A TRILOGY

DERIVATIVES OF HYDRAZINE AND OTHER HYDRONITROGENS HAVING N–N BONDS is part of a trilogy that comprises the second edition of OPEN-CHAIN ORGANIC NITROGEN COMPOUNDS which was originally published by W. A. Benjamin, Inc. in two volumes. The new three-part second edition has been revised, enlarged and reset.

DERIVATIVES OF HYDRAZINE AND OTHER HYDRONITROGENS HAVING N–N BONDS (1983)

DERIVATIVES OF AMMONIA (*in preparation*)

COMPOUNDS WITH N–O BONDS (*in preparation*)

DERIVATIVES OF
HYDRAZINE AND OTHER
HYDRONITROGENS
HAVING N–N BONDS

DERIVATIVES OF HYDRAZINE AND OTHER HYDRONITROGENS HAVING N–N BONDS

PETER A. S. SMITH

University of Michigan
Ann Arbor, Michigan

1983

THE BENJAMIN/CUMMINGS PUBLISHING COMPANY
Advanced Book Program
Reading, Massachusetts

London · Amsterdam · Don Mills, Ontario · Sydney · Tokyo

Library of Congress Cataloging in Publication Data

Smith, Peter Alan Somervail, 1920–
  Derivatives of hydrazine and other hydronitrogens
having N–N bonds.

  "Part of a trilogy that comprises the second edition
of Open-chain organic nitrogen compounds"—p.
  Bibliography: p.
  Includes index.
  1. Hydrazines. 2. Organonitrogen compounds.
I. Smith, Peter Alan Somervail, 1920–   Chemistry of
open-chain organic nitrogen compounds. II. Title.
QD305.A8S57   1983   547'.44   82-17735
ISBN 0-8053-8902-4

Dedication

I take pleasure in dedicating this volume to Royal Fred Sessions, who, when he taught chemistry at Santa Rosa Junior College, had a profound influence on my choice of career. His help, his guidance, his kindness, and the example he represented are remembered with appreciation.

CONTENTS

6. AZIDES

7. OTHER FUNCTIONS WITH CHAINS OF THREE OR MORE NITROGENS

PREFACE

In the roughly 17 years since publication of *The Chemistry of Open-Chain Organic Nitrogen Compounds*, of which this book is part of a second edition, the amount of published information on the subject seems to have doubled. This fact not only warrants a new edition, but makes it urgent that one be prepared before the sheer magnitude of the material puts the task beyond the capacity of a single author. The book that is before you now is the result of a frantic race to keep abreast of the torrent of new publications that constantly threaten to render the early chapters obsolete before the last ones are completed. The content is not intended to be encyclopedic, but it probably has gaping holes of which I am innocently unaware. I apologize for them. I have included information published as recently as early 1982, but my coverage of the most recent years is at best spotty.

I have completely rewritten the material in the original edition, saving hardly a line, but have tried to retain the same approach, in which phenomenology takes precedence over theory, and mechanism is considered conservatively and from the standpoint of its usefulness in correlating observed facts and allowing prediction of new ones. I have also maintained the emphasis on simple compounds, in the belief that in them the characteristics of a functional group are most clearly to be seen. As a result, alkaloids, steroids, and other naturally occurring compounds are almost totally ignored, fascinating though their chemistry may be. One can extrapolate from simple compounds to the complex ones more easily than the reverse. For similar reasons, I have given little attention to compounds with more than one functional group.

In addition to the new developments in subjects previously covered, some new subjects have been introduced. There are now short sections on analytical aspects—detection and determination. Nomenclature, spectroscopy, and photochemical transformations have been given greater attention. Some topics that previously were treated only lightly have been elevated to a position more befitting their current importance. Several chapters have been divided in order to make them more manageable. Hydrazine derivatives are now taken up in three chapters, according to oxidation state, and azides and diazoalkanes, never very congenial bedfellows, have been divorced.

This book has been written with advanced students and professional chemists in mind, with the aim of providing them with an integrated view of organic nitrogen chemistry in sufficient depth to be useful, yet sufficiently succinct and selective that one will be able to find what one wants reasonably efficiently. Although the style is not overtly didactic, I have tried to make the book usable for specialized courses. Its primary purpose, however, is to be used for reference in two ways: to obtain a working acquaintance of the properties and behavior of a particular type of nitrogen compound, and to obtain a guide to the literature in the form of key references.

The overall organization is still essentially that of the original work, but the need to produce this expanded edition as three volumes instead of two has led to the removal of azoxy compounds from their association with diazonium and azo compounds, to be placed in a separate chapter in a forthcoming volume, along with all the other functions having N-to-O bonds. As a concession to expediency, however, much diazoate chemistry is incorporated in the chapter on diazonium compounds, notwithstanding the presence of N—O bonds in them.

The three planned volumes of this edition should, God willing, be completed in approximately 3 years. The writing of this book has been a substantial drain of time and energy, and I am guilty of allowing it to entail neglect of family and friends, colleagues and correspondents, students and staff. I hope they will forgive me; without their forebearance, the writing would never have been finished. On the other hand, the writing would never have begun without the generous hospitality of Professor Leslie Hough and his colleagues at Queen Elizabeth College, Kensington, who provided congenial facilities and a haven from distraction during a sabbatical leave granted me by the University of Michigan. I thank them all.

Peter A. S. Smith

DERIVATIVES OF HYDRAZINE AND OTHER HYDRONITROGENS HAVING N–N BONDS

1.
HYDRAZINES

NOMENCLATURE

Alkyl and aryl derivatives of hydrazine are named as substitution products of the inorganic parent if no functional group of higher priority, such as carboxyl, is present. When there is more than one substituent, the locants 1,2 or N,N' are sanctioned by IUPAC. The compound p-$CH_3C_6H_4NHN(CH_3)_2$ is thus named *1,1-dimethyl-2-p-tolylhydrazine* (or N,N-, -N'-). The presence of a group of higher priority makes it necessary to use the prefix form *hydrazino,* the point of attachment being 1 or N, as in N'-methylhydrazinoacetic acid ($CH_3NHNHCH_2CO_2H$).

For disubstituted hydrazines it is sometimes convenient to use the prefixes *sym-* and *unsym-* to indicate 1,2- and 1,1-substitution, respectively, as in *sym*-dimethyl-hydrazine ($CH_3NHNHCH_3$). Although the foregoing method is not recommended by IUPAC, another alternative is permitted for *sym*-disubstituted hydrazines in which the two substituents are identical except for possible substitution on one or both: the prefix *hydrazo,* implying —NHNH—, may be used. Hydrazobenzene, for example, is an alternative for 1,2-diphenylhydrazine, and it may be treated as a parent structure when it bears substituents, as in 2,4'-dinitro-N-butylhydrazobenzene:

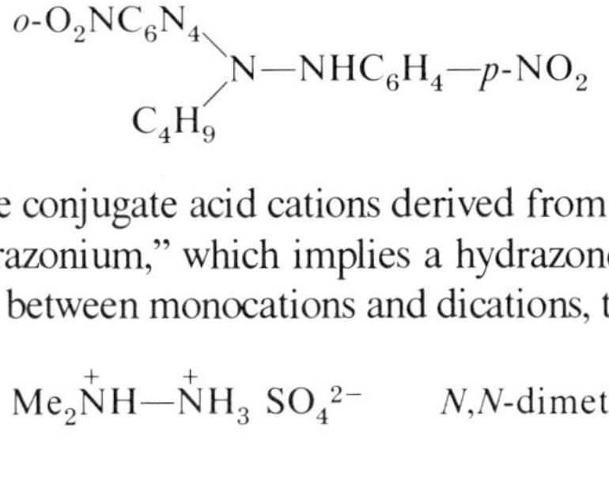

$$o\text{-}O_2NC_6N_4 \diagdown$$
$$\qquad\qquad N-NHC_6H_4-p\text{-}NO_2$$
$$C_4H_9 \diagup$$

The conjugate acid cations derived from hydrazines are named *hydrazinium* (*not* "hydrazonium," which implies a hydrazone derivative); if it is necessary to distinguish between monocations and dications, the suffixes $(1+)$ and $(2+)$ are included:

$$Me_2\overset{+}{N}H-\overset{+}{N}H_3 \; SO_4{}^{2-} \qquad N,N\text{-dimethylhydrazinium}(2+) \text{ sulfate}$$

PROPERTIES

The simple alkyl and aryl hydrazines are liquids with amine-like odors. They are reasonably stable to storage if protected from air, but the smaller ones, such as the dimethylhydrazines, are extremely flammable and may even be hypergolic in contact with oxidizing agents. They fume in moist air, dissolve exothermically in water, and are corrosive to flesh. Monoalkyl hydrazines boil at higher temperature than amines of the same molecular weight, and even tetraalkyl hydrazines, which have no NH for hydrogen bonding, boil at higher temperatures than the analogous hydrocarbons. Tetramethylhydrazine, for example, boils at 73°C, 15 degrees higher than tetramethylethane. Hydrogen bonding is nevertheless the major contributor to the boiling point of monoalkyl hydrazines, for successive methylation lowers the boiling point, which reaches a minimum with trimethylhydrazine (Table 1-1).

Table 1-1 Melting and Boiling Points and Basicities of Some Hydrazines

Hydrazine	Melting Point, °C	Boiling Point, °C (mm)	pK_a of Cation
NH_2NH_2		113.5	8.07
CH_3NHNH_2	−52.4	87	7.87
$CH_3NHNHCH_3$	−9	81	7.52
$(CH_3)_2NNH_2$	−57.2	63	7.21
$(CH_3)_2NNHCH_3$		60 (735)	6.56
$(CH_3)_2NH(CH_3)_2$		73 (730)	6.30
$C_2H_5NHNH_2$		99.5 (709)	7.99
$C_2H_5NHNHC_2H_5$		85	7.78
$(C_2H_5)_2NNH_2$		98	7.71
$C_6H_5NHNH_2$	19.6	243.5	5.1
$2,4,6\text{-}t\text{-}Bu_3C_6H_2NHNH_2$			3.66
$(CF_3)_2NN(CF_3)_2$		32	
$C_6H_5N(CH_3)NH_2$			5.0

N,N'-Diisopropyl hydrazine is miscible with warm water, whereas triethylhydrazine is "insoluble." *sec*-Butylhydrazine is "easily" soluble, and *N,N'*-diisobutylhydrazine is "difficulty" soluble. The solubility of phenylhydrazine is 10.9 percent at 19.6°C.

Hydrazines are weaker bases than amines by 1 to 3 powers of 10.[1] Alkylation lowers the base strength,[2] in contrast to the situation with ammonia, and the tetraalkyl hydrazines are the most weakly basic of the group. Most alkyl hydrazines have base strengths lying between those of ammonia and aniline and form salts readily with only one equivalent of acid. The tri- and tetraalkyl hydrazines with groups larger than butyl are nearly nonbasic and do not dissolve even in concentrated hydrochloric acid.[3] However, hydrazines show the pronounced nucleophilic character of the parent compound toward sp^2-hybridized carbon sites and react with carbonyl compounds more vigorously than amines.[4,5] The increased nucleophilicity that is generally observed when two atoms with unshared electron pairs are joined, known as the *alpha effect*,[6,7] thus includes hydrazines. One explanation is that this is a result of repulsion between the electron pairs, raising the ground-state energy. In the transition state, one of them becomes involved in bond formation with the electrophilic reagent, and the repulsion is lessened.[8] A similar phenomenon is observed with hydroxylamines.

Interpretation of the effect of structure on the basicity of hydrazines is complicated by the fact that there are two basic sites and they are not equivalent in most examples. Measured values, therefore, may not be unequivocally attributable to a specific nitrogen in the absence of other information. Methylhydrazine gives rise to substantial concentrations of both isomeric conjugate acids in solution,[9] whereas phenylhydrazine is protonated almost exclusively at the least-substituted nitrogen.[10] The fact that phenylhydrazine is a much weaker base than methylhydrazine suggests that electronic effects are transmitted in part from one nitrogen to the other. Support for this view is found in the effect of ring substituents on the basicity of phenylhydrazine and N-phenyl-N-methylhydrazine.[11] The effect of alkylation on the basic strength of hydrazine is clearly related to steric crowding and presumably is a consequence of changes in solvation. The sharply reduced basicity of 2,4,6-tri-*tert*-butylphenylhydrazine demonstrates this effect; simple steric interference with resonance between the ring and the exocyclic nitrogen would have an effect opposite to that observed here. The N—N bond distance[12] in both *sym*- and *unsym*-dimethylhydrazines, 1.45 Å, slightly shorter than the average N—C distance, implies that hydrazines should be especially sensitive to the effects of bulky substituents.

The acidity of the NH of hydrazines is very low; for a cyclic trialkyl hydrazine, a pK_a value of greater than 42 has been estimated from electrochemical data,[13] but the compound can be converted to a lithium derivative by reaction with t-butyllithium. The mono-substituted hydrazines are evidently more acidic, for both methylhydrazine and phenylhydrazine have been reported to form salts by reaction with sodamide or metallic sodium.[14] Quaternary hydrazinium hydroxides ($R_3\overset{+}{N}NH_2\ OH^-$), which are strong bases like the quaternary ammonium hydroxides,[15] show appreciable acidity at the NH site and equilibrate in aqueous solution with amine N-imides (R_3N^+—NH^-). N,N,N-Trimethylhydrazinium chloride has been converted by reaction with potassium t-butoxide to the simplest stable amine imide, isolated as an extremely hygroscopic solid having 2 moles of t-butyl alcohol of solvation.[16] If one substituent on the quaternary nitrogen is a benzyl group, it easily migrates to the other nitrogen irreversibly to form a trisubstituted hydrazine (Eq. 1-1).[17]

$$R_2\overset{+}{\underset{PhCH_2}{N}}-NH_2\ OH^- \rightleftharpoons R_2\overset{+}{\underset{PhCH_2}{N}}-NH^- \longrightarrow R_2N-NHCH_2Ph \qquad (1\text{-}1)$$

Allylic groups migrate similarly, with allylic rearrangement.[18]

If there are three different substituents on the quaternary nitrogen, it becomes a chiral center; two N-alkyl-N-methyl-N-phenylhydrazinium compounds have been resolved into their optical isomers.[19]

The infrared spectra of alkyl hydrazines resemble those of amines in the N—H stretching and bending regions. *Sym*- and *unsym*-disubstituted hydrazines may be distinguished by the fact that $\overset{+}{N}$—H stretching is always found well below $3100\ cm^{-1}$ in hydrazinium salts, and NH_2 stretching has symmetrical and unsymmetrical modes.[20] The nuclear magnetic resonance spectra of a variety of mono- to tetraalkyl hydrazines have been found to give signals at δ 2.26 to 2.73 for —CH_2—N$<$, 3.12 to 3.42 for $>$CH—N$<$, and 2.15 to 2.53 for CH_3—N.[21,22]

Tetraaryl hydrazines show some tendency to dissociated in solution into diarylamino radicals, in a manner similar to hexaarylethanes.[23] The radicals are colored, usually yellow to green, but in the crystalline state, in which dissociation does not take place, the hydrazines are colorless. The extent of dissociation is least when there are electron-withdrawing groups, such as nitro, in the *ortho* or *para* positions and greatest with electron-donating substituents, such as dimethylamino.

The oxidation potential for loss of an electron from tetraalkyl hydrazines to generate a cation radical has been found to show a linear correlation with vapor-phase ionization potentials and is much influenced by lone pair–lone pair interactions and steric strain, owing to the fact that substituents are eclipsed in the cation radical, and the N—N distance is shorter.[22,24]

The conformational analysis of hydrazines is surprisingly complex;[25,26] the *gauche* conformation is preferred to that with *trans* lone pairs.

The smaller alkyl hydrazines have been found to be carcinogenic in mammals as well as highly toxic in other ways. The toxic effects may be caused by inhalation as well as ingestion, and special care is called for in handling the highly volatile compounds having small alkyl groups. Phenylhydrazine has a long history of being harmful to chemists and other living things; it can be absorbed through the skin and can cause a cumulative poisoning.

REACTIONS

Heat. Hydrazines are in general reasonably stable to heat, and they can be distilled, although reduced pressure is advisable for these which would otherwise boil much above 100°C. Aryl hydrazines are more sensitive to heat than are alkyl hydrazines. Phenylhydrazine decomposes at its boiling point and slowly forms aniline, benzene, and nitrogen. The last two products probably arise from initial formation of phenyldiazene. *Sym*-diaryl hydrazines (hydrazobenzenes) have been more thoroughly studied;[27-29] their principal thermal reaction is dismutation into anilines and azobenzenes (Eq. 1-2). Unsymmetrically substituted hydrazobenzenes

$$ \text{ArNH—NHAr}' \xrightarrow{\Delta} \text{ArNH}_2 + \text{Ar}'\text{NH}_2 + \text{ArN}\!=\!\text{NAr}' \qquad (1\text{-}2) $$

give only the corresponding azobenzene and none of the symmetrical ones. Partial dissociation into anilino radicals, which then abstract hydrogen from the remaining hydrazobenzene, is a reasonable pathway. Thermolysis is kinetically first order, and there is only a minimal kinetic hydrogen-isotope effect.[30] Benzidine rearrangement (see below) is a competing process.[31,32] At 180°C, hydrazobenzene forms aniline and azobenzene in 70 percent yield, semidine in 14 percent yield, and smaller amounts of *o*-semidine, benzidine, and diphenyline.[30]

Tetraaryl hydrazines dissociate into free radicals more readily,[23,33] and through them undergo either rearrangement to semidines[34] (Eq. 1-3) or dismutation into

$$ \text{Ph}_2\text{N—NPh}_2 \;\rightleftharpoons\; 2\text{Ph}_2\text{N} \cdot \;\longrightarrow\; \text{Ph}_2\text{N}\!\!\diagdown\!\!\bigcirc\!\!=\!\text{NPh} \;\longrightarrow $$

$$ \text{Ph}_2\text{N}\!\!-\!\!\bigcirc\!\!-\!\!\text{NHPh} \qquad (1\text{-}3) $$

diarylamine and 9,10-diaryl-9,10-dihydrophenazine if the *para* position is blocked (Eq. 1-4). Polymeric substances derived from *p*-phenylenediamine have also been observed. Tetraalkyl hydrazines, however, appear to undergo elimination on thermolysis. When heated to 290°C, they form secondary amines in high yield, accompanied apparently by *N*-alkyl imines.[35]

$$(p\text{-}RC_6H_4)_2N\text{—}N(C_6H_4\text{—}p\text{-}R)_2 \rightleftarrows (p\text{-}RC_6H_4)_2N\cdot \longrightarrow$$

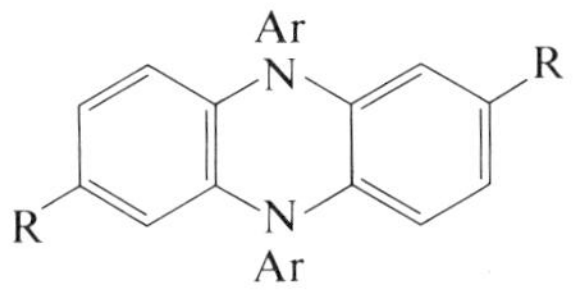

(1-4)

Acids. The simple alkyl hydrazines are extremely hygroscopic and fume in moist air, absorbing carbon dioxide readily to form hydrazinium salts of the corresponding carbazic acid (NH_2NHCO_2H). The salts of the common mineral acids are easily prepared, well crystallized, and suitable for storage. Aryl hydrazines behave similarly, but they are not hygroscopic. *tert*-Alkyl hydrazines form salts normally with acids, but on warming they are liable to fragmentation through loss of the alkyl group.[36,37] Hydrazines with two aryl groups, especially on the same nitrogen, are often cleaved by acid into amines and oxidation products in a manner resembling thermal decomposition (Eq. 1-3).[27-29] *p,p'*-Dinitrohydrazobenzene, for example, is converted by concentrated sulfuric acid into *p*-nitroaniline and dinitroazobenzene; *N,N*-diphenyl-*N',N'*-dimethylhydrazine is cleaved by dilute hydrochloric acid at room temperature. Some β-aminoalkyl and β-hydroxyalkyl hydrazines undergo *N,N* cleavage so easily (Eq. 1-5) that simple salts cannot ordinarily be prepared.[35,36]

$$(CH_3)_2N\text{—}NHCH_2\overset{\overset{\displaystyle OH}{|}}{\underset{\underset{\displaystyle CH_3}{|}}{C}}\text{—}Ph \xrightarrow[\text{ca. 20°C, 72 h}]{\text{aq.-alc. } H_2SO_4} (CH_3)_2NH + NH_3$$

$$+ \underset{29\%}{PhCOCH_3} + \underset{43\%}{PhCH(CH_3)CHO} \quad (1\text{-}5)$$

There are scattered reports of cleavage of other alkyl hydrazines under more drastic conditions.[40,41]

Benzidine rearrangement is the most prominent acid-catalyzed reaction of hydrazines; it is characteristic of hydrazobenzenes, which are very sensitive to acids.[42-46] Although salts, such as the dihydrobromides, can be isolated from nonpolar solvents,[47] aqueous acids usually cause the benzidine rearrangement to take place. The rearrangement is so named because the major products, *p,p'*-diaminobiphenyls, are commonly known as benzidines, but four other types of products may also be formed: diphenylines, semidines, *o*-semidines, and *o*-benzidines (observed only with naphthylhydrazines) (Eq. 1-6). Hydrazobenzene itself forms only benzidine and diphenyline, in the ratio 3:1; the other types of product shown in

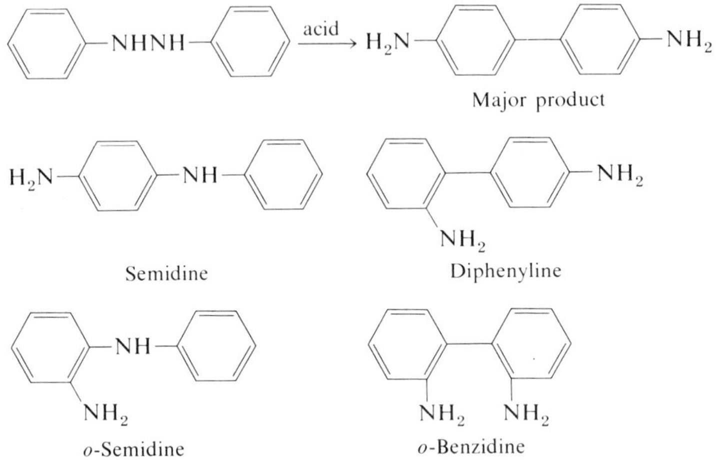

Major product

Semidine

Diphenyline

(1-6)

o-Semidine

o-Benzidine

Equation 1-6 may be formed from appropriately substituted hydrazobenzenes and are shown unsubstituted for reference.

Dismutation according to Equation 1-2 invariably accompanies acid-catalyzed benzidine rearrangement, in proportions varying with conditions and structure.[48] Reductive scission of the N—N bond may also occur; the o-semidine appears to be the reducing agent, being oxidized to phenazine derivatives.

Blocking the *para* positions generally promotes dismutation over benzidine rearrangement, but it sometimes happens that the interfering substituent is simply expelled, allowing the rearrangement to take place normally.[23] Substituents on the nitrogen atoms do not interfer, however, and even tetraaryl hydrazines undergo rearrangement.[29] It is apparently essential only that at least one aryl group be on each nitrogen. N-Acetylhydrazobenzene rearranges normally to monoacetylbenzidine, but the N,N'-diacetyl compound does not rearrange until an acetyl group has been hydrolyzed off.

The effect of structural factors in determining the predominant product from the rearrangement of a given hydrazobenzene may be determined by recourse to three rules proposed by Dewar:[42]

1. The normally preferred proportion of products is benzidine $>$ diphenyline $\gg$ semidine.
2. A diphenyline is the main product only from hydrazobenzenes in which the ring attached to the more basic nitrogen has a free *para* position; it is formed by linkage from this position to an *ortho* position of the other ring.
3. If a semidine is formed, the ring originally attached to the more basic nitrogen bears the primary amino group.

For the purpose of these rules, the relative basicities of the two nitrogens in a hydrazobenzene are deduced from those of the corresponding anilines. As examples of their application, p-chlorohydrazobenzene rearranges chiefly to a diphenyline (2,4'-diamino-5-chlorobiphenyl), but p-methylhydrazobenzene, in which the substituted ring is attached to the more basic nitrogen, gives chiefly an o-semidine (2-

anilino-*p*-toluidine). These rules do not apply to hydrazonaphthalenes, which give mainly *ortho*-benzidines.

The mechanism of the benzidine rearrangement has been the object of attention for many decades. It has been established that the reaction is intramolecular by experiments with unsymmetrical hydrazobenzenes, which produce only the corresponding unsymmetrical benzidine, and with mixtures of differently substituted hydrazobenzenes, which produce no benzidines arising from exchange of aryl groups[49] (Eq. 1-7). The reaction is subject to specific hydrogen-ion catalysis,[50,51]

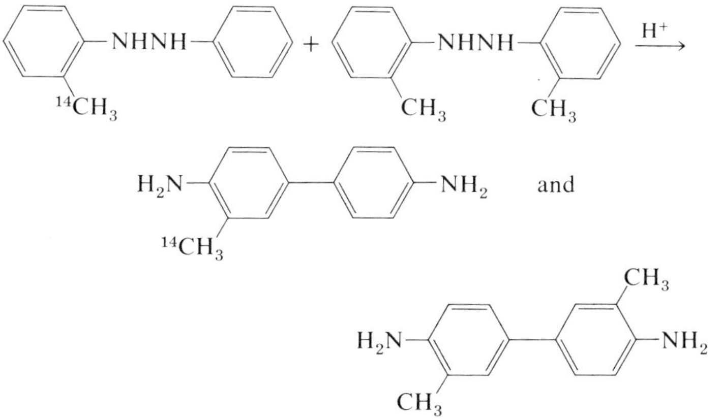

contrary to the earlier belief in general-acid catalysis. The dependence of the rate on acidity follows H_0 rather than concentration and is in most instances first-order at lower acidities, becoming second-order as acidity increases.[52,53] These facts speak for the existence of parallel paths of rearrangement, one through the monocation and one through the dication. At an earlier time, however, when the available evidence favored general-acid catalysis, it was convincingly argued that the addition of the second proton must be rate-controlling, rather than occurring as a prior equilibrium.[42] A pronounced kinetic isotope effect (in which substitution of deuterium oxide for water as the solvent doubles the velocity of the reaction that is first-order in acidity and quadruples that which is second-order) also implies normal, complete protonation prior to the transition state.[53] Deuteration of the ring positions of hydrazobenzene, however, does not affect the product ratio[54,55] and causes only a small kinetic isotope effect (k_H/k_D for deuteration in the *para* position is 0.962).[56] One can conclude that the loss of the protons from carbon at the new ring connection occurs after both the rate-determining transition state and the product-determining steps. Kinetic isotope effects have also been observed with ^{15}N, *p*-^{13}C, and *p*-^{14}C and imply that both N—N bond breaking and C—C bond formation are involved in the transition state in a concerted process in the case of benzidine formation. The evidence for formation of diphenyline, however, implies a dissociative rather than a concerted process.[56]

The foregoing facts point to a mechanism for benzidine formation in which the monocation or dication folds back upon itself so that the benzene rings are aligned face to face (Eq. 1-8). The resulting intermediate might be a pair of cation radicals

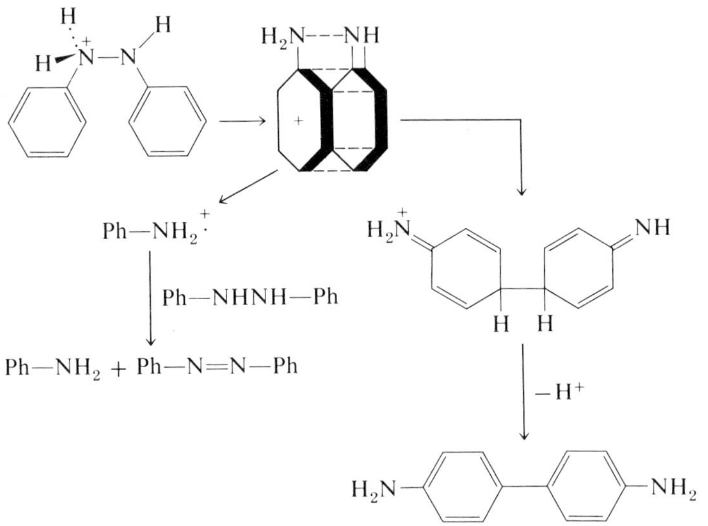

(1-8)

(π-complex mechanism through monocation)

held in a solvent cage, a π-complex derived from the species $ArNH_2$ and $ArNH^+$ or $ArNH_2$ and $ArNH_2^{2+}$, or the transition state for a symmetry-allowed [5 + 5] sigmatropic rearrangement. Through localization of a C—C bond between the 4 and 4′ positions, accompanied by unfolding and loss of protons from the same positions, benzidine is formed. Localization of the C—C bond at the 2 and 2′ positions would give rise to *ortho*-benzidine. Although it is possible to account for the other types of products by displacement or rotation of one ring with respect to the other, sufficient supporting evidence is lacking. Disruption of the complex (however it may be constituted) might give rise to separate cation radicals, which could be the source of other types of products, especially those of dismutation.[57]

The distribution of products in the thermally induced benzidine rearrangement is different from that of the acid-catalyzed reaction, and the mechanism may well be different. Photolysis of simple hydrazobenzenes dehydrogenates them to azobenzenes, but *N,N′*-dimethylhydrazobenzene forms semidine on photolysis. Since the quantum yield is not affected by triplet quenchers, the rearrangement is presumed to take place through a singlet-state intermediate.[58]

Studies of the reaction of Lewis acids with hydrazines appear to be confined to aliphatic hydrazines, and the possibility of benzidine rearrangement by this means has not been explored. Borane, trimethylborane, and boron halides have been found to form stable 1:1 adducts with the methylhydrazines, except for tetramethylhydrazine, which is unreactive.[59,60] The site of attachment appears to be the more highly methylated nitrogen (Eq. 1-9).

$$Me_2N—NH_2 + BF_3 \longrightarrow Me_2\overset{+}{N}\begin{smallmatrix} NH_2 \\ \diagdown \\ BF_3^- \end{smallmatrix}$$ (1-9)

Bases. Hydroxides do not attack hydrazines, and the simple hydrazines are sometimes distilled from solid potassium hydroxide in order to obtain them in an anhydrous state. Sodamide, alkyllithium reagents, and Grignard reagents can remove one or more protons to form hydrazine anions, but this is generally all that happens. Arylmethylhydrazines, however, undergo a slow rearrangement when converted to their dianions by reaction with butyllithium (Eq. 1-10).[61]

$$\underset{\underset{Me}{|}}{Ar-NNH_2} \xrightarrow{BuLi} \underset{\underset{Me}{|}}{Ar-NN^{2-}} \longrightarrow Me\bar{N}-\bar{N}Ar \xrightarrow{H_2O}$$

$$MeNHNHAr \qquad (1\text{-}10)$$

Quaternary hydrazinium salts can be deprotonated even by aqueous base to form amine N-imides. If a benzylic or allylic group is present on the quaternary nitrogen, it readily shifts to the anionic site to form a structure that no longer has a separation of charge (Eq. 1-1),[17] but if a group having a β hydrogen is present, an analogue of the Cope elimination may take place (Eq. 1-11).[62]

$$\underset{\underset{CH_3}{}}{\overset{CH_3CH_2}{\diagdown}} CH\overset{+}{N}Me_2NH_2 \xrightarrow[t\text{-BuOH}]{KO\text{-}t\text{-Bu}} \underset{\underset{CH_3}{}}{\overset{CH_3CH_2}{\diagdown}} CH\overset{+}{N}Me_2NH^- \xrightarrow[90 \text{ min}]{reflux}$$

$$CH_3CH_2CH{=}CH_2 + \underset{CH_3}{\diagup}CH{=}CH\underset{CH_3}{\diagdown} \qquad (1\text{-}11)$$

$$+ \underset{CH_3}{\overset{CH_3}{\diagdown}}CH{=}CH\underset{CH_3}{\diagdown} + Me_2NNH_2$$

(73% yield, ratio 64.5:10.5:25)

Alkylating agents. Monoaryl and monoalkyl hydrazines react readily with conventional alkylating agents (halides or sulfates). Multiple alkylation is likely to occur, for nucleophilicity is not greatly diminished until the tetralkyl stage. The more substituted nitrogen is preferentially attacked in the absence of strong steric or electronic effects, and methylhydrazine, for example, is converted easily and cleanly to a quaternary hydrazinium salt by methyl iodide (Eq. 1-12).[15]

$$CH_3NHNH_2 + CH_3I \longrightarrow (CH_3)_3\overset{+}{N}NH_2 \ I^- \qquad (1\text{-}12)$$

Phenylhydrazine is also methylated mostly at the substituted nitrogen, but the selectivity is not so pronounced.[2,17] Both methylation and phenylation lower the basic strength of hydrazines, but they evidently increase the nucleophilicity. However, tetraalkyl hydrazines are not only of low basicity, but are highly resistant toward alkylating agents, unless cyclic.[26]

Steric hindrance must be pronounced to shift the site of alkylation, for even *tert*-butylhydrazine is alkylated almost entirely on the substituted nitrogen by methyl iodide,[63] but a triphenylmethyl substituent forces introduction of a second triphenylmethyl group to take place at the other nitrogen (Eq. 1-13).

$$Ph_3CNHNH_2 + Ph_3CCl \longrightarrow Ph_3CNHNHCPh_3 \qquad (1\text{-}13)$$

Alkylating agents of intermediate size often balk at putting a third alkyl group on the same nitrogen, however, and N,N,N'-trialkyl and sym-tetraalkyl hydrazines may be formed instead (Eq. 1-14). In such cases, alkylation at the substituted nitrogen

$$(PhCH_2)_2NNHCH_2Ph + PhCH_2Cl \longrightarrow (PhCH_2)_2N-N(CH_2Ph)_2 \qquad (1-14)$$

occurs to the largest extent with alkyl iodides and least with chlorides.[3] This is explained by the presumption that the transition state is looser with a strong leaving group such as iodide and is therefore not so subject to destabilization by steric effects. N,N-Diphenylhydrazine is alkylated by methyl sulfate at the unsubstituted nitrogen (Eq. 1-15), probably as a result of combined steric and electronic effects.[27]

$$Ph_2NNH_2 + (CH_3)_2SO_4 \longrightarrow Ph_2NNHCH_3 \longrightarrow Ph_2NN(CH_3)_2 \qquad (1-15)$$

   Benzylation of N-(X-phenyl)-N-benzylhydrazine takes place mostly at the N position to give a quaternary salt; the substituent X has a weak effect on the rate and the site selectivity ($X = p$-CH$_3$, higher rate and increased ratio of attack at N/N'; $X = m$-Cl, lower rate and decreased ratio). The rate of alkylation is faster in ethanol than in toluene, as is to be expected of a reaction in which charge separation is greater in the transition state.[17]

   As implied by the foregoing facts, the site selectivity for alkylation is normally kinetically controlled. However, alkylation is reversible, especially easily in polar solvents, and reversal is favored at warmer temperatures. The reverse reaction involves displacement at carbon by the anion or other available nucleophile and is accordingly markedly favored by halides over tosylates. When reversibility comes into play significantly, the composition of the products shifts from the kinetically controlled ratio toward the thermodynamically controlled one; for example, benzylation of N-(p-chlorophenyl)-N-benzylhydrazine at 0°C takes place in a site ratio N/N' of 9:1, but at 40°C the product is exclusively N-(p-chlorophenyl)-N,N'-dibenzylhydrazine (Eq. 1-16). As is further implied, quaternary hydrazinium salts

$$p\text{-ClC}_6H_4N\begin{array}{l} {}^{CH_2Ph} \\ {}_{NH_2} \end{array} \xrightarrow{\ PhCH_2I\ } p\text{-ClC}_6H_4\overset{+}{N}(CH_2Ph)_2\ I^- \\ \qquad\qquad\qquad\qquad\qquad\qquad\qquad\quad \underset{NH_2}{|}$$

$$+\ p\text{-ClC}_6H_4N\begin{array}{l} {}^{CH_2Ph} \\ {}_{NH} \\ {}_{|} \\ {}_{CH_2Ph} \end{array} \qquad (1\text{-}16)$$

(at 0°C, ratio 9:1; at 40°C, ratio < 1:100)

can be made to rearrange via dealkylation/realkylation by heating (Eq. 1-17).[17] Another manifestation of this phenomenon is the formation of N,N,N-trimethyl-hydrazinium salt when N,N-dimethylhydrazine is heated with methylene chloride.[64] These reactions are the hydrazine analogs of the redistribution reaction of alkylam-monium salts.

$$PhN^{+}(CH_2Ph)_2 \ Cl^{-} \atop | \atop NH_2} \xrightarrow[\text{(2) 4 h HCl}]{\text{(1) 100°C}} PhN{\overset{\displaystyle CH_2Ph}{\underset{\displaystyle NHCH_2Ph \cdot HCl}{\Big<}}} \ + \ PhN{\overset{\displaystyle CH_2Ph}{\underset{\displaystyle NH_2 \cdot HCl}{\Big<}}}$$

$$ 31\% 55\%$$

$$ + \ PhCH_2Cl \qquad (1\text{-}17)$$

Hydrazines may also be alkylated with conjugated olefins, such as acrylonitrile (Eq. 1-18).[65,66] Such alkylation is commonly followed by ring closure to the con-

$$PhNHNH_2 + CH_2{=}CHCN \xrightarrow[\text{1 day}]{\text{MeOH}} PhCH_2N{\overset{\displaystyle NH_2}{\underset{\displaystyle CH_2CH_2CN}{\Big<}}} \qquad (1\text{-}18)$$

$$ 95\%$$

jugated substituent to form pyrazole derivatives.[67] Styrenes do not react with hydrazines unless they are converted to their anions (Eq. 1-19).[68] Anions of

$$CH_3N^{-}Na^{+} + ArCH{=}CHR \longrightarrow ArCH_2CH{\overset{\displaystyle R}{\underset{\displaystyle NCH_3}{\Big<}}} \qquad (1\text{-}19) \atop { \atop |\atop NH_2}}$$

hydrazines also react with alkyl halides and are useful in synthesis to increase site selectivity. The lithium derivative of phenylhydrazine, for example, prepared by reaction with butyllithium, reacts with methyl iodide to form N-phenyl-N-methyl-hydrazine exclusively,[61] whereas the neutral base gives rise to a significant amount of the N,N' isomer.

Bidentate alkylating agents, such as $X(CH_2)_n X$, generally react with hydrazines at both nitrogens to close a ring if n is suitably small; otherwise they react with two equivalents of hydrazine.[64]

Arylation of hydrazines may be accomplished by reaction with sufficiently activated aryl halides. Methylhydrazine reacts with 2,4-dinitrochlorobenzene at the methylated nitrogen,[69] whereas unsym-dimethylhydrazine and picrylhydrazine, being more hindered, undergo arylation at the unsubstituted nitrogen.

Acylating agents. Acylation of hydrazines somewhat resembles alkylation in that an alkylated nitrogen is often preferred to an unsubstituted one. However, the site of acylation is markedly more sensitive to steric effects as well as to the nature of the acylating agent. The usual products from acylation of phenylhydrazine with either esters or acid chlorides are N'-phenylhydrazides (accompanied by N,N'-diacyl-phenylhydrazines); simple alkyl hydrazines are acylated at the substituted nitrogen by acetic anhydride, but at the unsubstituted one by ethyl acetate.

Methylhydrazine has been the object of greater study than other alkyl hydrazines.[70-72] Acetic anhydride in pyridine gives a ratio of N-acetyl to N'-acetyl of 38, rising to over 100 in acetic acid solution; ethyl acetate, in contrast, produces a ratio of only 0.3 (Eq. 1-20).[73] Acetyl chloride in methylene chloride acylates

$$CH_3NHNH_2 + AcY \longrightarrow CH_3N\begin{array}{c} NH_2 \\ \diagdown \\ Ac \end{array} \quad \text{and} \quad CH_3NHNHAc \quad (1\text{-}20)$$

(Y = OAc, ratio 38:1 or greater)
(Y = OEt, ratio 1:3.3)

methylhydrazine at the N position,[74] as do carboxylic acids and dicyclohexylcarbo-diimide,[75] alkyl formates,[67,76] and dithio esters. Although the equilibrium ratio of N-acetyl and N'-acetyl, 0.39 at 27°C and 0.49 at 87°C, is not very different from the product ratio obtained with use of ethyl acetate, the evidence is convincing that in all the acylations the ratio of products is kinetically controlled.

The explanation of the foregoing observations has been attributed to differences in the extent of bond formation in the transition state. A more advanced transition state would be more crowded, and the steric repulsion of the N-substituent would be more significant in decelerating acylation at the substituted nitrogen. Esters, being much less active acylating agents, are presumed to involve a tighter transition state than do acid anhydrides. Steric effects are demonstrable with bulky substituents and bulky acylating agents, such as the reaction of *tert*-butylhydrazine with benzoyl chloride, which results largely in N'-acylation.[63]

Dimethylhydrazines are slightly less readily acylated, and *unsym*-dimethylhydrazine, for example, does not react with most esters unless sodium methoxide is used as a catalyst.[77] Formates and oxalates, however, do not require catalysis.[78] In either case, only monoacylation takes place, whereas monoalkyl hydrazines are likely to undergo diacylation when Schotten-Baumann conditions are used.

Cyanic acid, phenyl isocyanate, and phenyl isothiocyanate acylate the substituted nitrogen of methylhydrazine, and even isopropylhydrazine forms 2-alkyl semicar-bazide derivatives (Eq. 1-21).[79-81]

$$RNHNH_2 + ArNCO \longrightarrow ArNHCON\begin{array}{c} NH_2 \\ \diagdown \\ R \end{array} \quad (1\text{-}21)$$

Tripropylhydrazine has been formylated by successive reaction with butyllithium, carbon monoxide, and water.[82]

Nitrosation of hydrazines is of special interest because of the further transformations that the initially formed nitroso compounds may undergo. Alkyl hydrazines may be attacked at either nitrogen when treated with aqueous nitrous acid; the reaction is strongly catalyzed by halide or thiocyanate ions.[9] Methylhydrazine is converted to the extent of one third to N-methyl-N-nitrosohydrazine and two thirds to the presumed decomposition products of the N'-nitroso isomer (Eq. 1-22). With

$$CH_3NHNH_2 + HNO_2 \longrightarrow CH_3N\begin{array}{c} NH_2 \\ \diagdown \\ NO \end{array} + [CH_3NHNHNO] \xrightarrow{H^+}$$

$$CH_3NH_3^+ + N_2O \quad (1\text{-}22)$$

higher concentrations of nitrous acid, the *N,N'*-dinitroso compound appears to be formed, as deduced from the appearance of its logical decomposition products: N_2O, N_2, and CH_3OH. The *N*-alkyl-*N*-nitrosohydrazines are generally isolable if they are treated gently, especially if they are not exposed to strong acid. Their structure is established by the fact that they react with ketones to form *N*-nitroso-hydrazones, $RN(NO)N{=}CR_2$.

Some *N*-alkyl-*N*-nitrosohydrazines have been reported to lose water when exposed to strong acids, forming alkyl azides. Such a reaction involves some sort of rearrangement of the nitrogen chain, which has not been elucidated. As a general rule, however, azides are not obtained in significant yield from alkyl hydrazines.

Phenylhydrazine behaves similarly to alkyl hydrazines[83,84] and forms isolable *N*-nitroso-*N*-phenylhydrazine. The reaction is very fast and may be a diffusion-controlled process with the nitrosating species and *N'* cation.[85] In excess nitrous acid, much benzenediazonium ion is formed, perhaps through dinitrosation. In the presence of excess acid, *N*-nitroso-*N*-phenylhydrazine is converted in excellent yield into phenyl azide, either by prior rearrangement to the *N'*-nitroso isomer or by dehydration to a three-membered ring (Eq. 1-23). Heating causes decomposition into aniline and nitrous oxide instead.

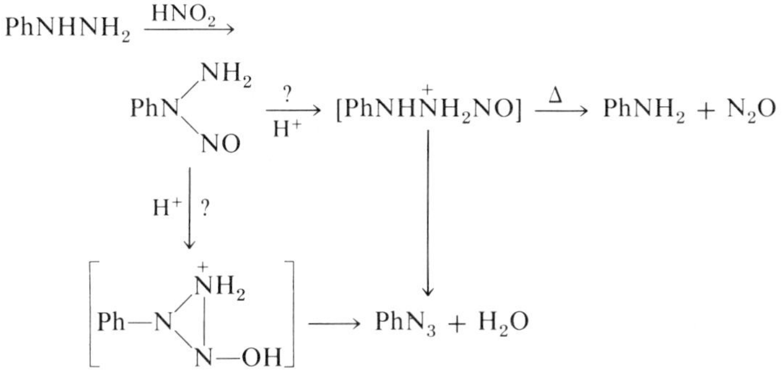

$$PhNHNH_2 \xrightarrow{HNO_2}$$

$$PhN{\Big\langle}{\substack{NH_2 \\ NO}} \xrightarrow[H^+]{?} [PhNH\overset{+}{N}H_2NO] \xrightarrow{\Delta} PhNH_2 + N_2O$$

$$H^+ \Big\downarrow ?$$

$$\left[Ph{-}N{\Big\langle}{\substack{\overset{+}{N}H_2 \\ N{-}OH}} \right] \longrightarrow PhN_3 + H_2O$$

$$(1\text{-}23)$$

Unsym-disubstituted hydrazines react with nitrous acid to form nitrous oxide and the corresponding secondary amine (Eq. 1-24) or its nitroso derivative if nitrous acid is used in excess.[86,87]

$$R_2NNH_2 + HNO_2 \longrightarrow R_2NH + N_2O \qquad (1\text{-}24)$$

N,N-Dimethylhydrazine initially gives a species in solution with ultraviolet absorption implying the presence of an *N*-nitroso structure, but it decomposes rapidly into dimethylammonium ion and nitrous oxide.[9] *Sym*-disubstituted hydrazines, however, form isolable nitroso compounds having either one or two nitroso groups. In the latter case, the overall effect is oxidation, since the *N,N'*-dinitroso compounds easily lose nitric oxide, leaving an azo compound (Eq. 1-25).[88,89]

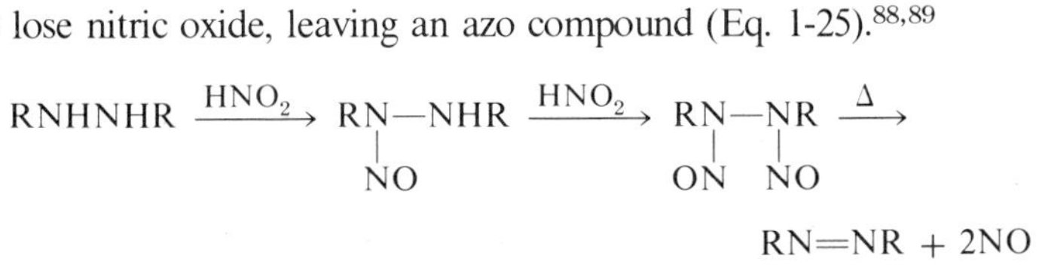

$$RNHNHR \xrightarrow{HNO_2} \underset{\substack{| \\ NO}}{RN}{-}NHR \xrightarrow{HNO_2} \underset{\substack{| \quad | \\ ON \quad NO}}{RN}{-}NR \xrightarrow{\Delta}$$

$$RN{=}NR + 2NO \qquad (1\text{-}25)$$

Acylating agents derived from sulfur and phosphorus acids, both inorganic and organic, react with hydrazines in much the same way as other acylating agents.[90,91] Thionyl chloride forms thionylhydrazines, $RNHN=SO$. Satisfactory yields are obtained only when thionyl chloride is added to an excess of the hydrazine; if phenylhydrazine is added slowly to thionyl chloride, phenyl azide is the main product, derived, apparently, from benzene-diazonium ion via diphenyltetrazene.[92] Thionylaniline also acts as a thionyl-transfer agent toward hydrazines. Sulfonyl hydrazides are generally formed by reaction of sulfonyl chlorides with hydrazines, but they cannot always be isolated in satisfactory yields, owing to the fact that many are easily cleaved into sulfinic acids and oxidation products of hydrazines. For example, when trimethylhydrazine is treated with p-toluenesulfonyl chloride, the product is *unsym*-dimethylhydrazine, formed via formaldehyde dimethylhydrazone (Eq. 1-26).[93,94] In some instances, the sulfonyl chloride is actually reduced all the way to a disulfide.

$$(CH_3)_2N-NHCH_3 \xrightarrow{ToSO_2Cl} \left[(CH_3)_2NN\begin{matrix} CH_3 \\ SO_2To \end{matrix} \right] \longrightarrow$$

$$ToSO_2H + (CH_3)_2NN=CH_2 \xrightarrow{H_2O} (CH_3)_2N-NH_2 \qquad (1\text{-}26)$$

Diazonium coupling.   Electrophilic attack on hydrazines by diazonium salts has been studied principally with phenylhydrazine and its derivatives.[95,96] In acetic acid solution, the major product is a 1,3-diaryl tetrazene, coupling having occurred at the substituted nitrogen, but when the reaction is carried out in strongly acidic solution, aryl azide and aniline (a pair of each if the aryl groups are different) are formed instead (Eq. 1-27). These products appear to be derived from the breakdown of a

$$ArN_2^+ + Ar'NHNH_2 \quad \begin{cases} \xrightarrow{AcOH} ArN=NN\begin{matrix} Ar' \\ NH_2 \end{matrix} \\ \\ \xrightarrow{strong\ HCl} [ArN=NNHNHAr'] \longrightarrow \\ \qquad\qquad ArN_3 + Ar'NH_2 \quad and \quad ArNH_2 + Ar'N_3 \end{cases} \qquad (1\text{-}27)$$

1,4-diaryl tetrazene, formed by coupling at the unsubstituted nitrogen; the 1,3-diaryl isomers do not decompose to azide and aniline when placed in strong acid. It has been suggested in explanation that the free arylhydrazines, which would be present in appreciable concentration in acetic acid solutions, couple rapidly at the substituted nitrogen, but in strongly acid solution, only the monocations are available for coupling. This enigmatic situation requires that the N-protonated cation, $Ar\overset{+}{N}H_2NH_2$, be very much more reactive toward diazonium ions than the more populous N' cation, $ArNHNH_3^+$, an ad hoc assumption. Alkyl hydrazines also couple at the substituted nitrogen, and *sym*-dimethylhydrazine couples twice to form hexazadienes (see Chap. 7).

Aldehydes and ketones.   The well-known reaction of hydrazone formation with aldehydes and ketones is typical of hydrazines having an NH_2 group (although difficult with quaternary hydrazinium salts). It occurs in two stages, addition and dehydration (Eq. 1-28), either of which may be the rate-determining step, depend-

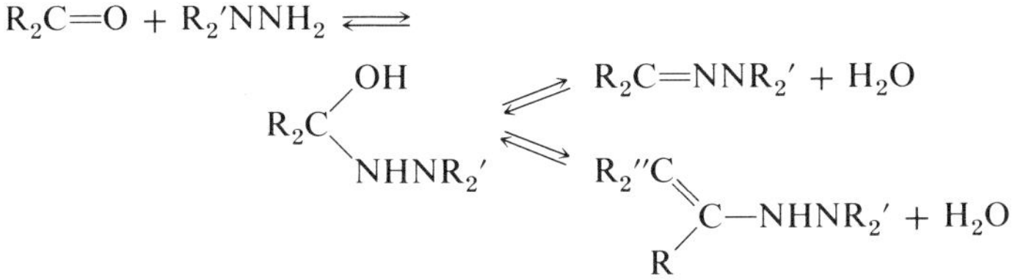

$$R_2C{=}O + R_2'NNH_2 \rightleftharpoons$$

(1-28)

ing on both structure and conditions.[97,98] With phenylhydrazine, for example, the addition step is rate-determining under acidic conditions, whereas the dehydration step is the slower under basic conditions.[99] Hydrazone formation varies in ease from spontaneous and exothermic with simple aldehydes and ketones to extremely difficult with highly hindered benzophenones. Vinylhydrazines may be formed initially.[100]

Hydrazone formation takes place without complication with aryl, acyl, and *unsym*-dialkyl hydrazines, but monoalkyl hydrazines can react with more than one equivalent of aldehyde, condensation occurring at the substituted nitrogen as well as at the unsubstituted one (Eq. 1-29).[101] *Sym*-disubstituted hydrazines have been re-

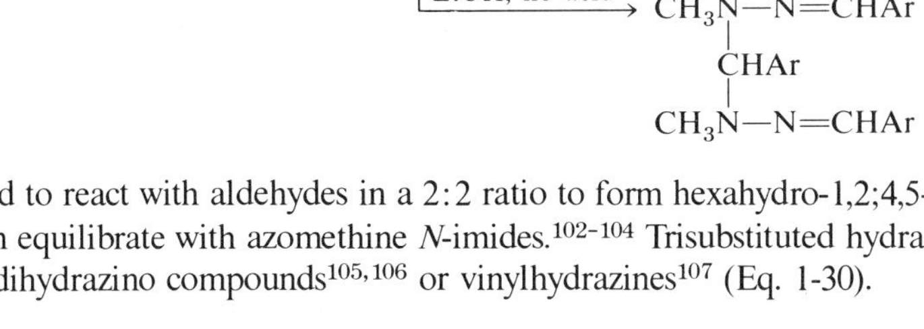

(1-29)

ported to react with aldehydes in a 2:2 ratio to form hexahydro-1,2;4,5-tetrazines, which equilibrate with azomethine *N*-imides.[102-104] Trisubstituted hydrazines form *gem*-dihydrazino compounds[105,106] or vinylhydrazines[107] (Eq. 1-30).

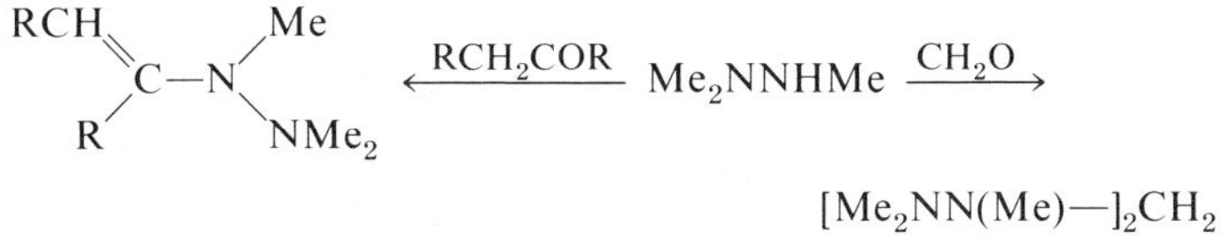

$$[Me_2NN(Me){-}]_2CH_2 \qquad (1\text{-}30)$$

In the presence of other reagents, products derived from hydrazones may be formed, such as α-hydrazino nitriles when HCN is present[108] and alkylated hydrazines when a reducing agent such as sodium cyanoborohydride is present[22] (Eq. 1-31).

$$R_2NNHCH_3 \xleftarrow{NaBH_3CN} R_2NNH_2 + CH_2{=}O \xrightarrow{HCN}$$
$$R_2NNHCH_2CN \qquad (1\text{-}31)$$

The presence of other functional groups in an α or β position to the carbonyl group may affect the nature of the ultimate product as a result of further reaction. β-Dicarbonyl compounds such as acetoacetic ester appear to undergo normal hydrazone formation as a first step, but the hydrazones undergo cyclization so easily (Eq. 1-32) that they are seldom isolable. These reactions are of much importance in the preparation of pyrazole derivatives.

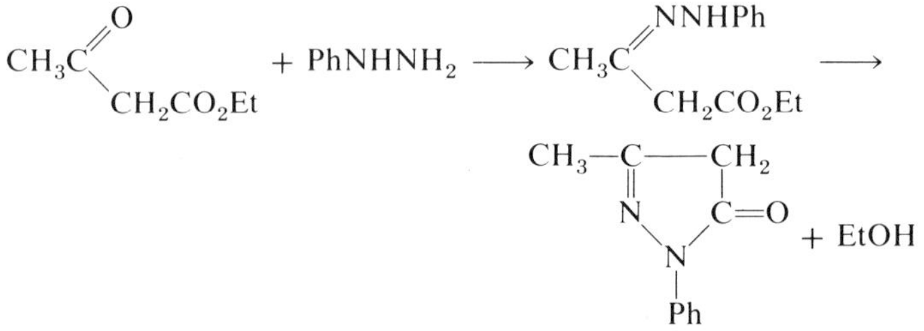

$$(1\text{-}32)$$

The most important secondary reaction between hydrazines and aldehydes and ketones is undoubtedly that leading to osazones. It occurs in general when an α-hydroxy aldehyde or ketone is treated with an excess of a monoaryl or N-aryl-N-alkyl hydrazine. Three equivalents of the hydrazine are consumed, of which two are incorporated into the osazone, and the third serves as an oxidizing agent, becoming reduced to ammonia and aniline (Eq. 1-33). Osazones are thus bis-hydrazones of

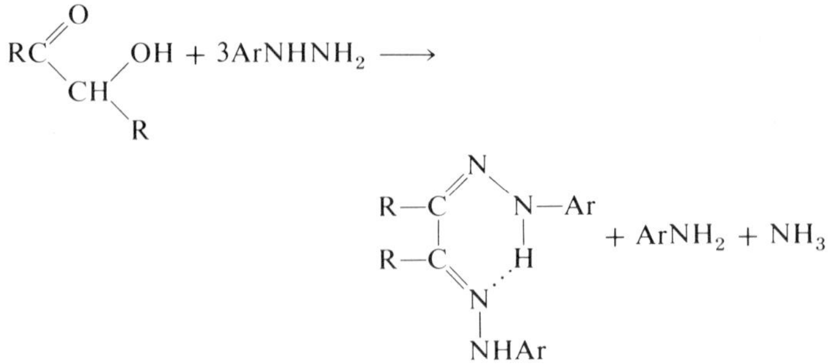

$$(1\text{-}33)$$

α-dicarbonyl compounds; those from monosubstituted hydrazines exist mainly as chelate hydrogen-bonded structures.

Osazone formation has received enormous application in sugar chemistry; it played a major role in the work of Emil Fischer in unraveling the confused skein of sugar chemistry. Its particular value is that it yields well-crystallized, easily isolated and identified products and destroys one specific chiral center if the carbon bearing the α-hydroxy group is not at the end of a chain.

The mechanism of osazone formation has been an object of concern for many decades. The evidence suggests[109,110] that the simple arylhydrazones react with a second equivalent of arylhydrazine by first tautomerizing to an α-hydrazino ketone, analogous to the Amadori rearrangement of α-hydroxy aldimines to α-amino ketones. The α-hydrazino hydrazone produced then tautomerizes to an enamine structure, which undergoes an acid-catalyzed cleavage to form aniline and an imine-hydrazone, which can react with a third equivalent of arylhydrazine to form the osazone with elimination of ammonia (Eq. 1-34).

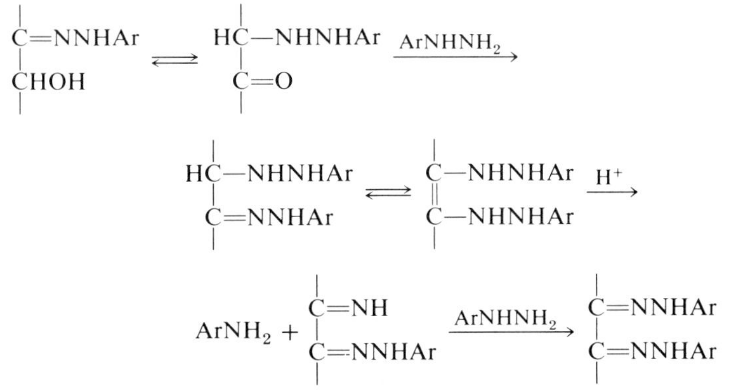

$$\begin{array}{c} \text{C=NNHAr} \\ | \\ \text{CHOH} \end{array} \rightleftharpoons \begin{array}{c} \text{HC—NHNHAr} \\ | \\ \text{C=O} \end{array} \xrightarrow{\text{ArNHNH}_2}$$

$$\begin{array}{c} \text{HC—NHNHAr} \\ | \\ \text{C=NNHAr} \end{array} \rightleftharpoons \begin{array}{c} \text{C—NHNHAr} \\ || \\ \text{C—NHNHAr} \end{array} \xrightarrow{\text{H}^+}$$

$$\text{ArNH}_2 + \begin{array}{c} \text{C=NH} \\ | \\ \text{C=NNHAr} \end{array} \xrightarrow{\text{ArNHNH}_2} \begin{array}{c} \text{C=NNHAr} \\ | \\ \text{C=NNHAr} \end{array} \qquad (1\text{-}34)$$

The process stops at the osazone stage, even if there is an oxidizable hydroxy group adjacent to the osazone structure, as there would be in most sugars. N-Methylphenylhydrazine, however, works down the sugar chain all the way to the hexahydrazone stage.[111] This fact has led to the suggested explanation[112] that osazones from monoaryl hydrazines are sufficiently stabilized by chelate hydrogen bonding to retard further reaction, but there are reservations about accepting this view, because osazones equilibrate with substantial amounts of their nonchelate forms.[113]

In a reaction formally analogous to hydrazone formation, o-nitroaryl hydrazines undergo a cyclic condensation on heating, losing water and forming N-hydroxy-benzotriazoles[114-116] (Eq. 1-35). 2,4-Dinitrophenylhydrazine reacts in this way only

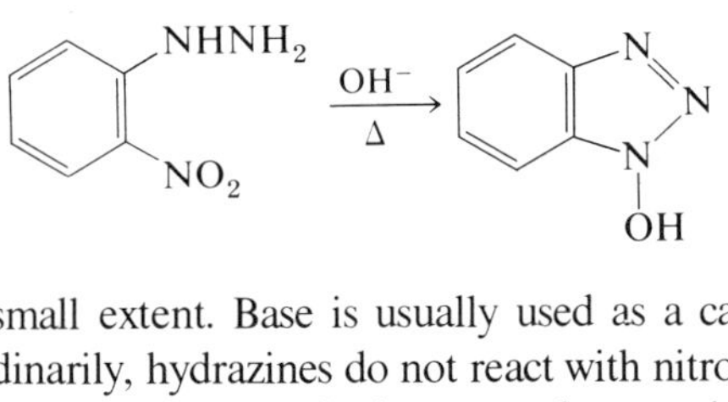

$$\xrightarrow[\Delta]{\text{OH}^-} \qquad (1\text{-}35)$$

to a small extent. Base is usually used as a catalyst, although it is not essential.

Ordinarily, hydrazines do not react with nitro groups except by oxidation-reduction. Nitroso compounds, however, do react intermolecularly with hydrazines to form analogous linear compounds (see Chap. 7), but phenylhydrazine and nitrsobenzene produce azoxybenzene, diphenylamine, benzene, and nitrogen in an apparent free-radical process.[117]

Oxidation. Hydrazines can act as both oxidizing and reducing agents, since their oxidation state stands between those of ammonia and elemental nitrogen; both behaviors are encountered to a significant extent, although the reducing properties are usually more pronounced. All classes of substituted hydrazines, including the tetrasubstituted hydrazines, are oxidizable under moderate conditions, except the quaternary hydrazinium salts, but the ease as well as the nature of the reactions vary considerably. Even very mild oxidizing agents, such as ammonical silver nitrate (Tollens' reagent) and Fehling's solution, are reduced by most hydrazines and can be used as qualitative test reagents for the hydrazine structure when other easily oxidized groups are known to be absent.

There is a difference of four electrons (or four hydrogen atoms) between the oxidation state of hydrazine and that of nitrogen. Oxidation can therefore be expected to occur in stages and the products to depend on both the quantity and the nature of the oxidizing agent. Although one-electron oxidations are well established with tri- and tetrasubstituted hydrazines, monosubstituted hydrazines usually give products resulting from two- and four-electron oxidations (loss of two or four hydrogen atoms). Two-electron oxidation of monosubstituted hydrazines would give rise to the diazene or tautomeric azamine structure: $RN{=}NH \rightleftarrows R\overset{+}{N}H{=}N^-$. These structures themselves have not been isolated from oxidations, but their expected decomposition products, nitrogen and a hydrocarbon (Eq. 1-36), comprise

$$RNHNH_2 \xrightarrow{[O]} [RN{=}NH] \longrightarrow RH + N_2 \qquad (1\text{-}36)$$

the major products from a variety of methods of oxidation.[69,115,116,118-120] Among the various oxidizing agents are copper sulfate, periodate, ferric chloride, ferricyanide, mercuric oxide, manganese dioxide, and molecular oxygen. The decomposition of phenylhydrazine in air, for example, gives principally benzene, along with small amounts of colored tars. Oxidative conversion of the hydrazino group to hydrogen is of considerable synthetic value, for it permits the replacement of oxygen or halogen by hydrogen (Eq. 1-37). It has been particularly useful in the synthe-

$$-CH{=}N- \qquad (1\text{-}37)$$

sis of heterocyclic parent systems, where the rings must often first be built up in cyclic amide form.

Alkyl hydrazines are smoothly oxidized to hydrocarbons by toluenesulfonyl azide under phase-transfer conditions,[121] but it is not clear whether the process takes place through formation of a sulfonhydrazide or by initial diazo transfer. Quinone oxidizes phenylhydrazine through intermediate ion radicals, detectable by CIDNP.[122]

Carbon tetrachloride and bromotrichloromethane also bring about oxidation of phenylhydrazine by a free-radical process; some halobenzene is formed as well as benzene.[123] When active manganese dioxide acts on aryl hydrazines in the presence of benzene, substantial amounts of arylbenzene are formed, presumably by substitution by aryl radicals.[124] Oxidation by O_2 is probably also free-radical in character; it occurs so easily that many known aryl hydrazines have never been obtained in an analytically pure form, but rather only in the form of derivatives. Autoxidation is nevertheless likely to be slow unless traces of heavy-metal compounds, particularly Cu^{II}, are present.[125] Electrochemical oxidation has been studied extensively and allows the separate stages of electron removal to be followed.[126,127]

Four-electron oxidation may take place if there is sufficient oxidizing agent and if it reacts fast enough to intercept the intermediate diazenes or free radicals. Halogens, such as iodine in the presence of triethylamine or bromine in acetic acid, give moderate to high yields of alkyl or heteroaryl halide, along with some alkane (Eq. 1-38).[128] Alkyl iodides largely retain the original configuration and may arise from

$$RNHNH_2 + 2I_2 \xrightarrow{Et_3N} RI + N_2 + 3Et_3NHI \tag{1-38}$$

unimolecular collapse of an iododiazene. In the presence of bicarbonate, the alcohol may become the major product; it is racemic. Aryl hydrazines are oxidizable to diazonium salts if the temperature is kept low.[118] Diazonium salts are seldom the only product, however, because even if they do not decompose, they can couple with unoxidized aryl hydrazine to form diaryl tetrazenes[118,129,130] (Eq. 1-39), which

$$PhNHNH_2 \cdot H_2SO_4 \xrightarrow{\begin{subarray}{c} HgO, \text{ excess} \\ \\ HgO, \text{ limited} \end{subarray}} \begin{array}{l} PhN_2^+ \; HSO_4^- \\ \\ [PhN{=}NNHNHPh] \longrightarrow \\ \qquad\qquad PhN_3 + PhNH_2 \end{array} \tag{1-39}$$

usually decompose into aryl azide and aniline. Under appropriate conditions, the usual decomposition products of diazonium salts can be obtained (Eq. 1-40). In

$$PhNHNH_2 \cdot HCl \xrightarrow[\text{Heat}]{\text{aq. } CuSO_4} PhCl + Cu \tag{1-40}$$

view of all the possibilities for reaction, it is not surprising that the oxidation of aryl hydrazines can give very complex mixtures.

Sym-disubstituted hydrazines lose two hydrogen atoms easily, and the azo compounds produced, being reasonably stable, can usually be isolated (Eq. 1-41). A

$$RNHNHR \xrightarrow{[O]} RN{=}NR \tag{1-41}$$

wide variety of oxidizing agents, including bromine, permanganate, mercuric oxide,[131] hydrogen peroxide,[132] cupric chloride,[133] and active manganese dioxide,[134] can be used. The last reagent gives good yields of cis-azo compounds if the temperature is kept below 70°C. Oxidation of hydrazines is perhaps the best general method for preparing aliphatic azo compounds.[135]

Peroxyacetic acid carries the oxidation all the way to azoxy compounds[136,137] (Eq. 1-42). Oxidation with molecular oxygen is particularly interesting, for it can give rise to hydrogen peroxide nearly quantitatively.[138]

$$\text{ArNHNHAr} + \text{AcOOH} \longrightarrow \text{ArN}\!\!=\!\!\overset{+}{\underset{\underset{O^-}{|}}{N}}\text{Ar} \tag{1-42}$$

With *unsym*-disubstituted hydrazines, the mildest oxidation removes one hydrogen atom and results in loss of nitrogen and formation of a secondary amine (Eq. 1-43). Even Fehling's solution can accomplish this, but ferric chloride or permanga-

$$\text{R}_2\text{NNH}_2 \xrightarrow[\text{or FeCl}_3]{\text{Cu(NH}_3)_4{}^{24}} \text{R}_2\text{NH} + \text{N}_2 \tag{1-43}$$

nate is easier to use. The reaction has value in structure proof, since the secondary amines are usually easily identified.[69,119]

The second stage of oxidation of *unsym*-disubstituted hydrazines, removal of two hydrogen atoms, leads to azamines (aminonitrenes, $\text{R}_2\overset{+}{\text{N}}\!\!=\!\!\text{N}^- \leftrightarrow \text{R}_2\text{N}\!-\!\overset{..}{\text{N}}\!:$), their dimers (tetrazenes) (Eq. 1-44), or their decomposition products (Eqs. 1-45 and 1-46).[124,139-146] What is isolated in a given experiment depends not only on how

$$\text{(CH}_3)_2\text{NNH}_2 \underset{\text{SnCl}_2}{\overset{\text{H}^+, \text{Br}_2}{\rightleftarrows}} \text{(CH}_3)_2\overset{+}{\text{N}}\!\!=\!\!\text{NH} \xrightarrow{\text{OH}^-}$$
$$\text{(CH}_3)_2\text{NN}\!\!=\!\!\text{NN(CH}_3)_2 \tag{1-44}$$

$$\text{R}_2\overset{+}{\text{N}}\!\!=\!\!\text{N}^- \longrightarrow \text{R}\!-\!\text{R} + \text{N}_2 \tag{1-45}$$

$$\text{R}_2\text{NN}\!\!=\!\!\text{NNR}_2 \longrightarrow \text{N}_2 + \text{R}_2\text{N}\!-\!\text{NR}_2 \tag{1-46}$$

much oxidizing agent is used, but also on the nature of the oxidizing agent and the conditions. Since the decomposition of N',N'-dialkylsulfonhydrazides, which is believed to pass through the azamine stage, sometimes gives products different from those of oxidation of the corresponding hydrazine,[145] it seems reasonable to attribute the differences to differences in the ease of formation of tetrazene under the respective reaction conditions.

It has been shown[142] that there is an optimum pH for tetrazene formation, which apparently requires the presence of both azamine and its conjugate acid. Tetrazenes might also be formed through reaction of the azamine with unreacted hydrazine to give an easily oxidized tetrazane.[139] Fragmentation of azamines, which apparently gives free radicals that either dimerize (Busch-Weiss reaction[147]) or disproportionate, appears to be favored by the presence on the α-carbons of groups that stabilize radicals or charges in the transition state. Oxidation of N-amino-2,6-dimethylpiperidine leads to tetrazene, but oxidation of N-amino-2,6-dicyanopiperidine gives mostly 1,2-dicyanocyclopentane.[148] When azamines are generated by reduction of nitrosamines with dithionite, the same products are obtained as when they are formed from hydrazines.

Trisubstituted hydrazines are more difficult to oxidize than mono- and disubstituted compounds and are generally not affected by Tollens' or Fehling's reagents. The difference in susceptibility is well illustrated by 2,4-dinitro-5-(dimethylhydrazino)-phenylhydrazine (Eq. 1-47); hot, neutral permanganate affects only the unsubstituted hydrazino group, giving 2,4-dinitrophenyl-N',N'-dimethylhydrazine.[69]

$$\text{O}_2\text{N}\underset{\underset{\text{Me}_2\text{NNH}}{|}}{\overset{\overset{\text{NO}_2}{|}}{\bigcirc}}\text{NHNH}_2 \xrightarrow[\text{aq. EtOH}]{\text{KMnO}_4} \text{O}_2\text{N}\underset{\underset{\text{Me}_2\text{NNH}}{|}}{\overset{\overset{\text{NO}_2}{|}}{\bigcirc}} + \text{N}_2 \qquad (1\text{-}47)$$

The immediate products of oxidation of trisubstituted hydrazines are free radicals (hydrazyls).[149-151] Some of them are reasonably stable and exist in equilibrium with tetrazanes, their dimers (Eq. 1-48). Trialkyl hydrazyls, formed either electrolyti-

$$\underset{R}{\overset{Ar}{>}}\text{N}-\text{NHR} \xrightarrow{[O]} \underset{R}{\overset{Ar}{>}}\text{N}-\overset{\cdot}{\text{N}}\text{R} \longrightarrow \underset{R}{\overset{Ar}{>}}\text{N}-\underset{\underset{R}{|}}{\text{N}}-\underset{\underset{R}{|}}{\text{N}}-\text{N}\overset{Ar}{\underset{R}{<}} \qquad (1\text{-}48)$$

cally[152a] or by photolysis in the presence of di-t-butyl peroxide,[152b] are less stable, in general, but some have been reported to be indefinitely stable in solution in the absence of air.

The simplest form of two-electron oxidation would produce diazenium ions. N-Allyl-N,N'-di-t-butylhydrazine, for example, is oxidized by silver fluoroborate to an isolable diazenium salt[152a] (Eq. 1-49). When the N'-substituent bears an α-hydrogen, however, a prototropic shift occurs to form a hydrazonium salt[93] (Eq. 1-50).

$$\underset{\underset{\underset{\text{CH}=\text{CH}_2}{|}}{\text{CH}_2}}{\overset{t\text{-Bu}}{>}}\text{N}-\text{NH}-t\text{-Bu} + \text{AgBF}_4 \longrightarrow \underset{\underset{\underset{\text{CH}=\text{CH}_2}{|}}{\text{CH}_2}}{\overset{t\text{-Bu}}{>}}\overset{+}{\text{N}}=\text{N}-t\text{-Bu}\ \text{BF}_4^- \qquad (1\text{-}49)$$

$$(\text{CH}_3)_2\text{N}-\text{NHCH}_3 \xrightarrow{\text{Br}_2} [(\text{CH}_3)_2\overset{+}{\text{N}}=\text{NCH}_3] \longrightarrow$$

$$(\text{CH}_3)_2\overset{+}{\underset{\underset{\text{H}}{|}}{\text{N}}}-\text{N}=\text{CH}_2\ \text{Br}^- \quad (1\text{-}50)$$

Oxygen-donating agents lead to oxygen-containing products, as shown by the reaction of t-butyl hydroperoxide with triaryl hydrazines[153] (Eq. 1-51), which is

$$\text{Ph}_2\text{NNHAr} + t\text{-BuOOH} \longrightarrow \text{Ph}_2\text{NH} + \text{ArN}=\text{O} \qquad (1\text{-}51)$$

believed to be a chain reaction. Tetraphenylhydrazine is not attacked, whereas N,N-diphenylhydrazine is oxidized to diphenylamine. Hydrazyls formed by other reagents are converted by oxygen to N-amino nitroxides (hydrazinoxyls)[151] and by NO_2 in benzene to N-hydroxyhydrazines.[154]

Tetrasubstituted hydrazines are converted electrochemically[24,155] or by strong oxidizing agents to cation radicals.[94] With tetraaryl hydrazines, these are stable enough for the reaction to be easily reversible (Eq. 1-52). The reaction is analogous to that of triphenylamine. Salts of the colored cation radicals have been isolated in a num-

$$Ar_2N-NAr_2 \underset{KI}{\overset{Br_2}{\rightleftharpoons}} [Ar_2NNAr_2]^{+\cdot} \tag{1-52}$$

ber of cases. They have unfortunately been called "hydrazinium salts" in some of the older literature, a totally unacceptable term because it is the systematic name for nonradical hydrazine salts in general. *Hydrazylium* seems to be an unambiguous alternative.

Amination may be formally considered to be a form of oxidation, analogous to hydroxylation. Hydroxylamine-*O*-sulfonic acid, a general aminating agent toward nucleophiles, reacts with *unsym*-dimethylhydrazine to form a stable triazanium salt[156] (Eq. 1-53).

$$Me_2NNH_2 + NH_2OSO_3H \longrightarrow Me_2\overset{+}{N}\underset{NH_2}{\overset{NH_2}{\diagup}} \tag{1-53}$$

Reduction. Reduction of hydrazines cleaves the N—N bond and produces two molecules of amine[157] (Eq. 1-54). It is not a rapid reaction, however, and many

$$\underset{Me}{\overset{Ph}{\diagdown}}N-NH_2 \xrightarrow[HCl]{Zn} PhNHMe + NH_3 \tag{1-54}$$

hydrazines may therefore be prepared by reductive methods. Hydrogen over Raney nickel,[17,158] diborane,[159] and metals in the presence of strong acid[17] seem to be the most effective reducing agents, although magnesium with methanol has been used to reduce quaternary hydrazinium salts. *p,p'*-Dinitrodimethylhydrazobenzene is an unusual case, in that it is reduced smoothly by sodium sulfide to *p*-nitro-*N*-methyl-aniline.[28] Pentafluorophenylhydrazine[160] and decafluorohydrazobenzene[161] have been reduced with aqueous hydrogen iodide. A different type of reduction is found in the reaction of tetrafluorohydrazine with propene[162] (Eq. 1-55), but it is not known whether it has any generality.

$$F_2N-NF_2 + CH_2=CHCH_3 \longrightarrow F_2NCH_2CH\underset{NF_2}{\overset{CH_3}{\diagup}} \tag{1-55}$$

PREPARATIVE METHODS[163-167]

With hydrazines as with amines, the simplest seeming preparative method, direct alkylation of the inorganic parent, is not the best method, because multiple alkylation is hard to prevent.[3,38,39,70-72,168] Alkylation of hydrazine with conventional alkylating agents may give usable yields of monoalkylhydrazine, but di-, tri-, and tetraalkyl hydrazines are formed as well. The yield of ethylhydrazine from ethyl sulfate, for example, is only 32%.[169,170] Results can be improved by using a large excess of hydrazine, and butylhydrazine has been obtained in 71% yield using a tenfold excess[171] (Eq. 1-56), but the cost in wasted hydrazine may not be trivial.

$$BuBr + 10N_2H_4 \longrightarrow BuNHNH_2 + N_2H_5Br \qquad (1\text{-}56)$$

Other types of alkylating agents have been used, such as epoxides, α,β-unsaturated carbonyl compounds or nitriles, and alkanediazoates.[172] Compounds bearing "positive" halogens, such as bromomalonic ester and α-bromoacetoacetic ester, do not alkylate hydrazine, but are reduced by it. Several β-phenylethylhydrazines have been made by alkylation of sodium hydrazide ($NaNHNH_2$) with styrenes.[173]

Aryl hydrazines, however, can be prepared quite satisfactorily by direct arylation if sufficiently active aryl halides, such as 2,4-dinitrochlorobenzene or 2-chloropyridine,[174] are used, and this is the best way to prepare the carbonyl reagent 2,4-dinitrophenylhydrazine.[175] If the aryl halide is unreactive, it may not be feasible to force the reaction, owing to oxidation-reduction reactions and condensation of the hydrazino group with a nitro group.[176] Arylation with phenols is applicable in the naphthalene series by use of the conditions of the Bucherer reaction[177] (Eq. 1-57),

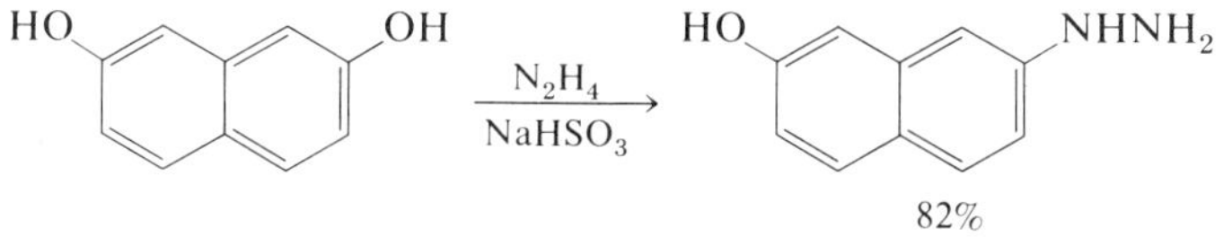

$$(1\text{-}57)$$

82%

but phenol itself gives only a poor yield of phenylhydrazine. Aryl hydrazines may also be prepared through a benzyne intermediate by reaction of unactivated aryl halides with sodium hydrazide; some diaryl hydrazine may be formed, and there is a problem with positional isomers in most instances.[173]

Multiple alkylation can in principle be prevented by means of blocking groups, as in the preparation of primary amines, but with hydrazine, there is only one generally used reagent, benzaldazine. It provides a good method for preparing methylhydrazine[178] (Eq. 1-58), but azines are poor nucleophiles, and only fairly

$$PhCH{=}NN{=}CHPh + (CH_3)_2SO_4 \longrightarrow$$

$$PhCH{=}\overset{+}{N}N{=}CHPh\ CH_3OSO_3^- \xrightarrow{H_2O,\ H^+}$$
$$\underset{CH_3}{|}$$

$$PhCH{=}O + CH_3NHNH_2 \text{ (as salt)} \qquad (1\text{-}58)$$

reactive alkylating agents will react with them. Ethylhydrazine has been prepared in 95% yield using triethyloxonium fluoroborate as the alkylating agent.[179]

Pyridyl, quinolyl, and isoquinolyl hydrazines have been prepared from the corresponding parent heterocycles by heating with sodium hydrazide[14,173] or hydrazine[128b] (Eq. 1-59) in a modification of the Tschitschibabin reaction; yields vary from 25 to greater than 90%.

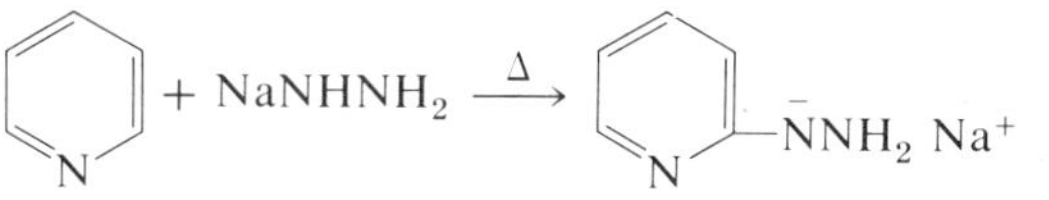

$$(1\text{-}59)$$

Many reductive methods are known for preparing monoalkyl hydrazines; in general, they involve reduction of hydrazone or hydrazide functions. The most general method for primary alkyl hydrazines is probably reduction of hydrazides with lithium aluminum hydride[77] (Eq. 1-60), although yields are commonly rather low.

$$RCONHNH_2 \xrightarrow{LiAlH_4} RCH_2NHNH_2 \tag{1-60}$$

Reduction of azines or the more difficultly prepared unsubstituted hydrazones also gives primary alkyl hydrazines if aldehyde derivatives are used; reduction is usually carried out with sodium amalgam or by catalytic hydrogenation over platinum or palladium.[180,181] With azines, the amount of reduction must be controlled so that only one imine bond is reduced, and the resulting hydrazone must then be hydrolyzed (Eq. 1-61). Sodium amalgam does not reduce ketazines, however. Some

$$R_2C{=}NN{=}CR_2 \xrightarrow{[H]} R_2CHNHN{=}CR_2 \xrightarrow{[H_2O]} R_2CHNHNH_2 \tag{1-61}$$

chemists have preferred to allow reduction to go all the way to *sym*-disubstituted hydrazines and then to oxidize these to azoalkanes,[182] which are readily isomerized to the hydrazone that would have been produced by partial reduction[180] (Eq. 1-62).

$$R_2C{=}NN{=}CR_2 \xrightarrow{[H]} R_2CHNHNHCHR_2 \xrightarrow[K_2Cr_2O_7]{HgO\ or}$$

$$R_2CHN{=}NCHR_2 \longrightarrow$$

$$R_2CHNHN{=}CR_2 \xrightarrow{[H_2O]} R_2CHNHNH_2 \tag{1-62}$$

A modification of the reduction of hydrazones is reductive alkylation of hydrazine[183] by treating a mixture of ketone and hydrazine hydrochloride with hydrogen and a platinum catalyst.

A related route is reduction of *N*-acylhydrazones (Eq. 1-63); semicarbazones[184]

$$R_2C{=}NNHCOX \xrightarrow[AcOH,\ 50\ psi]{H_2/PtO_2} R_2CHNHNHCOX \xrightarrow{[H_2O]}$$

$$R_2CHNHNH_2 \tag{1-63}$$

and *N*-carboethoxyhydrazones[185,186] are usually used. With aldehyde derivatives, some *N,N*-dialkyl hydrazine may be formed. There is a considerable advantage in this method over the use of azines or simple hydrazones, for all the original ketone or aldehyde is converted to hydrazine (rather than just half, as in the case of azines), and these substituted hydrazones are much easier to prepare and handle than are unsubstituted hydrazones. It has been amply demonstrated that lithium aluminum hydride may be used in the reducing step.[144,187]

Alkyl hydrazines can also be prepared by reduction of *N*-alkyl-*N*-nitroso ureas or urethans by catalytic hydrogenation and hydrolysis[188,189] (Eq. 1-64). Since the nitroso compounds are prepared from *N*-alkyl ureas or urethans, which are in turn easily prepared from amines, this route constitutes a synthesis of hydrazines by indirect amination of primary amines.

Aliphatic diazo compounds can be reduced to hydrazines, but the process is not

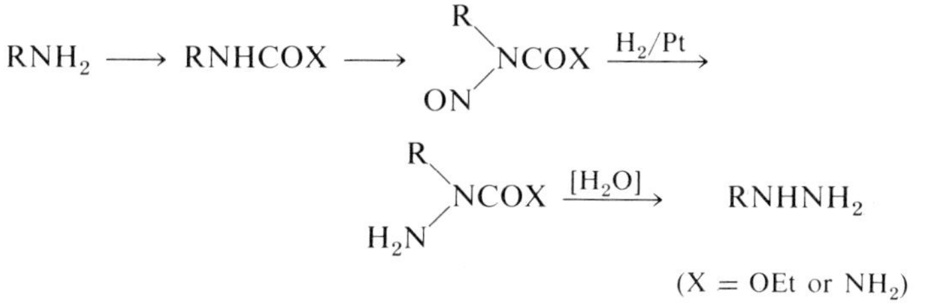

$$RNH_2 \longrightarrow RNHCOX \longrightarrow \underset{ON}{\overset{R}{\diagdown}}NCOX \xrightarrow{H_2/Pt}$$

$$\underset{H_2N}{\overset{R}{\diagdown}}NCOX \xrightarrow{[H_2O]} RNHNH_2 \qquad (1\text{-}64)$$

$$(X = OEt \text{ or } NH_2)$$

regarded as a useful synthesis for them, because the intermediates required to prepare most diazoalkanes can be converted directly to a hydrazine more efficiently. However, reduction of aromatic diazonium salts is a major preparative method for aryl hydrazines.[190-192] Reduction is usually accomplished by dissolving metals or stannous chloride (see Chap. 4). Since diazonium salts usually have low stability and cannot be heated to speed up otherwise slow reductions, it may be desirable to make use of Fischer's method, in which the diazonium salt is converted to a more stable arylazosulfonate salt by reaction with sodium bisulfite (Eq. 1-65). The stan-

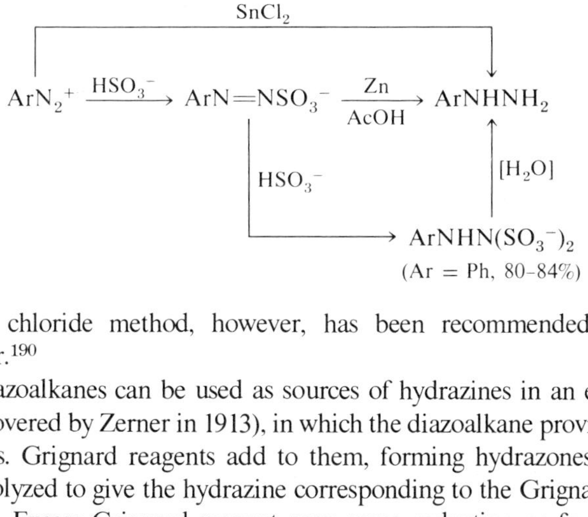

$$ArN_2^+ \xrightarrow{HSO_3^-} ArN{=}NSO_3^- \xrightarrow[AcOH]{Zn} ArNHNH_2 \qquad (1\text{-}65)$$

$$\begin{array}{c} SnCl_2 \\ \downarrow HSO_3^- \quad \uparrow [H_2O] \\ \longrightarrow ArNHN(SO_3^-)_2 \end{array}$$

$$(Ar = Ph, 80\text{-}84\%)$$

nous chloride method, however, has been recommended as being generally better.[190]

Diazoalkanes can be used as sources of hydrazines in an entirely different way (discovered by Zerner in 1913), in which the diazoalkane provides only the nitrogen atoms. Grignard reagents add to them, forming hydrazones, which can then be hydrolyzed to give the hydrazine corresponding to the Grignard reagent[193,194] (Eq. 1-66). Excess Grignard reagent may cause reduction or further addition, so the

$$Ph_2C{=}N_2 \xrightarrow[\text{(2) } H_2O]{\text{(1) } t\text{-BuMgCl}} \underset{60\%}{Ph_2C{=}NNH{-}t\text{-Bu}} \xrightarrow{[H_2O]}$$

$$\underset{87\%}{t\text{-BuNHNH}_2 + Ph_2CO} \qquad (1\text{-}66)$$

method is not always successful. A related method is the addition of Grignard reagents to diazirines, discussed later. Another manifestation of addition to N—N multiple bonds is shown in the preparation of t-butylhydrazine in about 25% overall yield by treating di-t-butyl azoformate with t-butylmagnesium chloride, followed by hydrolysis of the resulting hydrazide[195] (Eq. 1-67).

$$t\text{-BuO}_2\text{CN}{=}\text{NCO}_2\text{---}t\text{-Bu} \xrightarrow[\text{(2) H}_2\text{O}]{\text{(1) } t\text{-BuMgCl}} t\text{-BuO}_2\text{CNHN}\overset{\text{CO}_2\text{-}t\text{-Bu}}{\underset{t\text{-Bu}}{\diagdown}} \xrightarrow{\text{[H}_2\text{O]}}$$

$$t\text{-BuNHNH}_2 + \text{CO}_2 + t\text{-BuOH} \qquad (1\text{-}67)$$

Yet another way of preparing monosubstituted hydrazines with organometallic reagents is to allow them to add to one of the $C{=}N$ bonds of an azine, analogous to addition to carbonyl groups. The product is an N-substituted hydrazone,[196] from which the hydrazine can be obtained by hydrolysis (Eq. 1-68). The most serious

$$\text{Me}_2\text{C}{=}\text{N---N}{=}\text{CMe}_2 \xrightarrow{\text{RMgX or RLi}} \text{RC(Me}_2)\text{NHN}{=}\text{CMe}_2 \xrightarrow{\text{[H}_2\text{O]}}$$

$$\underset{\underset{R}{|}}{\text{Me}_2\text{C---NHNH}_2} + \text{O}{=}\text{CMe}_2 \qquad (1\text{-}68)$$

$$(\text{R} = \text{Ph, 30\%})$$

limitation of the method seems to be the incursion of a competing acid-base reaction when the azine has α-hydrogens, thus destroying the organometallic reagent.

There are several methods in which hydrazines are prepared by creating a new $N\text{---}N$ bond. The oldest is due to Shestakov, who showed that urea can be converted into hydrazine by treatment with hypochlorite and a base.[197] Although the reaction bears a formal analogy to the Hofmann rearrangement of amides, it probably proceeds through a diaziridinone, formed by an internal nucleophilic displacement in the anion of N-chlorourea. Aryl hydrazines have been obtained from aryl ureas in this way,[198,199] and surprisingly high yields of $tert$-alkyl hydrazines have been claimed[200] (Eq. 1-69).

$$\text{RNHCONH}_2 \xrightarrow[\text{H}_2\text{O, 15°C}]{\text{Cl}_2} \text{RNHCONHCl} \xrightarrow[\text{10–25°C}]{\text{NaOH}(aq)}$$

$$\text{RNHNH}_2 \qquad (1\text{-}69)$$

$$(\text{R} = t\text{-Bu, 90\%; R} = \alpha\text{-cumyl, 70\%})$$

Bimolecular nucleophilic displacement to form the hydrazo or hydrazino groups has been accomplished with a variety of reagents of the type NH_2X or, less satisfactorily, RNHX, in which X is halogen, sulfonyloxy, or even dinitrophenyl. Chloramine[201] and hydroxylamine-O-sulfonic acid are the most accessible reagents of this type and thus the most widely used.[202,203] Yields are usually moderate[204] (Eq. 1-70). Mesitylenesulfonyloxyamine has been used mostly with secondary and tertiary amines.[205] 2,4-Dinitrophenoxyamine gives high yields, but is somewhat troublesome to make.[206] O-Mesitoylhydroxylamine is also useful.[207]

$$\text{BuNH}_2 + \text{NH}_2\text{OSO}_3\text{H} \longrightarrow \text{BuNHNH}_2 \cdot \text{H}_2\text{SO}_4 \qquad (1\text{-}70)$$

$$49\text{–}56\%$$

A distinctive variant of the foregoing reactions is the reaction of chloramine with ketimines to form N-substituted diaziridines, which are easily hydrolyzable to hydrazines[208,209] (Eq. 1-71). The same diaziridines can also be obtained by addition of Grignard reagents to diazirines.

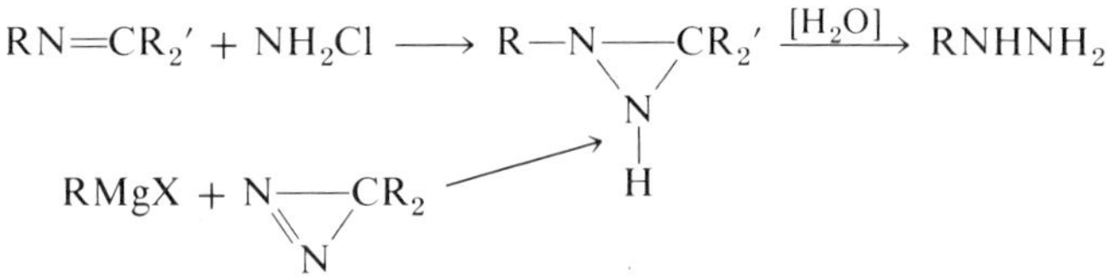

$$RN{=}CR_2' + NH_2Cl \longrightarrow R{-}N{-}CR_2' \xrightarrow{[H_2O]} RNHNH_2 \qquad (1\text{-}71)$$

$$RMgX + N{-}CR_2 \longrightarrow$$

Hydrolysis of sydnones is the heart of another method for preparing monoalkyl hydrazines from primary amines. The sydnones, mesoionic oxadiazole derivatives, are prepared from primary amines and chloroacetic acid in three steps, generally in good yield[210-212] (Eq. 1-72).

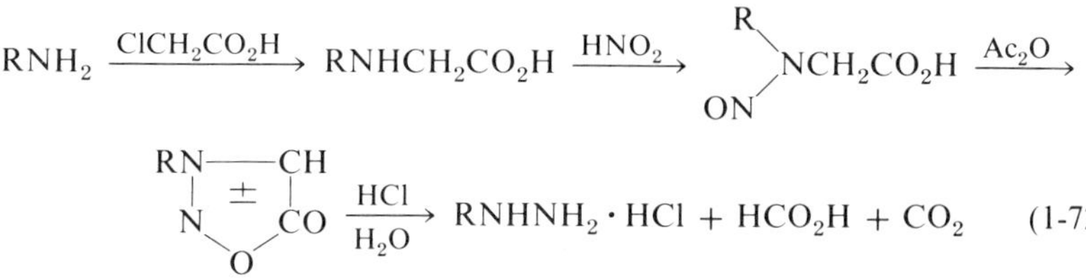

$$RNH_2 \xrightarrow{ClCH_2CO_2H} RNHCH_2CO_2H \xrightarrow{HNO_2} \underset{ON}{\overset{R}{\diagdown}}NCH_2CO_2H \xrightarrow{Ac_2O}$$

$$RN{-}CH \atop \underset{N}{|} \pm \underset{CO}{|} \xrightarrow[H_2O]{HCl} RNHNH_2 \cdot HCl + HCO_2H + CO_2 \qquad (1\text{-}72)$$

Olefins and aromatic hydrocarbons can be converted to hydrazines by reaction with azoformic esters[213] (see Chap. 4). The initial products are hydrazides (N,N'-dicarboethoxyhydrazines), which are easily hydrolyzed (Eq. 1-73). With olefins, the

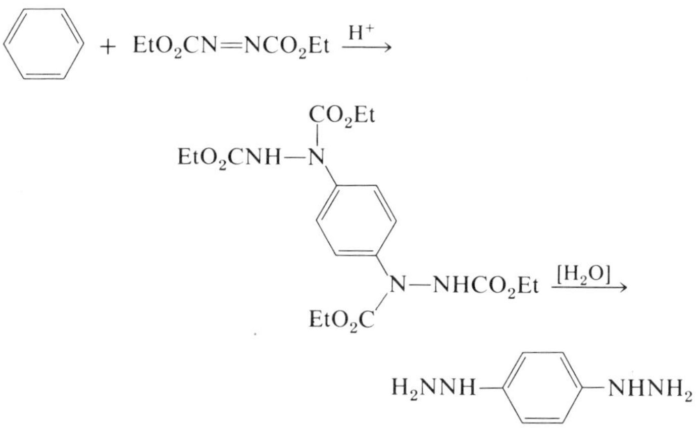

$$\bigcirc + EtO_2CN{=}NCO_2Et \xrightarrow{H^+}$$

$$EtO_2CNH{-}N{\overset{CO_2Et}{|}}$$

$$\underset{EtO_2C}{\diagup}N{-}NHCO_2Et \xrightarrow{[H_2O]}$$

$$H_2NNH{-}\bigcirc{-}NHNH_2 \qquad (1\text{-}73)$$

position of substitution is one of the trigonal carbons, and migration of the double bond occurs as a consequence of the ene reaction mechanism (Eq. 1-74). Either allylic hydrazines or their saturated analogs can be obtained. An advantage of the method is that primary alkyl hydrazines can be prepared without fear of contamination with di- and polyalkyl hydrazines.

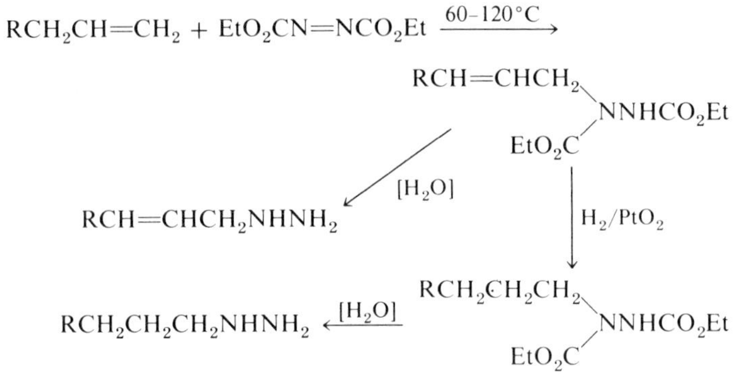

$$RCH_2CH{=}CH_2 + EtO_2CN{=}NCO_2Et \xrightarrow{60-120\,°C}$$

$$RCH{=}CHCH_2 \diagdown \atop EtO_2C \diagup \quad NNHCO_2Et$$

$$RCH{=}CHCH_2NHNH_2 \xleftarrow{[H_2O]} \qquad\qquad \downarrow H_2/PtO_2 \qquad\qquad (1\text{-}74)$$

$$RCH_2CH_2CH_2NHNH_2 \xleftarrow{[H_2O]} \quad RCH_2CH_2CH_2 \diagdown \atop EtO_2C \diagup \quad NNHCO_2Et$$

Disubstituted Hydrazines

Direct alkylation is not usually a useful process for the synthesis of *sym-* or *unsym-* disubstituted hydrazine, since on the one hand it is difficult to stop alkylation cleanly at a given stage and on the other hand it may give mixtures of *sym* and *unsym* isomers. Nevertheless, it may be a moderately successful method in specific instances, such as the reaction of benzyl chloride with a twofold excess of hydrazine, which affords *N,N*-dibenzylhydrazine in 35% yield.[214] When bulky groups are involved, *sym*-disubstituted hydrazines may be formed sufficiently cleanly that the method may have preparative value, as in the reaction of isopropylhydrazine with 2-bromo-1-phenylpropane.[215] An outstandingly successful use of alkylation is the preparation of *N-o*-chlorobenzyl-*N*-methylhydrazine from methylhydrazine and *o*-chlorobenzyl chloride in 92% yield.[216]

Sym-disubstituted hydrazines are more generally prepared by alkylating hydrazides; *N,N'*-di(ethoxycarbonyl)-,[217] *N,N'*-diformyl-,[218] and *N,N'*-dibenzoyl-hydrazine, which are easily available, are the usual substrates. They may be alkylated once on each nitrogen by most unhindered alkylating agents to form dialkyl hydrazides that can easily be hydrolyzed to *sym*-dialkyl hydrazines by hot aqueous mineral acid[219] (Eq. 1-75). By starting with a monosubstituted hydrazine, which can

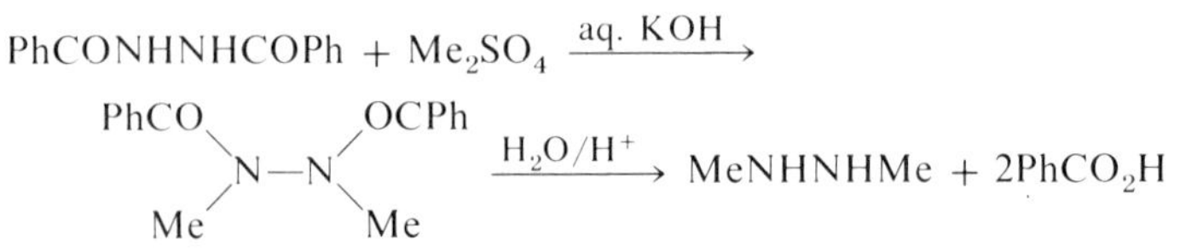

$$PhCONHNHCOPh + Me_2SO_4 \xrightarrow{aq.\ KOH}$$

$$PhCO \diagdown \quad OCPh \atop Me \diagup N{-}N \diagdown Me \xrightarrow{H_2O/H^+} MeNHNHMe + 2PhCO_2H \qquad (1\text{-}75)$$

be converted to the *N,N'*-dibenzoyl derivative by direct acylation, a second alkyl group of a different kind can be introduced, as exemplified by the preparation of *N*-ethyl-*N'*-methylhydrazine[179] and *N*-isopropyl-*N*-methylhydrazine.[220] However, it is not necessary to resort to diacylation for this purpose when starting with a monoalkyl hydrazine; *N*-methyl-*N*-acetylhydrazine, obtained by acetylation of methylhydrazine, may be alkylated cleanly on the unsubstituted nitrogen[73] (Eq. 1-76). Phthalhydrazide,[221] *N,N'*-di(*t*-butoxycarbonyl)hydrazine,[222] and *N,N'*-

$$\begin{matrix} Ac \\ \diagdown \\ \diagup \\ Me \end{matrix} NNH_2 \xrightarrow{RX} \begin{matrix} Ac \\ \diagdown \\ \diagup \\ Me \end{matrix} NNHR \xrightarrow{[H_2O]} MeNHNHR \qquad (1\text{-}76)$$

di(benzyloxycarbonyl)hydrazine[223] have also been used as substrates for the alkylative preparation of hydrazines.

An alternative way to utilize acyl groups to control alkylation is to generate the hydrazide anion by reduction of an azo compound rather than by treating the hydrazide with base. Thus N,N'-dibenzylhydrazine has been prepared in 60% yield by treating ethyl azoformate with metallic potassium to form the dianion, followed by alkylation with benzyl chloride.[224]

Blocking with acyl groups can also be used to prepare *unsym*-dialkyl hydrazines, for a single acyl group will direct alkylation to the unsubstituted nitrogen. *t*-Butoxycarbonylhydrazine has been successively alkylated with two different alkylating agents, for example[225] (Eq. 1-77).

$$t\text{-BuO}_2CNHNH_2 \xrightarrow[\text{(2) R'X}]{\text{(1) RX}} t\text{-BuO}_2CNHNRR' \xrightarrow{[H_2O]}$$

$$H_2NNRR' \qquad (1\text{-}77)$$

A large variety of reductive routes may be utilized for preparation of disubstituted hydrazines. They involve reduction at carbon, in the form of acyl or alkylidene groups, or at nitrogen, in the form of azo, nitroso, or nitro groups.

N-Substituted hydrazones of aldehydes or ketones may in general be reduced to *sym*-disubstituted hydrazines. Reagents that have been used successfully include hydrogen over a platinum catalyst,[226] sodium borohydride,[73] diborane,[227] lithium aluminum hydride,[38, 144, 228] and lithium in liquid ammonia.[229] Azines may be reduced similarly and have been used widely[71,230] (Eq. 1-78). Sodium amalgam is a

$$Me_2C=N-N=CMe_2 \xrightarrow{H_2/Pt} Me_2CHNHNHCHMe_2 \qquad (1\text{-}78)$$
$$90\%$$

useful reagent for the purpose, but it does not work well with purely aliphatic ketazines. Hydrogenation over platinum is more general, and lithium aluminum hydride has been used with success.[182]

Reduction of the acyl group of hydrazides to alkyl is commonly carried out with lithium aluminum hydride[228] (Eq. 1-79). It is an especially useful route to

$$RCONHNHPh \xrightarrow[\text{(2) H}_2O]{\text{(1) LiAlH}_4} RCH_2NHNHPh \qquad (1\text{-}79)$$

hydrazines with two different substituents. *Sym*-disubstituted hydrazines can be obtained by starting with *N'*-substituted hydrazides, and *unsym*-disubstituted hydrazines with *N*-substituted hydrazides. Methyl groups can be obtained by reduction of formyl or alkoxycarbonyl groups.[231] Although diacyl hydrazines can be reduced to dialkyl hydrazines,[77] it is sometimes possible to reduce an *N*-substituted hydrazide function selectively when a hydrazide function not substituted at N is also present. This makes possible the preparation of N-methyl-N-aryl hydrazines from the products of the reaction of an arene with ethyl azoformate[232] (Eq. 1-80).

$$\text{Arn}\underset{\text{CO}_2\text{Et}}{\overset{\text{NHCO}_2\text{Et}}{<}} \xrightarrow{\text{LiAlH}_4} \text{ArN}\underset{\text{CH}_3}{\overset{\text{NHCO}_2\text{Et}}{<}} \xrightarrow{[\text{H}_2\text{O}]} \underset{\text{CH}_3}{\overset{\text{Ar}}{>}}\text{NNH}_2 \qquad (1\text{-}80)$$

Combined reduction of hydrazide and hydrazone functions can be accomplished with lithium aluminum hydride and allows the preparation of N,N'-disubstituted hydrazines from N'-acyl hydrazones in useful yields[233] (Eq. 1-81).

$$\text{AcNHN}{=}\text{CMe}_2 \xrightarrow{\text{LiAlH}_4} \underset{61\%}{\text{EtNHNHCHMe}_2} \qquad (1\text{-}81)$$

The most widely used method for preparing *unsym*-disubstituted hydrazines is reduction of nitrosamines, which are easily prepared from secondary amines. Among the useful reducing agents are lithium aluminum hydride,[234-236] hydrogenation over a palladium catalyst,[237] sodium and alcohol,[234] aluminum under wet ether,[144] sodium and ammonia,[234] and zinc and acetic acid[238] (Eq. 1-82). The last-

$$t\text{-BuNHMe} \xrightarrow{\text{HNO}_2} \underset{\text{Me}}{\overset{t\text{-Bu}}{>}}\text{NNO} \xrightarrow{\text{Zn, AcOH}} \underset{\underset{24\text{-}67\%}{\text{Me}}}{\overset{t\text{-Bu}}{>}}\text{NNH}_2 \qquad (1\text{-}82)$$

named reagent succeeded where the others failed. Some investigators have reported difficulty with lithium aluminum hydride because it appears to have an induction period, after which it reacts so vigorously as to be potentially dangerous. One recommendation for avoiding this eventuality is to add the nitrosamine to a refluxing solution of the reducing agent in ether.[216] Electrolytic reduction has given good yields (i-Bu$_2$NNH$_2$, 78%).[239]

Diarylnitrosamines are difficult to reduce cleanly to hydrazines, owing to the ease with which the N—N bond is cleaved, and small to overwhelming amounts of the diarylamine, difficult to separate from the hydrazine, may be produced. The use of zinc dust with ammonium carbonate in ethanol at 0°C has succeeded where other methods have failed.[240] Alternatively, it has been recommended that the Curtius rearrangement of diarylcarbamoyl azides[241] (Eq. 1-83), which gives good overall yields of high purity, be used instead for this class of hydrazine.

$$\text{Ar}_2\text{NCOCl} \xrightarrow{\text{NaN}_3} \text{Ar}_2\text{NCON}_3 \xrightarrow{t\text{-BuOH}} \text{Ar}_2\text{NNHCO}_2\text{-}t\text{-Bu} \xrightarrow{\text{HCl (conc.)}}$$
$$\text{Ar}_2\text{NNH}_2 \qquad (1\text{-}83)$$

Reduction of aliphatic azo compounds is not commonly regarded as a useful preparative method for hydrazines, owing to the fact that hydrazines are frequently the starting materials for preparing the azo compounds. However, in certain instances it may have value. Reduction has been accomplished with zinc and alkali

(N,N'-dimethylhydrazine, 70%),[242] hydrazine and Raney nickel (N,N'-dibutylhydrazine, 84%),[243] and hydrogenation over palladium[244] (Eq. 1-84). Reduction of

$$t\text{-BuN}=\text{N-}t\text{-Bu} \xrightarrow{\text{H}_2/\text{Pd}} t\text{-BuNHNH-}t\text{-Bu} \qquad (1\text{-}84)$$
$$92\%$$

azobenzene derivatives, however, can be a practical route to hydrazobenzenes when the azo compounds can be obtained by diazonium coupling (see Chap. 4 for methods).

Hydrazobenzenes are available in wide variety through reduction of aromatic nitro compounds by various means under neutral or alkaline conditions, especially with zinc and sodium hydroxide[57,245] (Eq. 1-85).

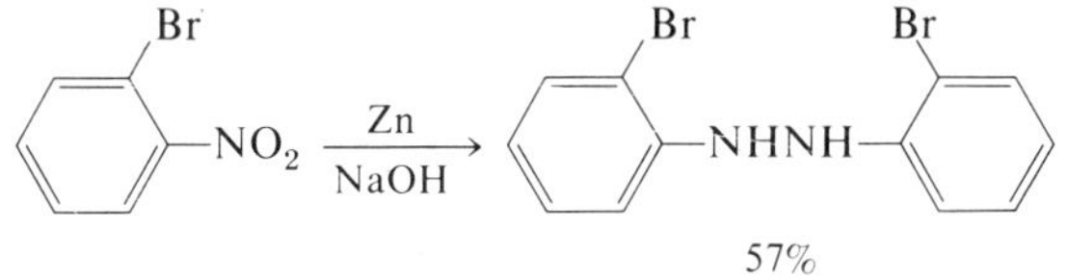

$$(1\text{-}85)$$
$$57\%$$

Displacement by a secondary amine on a reagent of the type NH_2X, analogous to Equation (1-70) has been used mostly with hydroxylamine-O-sulfonic acid[246] or chloramine.[201-203,247,248] Yields are generally moderate, but have run as high as 99 percent. Reagents of the type RNHX give only very low yields of N,N'-dialkyl hydrazines when treated with primary amines. Sym-disubstituted hydrazines can be prepared conveniently by intramolecular displacement of this sort, however. N,N'-Dialkyl ureas, readily prepared from isocyanates and primary amines, have been converted in high yields through their N-chloro derivatives to N,N'-dialkyl hydrazines[249] (Eq. 1-86). N,N'-Dialkyl sulfamides, which can be prepared from

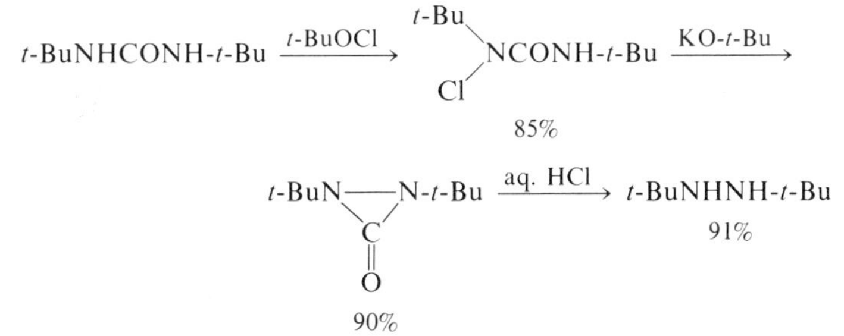

$$(1\text{-}86)$$

primary amines and sulfuryl chloride, also give good to high yields of N,N'-dialkyl hydrazines[250,251] (Eq. 1-87). The urea method is readily adapted to preparation of

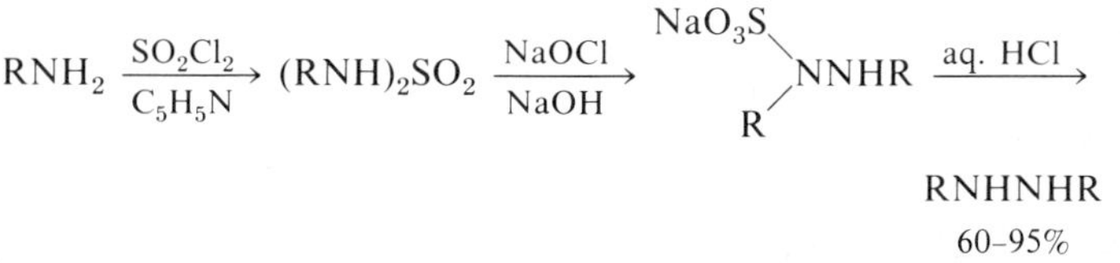

$$\text{RNHNHR} \qquad (1\text{-}87)$$
$$60\text{-}95\%$$

hydrazines with two different substituents, and the sulfamide route can be adapted by starting with chlorosulfonyl isocyanate.[252]

N,N'-Dialkyl hydrazines can be prepared from azines by reaction with one equivalent of Grignard reagent to form a monoalkyl hydrazone, which is then reduced[196] (Eq. 1-88). This scheme is suited to the preparation of hydrazines bearing two

$$R_2C{=}NN{=}CR_2 \xrightarrow[\text{(2) } H_2/PtO_2]{\text{(1) } R'MgX} R{-}\underset{\underset{R'}{|}}{\overset{\overset{R}{|}}{C}}{-}NHNHCHR_2 \tag{1-88}$$

different groups, and when a ketazine is used, one alkyl group will be tertiary. Addition of Grignard reagents to N-monosubstituted hydrazones, however, does not produce N,N'-disubstituted hydrazines (see Chap. 2).

The Tschitschibabin reaction (Eq. 1-59) may be adapted to the preparation of N-α-pyridyl-N'-alkyl hydrazines by use of sodium N-alkyl hydrazides.[14]

The formation of diaziridines from aldimines and alkylchloramines, analogous to Equation 1-71, followed by hydrolysis, makes a practical synthesis of N,N'-dialkyl hydrazines with differing alkyl substituents.[209,253]

Trisubstituted Hydrazines

Several of the reactions used to prepare less-substituted hydrazines can be adapted. Direct alkylation is of only limited use, such as in the case of N,N-diphenylhydrazine[27,29] (Eq. 1-15). Reduction of N,N-disubstituted hydrazones using lithium aluminum hydride or sodium cyanoborohydride has given good results[38,254,255] (Eq. 1-89). Both N,N'- and N',N'-dialkyl hydrazides can be reduced with lithium alumi-

$$CH_2{=}NNMe_2 \xrightarrow[\text{(2) } H_2O]{\text{(1) } LiAlH_4} CH_3NHNMe_2 \tag{1-89}$$

num hydride[228,231] or diborane[159] (Eq. 1-90). Yields may be poor to good, and diborane has in some cases given better results.

$$AcNHNMe_2 \xrightarrow[\text{(2) } H_2O]{\text{(1) } LiAlH_4} \underset{35\%}{EtNHNMe_2} \tag{1-90}$$

Addition of Grignard reagents to hydrazones is a general preparative method for trisubstituted hydrazines[35,256-258] (Eq. 1-91), but it is sensitive to crowding, and

$$Me_2NN{=}CHCH_3 \xrightarrow[\text{(2) } H_2O]{\text{(1) } EtMgBr} Me_2NNHC\overset{\overset{\displaystyle CH_3}{\diagup}}{\underset{\diagdown Et}{H}} \tag{1-91}$$

$$70\%$$

N—N cleavage may detract seriously from the yield. Addition of organometallic reagents to azobenzene derivatives does not give high yields,[259,260] but it may be useful in certain instances (triphenylhydrazine, 20%). Reduction of the azo group is a serious competing reaction.

Grignard reagents also add to nitrosamines to form hydrazines[261] (Eq. 1-92). A hydroxyhydrazine is presumably formed first, but more Grignard reagent reduces it.

$$
\begin{array}{c}
\text{Ar} \\
\diagdown \\
\text{N—N=O} + \text{R'MgX} \\
\diagup \\
\text{R}
\end{array}
\longrightarrow
\begin{array}{c}
\text{Ar} \qquad\qquad \text{OMgX} \\
\diagdown \qquad\qquad \diagup \\
\text{N—N} \\
\diagup \qquad\qquad \diagdown \\
\text{R} \qquad\qquad\quad \text{R'}
\end{array}
\xrightarrow{\text{R'MgX}}
$$

$$
\begin{array}{c}
\text{Ar} \\
\diagdown \\
\text{N—NHR'} \qquad (1\text{-}92)\\
\diagup \\
\text{R}
\end{array}
$$

The Tschitschibabin reaction with sodium dialkylhydrazides has been adapted to the preparation of N',N'-dialkyl-α-pyridyl hydrazines.[14]

The difficulties with direct alkylation, which is liable to give quaternary hydrazinium salts, have been circumvented by using iminophosphoranes (analogs of Wittig reagents), which can be prepared from N,N-dialkyl hydrazines. They undergo alkylation only at the imine nitrogen, and hydrolysis then gives trialkyl hydrazines.[262]

Tetrasubstituted Hydrazines

Tetrasubstituted hydrazines have been made in many instances by direct alkylation[3,27,29,214] (Eq. 1-93), starting with hydrazine or its mono-, di-, or trisubstituted

$$\text{BuNH—NH}_2 + \text{BuCl} \xrightarrow{\text{base}} \text{Bu}_2\text{N—NBu}_2 \qquad (1\text{-}93)$$

derivatives. The major limitations are that small, reactive alkylating agents, such as methyl iodide, favor continuing alkylation at the same nitrogen, producing quaternary hydrazinium salts, whereas alkylating agents that are secondary, even isopropyl, are usually incapable of adding a fourth alkyl group. Reductive alkylation using formaldehyde and sodium cyanoborohydride has been used with success in preparing many tetraalkyl hydrazines bearing methyl groups; one may start with either hydrazines or formaldehyde hydrazones.[22,255]

Reduction of N,N'-dialkyl-N,N'-diacyl hydrazines has been used to prepare a variety of tetraalkyl hydrazines (Eq. 1-94), both open-chain and cyclic, using lithium aluminum hydride or diborane.[22,77,159,263]

$$
\begin{array}{c}
\text{EtO}_2\text{C} \qquad\qquad \text{CO}_2\text{Et} \\
\diagdown \qquad\qquad\quad \diagup \\
\text{N—N} \\
\diagup \qquad\qquad\quad \diagdown \\
\text{CH}_3 \qquad\qquad\quad \text{CH}_3
\end{array}
\xrightarrow{\text{LiAlH}_4}
(\text{CH}_3)_2\text{NN}(\text{CH}_3)_2 \qquad (1\text{-}94)
$$

The foregoing methods are not adaptable to the preparation of tetraaryl hydrazines. These are instead prepared by oxidizing diarylamines with lead dioxide or permanganate[264,265] (Eq. 1-95). Tetraalkyl hydrazines can also be made by oxida-

$$\text{Ar}_2\text{NH} \xrightarrow{[O]} \text{Ar}_2\text{N—NAr}_2 \qquad (1\text{-}95)$$

tion, by electrolyzing lithium dialkylamides in tetrahydrofuran solution containing lithium perchlorate[266] (Eq. 1-96).

$$\text{Bu}_2\text{NLi} \xrightarrow[\text{THF/LiClO}_4]{\text{anodic ox.}} \underset{38\%}{\text{Bu}_2\text{N---NBu}_2} \tag{1-96}$$

An alkylarylamine has been converted to a dialkyldiaryl hydrazine by coupling its lithium derivative with its *N*-chloro derivative[267] (Eq. 1-97), but the generality of the

$$3,5\text{-}(t\text{-Bu})_2\text{C}_6\text{H}_3\text{N(Li)---}t\text{-Bu} + 3,5\text{-}(t\text{-Bu})_2\text{C}_6\text{H}_3\text{N} \begin{matrix} t\text{-Bu} \\ \diagdown \\ \text{Cl} \end{matrix} \longrightarrow$$

$$\begin{matrix} \text{Ar} & & \text{Ar} \\ \diagdown & & \diagup \\ & \text{N---N} \\ \diagup & & \diagdown \\ t\text{-Bu} & & t\text{-Bu} \end{matrix} \tag{1-97}$$

method appears not to have been explored. Another coupling method starts with *unsym*-disubstituted hydrazines, which are first oxidized to tetrazenes. These lose nitrogen on heating and give rise to hydrazines[268] (Eq. 1-98).

$$\begin{matrix} \text{Ar} \\ \diagdown \\ & \text{NNH}_2 \\ \diagup \\ \text{R} \end{matrix} \xrightarrow[\text{or HgO, FeCl}_3, \text{ HOCl}]{\text{EtO}_2\text{CN}{=}\text{NCO}_2\text{Et}} \begin{matrix} \text{Ar} & & \text{Ar} \\ \diagdown & & \diagup \\ & \text{NN}{=}\text{NN} \\ \diagup & & \diagdown \\ \text{R} & & \text{R} \end{matrix} \xrightarrow{120-140\,^{\circ}\text{C}}$$

$$\begin{matrix} \text{Ar} & & \text{Ar} \\ \diagdown & & \diagup \\ & \text{N---N} \\ \diagup & & \diagdown \\ \text{R} & & \text{R} \end{matrix} \tag{1-98}$$

Quaternary Hydrazinium Salts

Direct alkylation of less-substituted hydrazines is generally successful (see Eq. 1-12), so long as the groups involved are not too large.[15,17] Methylation is nearly always successful in producing quaternary salts, but with larger groups, alkylation usually stops short of quaternization. Tetrasubstituted hydrazines may be difficult to quaternize.[77] Quaternary salts can also be obtained by hydrolysis of quaternary hydrazonium salts[269] or quaternized hydrazides (Eq. 1-99). Amination of tertiary

$$\overset{+}{\text{R}_3\text{N}}\text{---N}{=}\text{CR}_2' \quad \text{or} \quad \overset{+}{\text{R}_3\text{N}}\text{---NHCOR}' \xrightarrow{\text{H}_2\text{O, H}^+} \overset{+}{\text{R}_3\text{N}}\text{---NH}_2 \tag{1-99}$$

amines with chloramine or hydroxylamine-*O*-sulfonic acid appears to be a general route to quaternary hydrazinium salts[15,246,247,270,271] (Eq. 1-100).

$$\text{R}_3\text{N} + \text{NH}_2\text{X} \longrightarrow \overset{+}{\text{R}_3\text{N}}\text{---NH}_2\text{X}^- \tag{1-100}$$

Amine Imides

Amine imides are generally formed by treating quaternary hydrazinium salts with a strong base[16] (Eq. 1-101). However, they are extremely difficult to isolate. Some

$$\overset{+}{\text{R}_3\text{N}}\text{---NH}_2 \xrightarrow{\text{base}} \overset{+}{\text{R}_3\text{N}}\text{---NH}^- \tag{1-101}$$

β-hydroxyalkyl derivatives, which are perhaps stabilized by hydrogen bonding, have been prepared by the reaction of oxiranes with *unsym*-dialkyl hydrazines[272,273] (Eq. 1-102).

$$R_2NNH_2 + R'CH\underset{O}{\diagdown\!\!\!\diagup}CH_2 \longrightarrow R_2\overset{+}{N}-NH^- \atop \underset{CH_2CHOHR'}{|} \qquad (1\text{-}102)$$

ANALYTICAL METHODS[274]

Qualitative and quantitative analytical methods for hydrazines depend mostly on the pronounced susceptibility of hydrazines to oxidation. Tollens' reagent is frequently used for qualitative detection, when it is known that other easily oxidizable functions are absent. Quantitative methods have been developed primarily for hydrazobenzenes. Oxidation with excess standard potassium permanganate followed by iodimetric titration of the excess permanganate is one such method. Another is based on the capacity of the dye Bindschedler's green to oxidize hydrazo compounds. When the oxidation is fast enough, the highly colored oxidant serves as its own indicator; otherwise, excess standard dye solution can be used and the excess titrated with titanous chloride.[275,276] Iodate can be used to titrate alkyl hydrazines,[277] and periodate to titrate aryl hydrazines.[278] Cupric ion liberates nitrogen quantitatively from monosubstituted hydrazines.[279] Very strong reducing agents, especially titanous chloride, can be used to titrate hydrazo compounds by reducing them to amines.[274,280] A colorimetric method has been developed on the basis of the benzidine rearrangement. The benzidine can be diazotized and coupled to a naphthylamine to form an intensely colored azo dye.[281]

Hydrazines having an unsubstituted NH_2 group react essentially quantitatively with aldehydes and ketones. Qualitative detection and even assay may be accomplished with choice of a suitable reagent, preferably an aldehyde. Solid aldehydes (e.g., nitrobenzaldehydes) would be suitable for gravimetric assay, whereas highly conjugated aldehydes (cinnamaldehyde or dimethylaminobenzaldehyde) are more suited to colorimetric detection of low concentrations.[274] A volumetric method has been developed on the basis of hydrazone formation using a standard solution of benzaldehydesulfonic acid and an external indicator.[282]

If amines are known to be absent, hydrazines can be titrated with standard acid, using the techniques developed for amines in aqueous or nonaqueous medium.

REFERENCES

1. Reviewed in detail by P. J. Krueger, in *The Chemistry of the Hydrazo, Azo and Azoxy Groups*, S. Patai (ed.), Wiley, New York, 1975, pp. 159–164.
2. R. L. Hinman, *J. Org. Chem.* **23**, 1587 (1958).
3. O. Westphal, *Ber.* **74**, 759, 1365 (1941).
4. M. J. Gregory and T. C. Bruice, *J. Am. Chem. Soc.* **89**, 4400 (1967).
5. J. E. Dixon and T. C. Bruice, *J. Am. Chem. Soc.* **94**, 2052 (1972).
6. J. D. Aubort, R. F. Hudson, and R. C. Woodcock, *Tetrahedron Lett.*, 2229 (1973).

7. T. C. Bruice, A. Denzel, R. W. Huffman, and A. R. Butler, *J. Am. Chem. Soc.* **89**, 2106 (1967).

8. J. O. Edwards and R. G. Pearson, *J. Am. Chem. Soc.* **84**, 16 (1962).

9. J. R. Perrott, G. Stedman, and N. Uysal, *J. Chem. Soc.* [*Perkin II*], 274 (1977).

10. F. E. Condon, R. T. Reece, D. G. Shapiro, D. C. Thokker, and T. B. Goldstein, *J. Chem. Soc.* [*Perkin II*], 1112 (1974).

11. H. H. Stroh and O. Westphal, *Chem. Ber.* **96**, 184 (1963); **97**, 83 (1964).

12. W. H. Beamer, *J. Am. Chem. Soc.* **70**, 2979 (1948).

13. S. F. Nelsen and R. T. Landis, Jr., *J. Am. Chem. Soc.* **95**, 5422 (1973).

14. T. Kauffmann, J. Hansen, C. Kosel, and W. Schoeneck, *Liebig's Ann. Chem.* **656**, 103 (1962).

15. H. H. Sisler, G. M. Omeitanski, and B. Rudner, *Chem. Rev.* **57**, 1021 (1957).

16. R. Appel, N. Heinen, and R. Schöllhorn, *Chem. Ber.* **99**, 3118 (1966).

17. P. A. S. Smith and G. L. DeWall, *J. Am. Chem. Soc.* **99**, 5751 (1977).

18. R. F. Smith, R. D. Blondell, R. A. Abgott, K. B. Lipkowitz, J. A. Richmond, and K. A. Fountain, *J. Org. Chem.* **39**, 2036 (1974).

19. B. K. Singh, *J. Chem. Soc.* **103**, 604 (1913); **105**, 1972 (1914).

20. J. A. Blair and R. J. Garder, *J. Chem. Soc.* [*C*], 2707 (1970).

21. F. E. Condon, *Org. Mag. Res.* **6**, 517 (1974).

22. S. F. Nelsen, V. Peacock, and G. R. Weisman, *J. Am. Chem. Soc.* **98**, 5269 (1976).

23. H. Musso, *Chem. Ber.* **92**, 2881 (1959).

24. S. F. Nelsen, L. A. Grazzo, and V. E. Peacock, *J. Org. Chem.* **46**, 2402 (1981).

25. Y. Shvo, in *The Chemistry of the Hydrazo, Azo, and Azoxy Groups,* S. Patai (ed.), Wiley, New York, 1975, Chap. 21.

26. S. F. Nelsen and P. M. Gannett, *J. Am. Chem. Soc.* **103**, 3300 (1981); C. Van Alsenoy, J. O. Williams, J. N. Scarsdale, H. J. Geise, G. Smits, and L. Schäfer, *Bull. Soc. Chim. Belg.* **89**, 737 (1980).

27. H. Wieland and E. Schamberg, *Ber.* **53**, 1329 (1920).

28. O. N. Witt and A. Kopetchni, *Ber.* **45**, 1134 (1912).

29. G. S. Hammond, B. Seidel, and R. E. Pincock, *J. Org. Chem.* **28**, 3275 (1963).

30. A. Heesing and U. Schinke, *Chem. Ber.* **110**, 2867 (1977).

31. H. J. Shine, F.-T. Huang, and R. L. Snell, *J. Org. Chem.* **26**, 380 (1961).

32. Y. O. Lukashevich and L. G. Krolik, *Dokl. Akad. Nauk S.S.S.R.* **147**, 1090 (1962) (English trans. p. 1080).

33. H. Wieland, *Ber.* **48**, 1091 (1915).

34. P. Wetzel, C. Dietz, and G. Eckhardt, *Chem. Ber.* **108**, 3550 (1975).

35. O. Westphal and M. Eucken, *Ber.* **76**, 1137 (1943).

36. D. J. Cram and M. C. V. Sahyun, *J. Am. Chem. Soc.* **85**, 1257 (1963).

37. J. Thiele and K. Heuser, *Liebig's Ann. Chem.* **290**, 1 (1896).

38. E. F. Elslager, E. A. Weinstein, and D. F. Worth, *J. Med. Chem.* **7**, 493 (1964).

39. W. H. Urry, P. Szecsi, C. Ikoku, and D. W. Moore, *J. Am. Chem. Soc.* **86**, 2224 (1964).

40. J. R. Bailey and W. T. Read, *J. Am. Chem. Soc.* **36**, 1747 (1914).

41. S. Gabriel, *Ber.* **47**, 3028 (1914).

42. M. J. S. Dewar, in *Molecular Rearrangements,* P. de Mayo (ed.), Wiley, New York, 1963, pp. 323 *et seq.*

43. H. J. Shine, *Aromatic Rearrangements.* Elsevier, Amsterdam, 1967.

44. MTP Internat. Rev. Sci., *Organic Chemistry,* Ser. 1, Vol. 3, H. Zollinger (ed.), Butterworths, London, 1973, pp. 79–84.

45. D. V. Banthorpe, E. D. Hughes, C. K. Ingold, R. Bramley, and J. A. Thomas, *J. Chem. Soc.,* 2864 (1964).

46. R. A. Cox and E. Buncel, in *The Chemistry of the Hydrazo, Azo, and Azoxy Groups,* S. Patai (ed.), Wiley, New York, 1975, pp. 776–805.

47. A. Pongratz and K. Scholtis, *Ber.* **75,** 138 (1942).

48. J.-D. Cheng and H. J. Shine, *J. Org. Chem.* **40,** 703 (1975).

49. D. V. Banthorpe, *J. Chem. Soc.,* 2413 (1962).

50. D. V. Banthorpe, E. D. Hughes, C. K. Ingold, and J. Roy, *J. Chem. Soc.,* 3294 (1962).

51. M. D. Cohen and G. S. Hammond, *J. Am. Chem. Soc.* **75,** 880 (1953).

52. D. V. Banthorpe, C. K. Ingold, J. Roy, and S. M. Somerville, *J. Chem. Soc.,* 2436 (1962).

53. H. J. Shine and J. T. Chamness, *J. Org. Chem.* **28,** 1232 (1963).

54. D. V. Banthorpe and E. D. Hughes, *J. Chem. Soc.,* 3308 (1962).

55. D. V. Banthorpe, E. D. Hughes, C. K. Ingold, and R. Humberlin, *J. Chem. Soc.,* 3299 (1962).

56. H. J. Shine, H. Zmuda, K. H. Park, H. Kwart, A. G. Horgan, and M. Brechbiel, *J. Am. Chem. Soc.* **104,** 2501 (1982).

57. G. S. Hammond and J. S. Clovis, *J. Org. Chem.* **28,** 3283, 3290 (1963).

58. J.-D. Cheng and H. J. Shine, *J. Org. Chem.* **39,** 2835 (1974).

59. L. K. Peterson and G. L. Wilson, *Can. J. Chem.* **49,** 3171 (1971).

60. L. Belinski, G. Franceis, C. Horney, and F. X. Lalau-Kiraly, *Compt. Rend. Acad. Sci.* [D] (*Paris*) **259,** 3737 (1964).

61. R. West and H. F. Stewart, *J. Am. Chem. Soc.* **92,** 853 (1970).

62. H. Posvic and D. Rogers, *J. Org. Chem.* **39,** 1588 (1974).

63. J. Lakritz, doctoral dissertation, University of Michigan, 1960.

64. R. F. Evans, W. Kynaston, and J. I. Jones, *J. Chem. Soc.,* 4031 (1963).

65. S. Pietro, *Boll. Sci. Fac. Chim. Ind. Bologna* **11,** 78, 83 (1953); *Chem. Abstr.* **49,** 13975 (1955).

66. H. Dorn and K. Walter, *Z. Chem.* **7,** 151 (1967).

67. H. Dorn, A. Zubek, and G. Hilgetag, *Chem. Ber.* **98,** 3377 (1965).

68. T. Kauffmann, K. Lötzsch, and D. Wolf, *Chem. Ber.* **99,** 3148 (1966).

69. B. Vis, *Rec. Trav. Chim. Pays Bas* **58,** 387 (1939).

70. R. L. Hinman and D. Fulton, *J. Am. Chem. Soc.* **80,** 1895 (1958).

71. K. Jensen, H. R. Baccaro, O. Ruchardt, G. E. Olsen, C. Pedersen, and J. Toft, *Acta Chem. Scand.* **15,** 1109 (1961).

72. W. J. Theuer and J. A. Moore, *J. Org. Chem.* **29,** 3734 (1964).

73. F. E. Condon, *J. Org. Chem.* **37,** 3608 (1972).

74. N. P. Peet, S. Sunder, and R. J. Cregge, *J. Org. Chem.* **41,** 2733 (1976).

75. R. F. Smith, A. C. Bates, A. J. Battisti, P. G. Byrnes, C. T. Mroz, T. J. Smearing, and F. X. Albrecht, *J. Org. Chem.* **33,** 851 (1968).

76. H. Dorn, A. Zubek, and K. Walter, *Liebig's Ann. Chem.* **707,** 100 (1967).

77. R. L. Hinman, *J. Am. Chem. Soc.* **78,** 1645 (1956).

78. G. Zinner and U. Gebhardt, *Arch. Pharm.* (*Weinheim*) **304,** 706 (1971).

79. G. Zinner and K. Doerschner, *Arch. Pharm.* (*Weinheim*) **306,** 35 (1973).

80. G. Kempter and H. Liehr, *Z. Chem.* **9,** 339 (1969).

81. S. Veibel, *Dan. Tidsskr. Farm.* **17,** 42 (1943).

82. V. Rautenstrauch and F. Delay, *Angew. Chem. Int. Ed.* **19,** 726 (1980).

83. K. Clusius and K. Schwarzenbach, *Helv. Chim. Acta* **42,** 739 (1959).

84. T. Taguchi, T. Matsuo, and M. Kojima, *J. Org. Chem.* **29**, 1104 (1964).

85. G. Stedman and N. Uysal, *J. Chem. Soc.* [*Perkin II*], 667 (1977).

86. M. N. Sheng and A. R. Day, *J. Org. Chem.* **28**, 736 (1963).

87. M. De Rosa and P. Haberfield, *J. Org. Chem.* **46**, 2639 (1981).

88. J. Thiele, *Liebig's Ann. Chem.* **376**, 264 (1910).

89. S. G. Cohen, F. Cohen, and C.-H. Wang, *J. Org. Chem.* **28**, 1479 (1963).

90. A. Michaelis, *Liebig's Ann. Chem.* **270**, 117 (1892).

91. D. Klamann, U. Krämer, and P. Weyerstahl, *Chem. Ber.* **95**, 2694 (1962).

92. L. B. Pearce, M. H. Feingold, K. F. Cerny, and J.-P. Anselme, *J. Org. Chem.* **44**, 1881 (1979).

93. S. Wawzonek and W. McKillip, *J. Org. Chem.* **27**, 3946 (1962).

94. G. Zinner and W. Ritter, *Arch. Pharm.* (*Weinheim*) **296**, 681 (1963).

95. J. P. Horwitz and V. A. Grakauskas, *J. Am. Chem. Soc.* **80**, 926 (1958).

96. H. Minato, M. Oku, and S. H.-P. Chan, *Bull. Chem. Soc.* Japan **39**, 1049 (1966).

97. E. H. Cordes and W. P. Jencks, *J. Am. Chem. Soc.* **84**, 825 (1962).

98. W. P. Jencks, *Catalysis in Chemistry and Enzymology*, McGraw-Hill, New York, 1969, Chap. 10.

99. R. Moscovici, J. P. Ferraz, E. A. Neves, J. O. Tognoli, M. I. El-Seoud, and L. do Umaral, *J. Org. Chem.* **42**, 4093 (1977).

100. Yu. P. Kitaev and T. V. Troepol'skaya, *Izv. Akad. Nauk S.S.S.R., Otd. Khim. Nauk,* 454, 465 (1963) (English trans. pp. 408, 418).

101. J. C. Howard, G. Gever, and P. H.-L. Wei, *J. Org. Chem.* **28**, 868 (1963).

102. W. Oppolzer, *Angew. Chem. Int. Ed.* **16**, 10 (1977).

103. R. Grashey, R. Huisgen, K. K. Sun, and R. M. Moriarty, *J. Org. Chem.* **30**, 74 (1965).

104. B. Rassow and M. Lummerzheim, *J. Prakt. Chem.* [2]**64**, 138 (1901).

105. G. Zinner, *Angew. Chem.* **75**, 687 (1963).

106. B. Rassow and R. Külke, *J. Prakt. Chem.* [2]**65**, 97 (1902).

107. H. Plieninger and I. Nogradi, *Chem. Ber.* **88**, 1964 (1955).

108. M. Göotz and K. Zeile, *Tetrahedron* **26**, 3185 (1970).

109. G. J. Bloink and K. H. Pausacker, *J. Chem. Soc.*, 661 (1952).

110. E. Hardegger and G. Schreier, *Helv. Chim. Acta* **35**, 232 (1952).

111. O. L. Chapman, W. J. Welstead, Jr., T. J. Murphy, and R. W. King, *J. Am. Chem. Soc.* **86**, 732 (1964).

112. L. Mester, E. Moczar, and J. Parello, *J. Am. Chem. Soc.* **87**, 596 (1965).

113. O. L. Chapman, W. J. Welstead, Jr., T. J. Murphy, and R. W. King, *J. Am. Chem. Soc.* **89**, 7005 (1967).

114. H. Goldstein and R. Voegeli, *Helv. Chim. Acta* **26**, 475 (1943).

115. L. Maskaant, *Rec. Trav. Chim. Pays Bas* **56**, 211 (1937).

116. D. J. Cram and J. S. Bradshaw, *J. Am. Chem. Soc.* **85**, 1108 (1963).

117. H. Minato and A. Kusuoka, *J. Org. Chem.* **39**, 3419 (1974).

118. F. D. Chattaway, *J. Chem. Soc.* **93**, 852 (1908); **91**, 1323 (1907).

119. L. A. Carpino, *J. Org. Chem.* **35**, 3971 (1970).

120. T. G. Back, S. Collins, and R. G. Kerr, *J. Org. Chem.* **46**, 1564 (1981).

121. B. Stanovnik, M. Tišler, M. Kunaver, D. Gabrijelcic, and K. Kocevar, *Tetrahedron Lett.*, 3059 (1978).

122. T. G. Sterleva, L. A. Kiprianova, A. F. Levit, and I. P. Gregorova, *Zh. Org. Khim.* **12**, 1927, 2034 (1976).

123. E. S. Huyser and R. H. S. Wang, *J. Org. Chem.* **33**, 3901 (1968).

124. I. Bhatnagar and M. V. George, *J. Org. Chem.* **33,** 2407 (1968).

125. H. Aebi, B. Dewald, and H. Suter, *Helv. Chim. Acta* **48,** 656 (1965).

126. S. D. Ross, M. Finkelstein, and E. J. Rudd, *Anodic Oxidation,* Academic Press, New York, 1975.

127. H. Eisner and E. Kirowa-Kisner, in *Encyclopedia of the Electrochemistry of the Elements,* Vol. XIII, A. J. Bard and H. Lund (eds.), Marcel Dekker, New York, 1979, pp. 220–362.

128. (a) D. M. Brown, G. H. Jones, B. E. Looker, C. D. McLean, and S. Middleton, *J. Chem. Soc.* [*Perkin I*], 2052 (1977); (b) A. Courotte-Potman, H. C. van der Plas, B. van Veldhuizen, and C. A. Landheer, *J. Org. Chem.* **46,** 5102 (1981); (c) V. A. Ershov and I. Y. Postovskii, *Khim. Geterotsikl. Soedin.* **4,** 571 (1971).

129. R. Stollé, *J. Prakt. Chem.* **[2]66,** 332 (1902).

130. L. F. Audrieth, *Chem. Rev.* **15,** 169 (1934).

131. J. R. Shelton and C. K. Liang, *Synthesis,* 204 (1971).

132. C. G. Overberger and H. Gainer, *J. Am. Chem. Soc.* **80,** 4556 (1958).

133. E. L. Allred and J. G. Henshaw, *J. Chem. Soc. Chem. Commun.,* 1021 (1969).

134. J. A. Hyatt, *Tetrahedron Lett.,* 141 (1977).

135. A. M. Blackham and N. L. Eatough, *J. Am. Chem. Soc.* **84,** 2922 (1962).

136. B. T. Newbold, *J. Org. Chem.* **27,** 3919 (1962).

137. H. Bock, G. Rodolph, and E. Baltin, *Chem. Ber.* **98,** 2054 (1965).

138. J. H. Walton and G. W. Filson, *J. Am. Chem. Soc.* **54,** 3228, (1932).

139. C. G. Overberger, *Rec. Chem. Prog.* **21,** 21 (1960).

140. R. A. Abramovitch and B. A. David, *Chem. Rev.* **64,** 149 (1964).

141. R. L. Hinman and K. L. Hamm, *J. Am. Chem. Soc.* **81,** 3294 (1959).

142. W. R. McBride and E. M. Bens, *J. Am. Chem. Soc.* **81,** 5546 (1959).

143. C. G. Overberger and L. P. Herin, *J. Org. Chem.* **27,** 417, 2423 (1962).

144. L. A. Carpino, A. A. Santelli, and R. W. Murray, *J. Am. Chem. Soc.* **82,** 2728 (1960).

145. D. M. Lemal, T. W. Rave, and S. D. McGregor, *J. Am. Chem. Soc.* **85,** 1944 (1963).

146. L. Horner and H. Ferkeness, *Chem. Ber.* **94,** 712 (1961).

147. L. A. Carpino, *J. Am. Chem. Soc.* **79,** 4427 (1957).

148. C. G. Overberger and S. Altscher, *J. Org. Chem.* **31,** 1728 (1966).

149. A. E. Arbuzov, F. G. Valitova, A. V. Il'yanov, B. M. Kozyrev, and Yu. B. Yablokov, *Dokl. Akad. Nauk S.S.S.R.,* **147,** No. 1, 99 (1962) (English trans. p. 949).

150. A. T. Balaban, M. T. Caproiu, N. Negoita, and R. Baican, *Tetrahedron* **33,** 2249 (1977).

151. R. H. Poirier and F. Benington, *J. Org. Chem.* **19,** 1157 (1954).

152. (a) S. F. Nelsen and R. T. Landis, II, *J. Am. Chem. Soc.* **96,** 1788 (1974); (b) L. Lunazzi and K. U. Ingold, *J. Am. Chem. Soc.* **96,** 5558 (1974).

153. K. Maruyama, T. Otsuki, and T. Iwao, *J. Org. Chem.* **33,** 82 (1968).

154. R. H. Poirier, E. J. Kahler, and F. Benington, *J. Org. Chem.* **17,** 1437 (1952).

155. S. F. Nelsen and J. Hintz, *J. Am. Chem. Soc.* **94,** 7108 (1972).

156. K. H. Linke, R. Turley, and W. Gossel, *Z. Naturforsch.* [*C*] **27B,** 1005 (1972).

157. E. Fischer, *Liebig's Ann. Chem.* **239,** 248 (1887).

158. S. Bozzini and A. Stetner, *Ann. Chim.* (*Rome*) **58,** 169 (1968).

159. H. Feuer and F. Brown, Jr., *J. Org. Chem.* **35,** 1468 (1970).

160. J. M. Birchall, R. N. Haszeldine, and A. R. Parkinson, *J. Chem. Soc.,* 4966 (1962).

161. J. M. Birchall, R. N. Haszeldine, and J. E. G. Kemp, *J. Chem. Soc.* [*C*], 449 (1970).

162. R. Shaw, *Int. J. Chem. Kinet.* **9,** 689 (1977).

163. J. W. Timberlake and J. Stowell, in *The Chemistry of the Hydrazo, Azo, and Azoxy Groups*, S. Patai (ed.), Wiley, New York, 1975, pp. 69–82.

164. Reviewed in *Methodicum Chemicum*, Vol. 6: *C—N Compounds* F. Zymalkowski (ed.), Academic Press, New York, 1975, pp. 73–126 and 153–158.

165. Reviewed by E. Müller and E. Enders, in *Houben-Weyl Methoden der Organischen Chemie*, Vol. 10/2, R. Stroh (ed.), Georg Thieme Verlag, Stuttgart, 1967, pp. 1–63, 169–497.

166. S. R. Sandler and W. Karo, *Organic Functional Group Preparations*, Vol. 1, Academic Press, New York, 1968, pp. 363–379.

167. R. Ohme and A. Zubek, *Z. Chem.* **8**, 41 (1968).

168. F. Klages, G. Nober, F. Kirchner, and M. Bock, *Liebig's Ann. Chem.* **547**, 1 (1941).

169. R. D. Brown and R. A. Kearley, *J. Am. Chem. Soc.* **72**, 2762 (1950).

170. A. N. Kost and R. S. Sagitullin, *Zh. Obshch. Khim.* **33**, 867 (1963).

171. H.-H Stroh and H.-G. Scharnow, *Chem. Ber.* **98**, 1588 (1965).

172. R. A. Moss and C. E. Powell, *J. Org. Chem.* **40**, 1213 /(1975).

173. T. Kauffmann, in *Newer Methods of Synthetic Organic Chemistry*, Vol IV, W Foérst (ed.), Academic Press, New York, 1968.

174. J. S. McFadyen and T. S. Stevens, *J. Chem. Soc.*, 584 (1936).

175. C. F. H. Allen, *Org. Syntheses*, Coll. **II**, 228 (1943).

176. A. K. Macbeth and J. R. Price, *J. Chem. Soc.*, 982 (1937).

177. N. L. Drake, *Org. Reactions* **1**, 114 (1942).

178. H. H. Hatt, *Org. Syntheses* Coll. **II**, 395 (1943).

179. T. J. Curphey and K. S. Prasad, *J. Org. Chem.* **37**, 2259 (1972).

180. F. P. Jahn, *J. Am. Chem. Soc.* **59**, 1761 (1937).

181. A. Wohl and C. Oesterlin, *Ber.* **33**, 2736 (1900).

182. R. Renaud and L. C. Leitch, *Can. J. Chem.* **32**, 545 (1954).

183. H. L. Lochte, W. A. Noyes, and J. R. Bailey, *J. Am. Chem. Soc.* **44**, 2556 (1922).

184. E. J. Poth and J. R. Bailey, *J. Am. Chem. Soc.* **45**, 3001 (1923).

185. M. C. Chaco and N. Rabjohn, *J. Org. Chem.* **27**, 2765 (1962).

186. M. C. Chaco, P. R. Stapp, J. A. Ross, and N. Rabjohn, *J. Org. Chem.* **27**, 3371 (1962).

187. A. Ebnöther, E. Jucker, A. Lindenmann, E. Rissi, R. Steiner, R. Süess, and A. Vogel, *Helv. Chim. Acta* **42**, 533 (1959).

188. H. J. Backer, *Rec. Trav. Chim. Pays Bas* **31**, 193 (1912).

189. C. Weygand, *Organic Preparations*, Interscience Publishers, New York, 1945, p. 241.

190. H. Zollinger, *Diazo and Azo Chemistry*, Interscience Publishers, New York, 1961, p. 169.

191. G. H. Coleman, *Org. Syntheses*, Coll. **I**, 2d ed., 442 (1941).

192. I. M. Hunsberger, E. R. Shaw, J. Fugger, R. Ketcham, and F. Lednicer, *J. Org. Chem.* **21**, 394, 2262 (1956).

193. G. H. Coleman, H. Gilman, C. E. Adama, and P. E. Pratt, *J. Org. Chem.* **3**, 99 (1938).

194. P. A. S. Smith, J. M. Clegg, and J. Lakritz, *J. Org. Chem.* **23**, 1595 (1958).

195. L. A. Carpino, P. H. Terry, and P. J. Crowley, *J. Org. Chem.* **26**, 4336 (1961).

196. C. G. Overberger and A. V. Di Giulio, *J. Am. Chem. Soc.* **80**, 6562 (1958).

197. P. Shestakov, German Patent No. 164, 755; *Chem. Zentr.* **1905II**, 1703.

198. A. Darapsky, *J. Prakt. Chem.* **[2]76**, 438 (1907).

199. G. R. Elliot, *J. Chem. Soc.* **123**, 804 (1923).

200. C. J. Sheppard and L. E. Korczykowski, U.S. Patent No. 3,956,366 (1976); *Chem. Abstr.* **85**, 176881 (1976). (see German Patent No. 1,818,020; *Chem. Abstr.* **87**, 38848 (1977))

201. L. H. Diamond and L. F. Audrieth, *J. Am. Soc.* **77**, 3131 (1955).

202. H. H. Sisler, H. S. Ahuja, and N. L. Smith, *J. Org. Chem.* **26**, 1819 (1961).

203. F. N. Collier, Jr., W. W. Horner, P. M. Dickens, J. G. Hull, and W. T. Layton, *J. Am. Chem. Soc.* **83**, 2235 (1961).

204. R. Gösl and H. Meuwsen, *Org. Syntheses* **43**, 1 (1963).

205. Y. Tamura, J. Minamikawa, and M. Ikida, *Synthesis*, 1 (1977).

206. T. Sheradsky, G. Salemnick, and M. Nir, *Tetrahedron* **28**, 3833 (1972).

207. L. A. Carpino, *J. Am. Chem. Soc.* **82**, 3133 (1960).

208. E. Schmitz, *Angew. Chem.* **76**, 197 (1964).

209. E. Schmitz, *Dreiringe mit Zwei Heteroatomen,* Springer-Verlag, New York, 1967, pp. 67–112.

210. J. Fugger, J. M. Tien, and I. M. Hunsberger, *J. Am. Chem. Soc.* **77**, 1843 (1955).

211. J. M. Tien and I. M. Hunsberger, *J. Am. Chem. Soc.* **77**, 6604, 6696 (1955).

212. F. H. C. Stewart, *Chem. Rev.* **64**, 129 (1964).

213. R. B. Carlin and M. S. Moores, *J. Am. Chem. Soc.* **84**, 4107 (1962).

214. M. J. S. Dewar and W. B. Jennings, *J. Am. Chem. Soc.* **95**, 1562 (1973).

215. J. H. Biel, A. E. Drukker, T. F. Mitchell, E. P. Sprengler, P. A. Nuhfer, A. C. Conway, and A. Horita, *J. Am. Chem. Soc.* **81**, 2805 (1959).

216. D. E. Butler, S. M. Alexander, J. W. McLean, and L. B. Strand, *J. Med. Chem.* **14**, 1052 (1971).

217. C. D. Hurd and F. F. Cesark, *J. Am. Chem. Soc.* **89**, 1417 (1967).

218. F. Hrabak, *Coll. Czech. Chem. Commun.* **34**, 4010 (1969).

219. H. H. Hatt, *Org. Syntheses,* Coll. **II**, 208 (1943).

220. H. C. Ramsperger, *J. Am. Chem. Soc.* **51**, 918 (1929).

221. E. Schmitz and R. Ohme, *Chem. Ber.* **95**, 2012 (1962).

222. L. A. Carpino, *J. Am. Chem. Soc.* **85**, 2144 (1963).

223. P. Zeller, H. Gutmann, B. Hegedus. A. Kaiser, A. Langemann, and M. Müller, *Experientia* **19**, 129 (1963).

224. A. Zweig and A. K. Hoffmann, *J. Am. Chem. Soc.* **85**, 2736 (1963).

225. W. D. Ollis, I. O. Sutherland, and Y. Thebtaranonth, *J. Chem. Soc. Chem. Commun.*, 1199 (1970).

226. H. Dorn and A. Zubek, *Z. Chem.* **12**, 129 (1972).

227. J. A. Blair and R. J. Gardner, *J. Chem. Soc. [C]*, 1714 (1970).

228. K. Kratzl and K. P. Berger, *Monatsh. Chem.* **89**, 83 (1958).

229. F. E. Henoch and C. R. Hauser, *Can. J. Chem.* **47**, 157 (1969).

230. K. Taipale, *J. Russ. Phys.-Chem. Soc.* **62**, 1241 (1930); *Chem. Abstr.* **25**, 2415 (1931).

231. R. T. Beltrami and E. R. Bissell, *J. Am. Chem. Soc.* **78**, 2467 (1956).

232. R. Huisgen, F. Jakob, W. Siegel, and A. Cadus, *Liebig's Ann. Chem.* **590**, 1 (1954).

233. L. Spialter, D. H. O'Brien, G. L. Untereiner, and W. A. Reusch, *J. Org. Chem.* **30**, 3278 (1965).

234. H. Zimmer, L. F. Audrieth, M. Zimmer, and R. V. Rowe, *J. Am. Chem. Soc.* **77**, 790 (1955).

235. R. L. Hinman and K. L. Hamm, *J. Org. Chem.* **23**, 529 (1958).

236. C. Hanna and F. W. Schueler, *J. Am. Chem. Soc.* **74**, 3693 (1952).

237. K. Klages, E. M. Wilson, and G. K. Helmkamp, *Ind. Eng. Chem.*, **52**, 119 (1960).

238. D. C. Iffland and E. Cordes, *J. Org. Chem.* **28**, 2769 (1963).

239. P. E. Iversen, *Acta Chem. Scand.* **25**, 2337 (1971).

240. B. T. Hayes and T. S. Stevens, *J. Chem. Soc. [C]*, 1088 (1970).

241. N. Koga and J.-P. Anselme, *J. Org. Chem.* **33**, 3963 (1968).

242. J. Thiele, *Ber.* **42**, 2575 (1909).

243. B. V. Ioffe, Z. I. Sergeeva, and Y. Y. Dumpis, *Zh. Org. Khim.* **5**, 1735 (1969) (English trans. p. 1683).

244. J. C. Stowell, *J. Org. Chem.* **32**, 2360 (1967).

245. H. R. Snyder, C. Weaver, and C. D. Marshall, *J. Am. Chem. Soc.* **71**, 289 (1949).

246. H. Gösl and A. Meuwsen, *Chem. Ber.* **92**, 2521 (1959).

247. W. Theilacker and E. Wegner, in *Newer Methods of Preparative Organic Chemistry,* Vol. 3, W. Foerst (ed.), Academic Press, New York, 1964, p. 303.

248. H. Schubert, F. Fohringen, and H. Noack, *Z. Chem.* **10**, 68 (1970).

249. F. D. Greene, J. C. Stowell, and W. R. Bergmark, *J. Org. Chem.* **34**, 2254 (1969).

250. R. Ohme, H. Preuschhof, and H.-U. Heyne, *Org. Syntheses* **52**, 11, 15 (1972).

251. R. Ohme and H. Preuschhof, *Liebig's Ann. Chem.* **713**, 74 (1968).

252. J. W. Timberlake, J. Alender, A. W. Garner, M. L. Hodges, C. Özmeral, S. Szilagyi, and J. O. Jacobus, *J. Org. Chem.* **46**, 2082 (1981).

253. E. Schmitz and K. Schinkowski, *Chem. Ber.* **97**, 49 (1964).

254. G. N. Walker, M. A. Moore, and B. N. Weaver, *J. Org. Chem.* **26**, 2740 (1961).

255. S. F. Nelsen and G. R. Weisman, *Tetrahedron Lett.,* 2321 (1973).

256. B. V. Ioffe and L. E. Poroshin, *Zh. Obshch. Khim.* **29**, 3154 (1959); *Chem. Abstr.* **54**, 11975e (1960).

257. A. Marxer and M. Horvath, *Helv. Chim. Acta* **47**, 1101 (1964).

258. D. Bar, F. Erb-Debruyne, P. Marcincal, A. Marcincal-Lefebvre, and O. Labian, *Bull. Soc. Chim. France*, 1783 (1971).

259. P. F. Holt and B. P. Hughes, *J. Chem. Soc.*, 764 (1954).

260. E. M. Kaiser and G. J. Bartling, *Tetrahedron Lett.,* 4357 (1969).

261. H. Wieland and H. Fressel, *Ber.* **44**, 901 (1911).

262. H. Zimmer and G. Singh, *J. Org. Chem.* **29**, 1579 (1964).

263. K.-H. Linke, R. Turley, and E. Flaskamp. *Chem. Ber.* **106**, 1052 (1973).

264. H. Wieland and S. Gumbarjan, *Ber.* **39**, 1500 (1906).

265. F. A. Neugebauer and P. H. H. Fischer, *Chem. Ber.* **98**, 844 (1965).

266. R. Bauer and H. Wendt, *Angew. Chem. Int. Ed.* **17**, 202 (1978).

267. S. F. Nelsen and R. T. Landis, II, *J. Am. Chem. Soc.* **95**, 8707 (1973).

268. H. Wieland and H. Fressel, *Liebig's Ann. Chem.* **392**, 135 (1912).

269. P. A. S. Smith and E. E. Most, Jr., *J. Org. Chem.* **22**, 358 (1957).

270. E. Appel and D. Hänssgen, *Chem. Ber.* **103**, 3733 (1970).

271. H. W. Schiessl and R. Appel, *J. Org. Chem.* **33**, 322 (1968).

272. R. C. Slagel, *J. Org. Chem.* **33**, 1374 (1968).

273. B. M. Culbertson, E. A. Sedor, and R. C. Slagel, *Macromolecules* **1**, 254 (1968).

274. S. Siggia, *Quantitative Analysis via Functional Groups,* 4th ed. Wiley, New York, 1979, pp. 654–679.

275. D. V. Banthorpe, E. D. Hughes, and C. K. Ingold, *J. Chem. Soc.*, 2386 (1962).

276. M. J. S. Dewar, *J. Chem. Soc.*, 777 (1946).

277. R. A. Penman and L. F. Audrieth, *Anal. Chem.* **20**, 1058 (1948).

278. A. Berka and J. Zyka, *Chem. Listy* **50**, 314 (1956).

279. S. Siggia and L. J. Lohr, *Anal. Chem.* **21**, 1202 (1949).

280. R. S. Bottei and N. H. Furman, *Anal. Chem.* **29**, 119 (1957).

281. M. Večera and J. Petránek, *Chem. Listy* **48**, 1351 (1954); *Chem. Abstr.* **49**, 4461 (1955).

282. R. Meyer, *Z. Anal. Chem.* **140**, 124 (1943).

2.
HYDRAZONES

NOMENCLATURE

In the absence of functional groups with higher seniority, hydrazones are named according to the IUPAC recommendations as derivatives of the corresponding aldehyde or ketone by adding the separate word *hydrazone*, which may have prefixes if there are substituents on the nitrogen. Examples are benzophenone hydrazone, Ph_2C=NNH_2, and pentan-3-one *N*-acetyl-*N*-methylhydrazone, Et_2C=$NN(Me)$-COMe. They are listed under the parent ketone or aldehyde in *Chemical Abstracts*. In situations where it is necessary to name a hydrazone as a substituted hydrazine, the substituent is named by the *-ylidene* convention, as in *p*-(isopropylidenehydrazino)benzenesulfonic acid, *p*-$(Me_2C$=$NNH)C_6H_4SO_3H$. Where it is necessary to name the divalent hydrazine function as a prefix, the term *hydrazono* is used, as in 2-hydrazono-2-phenylacetic acid, H_2NN=$C(Ph)CO_2H$.

Carbamylhydrazones are a special case and are by custom called *semicarbazones*. Bis-hydrazones of α-dicarbonyl compounds are commonly called *osazones* (an *N*-phenyl substituent is usually implied).

The hydrazones formed by reaction of hydrazine with two equivalents of carbonyl compound, one at each end, are called *azines* (aldazines or ketazines). They are named by following the name of the carbonyl compound with the word *azine*, as in acetone azine, Me_2C=NN=CMe_2. They also have been called by names derived outside the IUPAC recommendations, such as "benzaldazine" (benzaldehyde azine). When the azine function must defer to a functional group of higher seniority and appear as a prefix, the form *azino* is recommended. It stands for the group =NN=, as in 2,2'-azinodi(cyclopentanecarboxylic acid). Unsymmetrical azines must be named as alkylidene hydrazones, as in ethylidenehydrazonoacetonitrile, $MeCH$=NN=$CHCN$.

Compounds derived from the hypothetical tautomers of hydrazones, R_2C=$\overset{+}{N}H$—NH^-, are known as *azomethine imides*.

PROPERTIES

The simple hydrazones are mostly liquids, soluble in water if the carbon content is small. Dialkylhydrazones are reasonably stable to storage, but the unsubstituted and monoalkyl hydrazones of small aldehydes are prone to disproportionation to azine and hydrazine, or dimerization.[1] Formaldehyde dimethylhydrazone melts at $-103°C$ and boils at $72°C$ (730 mm); acetone methylhydrazone boils at $116-118°C$, and acetaldehyde dimethylhydrazone boils at $89-94°C$. Phenylhydrazones are usually solids, but many of the aliphatic ones have quite low melting points; for this reason, the higher-melting *p*-nitro- and 2,4-dinitrophenylhydrazones are preferred as derivatives for identification of aldehydes and ketones. Azines are mostly solids.

Conversion to a hydrazone lowers the basicity of the nitrogen atom that forms the double bond, analogous to the drop in basicity when an amine is converted to an imine. Phenylhydrazones and semicarbazones, which are derived from hydrazines in which the erstwhile —NH$_2$ group is the more basic site, are noticeably weaker bases; semicarbazones are about 5 powers of 10 weaker than semicarbazide.[2] Azines, in which both nitrogens are double-bonded, are usually insoluble in dilute aqueous acids; acetone azine, however, is reported[3] to have a pK_b value of 9. Hydrazones of benzaldehydes and aryl ketones have the possibility for π-electron delocalization spanning the C=N bond, and benzophenone hydrazone, for example, has a pK_a value of 3.85 for the conjugate acid in anhydrous methanol and is a much weaker base than hydrazine (p$K_a = 8.07$).[4] The effects of substituents on the benzene ring are transmitted to the hydrazone function, and p,p'-dimethoxybenzophenone hydrazone, (p$K_a = 4.70$) is a stronger base than the parent compound and p,p-dichlorobenzophenone hydrazone (p$K_a = 3.13$) is significantly weaker. These substances thus behave as vinylogs of aniline. Electronic interaction between the amino group and the C-phenyl group is best described in terms of a limiting structure that represents the hydrazonomethyl group as electron-donating[5,6] (Eq. 2-1).

$$\text{Ph}-\underset{\underset{R}{|}}{C}=N-NH_2 \longleftrightarrow \text{(ring)}=\underset{\underset{R}{|}}{C}-N=\overset{+}{N}H_2 \qquad (2\text{-}1)$$

Quaternary hydrazonium hydroxides are evidently strong bases, for their salts are neutral.

The α-hydrogens of ketone hydrazones have a feeble acidity sufficient to react with lithium diisopropylamide.[7,8] p-Nitrophenylhydrazones and 2,4-dinitrophenylhydrazones are acidic enough to dissolve in aqueous sodium hydroxide solution, presumably by reaction at the NH site.[9] Benzaldehyde 2,4-dinitrophenylhydrazone and its simple derivatives[10] have pK_a values of about 11. The hydrazone anions so derived are identical with those of the tautomeric azo compound (Eq. 2-2).

$$R_2C=N-\bar{N}R \longleftrightarrow R_2\bar{C}-N=NR \qquad (2\text{-}2)$$

The possibility of tautomerism in hydrazones was debated for a very long time,[11] but eventually physical methods became available that could distinguish between the possibilities unambiguously.[12-14] Monosubstituted hydrazones having at least one α-hydrogen can in principle tautomerize to azo compounds or to vinylhydrazines ("enehydrazines") (Eq. 2-3), and it was often asserted that one or

$$\underset{/}{\overset{\backslash}{}}CH-\underset{\underset{|}{}}{CH}-N=NR \rightleftharpoons \underset{/}{\overset{\backslash}{}}CH-\underset{\underset{|}{}}{C}=N-NHR \rightleftharpoons$$

$$\underset{/}{\overset{\backslash}{}}C=\underset{\underset{|}{}}{C}-NH-NHR \qquad (2\text{-}3)$$

another of the several forms in which hydrazones may be encountered was an azo tautomer. Nuclear magnetic resonance evidence is convincing, however, that these earlier claims are incorrect and that the observed substances were either geometric isomers, polymorphic modifications (common among hydrazines), dimers, or trimers.[12-15] In other cases, earlier investigators were misled by atmospheric oxidation of arylhydrazones of aldehydes. However, equilibration with appreciable amounts of the vinylhydrazine structure has been detected in derivatives of β-keto esters.[16] The adducts of phenylhydrazine and of N-phenyl-N-methylhydrazine with acetylenedicarboxylic ester exist as mixtures of hydrazone and vinylhydrazine, with the former predominating, and the pure tautomers have actually been isolated by crystallization.[17] Stable trisubstituted vinylhydrazines, such as 1-butenyltrimethylhydrazine, are well known.[18,19] Polarographic evidence shows that the reaction of phenylhydrazine with aliphatic ketones leads first to vinylhydrazines, which only slowly tautomerize to the more stable hydrazone structure.[20]

Hydrazones also equilibrate with detectable amounts of the azo tautomers, which are less stable by $\Delta G \cong 3$ kcal/mol.[20-22] The position of the equilibrium can be shifted toward formation of the azo tautomer by slow distillation with steam, which selectively removes the more volatile azo compound; in the absence of a strong base as catalyst, reversion to the hydrazone structure is extremely slow.

The hydrazone structure gives rise to geometric isomerism in derivatives of aldehydes and unsymmetrical ketones, owing to restricted rotation about the $C=N$ bond.[15] Only in a few instances was it possible to establish that different forms of hydrazones were a result of such isomerism before the advent of nuclear magnetic resonance.[12-14,23] An early example in which geometric isomerism was clearly demonstrated is the phenylbenzoylhydrazone of cyclohexanone-4-carboxylic acid. The isomers are mirror images, and in 1914, Mills and Bain[24,25] succeeded in resolving them. Furthermore, the diphenylhydrazones of p-methyl- and p-methoxybenzophenone,[26] with which tautomerism cannot occur, have been separated into isomeric pairs, and the carboethoxyhydrazones of benzil and benzoin also have been separated.[27] All three possible isomers, syn-syn, syn-anti, and anti-anti, of p-nitroacetophenone azine have been isolated.[28]

Separation of geometric isomers is not always feasible, owing to easy interconversion, and in other cases, only one isomer may be present. The m- and p-substituted acetophenones, for example, form only one isomer with dimethylhydrazine, but the o-substituted ketones give mixtures in which the isomers are detectable spectroscopically, but they equilibrate too fast to permit separation.[29] However, both geometric isomers of the semicarbazones of some phenyl alkyl ketones have been isolated from mixtures prepared by irradiating the stabler anti-phenyl isomers.[30] The ultraviolet spectra are appreciably different, owing to restricted rotation of the phenyl group in the syn-isomers.

The hydrazone function does not of itself give color to a molecule. When there is no conjugation, as in dimethylhydrazones of aliphatic ketones, there is weak absorption in the near ultraviolet range[6,31,32] (232-234 mμ) The position of the UV maximum for 2,4-dinitrophenylhydrazones and semicarbazones is sufficiently sensitive to structural variations in the ketone or aldehyde part as to be a useful diagnostic tool.[33-35] Absorption is shifted toward the visible by aryl groups on N or C (benzaldehyde ethylhydrazone, $\lambda_{max} = 291$ and 366 mμ).[36] The bathochromic shift is even further pronounced when electron-withdrawing para substituents are present, as a result of an extended delocalized electronic system.[6,14,31] In quaternary

hydrazonium salts, where the terminal hydrazone nitrogen can no longer participate, these effects are greatly altered.[31]

The infrared stretching vibration of the C=N group of hydrazones is weak and is
found[15,37,38] in the range 1588 to 1650 cm^{-1}; that of azines has been reported near
1610 cm^{-1}.

The ^{1}H NMR of aldehyde hydrazones is not so far downfield as that of the parent
aldehydes, but it may still fall below the aryl-H region. Benzaldehyde azine, for
example, has a signal at δ 8.75 for CH=N, and acetophenone benzylidenehydrazone shows the CH=N resonance at δ 8.41 and the α-methyl resonance at
2.55.[39] The NH resonance of 2,4-dinitrophenylhydrazones shows linear correlation
with the electron density as calculated by CNDO and can be used in the assignment
of geometric configuration.[40]

Some phenylhydrazones are phototropic and turn pink on exposure to light, but
they revert to colorless forms again in the dark.[33] Phenylhydrazones with long,
straight-chain *para* substituents form liquid crystals between the apparent melting
point and the formation of a clear melt.

The mass spectra of dimethylhydrazones and dinitrophenylhydrazones show
fragmentation cleavages analogous to the McLafferty rearrangement of ketones and
aldehydes.[41]

Relatively little attention seems to have been paid to the determination of hydrazone structures by x-ray crystallography, and the reported examples do not include
very simple compounds.[42-44] The C=N distances fall in the range 1.27 to 1.35 Å
and are slightly longer than the distance in imines. The N—N distances (1.38 to
1.41 Å) are shorter than that in hydrazine (1.46 Å), and the N-substituents are close
to coplanar with the C-substituents. These characteristics all point to an appreciable
contribution from a limiting structure with an N=N bond, as shown for the example of cyclopentadienone dimethylhydrazone. The implied higher bond order between the nitrogens implies restricted rotation, and indeed, the methyl groups appear as an NMR doublet at low temperatures.[45] The coalescence temperature,
$-65°$C, implïes a barrier to rotation of 11 kcal/mol. This example is an exception-

ally favorable one for the dipolar structure, and ordinary hydrazones should have a
lower barrier; indeed, benzaldehyde dimethylhydrazone shows a single resonance
for the methyl groups, with no broadening down to $-65°$C.

Nitrile imides are known only as highly reactive intermediates, and have not been
isolated in substance. They are known principally as *in situ* partners in cycloaddition
reactions, but a representative example, benzonitrile *N*-methylimide, has been
shown to undergo two intramolecular reactions: tautomerization to an azine, and
rearrangement to methylphenylcarbodiimide, presumably through successive formation of a diazirine and *N*-methylbenzimidoylnitrene.[45A]

The potential for toxicity in hydrazones is high, in view of the fact that they are more or less readily hydrolyzable to hydrazines, most of which are quite toxic. Hydrazones are widely represented among both medicinal agents and agricultural pesticides.

REACTIONS

Heat and light. Thermally induced reactions of hydrazones have not attracted study, probably because they are commonly not clean reactions. Quaternary hydrazonium salts of aldehydes, however, decompose at 240 to 250°C to form nitriles[46] (Eq. 2-4). Yields may be as high as 95%, but in some cases are low, presum-

$$RCH{=}\overset{+}{N}NR_3 \; X^- \xrightarrow{240-250°C} RCN + [R_3NH^+ \; X^-] \qquad (2\text{-}4)$$
$$(X = I \text{ or } BF_4)$$

ably owing to a competing dealkylation reaction. With hydrazonium salts of ketones, pyrolysis gives a quite different result.[47] The trimethylhydrazonium fluoroborates of some aryl alkyl ketones, when heated to approximately 200°C, undergo an astonishing conversion to pyridines, generally in high yields (Eq. 2-5). The γ carbon comes from one of the methyl groups.

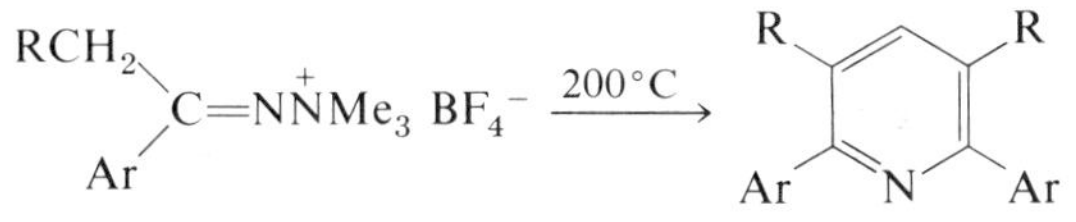

$$(2\text{-}5)$$

Strong heating of azines initiates two types of reaction: loss of nitrogen with formation of an olefin, and cleavage of the N—N bond with formation of imines, nitriles, and phenanthridines. The former reaction, exemplified by benzaldehyde azine,[48] is an ionic, chain process in which a diazoalkane is a chain carrier (Eq. 2-6);

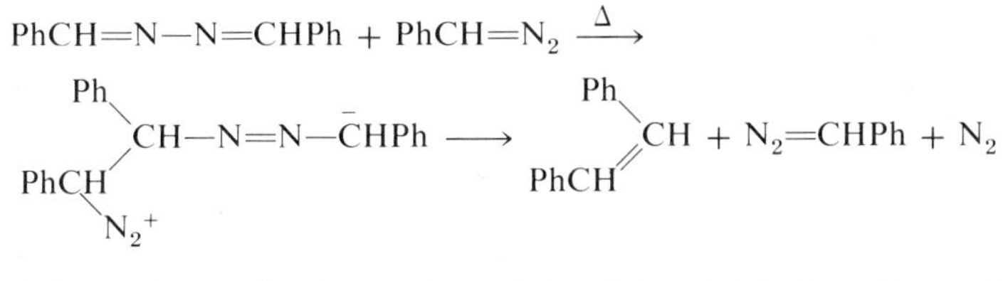

$$(2\text{-}6)$$

the reaction is intermolecular and is catalyzed by added diazoalkane. The other reaction, exemplified by benzophenone azine, is of unclear stoichiometry and is probably a free-radical process[49] (Eq. 2-7). The unsymmetrical azine, benzophe-

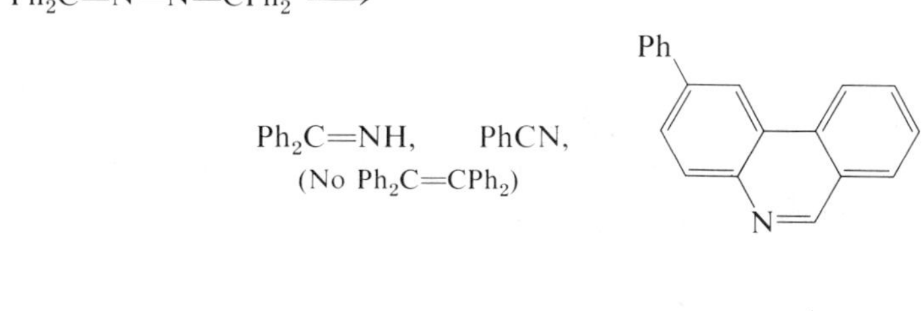

$$Ph_2C{=}N{-}N{=}CPh_2 \xrightarrow{\Delta}$$

$$Ph_2C{=}NH, \quad PhCN,$$
$$(No \; Ph_2C{=}CPh_2)$$

$$(2\text{-}7)$$

none benzylidenehydrazone, thermolyzes by both paths (the olefin formed is triphenylethylene). Cinnamaldehyde azines give pyrazoles in high yields.[50]

Photolysis of hydrazones leads to somewhat similar results. One type of reaction is a photochemical analog of the Wolff-Kishner reaction; benzophenone hydrazone, for example, gives mostly diphenylmethane[51] (Eq. 2-8). In the presence of air, how-

$$Ph_2C=NNH_2 \xrightarrow{h\nu} Ph_2CH_2 + N_2$$

$$(+ Ph_2C=O, Ph_2C=NH, \text{azine}) \qquad (2-8)$$

ever, benzophenone becomes the main product. With benzaldehyde phenylhydrazone, a formally analogous reaction is accompanied by N—N cleavage[52] (Eq. 2-9).

$$PhCH=NNHPh \xrightarrow[\text{MeOH}]{h\nu} PhCH_2Ph, \qquad PhCN, \qquad PhCHO,$$
$$\phantom{PhCH=NNHPh \xrightarrow[\text{MeOH}]{h\nu} } 14\% \qquad\qquad 8\% \qquad\quad 15\%$$

$$PhNH_2 \qquad (2-9)$$
$$17\%$$

Benzophenone phenylhydrazone undergoes only reductive cleavage to benzophenone imine and aniline when photolyzed in methanol.

Cleavage of the N—N bond is the usual result of photolysis of azines, but loss of N_2 and formation of olefin, analogous to Equation 2-6 but not involving a diazo compound, also may be important.[53] Acetone azine is photolyzed to acetone methylimine and acetonitrile[54] (Eq. 2-10). Benzaldehyde azine principally forms

$$Me_2C=N-N=CMe_2 \xrightarrow{h\nu} 2\ Me_2C=N\cdot \longrightarrow$$
$$Me_2C=NMe + MeCN \qquad (2\text{-}10)$$

benzaldimine and benzonitrile (up to 90% yield),[55,56] accompanied by small amounts of stilbene. Unsymmetrical azines give rise to all three possible stilbenes, as do mixtures of two different azines.[53] With photolysis in methanol solution, products of reduction are formed as well (diphenylmethane and tetraphenylethane from benzophenone azine,[57] diphenylmethane from benzaldehyde phenylhydrazone[52]). Photolysis of benzaldazine in the presence of sufficient oxygen oxidizes it largely to benzonitrile. The same result has been reported for photolysis in the presence of benzophenone, which was reduced to benzpinacol,[58] but the claim has been disputed.[59] Photolysis of aldehyde diphenylhydrazones also gives nitriles (benzonitrile, 30%)[60]; the diphenylamine to be expected as the other product is converted to carbazole under the conditions.

Solvolysis and related reactions. Hydrolysis of hydrazones, catalyzed by acid or base, leads to the parent aldehyde or ketone. However, hydrazones are more difficult to hydrolyze than imines and in general are inert to aqueous acid or base without heating; prolonged boiling may be necessary to achieve hydrolysis, and not infrequently even this may be insufficient. This resistance is substantially an equilibrium phenomenon, for hydrazones are formed in high yield in dilute water solution (K for formation of p-chlorobenzaldehyde semicarbazone, for example, is 10^6).

Simple aliphatic hydrazones hydrolyze more readily than arylhydrazones,[3,61] and some hydrazones, such as acetophenone 1-phenylethylhydrazone, are hydrolyzed by standing in moist air (oxygen is apparently involved, however).[62] Quaternary hydrazonium salts can be hydrolyzed rather readily in the presence of base, a bit less so in acid.[31] In fact, alkylation of an ordinary hydrazone to a quaternary salt followed by base-catalyzed hydrolysis (Eq. 2-11) has been recommended as a convenient method for hydrolyzing recalcitrant hydrazones.[7,63]

$$R_2C{=}NNMe_2 \xrightarrow{\ MeI\ } R_2C{=}N\overset{+}{N}Me_3\ I^- \xrightarrow[H_2O]{\ OH^-\ } R_2C{=}O \qquad (2\text{-}11)$$

Hydrolysis of hydrazones and its reverse, the reaction of ketones and aldehydes with hydrazine derivatives, may be pictured as a series of additions and eliminations (Eq. 2-12), analogous to the hydrolysis and formation equilibria of imines, which

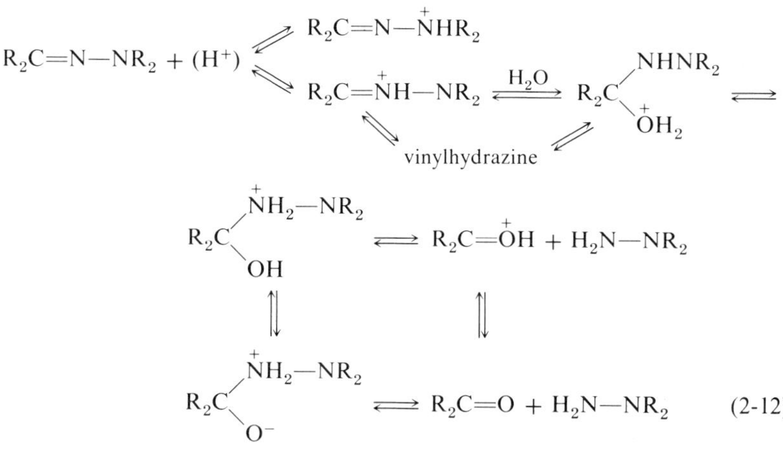

$$R_2C{=}N{-}NR_2 + (H^+) \rightleftharpoons \begin{array}{c} R_2C{=}N{-}\overset{+}{N}HR_2 \\ R_2C{=}\overset{+}{N}H{-}NR_2 \xrightleftharpoons{H_2O} R_2C\overset{NHNR_2}{\underset{\overset{+}{O}H_2}{\diagdown}} \rightleftharpoons \end{array}$$

vinylhydrazine

$$R_2C\overset{\overset{+}{N}H_2{-}NR_2}{\underset{OH}{\diagdown}} \rightleftharpoons R_2C{=}\overset{+}{O}H + H_2N{-}NR_2$$

$$R_2C\overset{\overset{+}{N}H_2{-}NR_2}{\underset{O^-}{\diagdown}} \rightleftharpoons R_2C{=}O + H_2N{-}NR_2 \qquad (2\text{-}12)$$

have been more thoroughly studied. Owing to the one-sided equilibrium for hydrazones, investigations of the mechanism have been much more concerned with formation[64,65] than with hydrolysis. No attempt is made in Equation 2-12 to deal with the details of proton transfer and its catalysis (see Chap. 1).

Because of the difficulty in shifting the equilibria in the direction of hydrolysis products, use of a trapping reagent to remove the liberated hydrazine is advisable for efficient hydrolysis of hydrazones to ketones or aldehydes. Pyruvic acid is especially effective,[66] for it forms very stable hydrazones, but even acetone has been used to generate ketones from toluenesulfonylhydrazones.[67,68] Acetylacetone is even more effective, for it converts the liberated hydrazine to a very stable pyrazole.[66] Exchange with a high-boiling ketone, such as benzophenone, with concurrent distillation of the more volatile aldehyde or ketone also has been recommended[69] (Eq. 2-13). Copper acetate strongly promotes hydrolysis by complexing the released hydrazine.[70]

As a result of the increasing interest in hydrazones as protecting groups for carbonyl compounds,[71] a number of alternative means of regenerating aldehydes and

$$RCH=NNHR' + Ph_2C=O \xrightarrow[\text{1 drop HCl}]{120\,^{\circ}C}$$

$$Ph_2C=NNHR' + RCHO \qquad (2\text{-}13)$$
$$66\text{--}74\%$$

ketones from them have been devised; they are to be found later in this chapter under the headings Nitrosation, Oxidation, and Reduction.

Unsubstituted hydrazones undergo self-solvolysis, resulting in disproportionation between azine and free hydrazine[72] (Eq. 2-14). The reaction is catalyzed by traces of

$$2R_2C=NNH_2 \rightleftharpoons R_2C=NN=CR_2 + N_2H_4 \qquad (2\text{-}14)$$

acid, or even water, and its occurrence may make the preparation of hydrazones difficult. However, since the reaction is reversible, an excess of anhydrous hydrazine effectively prevents formation of azine. Unsymmetrical azines undergo disproportionation into a pair of symmetrical ones on standing in solution over several days[39] (Eq. 2-15). The reaction is greatly accelerated by a drop of acetic acid and presumably proceeds by a solvolytic mechanism.

$$2PhCH=N-N=C\overset{\displaystyle Ph}{\underset{\displaystyle Me}{\Big\langle}} \rightleftharpoons PhCH=N-N=CHPh +$$

$$(K = 0.25)$$

$$\overset{\displaystyle Ph}{\underset{\displaystyle Me}{\Big\rangle}}C=N-N=C\overset{\displaystyle Me}{\underset{\displaystyle Ph}{\Big\langle}} \qquad (2\text{-}15)$$

Strong acids. Alkylhydrazones form water-soluble salts with mineral acids and can be separated from neutral substances by this means.[61] The most important reaction of hydrazones with acids, however, is the Fischer indole synthesis.[73]

Arylhydrazones of aliphatic ketones undergo acid-catalyzed cyclization that competes very successfully with hydrolysis and is known as the *Fischer indole synthesis.* It becomes almost quantitative when anhydrous catalysts, such as zinc chloride or alcoholic hydrogen chloride, are used; an ammonium salt is the other product (Eq. 2-16). Since its discovery in 1886, it has become the most widely used preparative method for this important heterocyclic system.

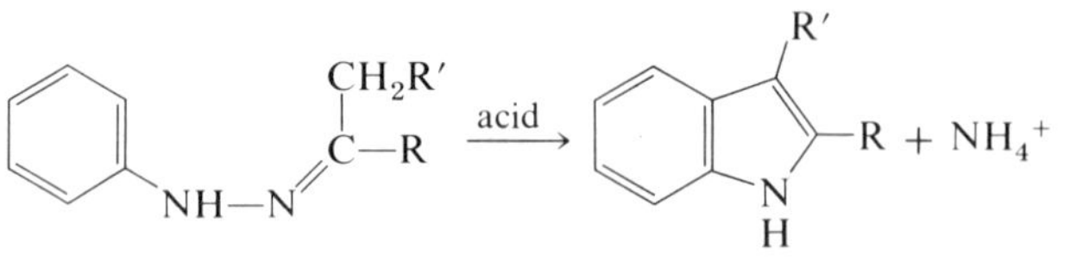

$$(2\text{-}16)$$

With *N,N*-disubstituted hydrazones, the cyclization occurs even more easily, forming *N*-substituted indoles. When the ketone moiety holds a secondary alkyl group, a true indole may be formed, if there is also a primary group, along with an indolenine[74] (Eq. 2-17), which is the only product if no primary group is present. Higher acidities favor formation of an indole over indolenine, presumably by shift-

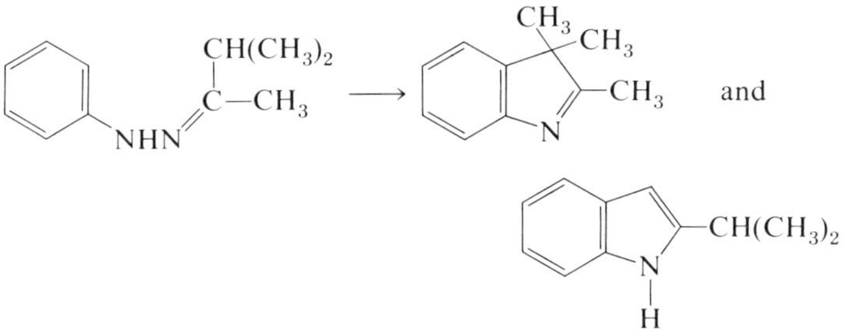

$$(2\text{-}17)$$

ing the reaction from kinetic to thermodynamic control. With disubstituted hydrazones of ketones bearing a secondary alkyl group, 2-methyleneindolines are formed (Eq. 2-18). When an *ortho*-substituent on the N-phenyl group blocks cycliza-

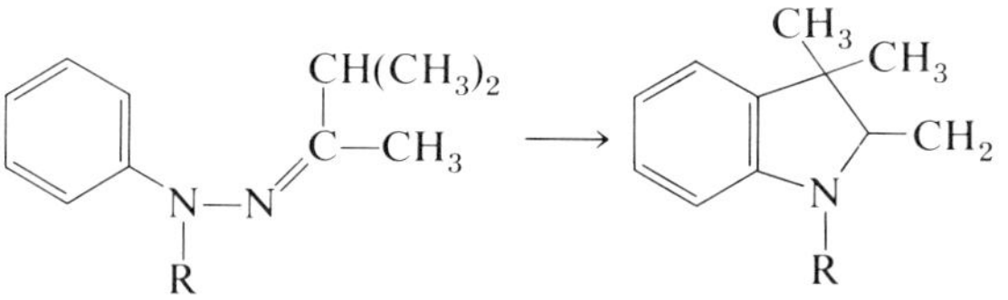

$$(2\text{-}18)$$

tion to form an indole, the group may migrate or, if it can form a stabilized cation, it may be ejected.[75,76]

The most plausible mechanism for the Fischer indole synthesis comes from Carlin,[77] who pointed out the similarity in form to the benzidine rearrangement. If the vinylhydrazine tautomer of the hydrazone is first formed, a benzidine type of rearrangement without rotation (the so-called *ortho*-benzidine rearrangement) would give rise to an o-aminophenyl enamine (or the tautomeric imine), which could easily form an indole by elimination of ammonia (Eq. 2-19). In one instance, such

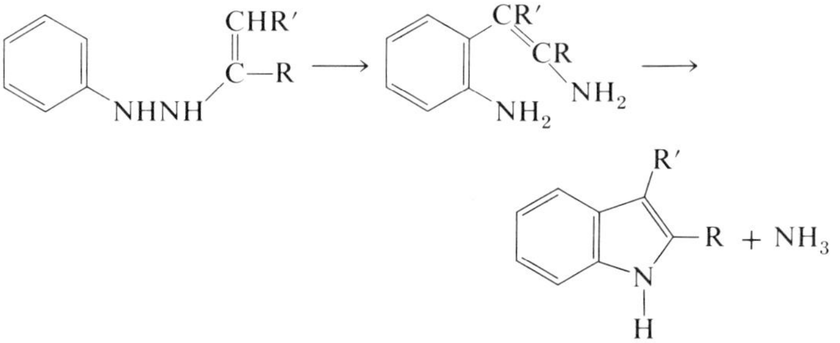

$$R + NH_3 \qquad (2\text{-}19)$$

an intermediate has actually been isolated and been shown to cyclize on more drastic treatment,[78] and in another, an imine intermediate has been characterized spectroscopically.[79] The role of the acid catalyst is presumably to accelerate formation of the vinylhydrazine and should be otherwise parallel to that in the benzidine rearrangement. Experiments with isotopic labeling have established that it is the imine nitrogen of the phenylhydrazone that is eliminated as ammonia. Structurally fixed vinylhydrazines have been demonstrated to undergo conversion to indoles when treated with acid[80,81] (Eq. 2-20).

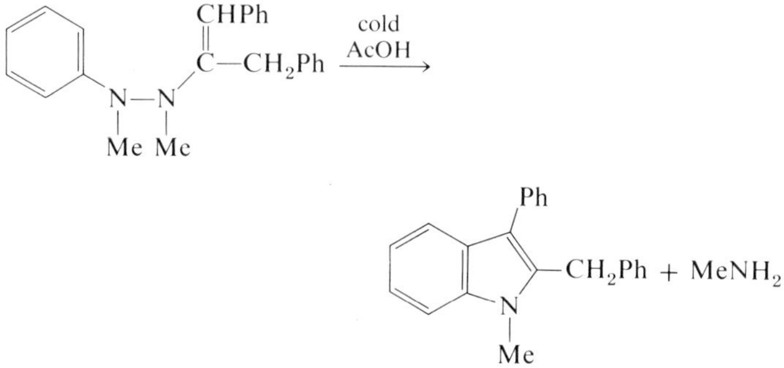

$$(2\text{-}20)$$

An unusual side reaction, observed with camphor phenylhydrazone, leads to o-phenylenediamine and benzimidazole in addition to the expected indole[82] (Eq. 2-21).

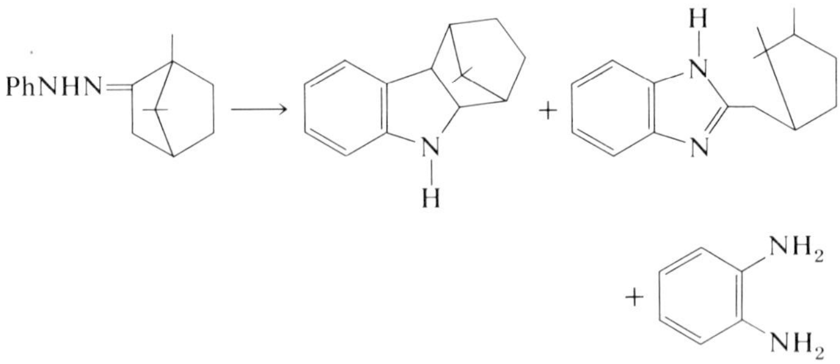

$$(2\text{-}21)$$

Azines undergo a cyclization analogous to the Fischer indole synthesis, known as the *Piloty pyrrole synthesis*[83a,83b] (Eq. 2-22); it has received little attention.

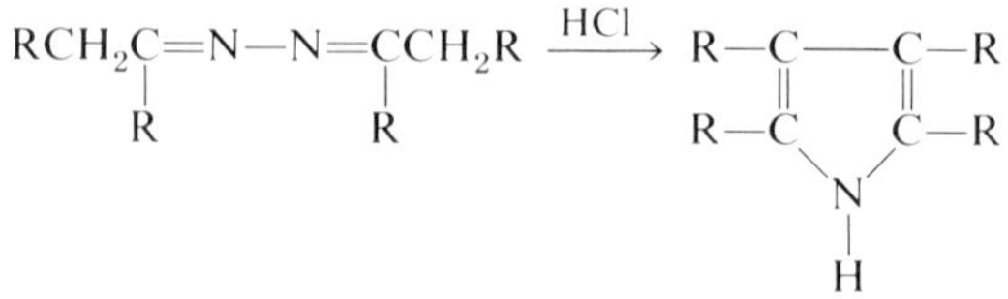

$$(2\text{-}22)$$

Aldehyde hydrazones, which are not good substrates for the Fischer indole synthesis, undergo an acid-catalyzed cleavage to form nitriles[84] (Eq. 2-23). Some

$$RCH{=}NNHR' \xrightarrow[\text{or HCl}]{ZnCl_2} RCN + (H_2NR')$$

$$(2\text{-}23)$$

aldazines have been observed to undergo an analog of the aldol condensation when treated with strong acid and thereby form pyrazolines[85a,85b] (Eq. 2-24).

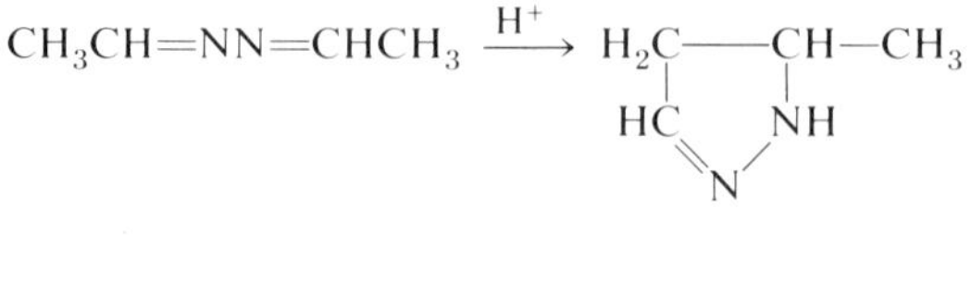

$$(2\text{-}24)$$

Arylhydrazones of benzaldehydes and diaryl ketones, which are structurally incapable of conversion to indoles, have been found to undergo rearrangement to biphenyl derivatives when heated with polyphosphoric acid[86] (Eq. 2-25). An electron-releasing group appears to be required on the carbon-attached ring, and yields are moderate to good. When a phenolic hydroxyl group is present, the isomeric diaryl ether may be formed as well.[87] Formation of biphenyls can compete with indole formation when favorable substituents are present, and p-aminoacetophenone 2,6-dimethylphenylhydrazone, for example, gives an indole derivative (mostly) with shift of a methyl group, the biphenyl according to Equation 2-25, and

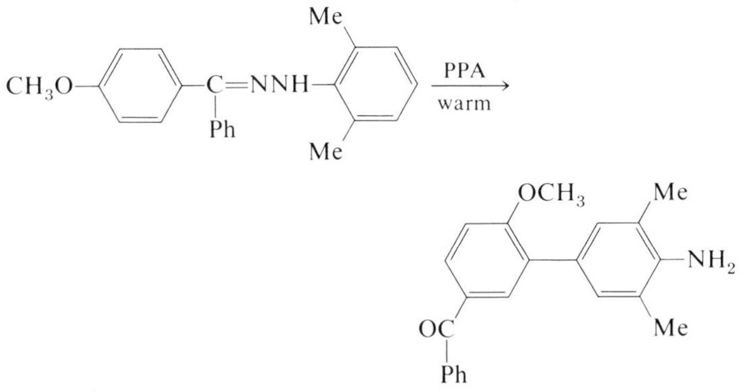

$$(2\text{-}25)$$

still another type of rearrangement product, p-amino-α-(4-amino-3,5-dimethylphenyl)acetophenone. The reaction leading to biphenyl derivatives has been interpreted as a [5,5] sigmatropic shift of the conjugate acid protonated at the imine nitrogen.

Formaldehyde hydrazones are special cases in various respects, including their behavior toward acids. The most carefully studied example appears to be formaldehyde dimethylhydrazone, which oxidizes itself to the bis-dimethylhydrazone of glyoxal; one-third of the original hydrazone is reduced to dimethylamine.[88] Even water catalyzes the reaction slowly, but acetic acid is very effective. The reaction proceeds via N,N'-dimethylhydrazinoacetaldehyde dimethylhydrazone (the heat-to-head dimer) followed by elimination of dimethylamine in overall conversion as high as 77%.

Strong bases.   The amino hydrogen on hydrazones having only one substituent or none is feebly acidic, although much less acidic than water. Alkali metal hydroxides are capable of forming low but kinetically significant concentrations of the conjugate base. The simplest consequence is reprotonation at carbon to form a small equilibrium concentration of the isomeric azo compound (see the section Properties).[36] Stronger bases, such as sodamide, can convert hydrazones completely to anions,[89] and butyllithium is strong enough to remove a proton from an α-carbon, forming a dianion if there was also an amino hydrogen[90] (Eq. 2-26). Disubstituted hydrazones also have been converted to α-carbanions with lithium diisopropylamide.[7,8,91,92]

$$\underset{\text{CH}_3}{\overset{\text{Ar}}{>}}\text{C}=\text{N}-\text{NHAr}' \xrightarrow{\text{BuLi}} \underset{-\text{CH}_2}{\overset{\text{Ar}}{>}}\text{C}=\text{N}-\overset{-}{\text{N}}\text{Ar}' \qquad (2\text{-}26)$$

The most important reaction of strong bases with hydrazones is the Wolff-Kishner reaction, in which an unsubstituted hydrazone is treated with an alkali hydroxide or alkoxide, resulting in loss of nitrogen and formation of a hydrocarbon[93,94] (Eq. 2-27). Semicarbazones ultimately react in the same way, for they are first hydrolyzed

$$R_2C=NNH_2 \xrightarrow[\text{KOH}]{150-200\,^\circ C} R_2CH_2 + N_2 \qquad (2\text{-}27)$$

to unsubstituted hydrazones. The reaction can be carried out on an aldehyde or ketone without actual isolation of the hydrazone, which may be formed in situ. Originally the reaction was carried out either with solid sodium hydroxide in the presence of platinum or with strong alcoholic potassium hydroxide in sealed tubes, for the temperatures required were usually above the boiling point of the reaction medium. The introduction of ethylene glycol as solvent by Huang Min-lon did away with the need for sealed tubes and thereby greatly increased the usefulness of the reaction. A further improvement is the use of potassium *tert*-butoxide in dimethyl sulfoxide, which allows the reaction to be carried out efficiently at room temperature.[95] In special cases, however, such as α-keto hydrazones, even warm aqueous sodium hydroxide is sufficient to bring about reaction.[96,97]

N-Alkylhydrazones undergo the Wolff-Kishner reaction, albeit in low yield, and give rise to the corresponding alkyl hydrocarbons[98] (Eq. 2-28). Dialkylhydrazones

$$PhCH=NNHEt \xrightarrow{\text{NaOH/Pt}} \underset{25\%}{PhCH_2Et + N_2}$$

$$(\text{via } PhCH_2N=NEt \ ?) \qquad (2\text{-}28)$$

are quite inert, but some dinitrophenylhydrazones undergo the reaction with loss of dinitrophenol.[98,99]

The kinetics of the reaction are first-order in hydrazone and in base.[98,100] The rate is independent of the cation (Li^+, Na^+, K^+, Mg^{2+}).[101] The rate-limiting step involves the hydrazone anion and at least two ROH solvent molecules, one to accept the remaining proton from nitrogen and the other to supply a proton to carbon, generating R_2CH^- upon loss of nitrogen.[101] A mild alternative to Wolff-Kishner reduction, involving tosylhydrazones, is described in Chapter 3.

Aldehyde arylhydrazones, which cannot undergo the Wolff-Kishner reaction, are susceptible to rearrangement to amidines by strong bases[102] (Eq. 2-29). It is not

$$PhCH=NNHAr \xrightarrow[\text{hot xylene}]{\text{NaNH}_2} PhC\overset{\displaystyle \nearrow NAr}{\underset{\displaystyle \searrow NH_2}{}} \qquad (2\text{-}29)$$

$$30\text{-}63\%$$

unreasonable to suppose that the hydrazone is first cleaved to a nitrile and anilide ion, which are known to be able to react to form an amidine. A route through a diaziridine seems less likely, in view of the stability of diaziridines and of their magnesium derivatives when prepared in other ways. A nitrile and aniline have indeed been obtained in addition to or instead of amidine under not very different conditions.[103] This cleavage is strongly promoted by oxygen and may involve a free-radical-chain process.

Some benzoylhydrazones are cleaved to amides by alumina or hot alkali.[104]

Some ketazines have been converted to symmetrical dianions by removal of α-protons by lithium diisopropylamide. The α,α'-dilithio derivatives are stable for a while in tetrahydrofuran, but within 24 hours they isomerise to the dianions of tetrahydropyridazines or γ-diketone diimines[105] (Eq. 2-30). The latter form pyrroles

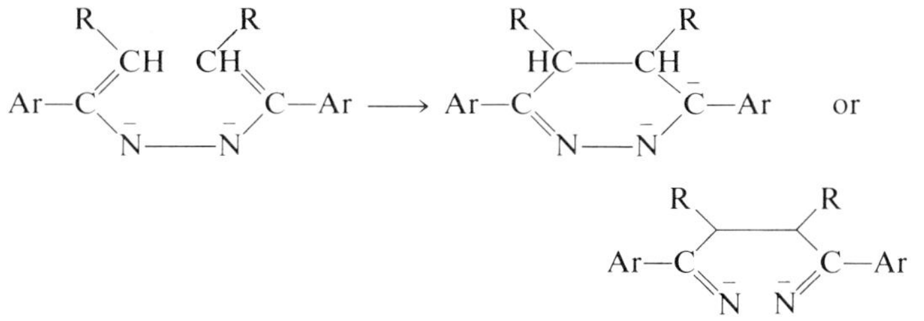

$$\text{(2-30)}$$

by elimination of ammonia during workup. Yields are poor (17–52%), and the factors that determine the type of product have not been elucidated.

Although quaternary hydrazonium salts are hydrolyzed by aqueous base, strictly anhydrous sodium ethoxide removes an α-hydrogen and thereby brings about closure of an azirine ring with ejection of a tertiary amine, analogous to the Neber reaction of sulfonyl oximes[31, 106] (Eq. 2-31). This reaction is an important prepara-

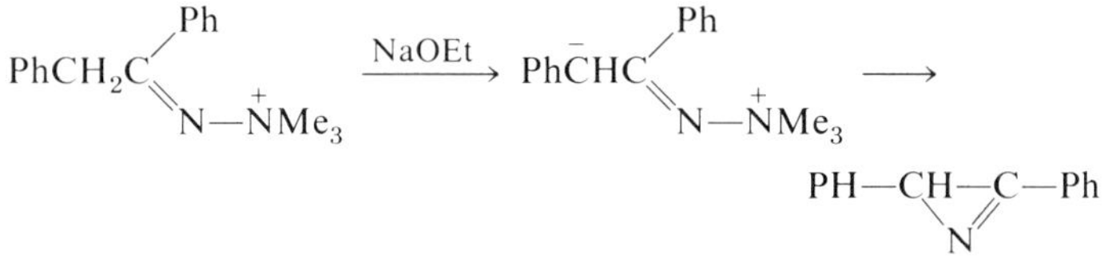

$$\text{(2-31)}$$

tive method for azirines,[107, 108] which are readily hydrolyzed to α-amino ketones. Quaternary hydrazonium salts of aldehydes under similar treatment undergo smooth elimination to give good yields of nitriles[109, 110] (Eq. 2-32). So easy is this

$$RCH{=}\overset{+}{N}NMe_3\ I^-\ \xrightarrow[CH_3OH]{NaOCH_3}\ RCN + (CH_3)_3N \qquad \text{(2-32)}$$

elimination that it is difficult to avoid it when one tries to hydrolyze such hydrazones, unless the medium is quite acidic (for example, 5% HCl).[7] A competing reaction observed with some quaternary hydrazones of ketones is elimination of an iminium salt and formation of the corresponding ketimine; subsequent reactions may lead to products of Mannich condensation.[111]

Carbon nucleophiles.   Carbon nucleophiles, such as Grignard reagents and cyanide ion, do not react with hydrazones so readily as they do with aldehydes and ketones, and removal of a proton from an α-carbon may predominate. Nevertheless, good yields of hydrazines and derivatives have been obtained in many instances[112,113] (Eq. 2-33). Azines may undergo addition at either one or both C=N bonds (Eq. 2-34), but aldazines are sometimes reduced by Grignard reagents instead[114] (Eq. 2-35). Quaternary hydrazonium salts undergo reductive cleavage of the N—N

$$\text{Me}_2\text{N}-\text{N}=\text{CH}_2 + \text{EtMgBr} \longrightarrow \text{Me}_2\text{N}-\bar{\text{N}}\text{CH}_2\text{Et} \xrightarrow{\text{H}_2\text{O}}$$

$$\text{Me}_2\text{NNHCH}_2\text{Et} \qquad (2\text{-}33)$$

$$\text{R}_2\text{C}=\text{N}-\text{N}=\text{CR}_2 + \text{R}'\text{MgX} \longrightarrow \underset{\overset{|}{\text{R}'}}{\text{R}_2\text{C}}-\text{NHN}=\text{CR}_2 \xrightarrow{\text{R}'\text{MgX}}$$

$$\underset{\overset{|}{\text{R}'}}{\text{R}_2\text{C}}-\text{NHNH}-\underset{\overset{|}{\text{R}'}}{\text{CR}_2} \qquad (2\text{-}34)$$

$$\text{PhCH}=\text{N}-\text{N}=\text{CHPh} + \text{EtMgI} \longrightarrow$$

$$\text{PhCH}=\text{N}-\text{NHCH}_2\text{Ph} + \text{C}_2\text{H}_6 \qquad (2\text{-}35)$$

bond, with little or no addition to carbon.[115] The imine nitrogen is substituted to some extent, however, a reaction that has not been observed with other hydrazones (Eq. 2-36). Many hydrazones will add the elements of HCN when treated with

$$\text{Ar}_2\text{C}=\text{N}-\overset{+}{\text{N}}\text{R}_3 + \text{Ar}'\text{MgX} \underset{(\text{or Ar}'\text{Li})}{\Big\langle} \overset{\displaystyle \text{Ar}'-\text{Ar}' + \text{Ar}_2\text{C}=\text{NH} + \text{R}_3\text{N}}{\underset{\displaystyle \text{Ar}_2\text{C}=\text{NAr}' + \text{R}_3\text{N}}{}} \qquad (2\text{-}36)$$

potassium cyanide and aqueous ammonium chloride or liquid HCN;[116] the α-hydrazino nitriles that are produced are a useful source of α-hydrazino acids[117] (Eq. 2-37). Isocyanides also will add if a molecule of acid is included in the reaction.[118]

$$\text{R}_2\text{C}=\text{NNR}_2' + \text{HCN} \longrightarrow \text{R}_2\text{C} \overset{\displaystyle \text{CN}}{\underset{\displaystyle \text{NHNR}_2'}{\Big\langle}} \xrightarrow{[\text{H}_2\text{O}]}$$

$$\text{R}_2\text{C} \overset{\displaystyle \text{COOH}}{\underset{\displaystyle \text{NHNR}_2'}{\Big\langle}} \qquad (2\text{-}37)$$

The reaction presumably takes place between the isocyanide and the conjugate acid of the hydrazone; the immediate product is then a cation, which adds the anion of the acid (Eq. 2-38). In many cases, a further change, isomerization or cyclization, follows. Only hydrazones with base-weakening substituents on the saturated nitrogen atom, such as semicarbazones, take part readily in this reaction, presumably because ordinary hydrazones are not significantly protonated on the imine nitrogen.

$$R_2C \overset{+}{=\!\!=} NHNR_2' + R''NC \longrightarrow R_2C \overset{NHNR_2'}{\underset{C \equiv \overset{+}{N}R''}{\Big\langle}} \xrightarrow{\quad AcO^- \quad}$$

$$R_2C \overset{NHNR_2'}{\underset{AcO}{\Big\langle}} \overset{}{C=NR''} \qquad (2\text{-}38)$$

Alkylation. The reaction of neutral hydrazones with alkylating agents has not received much attention, except for the quaternization of dialkylhydrazones, especially dimethylhydrazones. Alkylation occurs exclusively at the amine nitrogen, the
more basic site, yielding quaternary hydrazonium salts, the structure of which is
shown by their hydrolysis to quaternary hydrazinium salts[31,119,120] (Eq. 2-39).

$$R_2C = NNMe_2 + MeI \longrightarrow R_2C = N\overset{+}{N}Me_3 \; I^- \xrightarrow{\;[H_2O]\;}$$
$$R_2CO + H_2\overset{+}{N}NMe_3 \; I^- \qquad (2\text{-}39)$$

Arylhydrazones are presumably too weakly nucleophilic to be readily alkylated in
the neutral state, but their monoanions, generated with sodamide, lithium diisopropylamide, or even sodium hydroxide in favorable cases, undergo alkylation with
primary alkyl halides at the saturated nitrogen atom[89,121,122] (Eq. 2-40). Acetone

$$PhCH = N\overset{-}{N} - C_6H_4NO_2\text{-}p \; K^+ \xrightarrow{\;CH_3I\;} PhCH = NN \overset{CH_3}{\underset{C_6H_4NO_2\text{-}p}{\Big\langle}} \qquad (2\text{-}40)$$

acetylhydrazone has been benzylated in the presence of 40% sodium hydroxide, and
a number of phenylhydrazones have been alkylated in 69 to 72% yields after treatment with sodamide in liquid ammonia.[89]

Much recent attention has been given to alkylation of *N,N*-disubstituted
hydrazones through lithiation with lithium diisopropylamide. Much of the work has
been done with dimethylhydrazones,[7,8,92] but hydrazones derived from *N*-amino-
2-methoxymethylpyrrolidine, which can be obtained in optically active form from
proline, are valuable for enantioselective alkylation.[123] Alkylation occurs exclusively at an α-carbon and in very high yields, and when followed by hydrolysis, it is
a valuable process for the synthesis of ketones (Eq. 2-41). However, it has been

$$Me_2NN = C \overset{CH_2R}{\underset{CHR_2}{\Big\langle}} \xrightarrow[-78°C]{Li^+i\text{-}Pr_2N^-} Me_2NN - C \overset{CH_2R}{\underset{\underset{Li}{\overset{|}{}} \; CR_2}{\Big\langle}} \xrightarrow[-78°C]{R'X} Me_2NN = C \overset{CH_2R}{\underset{\underset{R'}{\overset{|}{CR_2}}}{\Big\langle}}$$

$$\Big\uparrow \qquad\qquad\qquad\qquad\qquad\qquad\qquad\qquad\qquad\qquad \Big\downarrow \qquad\qquad (2\text{-}41)$$

$$O = C \overset{CH_2R}{\underset{CHR_2}{\Big\langle}} \qquad\qquad\qquad\qquad\qquad\qquad\qquad\qquad O = C \overset{CH_2R}{\underset{\underset{R'}{\overset{|}{CR_2}}}{\Big\langle}}$$

claimed that acetone N-methyl-N-phenylhydrazone becomes alkylated at the imine nitrogen when treated with magnesium oxide in hot benzene and then methyl sulfate, or when acetone phenylhydrazone is converted to a sodium salt with excess metallic sodium dispersed in hot toluene and then treated with methyl iodide; the product was stated to be N-pheny-N,N'-dimethyl-N'-isopropenylhydrazine.[20] However, the product was not characterized spectroscopically, and the properties reported for it were not significantly different from those of butan-2-one N-methyl-N-phenylhydrazone, the product to be expected of methylation at carbon.

In a totally different type of process, propionaldehyde methylhydrazone has been alkylated at the trigonal carbon by butadiene in the presence of a nickel-triphenylphosphine complex.[124] The product is an azo compound and is formally the result of an ene reaction with two equivalents of diene (Eq. 2-42). Methyl acetylene-

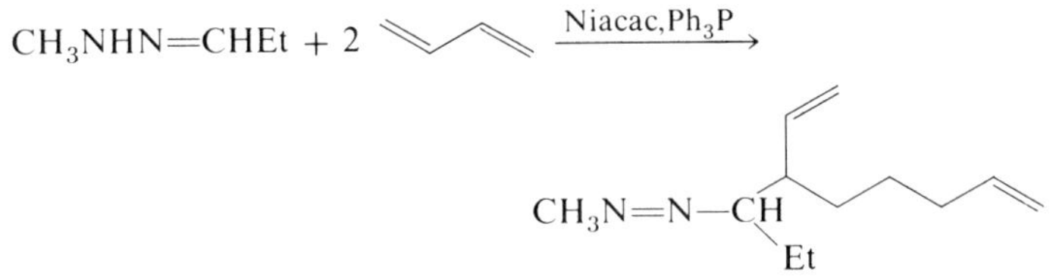

$$CH_3NHN{=}CHEt + 2 \quad \diagdown\diagup\diagup \quad \xrightarrow{\text{Niacac,Ph}_3\text{P}}$$

$$CH_3N{=}N{-}\underset{\underset{Et}{|}}{CH}\diagup\diagdown\diagup\diagdown\diagup \qquad (2\text{-}42)$$

(and isomers)

dicarboxylate, however, reacts with acetone dimethylhydrazone to form products derived from electrophilic attack at either nitrogen atom.[125]

Azomethine imides are also attacked by unsaturated compounds. 1,3-Cycloaddition occurs, involving attachment at both C and N, when the substrates are generated in situ[126] (Eq. 2-43).

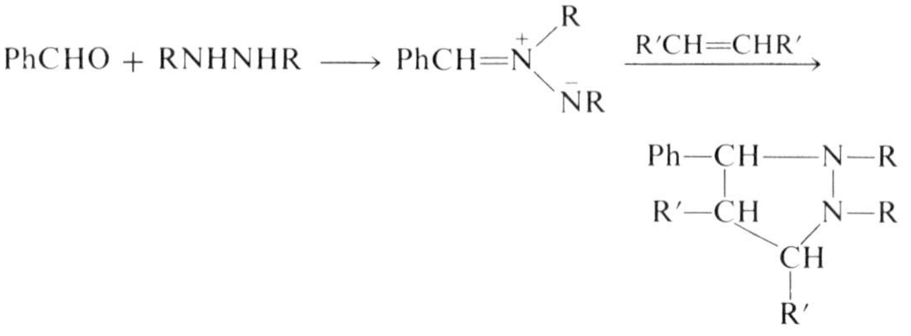

$$PhCHO + RNHNHR \longrightarrow PhCH{=}\overset{+}{N}\overset{\diagup R}{\underset{\diagdown NR}{{}^{-}}} \xrightarrow{R'CH{=}CHR'}$$

$$\begin{array}{c} Ph{-}\underset{|}{CH}{-\!-\!-}N{-}R \\ R'{-}\underset{}{CH}\quad N{-}R \\ \underset{|}{CH} \\ R' \end{array} \qquad (2\text{-}43)$$

Less attention has been paid to azines, but alkylation at one nitrogen by means of powerful primary alkylating agents to form an intermediate hydrazonium salt is a good preparative route to simple monoalkyl hydrazines.[127]

Aldehydes and ketones. In addition to the exchange of alkylidene groups shown in Equation 2-13, hydrazones can react with carbonyl compounds at the saturated nitrogen or an α-carbon atom. Unsubstituted hydrazones react very easily at the amino group to form azines (Eq. 2-44). This type of reaction may occur intra-

$$R_2C{=}NNH_2 + R_2'C{=}O \longrightarrow R_2C{=}NN{=}CR_2' + H_2O \qquad (2\text{-}44)$$

molecularly when a hydrazone function is formed in a position β to a carbonyl group. Cyclization usually ensues rapidly, even with monsubstituted hydrazones if enolization is possible; the products are pyrazoles (Eq. 2-45).

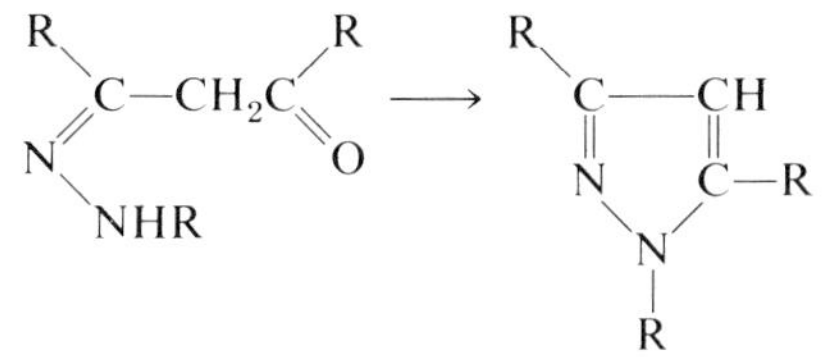

(2-45)

Methylhydrazones (and presumably others) of enolizable ketones react with aldehydes at both N and α-C to close to a pyrazoline ring[128] (Eq. 2-46). Ketones having

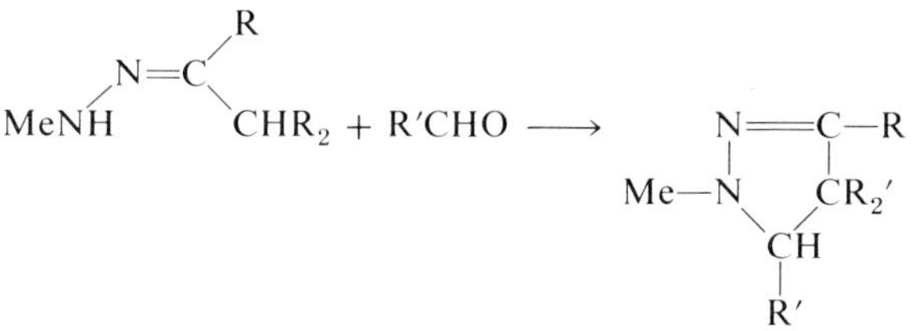

(2-46)

an α-CH_2 group react in a different way in the presence of a catalytic amount of toluenesulfonic acid; N-methyl pyrroles are formed in a reaction analogous to the Piloty pyrrole synthesis[129] (Eq. 2-47).

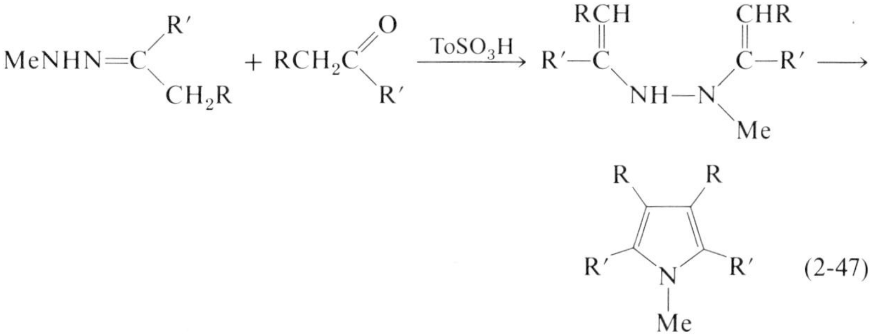

(2-47)

40–90%

Exchange of hydrazono and oxo groups between a hydrazone and an aldehyde or ketone through the hydrolysis equilibrium has already been mentioned. There is a process for cleaving tosylhydrazones and dinitrophenylhydrazones that is closely related and may be the same mechanistically: prolonged (18 to 100 hours) heating with acetone in a sealed tube at 50 to 80°C[67] (Eq. 2-48). The analogous reaction

$$R_2C=NNC_6H_3(NO_2)_2\text{-}2,4 + Me_2C=O \longrightarrow$$

$$R_2C=O + Me_2C=NNC_6H_3(NO_2)_2\text{-}2,4 \quad (2\text{-}48)$$

apparently occurs with formaldehyde, but it is accompanied by a bewildering variety of competing and secondary reactions that are incompletely understood, having been investigated at a time when neither chromatographic separation nor spectro-scopic identification was available.[130]

Lithiated dialkylhydrazones react readily with ketones by simple addition, giving rise to aldol hydrazones[63] (Eq. 2-49). Addition can be directed to the β-carbon of

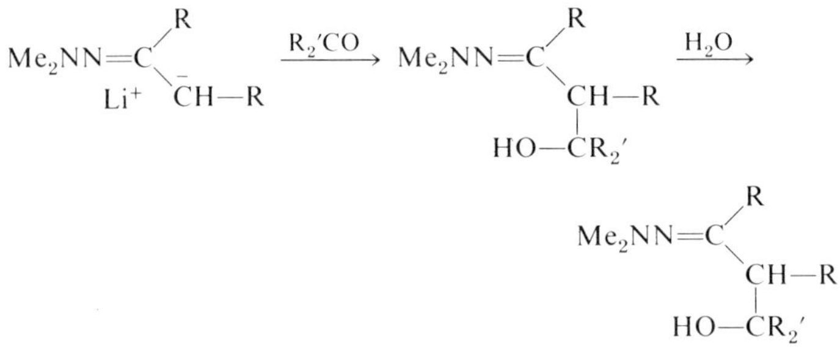

$$\underset{\substack{| \\ Li^+}}{Me_2NN=C}\overset{R}{\underset{CH-R}{\diagdown}} \xrightarrow{R_2'CO} \underset{\substack{| \\ HO-CR_2'}}{Me_2NN=C}\overset{R}{\underset{CH-R}{\diagdown}} \xrightarrow{H_2O}$$

$$\underset{\substack{| \\ HO-CR_2'}}{Me_2NN=C}\overset{R}{\underset{CH-R}{\diagdown}} \qquad (2\text{-}49)$$

α,β-unsaturated carbonyl compounds by first converting the lithiated hydrazone to the dialkylcuprate by reaction with cuprous iodide.

Whereas the reactions with carbonyl compounds so far considered have entailed formation of a bond from the carbonyl carbon to either of the hydrazone nitrogens or to the α-carbon, a further possibility is bond formation at the trigonal carbon of the hydrazone. This has been reported in situations in which the aldehydic hydro-gen of an aldehyde hydrazone is activated by another functional group. Phenylhy-drazones of α-keto aldehydes provide such a situation; they condense rapidly with formaldehyde to produce α-hydroxy hydrazones[131] (Eq. 2-50), perhaps by way of the azo tautomer.

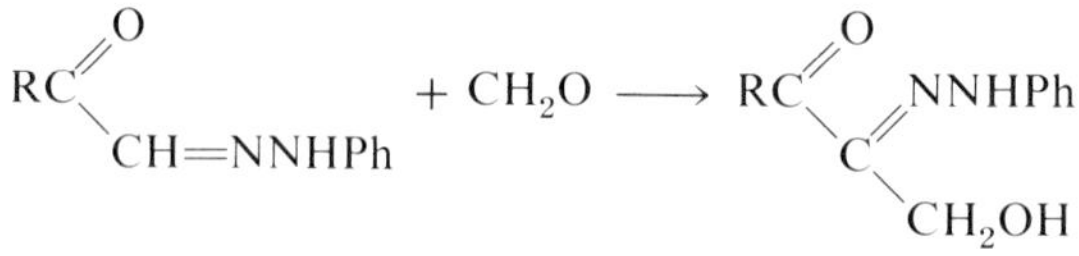

$$\underset{\substack{\diagdown \\ CH=NNHPh}}{RC}\overset{O}{\diagup} + CH_2O \longrightarrow \underset{\substack{\diagdown \\ CH_2OH}}{RC}\overset{O}{\diagup}\underset{C}{\diagdown}{NNHPh} \qquad (2\text{-}50)$$

Acylation. Hydrazones react normally with acylating agents if there is an unsubsti-tuted position on nitrogen, and a variety of acetyl, benzoyl, and arenesulfonyl hydrazones have been prepared from alkyl hydrazones under Schotten-Baumann conditions[132,133] (Eq. 2-51). In some cases, cyclization to an active methylene group

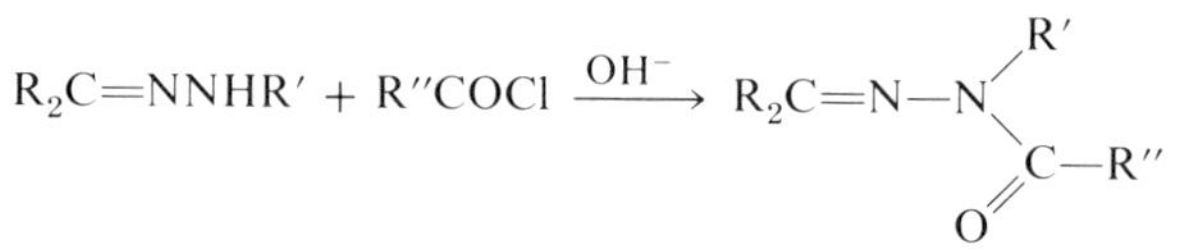

$$R_2C=NNHR' + R''COCl \xrightarrow{OH^-} \underset{\substack{| \\ O^{\diagup}}}{R_2C=N-N}\overset{R'}{\underset{C-R''}{\diagdown}} \qquad (2\text{-}51)$$

ensues, as in the acetylation of acetone phenylhydrazone[134] (Eq. 2-52). Acylation of tosylhydrazones to ditosylhydrazones is described in Chapter 3.

In a rather special case, formaldehyde dimethylhydrazone has been acylated at carbon by 1,1-thiocarbonylbistriazole.[135] The product is the dimethylhydrazone of thioglyoxalyltriazole, from which the remaining triazolyl group is easily displaced

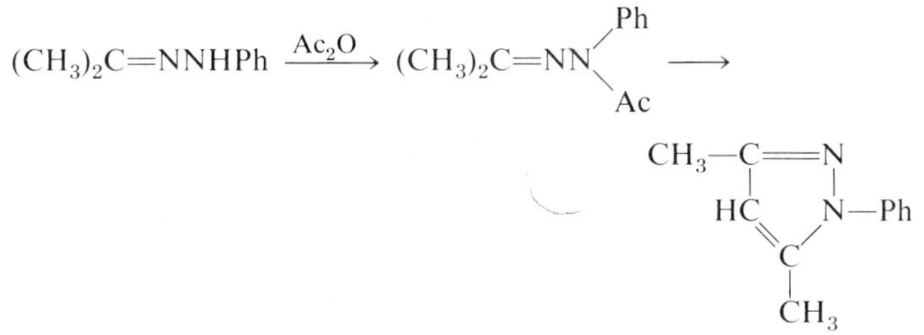

$$(CH_3)_2C=NNHPh \xrightarrow{Ac_2O} (CH_3)_2C=NN\begin{smallmatrix}Ph\\ \\Ac\end{smallmatrix} \longrightarrow$$

$$\begin{array}{c} CH_3-C=\!\!=\!\!N \\ \mid \qquad \mid \\ HC \qquad N-Ph \\ \diagdown C \diagup \\ \mid \\ CH_3 \end{array} \qquad (2\text{-}52)$$

by nucleophiles, such as ammonia, to form such compounds as dimethylhydrazonothioacetamide, $Me_2NN=CHCSNH_2$.

Both simple hydrazones and their monosubstituted derivatives can be nitrosated by nitrous acid. In the presence of water, unsubstituted hydrazones and semicarbazones are converted to the corresponding aldehyde or ketone (Eq. 2-53). So smooth

$$R_2C=NNH_2 \text{ (or } R_2CNNHCONH_2) \xrightarrow[H_2O]{HNO_2} R_2C=O + HN_3 \qquad (2\text{-}53)$$

is the reaction that it has been recommended as the method of choice for regenerating aldehydes and ketones from their semicarbazones; aqueous sodium nitrite is added to a solution of the semicarbazone in glacial acetic acid.[136] In concentrated sulfuric acid, however, rearrangement takes place, giving amides[137,138] (Eq. 2-54)

$$R_2C=NNH_2 \text{ (or } R_2CNNHCONH_2) \xrightarrow[\text{conc. } H_2SO_4]{NaNO_2} [R_2C=N-N_2^+] \longrightarrow$$

$$RCONHR + N_2 \qquad (2\text{-}54)$$

(lactams from hydrazones of cyclic ketones). This is a rearrangement of the Schmidt type, in which an intermediate iminodiazonium ion, also believe to be formed in the reaction of ketones with hydrogen azide under similar conditions, is generated.

Monosubstituted hydrazones are readily nitrosated, giving reasonably stable N-nitrosohydrazones[98,139,140] (Eq. 2-55). Aldehyde phenylhydrazones, however, undergo C-nitrosation in the presence of base to form azo oximes[141] (Eq. 2-56), but

$$PhCH=NNHCH_3 \xrightarrow[\text{hexane}]{RONO} PhCH=NN\begin{smallmatrix}CH_3\\ \\NO\end{smallmatrix} \qquad (2\text{-}55)$$

$$35\%$$

$$PhCH=NNHPh \xrightarrow[\text{NaOEt}]{RONO} PhC\begin{smallmatrix}NOH\\ \\N=NPh\end{smallmatrix} \qquad (2\text{-}56)$$

N-nitrosation in acetic acid.[140] This effect of the medium on the site of nitrosation appears to have some generality, but in neutral medium, isoamyl nitrite is reported to convert benzaldehyde phenylhydrazone to the C-nitro derivative rather than nitroso.[140,142] C-Nitro hydrazones are obtained directly from aldehyde phenylhydrazones and propyl nitrate in the presence of strong bases[142] (potassium amide or *t*-butoxide) in yields ranging up to 94%.

Electrophilic nitrogen (diazonium, nitroso, azide).   Arylhydrazones of benzaldehydes couple with diazonium salts at N, forming tetrazene derivatives. These rearrange very easily to *gem*-bis(azo) compounds, which are isolable, but tautomerize readily to formazans[143] (Eq. 2-57). Coupling in basic medium proceeds directly to

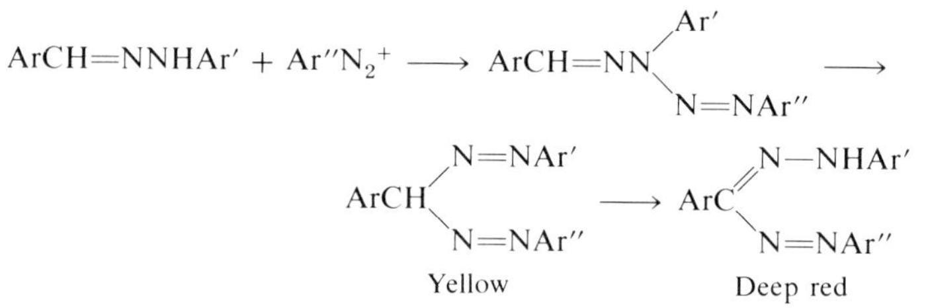

$$\text{ArCH=NNHAr' + Ar''N}_2{}^+ \longrightarrow \text{ArCH=NN} \overset{\text{Ar'}}{\underset{\text{N=NAr''}}{}} \longrightarrow$$

$$\text{ArCH} \overset{\text{N=NAr'}}{\underset{\text{N=NAr''}}{}} \longrightarrow \text{ArC} \overset{\text{N—NHAr'}}{\underset{\text{N=NAr''}}{}} \qquad (2\text{-}57)$$

$$\quad\text{Yellow} \qquad\qquad\qquad \text{Deep red}$$

formazans.[144] *N*-Methyl-*N*-phenylhydrazones of aldehydes do not couple at the aldehyde carbon, presumably because they cannot first accept the diazonium group at *N*.

Aryl azides also couple at the aldehyde carbon of phenylhydrazones with base catalysis and form *N*-aminoformazans[145] (Eq. 2-58). Benzaldehyde phenylhydra-

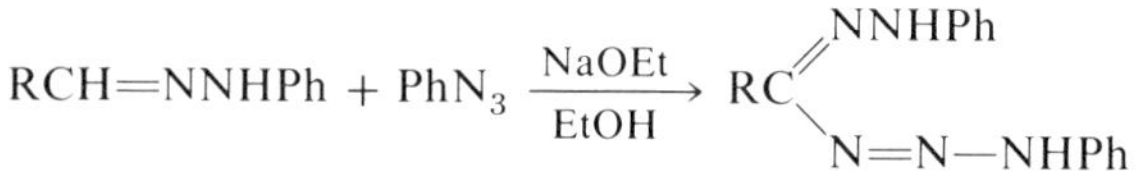

$$\text{RCH=NNHPh + PhN}_3 \xrightarrow[\text{EtOH}]{\text{NaOEt}} \text{RC} \overset{\text{NNHPh}}{\underset{\text{N=N—NHPh}}{}} \qquad (2\text{-}58)$$

zone, however, gives a different type of product, perhaps by a secondary reaction of an aminoformazan first produced. The N—N bond of the hydrazone is cleaved and a 2,5-disubstituted tetrazole results[146] (Eq. 2-59).

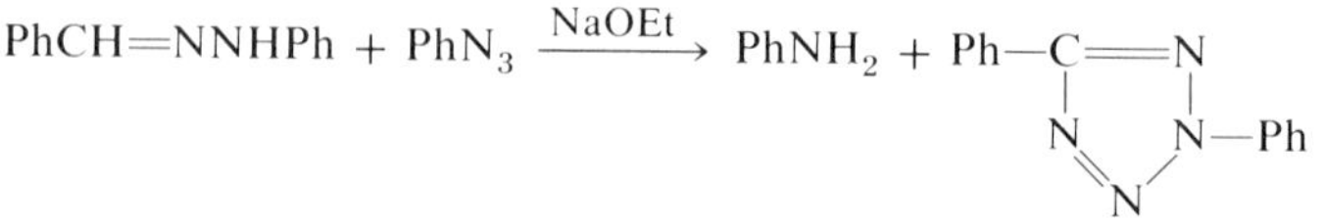

$$\text{PhCH=NNHPh + PhN}_3 \xrightarrow{\text{NaOEt}} \text{PhNH}_2 + \text{Ph—C=N} \qquad (2\text{-}59)$$

Nitrosobenzene attacks the aldehyde carbon of benzaldehyde phenylhydrazone and gives rise first to a hydroxylamino azo compound, which loses the elements of phenyldiazene and leaves benzaldehyde *N*-phenylnitrone[147,148] (Eq. 2-60). Some

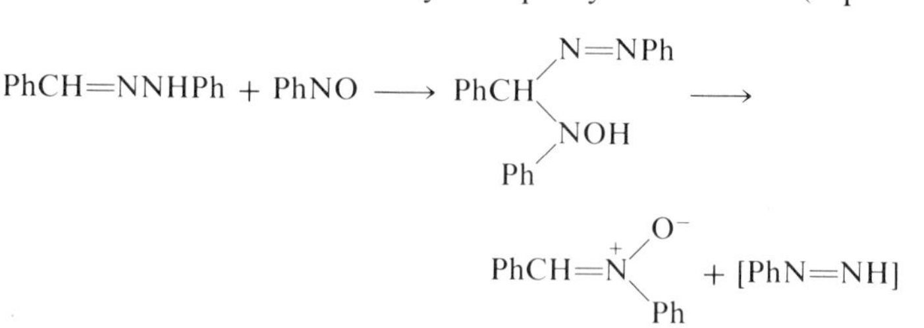

$$\text{PhCH=NNHPh + PhNO} \longrightarrow \text{PhCH} \overset{\text{N=NPh}}{\underset{\underset{\text{Ph}}{}}{\text{NOH}}} \longrightarrow$$

$$\text{PhCH=N} \overset{\text{O}^-}{\underset{\text{Ph}}{}} + [\text{PhN=NH}] \qquad (2\text{-}60)$$

benzophenone *N*-phenylnitrone and phenylhydrazone are also formed, and phenyl radicals have been detected.

Halogenation. Halogens attack the trigonal carbon of hydrazones and azines, replacing the aldehyde hydrogen if one is present. Ketone hydrazines and ketazines react with chlorine to form α-chloro azo compounds[149,150] (Eq. 2-61); these are

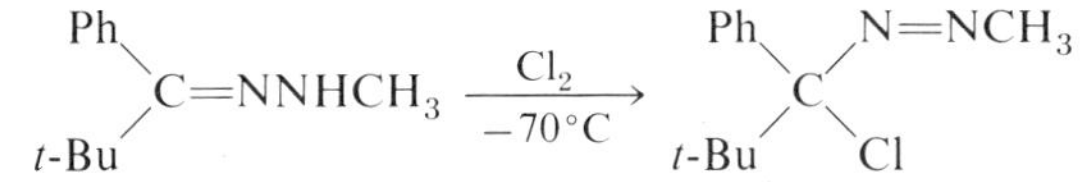

$$\underset{t\text{-Bu}}{\overset{Ph}{>}}C{=}NNHCH_3 \xrightarrow[-70°C]{Cl_2} \underset{t\text{-Bu}}{\overset{Ph}{>}}C\underset{Cl}{\overset{N=NCH_3}{<}} \qquad (2\text{-}61)$$

useful intermediates because of the ease with which the chlorine may be replaced by hydrogen or nucleophiles. Aldehyde hydrazones are halogenated to hydrazonoyl halides[151-154] (Eq. 2-62) or their cyclization products (oxadiazoles in the case of

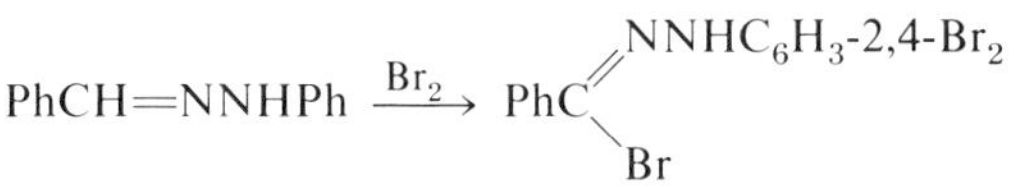

$$PhCH{=}NNHPh \xrightarrow{Br_2} PhC\underset{Br}{\overset{NNHC_6H_3\text{-}2,4\text{-}Br_2}{<}} \qquad (2\text{-}62)$$

semicarbazones), and they may be oxidized to nitriles at higher temperatures (160°C).[155] The rate of halogenation of aldehyde hydrazones appears to be controlled by conversion of the *syn*-H stereoisomer to the more reactive *anti*-H form.[151] Treatment with excess chlorine may convert aldehyde arylhydrazones to α,α-dichloro azo compounds[150] (Eq. 2-63). In contrast to the foregoing reactions is the

$$RCH{=}NNHAr \xrightarrow{Cl_2} RCCl_2N{=}NAr \qquad (2\text{-}63)$$

bromination of certain dinitrophenylhydrazones of aliphatic ketones in the steroid field. They are converted to the hydrazones of α-bromo ketones, which eliminate HBr spontaneously to form α,β-unsaturated hydrazones[156] (Eq. 2-64). The scope and limitations of this reaction do not appear to be known.

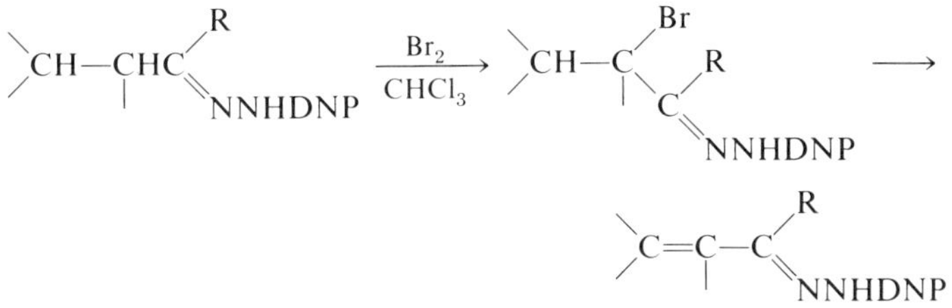

$$\underset{}{>}CH{-}CHC\underset{NNHDNP}{\overset{R}{<}} \xrightarrow[CHCl_3]{Br_2} \underset{}{>}CH{-}C\underset{\underset{NNHDNP}{C}}{\overset{Br}{|}}R \longrightarrow$$

$$\underset{}{>}C{=}C{-}C\underset{NNHDNP}{\overset{R}{<}} \qquad (2\text{-}64)$$

Oxidation. Unsubstituted hydrazones are oxidized to diazo alkanes by a wide variety of oxidizing agents, and the reaction is an important preparative method (see Chap. 5 for a discussion of conditions and choice of oxidizing agent). Besides diazo alkane, some azine is almost always produced (Eq. 2-65), as well as products of

$$R_2C{=}NNH_2 \xrightarrow[OH^-]{HgO} R_2C{=}N_2 \quad (\text{and } R_2C{=}N{-}N{=}CR_2) \qquad (2\text{-}65)$$

decomposition of the diazo compound. Much of the earlier work was concerned with mercuric oxide as the reagent, but in spite of much attention, the detailed nature of the oxidation remains uncertain. It is catalyzed by base, and a limited amount of water or alcohol is helpful.[157] Decomposition of the diazo compound is catalyzed by mercury salts, and in sensitive cases, no diazo compound can be isolated at all.[158] Such a case is camphor hydrazone, the oxidation of which gives tricyclene instead of diazocamphane. It has been proposed that azines are formed by the loss of nitrogen from dimers of diazo compounds, which might be formed by coupling between diazo compound and hydrazone followed by oxidation. Azines are also known to be formed from diazo compounds alone (see Chap. 5). However, it has been shown that oxidation of hydrazones by radical-producing reagents (for example, I_2, NBS) gives azines without going through intermediate diazo compounds.[159] Mercuric trifluoroacetate is responsible for a curiously specific reaction, oxidation to an acetylene[160] (Eq. 2-66); this reaction is not observed with aldehyde

$$ArCH_2C \overset{R}{\underset{NNH_2}{\Big\backslash}} \quad \xrightarrow{Hg(O_2CCF_3)_2} \quad \underset{15-55\%}{ArC\equiv CR} \tag{2-66}$$

hydrazones, nor with silver trifluoroacetate. In some solvents (for example, ethanol), mercuric oxide oxidizes benzaldehyde hydrazones to nitriles,[161] via diazo compound.

The anions of hydrazones, prepared by reaction with alkyllithium reagents, appear to be particularly sensitive to oxidation. Lithiated benzophenone hydrazone reacts even with oxygen, forming diphenyldiazomethane and hydroperoxide.[160]

Oxidation of unsubstituted hydrazones with excess halogen replaces the hydrazono group with two halogen atoms[162] (Eq. 2-67). This is a known reaction of diazoalkanes, which are presumably intermediates; yields are only moderate.

$$t\text{-BuCH}{=}NNH_2 + ICl \longrightarrow t\text{-BuCH}\overset{I}{\underset{Cl}{\Big\backslash}} \tag{2-67}$$

Monosubstituted hydrazones may be oxidized by a variety of reagents; the products may be dimeric (joined by C—C, C—N, or N—N bonds), oxygen-containing, or both. Oxidation of phenylhydrazones of aldehydes, especially benzaldehydes, by mercuric oxide, iodine, alkyl nitrites, etc. produces dimeric substances whose number and identity have for a long time been controversial.[163, 164] Unraveling the problem was hampered by the inherent imprecision of hydrogen analyses, by the occurrence of easy further transformations, and by the existence of solid solutions. The combined evidence of nuclear magnetic resonance and infrared and ultraviolet spectroscopy eventually brought order to the area in the representative instance of benzaldehyde phenylhydrazone,[165, 166] with which most investigators have been concerned.

An initial oxidation product of benzaldehyde phenylhydrazone is a bright-yellow substance with a melting point reported variously from 180 to 186°C. Although it was for a long time thought to be a tetrazane, it is now firmly established as a bis-azo compound (Eq. 2-68). It is accompanied by an isomer (mp 203°C) that has a hydrazidine structure and by an osazone, the *anti*-bis(phenylhydrazone) of benzil (mp 234°C). The bis-azo compound is easily tautomerized by treatment with base to the osazone, and all of these substances can be oxidized further. When manganese

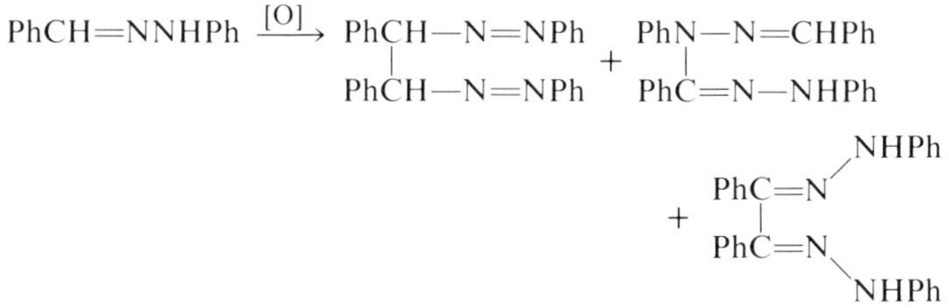

$$PhCH=NNHPh \xrightarrow{[O]} \begin{matrix} PhCH-N=NPh \\ | \\ PhCH-N=NPh \end{matrix} + \begin{matrix} PhN-N=CHPh \\ | \\ PhC=N-NHPh \end{matrix}$$

$$+ \begin{matrix} PhC=N^{\diagup NHPh} \\ | \\ PhC=N_{\diagdown NHPh} \end{matrix} \qquad (2\text{-}68)$$

dioxide is used as the oxidant, an osotriazole (1,3,4-triaryl-1,2,5-triazole) may be formed as well, in yields up to 50% in refluxing benzene, lower at lower temperatures. Biphenyl accompanies it.[167] The initial step is believed to be formation of a free radical delocalized over the CNN triad.[165-167] Conditions can be chosen (for example, oxygen and alcoholic alkali) so as to make the osazone the predominant product in preparatively useful yield.[163]

Ketone phenylhydrazones[168] are apparently oxidized to an analogous free radical, but dimerization takes place predominantly if not entirely by C—N bond formation (Eq. 2-69). The possibility for tautomerism inherent in the aldehyde hydra-

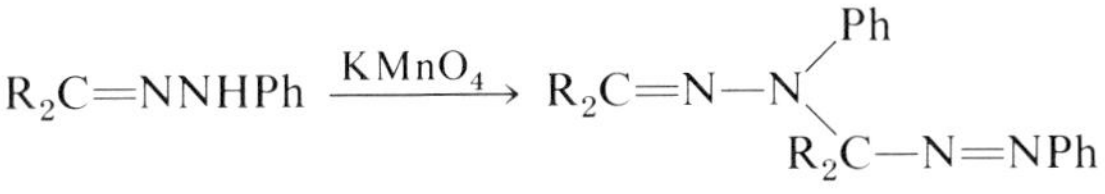

$$R_2C=NNHPh \xrightarrow{KMnO_4} \begin{matrix} Ph \\ R_2C=N-N^{\diagup} \\ \diagdown R_2C-N=NPh \end{matrix} \qquad (2\text{-}69)$$

zone oxidation products is, of course, not present, and the products from ketones are fixed in the azo hydrazine structure. In contrast, oxidation of some ketone phenylhydrazones under different conditions (MnO_2 suspended in benzene) has been found to give the parent ketone[167] (Eq. 2-70).

$$Ph_2C=NNHPh \xrightarrow{MnO_2} Ph_2C=O + Ph-Ph + N_2 \qquad (2\text{-}70)$$

Oxidizing agents that are donors of oxygen may attack either the carbon or the nitrogen of the hydrazone structure. Arylhydrazones of aldehydes react easily with atmospheric oxygen, forming what are apparently azo hydroperoxides[14] (Eq. 2-71).

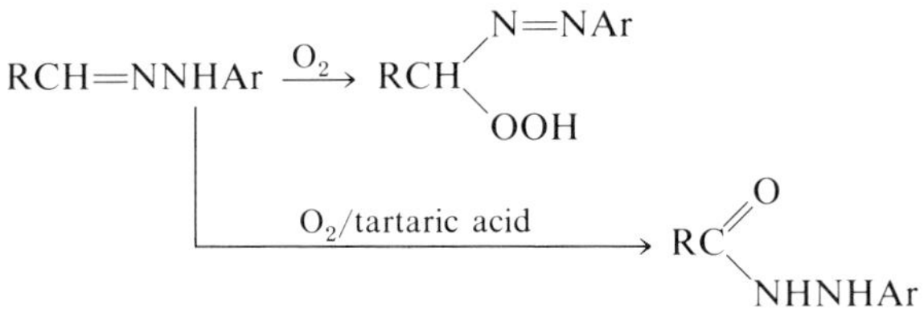

$$RCH=NNHAr \xrightarrow{O_2} RCH \begin{matrix} \diagup N=NAr \\ \diagdown OOH \end{matrix}$$

$$\downarrow \xrightarrow{O_2/\text{tartaric acid}} RC \begin{matrix} \diagup O \\ \diagdown NHNHAr \end{matrix} \qquad (2\text{-}71)$$

Although there are several reports that ketone phenylhydrazones are inert to autoxidation,[169,170] some aliphatic phenylhydrazones react with oxygen when heated and form analogous hydroperoxy azo compounds. The reaction is a radical-chain process and is faster in nonpolar solvents.[171-173]

In the presence of tartaric acid, benzaldehyde phenylhydrazone gives considerable amounts of N'-phenylbenzhydrazide, which is probably a transformation product of the hydroperoxide. In hot alcoholic solution, dimerization as well as addition of oxygen ensues,[174] giving products of imperfectly established structure that appear to be *N*-benzoyl hydrazidines (or an azo tautomer) (Eq. 2-72). A tetrazane structure

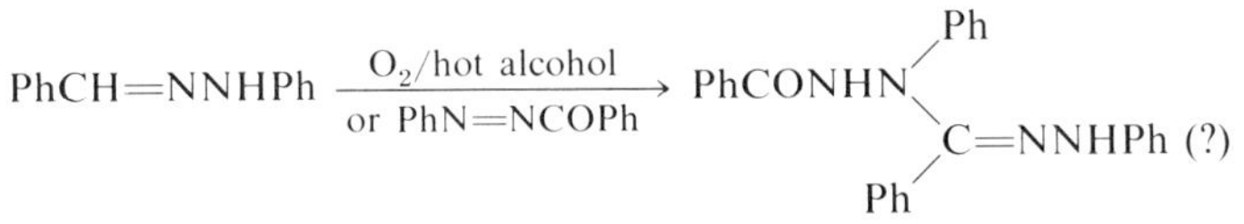

$$PhCH=NNHPh \xrightarrow[\text{or } PhN=NCOPh]{O_2/\text{hot alcohol}} PhCONHN\underset{Ph}{\overset{Ph}{<}}C=NNHPh\ (?) \qquad (2\text{-}72)$$

has been proposed for such products on the basis of their known formation from the hydrazone and *N*-benzoyl-*N'*-phenyldiazene, but the argument is unconvincing. The easy conversion to formazan derivatives and other hydrazidines implies a dimer formed through a C—N bond, and the reaction of ethyl azoformate with aldehyde phenylhydrazones, which gives analogous substances, has been shown to involve C—N rather than N—N bond formation[175] (see Chap. 4).

Peroxy acids convert benzaldehyde alkyl- and arylhydrazones to dimeric azoxy compounds[176] (Eq. 2-73). These are sparingly soluble, relatively high-melting sub-

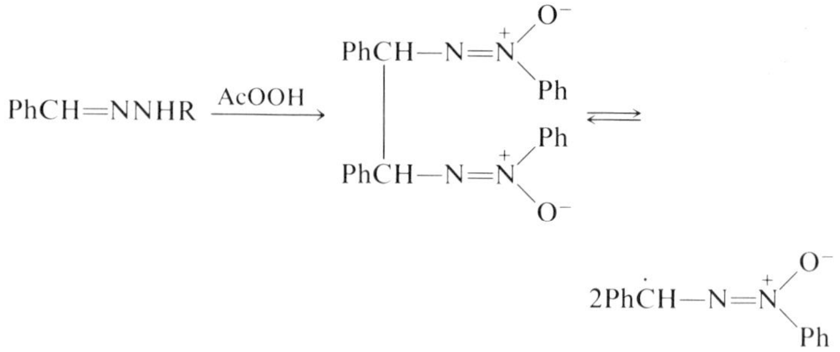

$$PhCH=NNHR \xrightarrow{AcOOH} \begin{matrix} PhCH-\overset{+}{N}=N\overset{O^-}{<}_{Ph} \\ | \\ PhCH-\overset{+}{N}=N\underset{O^-}{<}^{Ph} \end{matrix} \rightleftharpoons$$

$$2Ph\overset{\cdot}{C}H-\overset{+}{N}=N\overset{O^-}{<}_{Ph} \qquad (2\text{-}73)$$

stances, the structure of which was for a long time enigmatic and controversial. They were eventually recognized as azoxy compounds,[133] but their dimeric nature remained undetected still longer, owing to the difficulty of determining molecular weights and the ambiguity of hydrogen analyses. They exist in meso and racemic forms, which slowly interconvert by dissociation into allylically delocalized radicals.[176]

Monosubstituted hydrazones of aldehydes other than benzaldehydes and of ketones do not form dimeric azoxy compounds but instead are converted by peroxy acids into hydrazides[177] or α-acetoxy azoxy compounds,[178] respectively (Eqs. 2-74, 2-75). Cinnamaldehyde methylhydrazone gives a product like that from ketone

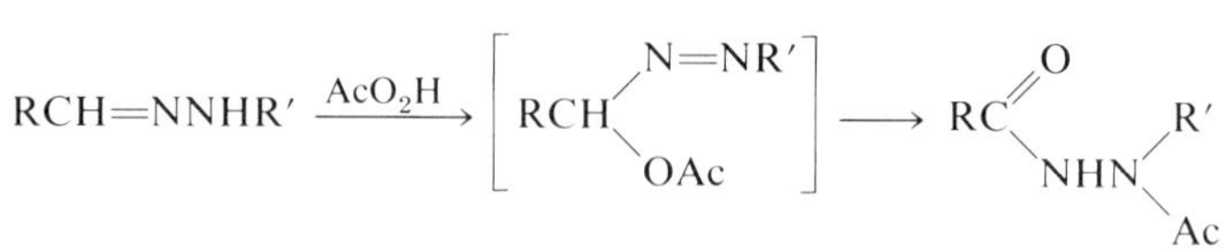

$$RCH=NNHR' \xrightarrow{AcO_2H} \left[RCH\underset{OAc}{\overset{N=NR'}{<}} \right] \longrightarrow RC\underset{NHN-Ac}{\overset{O}{<}}R' \qquad (2\text{-}74)$$

$$R_2C=NNHR' \xrightarrow{AcO_2H} R_2C\underset{\overset{+}{N}=N-R'}{\overset{OAc}{<}}\underset{O^-}{} \qquad (2\text{-}75)$$

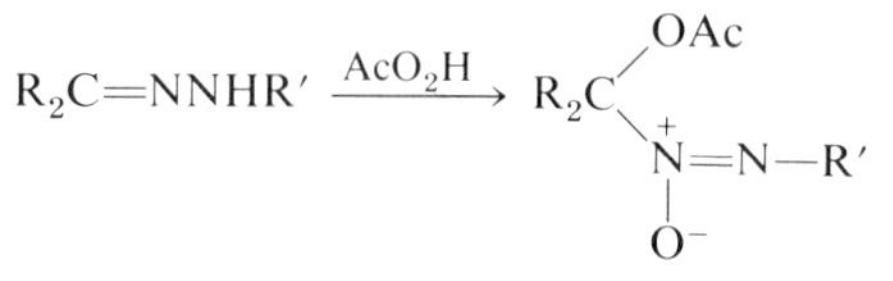

methylhydrazones.[179] Ketone acylhydrazones, such as semicarbazones, however, are cleaved smoothly to ketones and a *sym*-diacylhydrazine.[180]

The action of peroxy acids in disubstituted hydrazones has not been completely explored. Ketone derivatives have been converted to the parent ketone and tetrazenes by peroxyacetic acid (Eq. 2-76), perhaps through intermediate azamines

$$R_2C\!=\!NNR_2' \xrightarrow{\text{AcO}_2\text{H}} R_2CO + R_2'N\!-\!N\!=\!N\!-\!NR_2' \qquad (2\text{-}76)$$

(under similar conditions, unsubstituted hydrazones are converted to ketone and N_2 via intermediate diazoalkane). Ketazines, however, are converted to N-oxides,[181,182] which are oxidized by excess peroxy acid to ketones and nitrogen (Eq. 2-77). The

$$Ph_2C\!=\!NN\!=\!CPh_2 \xrightarrow{\text{AcO}_2\text{H}}$$

$$Ph_2C\!=\!N\!-\!\overset{+}{N}\!\!\begin{array}{c}\text{O}^-\\ \diagup \\ \diagdown \\ \text{CPh}_2\end{array}\quad\begin{array}{c}\xrightarrow{\text{AcO}_2\text{H}} Ph_2CO + N_2 \qquad (2\text{-}77)\\[4pt] \xrightarrow[h\nu]{\Delta\ \text{or}} Ph_2\!=\!N_2 + Ph_2CO\end{array}$$

isolable N-oxides are converted back to the azine by reducing agents and are cleaved by photolysis or thermolysis to ketone and diazoalkane. So easy can the latter reaction be that N-oxides of aliphatic azines appear to decompose as fast as they are formed; the diazoalkane immediately reacts with the acid present to form an ester, which is the product isolated. The mixed azine benzophenone isopropylidenehydrazone does not form an isolable N-oxide when treated with trifluoroperoxyacetic acid, but the intense color of the transient diazodiphenylmethane can be seen. Another mixed azine, benzophenone benzylidenehydrazone, is selectively oxidized at the ketazine nitrogen[181] (Eq. 2-78). Aldazines (for example, of furfural) are oxidized to nitriles.[181]

$$PhCH\!=\!NN\!=\!CPh_2 \xrightarrow{\text{CF}_3\text{CO}_3\text{H}} PhCH\!=\!N\!-\!\overset{+}{N}\!\!\begin{array}{c}\text{O}^-\\ \diagup \\ \diagdown \\ \text{CPh}_2\end{array} \xrightarrow{\text{H}_2\text{O/H}^+}$$

$$75\%$$

$$Ph_2CO\ (+\ PhCH_2OH) \qquad (2\text{-}78)$$

Methanolic hydrogen peroxide has been used to cleave dialkylhydrazones of aldehydes to nitriles and dialkylhydroxylamine[183] (Eq. 2-79); an intermediate N-oxide may be presumed.

$$RCH\!=\!NNMe_2 + H_2O_2\ (30\%) \xrightarrow{\text{MeOH}} RCN + Me_2NOH \qquad (2\text{-}79)$$

Singlet oxygen attacks dimethylhydrazones of ketones by an apparent ene reaction; the isolated products are ketones, in 48 to 88% yields[184] (Eq. 2-80).

$$\begin{array}{c}RCH_2\\ \diagdown\\ \diagup\\ R\end{array}\!\!C\!=\!NNMe_2 + O_2 \xrightarrow[\text{(2) Ph}_3\text{P or Me}_2\text{S}]{\text{(1) }h\nu^-,\ \text{sens.}} RCH_2COR \qquad (2\text{-}80)$$

Adamantanone dimethylhydrazone, in which the ene reaction would generate a bridgehead double bond, does not react.

Lead tetraacetate attacks hydrazones at the trigonal carbon, forming α-acetoxy-azoalkanes from ketone derivatives[185-187] (Eq. 2-81). Aldazines give 1,3,4-oxadia-

$$R_2C{=}NNHR' \xrightarrow{Pb(OAc)_4} R_2C\begin{smallmatrix} OAc \\ \diagup \\ \diagdown \\ N{=}NR' \end{smallmatrix} \qquad (2\text{-}81)$$

zolines, perhaps derived from intermediate α-acetoxy compounds, and ketazines give α,β-unsaturated azoalkyl acetates[179] (Eq. 2-82). Urethans are formed from

$$\begin{smallmatrix} CH_3 \\ \diagdown \\ \diagup \\ R \end{smallmatrix}C{=}NN{=}C\begin{smallmatrix} CH_3 \\ \diagup \\ \diagdown \\ R \end{smallmatrix} \xrightarrow{Pb(OAc)_4} CH_2{=}C\begin{smallmatrix} R \\ \diagup \\ \diagdown \\ N{=}N{-}\underset{\underset{CH_3}{|}}{\overset{\overset{R}{|}}{C}}{-}OAc \end{smallmatrix} \qquad (2\text{-}82)$$

semicarbazones, presumably through intermediate acetoxy azo compounds[188] (Eq. 2-83).

$$Ph_2C{=}NNHCONEt_2 \xrightarrow{Pb(OAc)_4}$$

$$Ph_2C\begin{smallmatrix} OAc \\ \diagup \\ \diagdown \\ OCONEt_2 \end{smallmatrix} \quad \text{and} \quad Ph_2C\begin{smallmatrix} OAc \\ \diagup \\ \diagdown \\ CONEt_2 \end{smallmatrix} \qquad (2\text{-}83)$$

(63% yield, ratio 20:1)

Ozone evidently attacks at carbon rather than nitrogen, for ozonolysis of dinitrophenylhydrazones gives rise principally to ketone and dinitrobenzene[189] (Eq. 2-84) (the expected product of decomposition of dinitrophenyldiazene). Dimethyl-

$$Me_2C{=}NNHC_6H_3\text{-}2,4\text{-}(NO_2)_2 \xrightarrow{O_3}$$
$$Me_2CO + N_2 + 1,3\text{-}(NO_2)_2C_6H_4 \qquad (2\text{-}84)$$
$$50\text{-}100\% 35\text{-}50\%$$

hydrazones are comparable to alkenes in reactivity toward ozone, and a variety of ketone derivatives have been cleaved in preparatively useful yields (Eq. 2-85). The

$$\begin{smallmatrix} Pr \\ \diagdown \\ \diagup \\ Et \end{smallmatrix}C{=}NNMe_2 + 2O_3 \xrightarrow{CH_2Cl_2} \begin{smallmatrix} Pr \\ \diagdown \\ \diagup \\ Et \end{smallmatrix}C{=}O + O{=}N{-}NMe_2 + O_2 \qquad (2\text{-}85)$$
$$93\%$$

fact that two equivalents of ozone are required suggests that the reaction does not take place by cycloaddition, but by attack at carbon only, followed by cleavage to ketone, oxygen, and azamine, which is oxidized to a nitrosamine by more ozone. However, ring substituents on acetophenone dimethylhydrazones do not affect the reactivity appreciably.[189]

Among other oxidizing agents useful for regenerating ketones from dialkylhydrazones should be mentioned cobalt trifluoride,[190] which is reported to give yields of 46 to 94%, and periodate,[7] which also gives high yields and permits a choice of pH between 4.5 and 7. Acidic conditions are desirable with aldehyde dimethylhydrazones, for there is otherwise a competing reaction leading to nitrile. Periodate oxidation of ketone dimethylhydrazones proceeds with gas evolution; the gas shows an m/e value of 44 in the mass spectrometer and may be nitrous oxide. Photolytic oxidation has been noted under the section entitled Heat and light.

Finally, the oxidation of osazones should be mentioned. They are very easily converted to 2-aryl-1,2,3-triazoles by copper sulfate, one phenyl group and its attached nitrogen being lost[191] (Eq. 2-86). This reaction is often carried out because

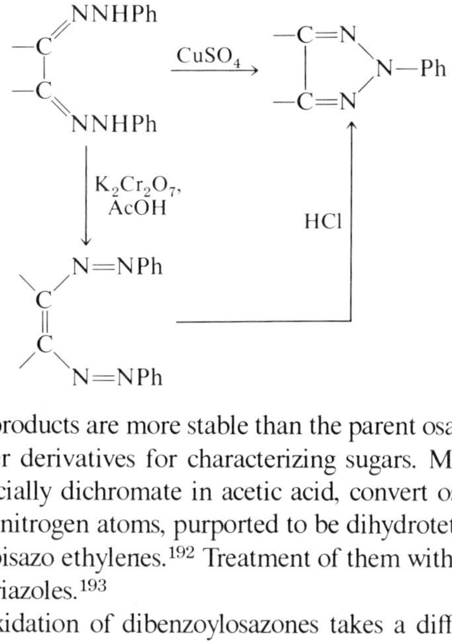

(2-86)

the products are more stable than the parent osazones; these osotriazoles thus make better derivatives for characterizing sugars. Most oxidizing agents, however, and especially dichromate in acetic acid, convert osazones to products still containing four nitrogen atoms, purported to be dihydrotetrazines, but which are more probably bisazo ethylenes.[192] Treatment of them with hydrochloric acid converts them to osotriazoles.[193]

Oxidation of dibenzoylosazones takes a different course, however, involving a rearrangement, and an aminotriazole derivative is formed[194] (Eq. 2-87).

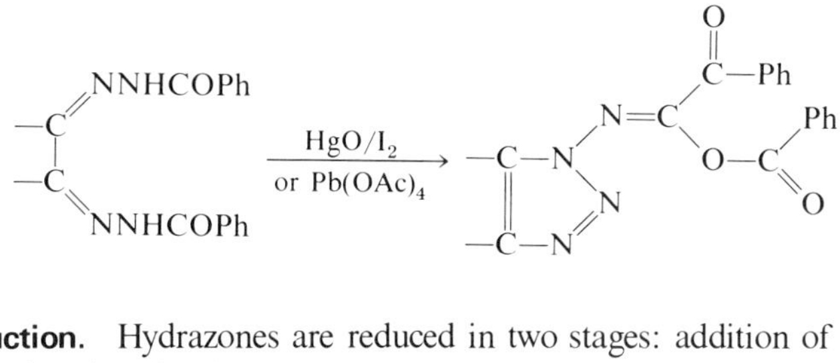

(2-87)

Reduction.   Hydrazones are reduced in two stages: addition of hydrogen to the C=N bond and reductive cleavage of the N—N bond. Catalytic hydrogenation with platinum, platinum oxide, or palladium catalyst at or near room temperature and with mild pressure usually hydrogenates the double bond only, giving hydrazines in preparatively useful yields[195-197] (Eq. 2-88). This type of reduction

$$Me_2C=NNHCOPh \xrightarrow{H_2/5\% \text{ Pt on } C} Me_2CHNHNHCOPh \qquad (2\text{-}88)$$

does not affect the carbonyl groups of acylhydrazones, which are reduced to hydrazides. Azines are reduced by the same means.[198] Azo compounds appear to be the result of the first stage of reduction; they may tautomerize to a hydrazone if further reduction does not intervene[199,200] (Eq. 2-89).

$$R_2C=N-N=CR_2 \xrightarrow{H_2/Pt}$$

$$R_2CHN=NCHR_2 \longrightarrow R_2CHNHNHCHR_2 \qquad (2\text{-}89)$$

$$R_2C=NNHCHR_2$$

Hydrogenation with a nickel catalyst generally brings about cleavage of the N—N bond with formation of amines[131,201,202] (Eq. 2-90). The same result oc-

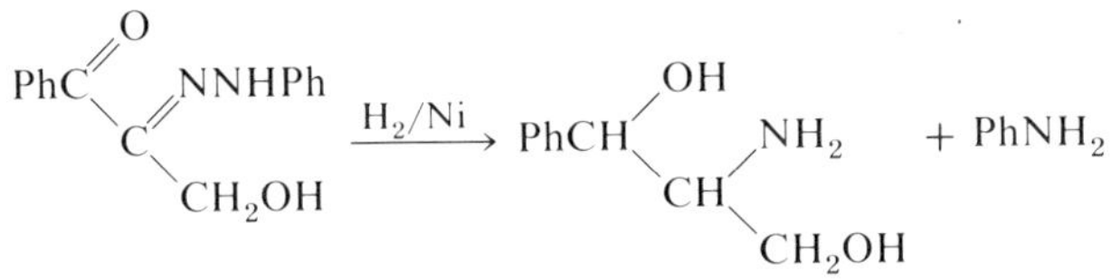

$$(2\text{-}90)$$

curs with phenylhydrazones of α-keto acids even with palladium or platinum catalysts.[203]

Diborane by itself is inert toward hydrazones, but when HCl is present, they are reduced to hydrazines in high yields at room temperature.[204] Sodium borohydride reduces disubstituted hydrazones to hydrazines, but gives poor yields when an NH is still present.[205] Lithium aluminum hydride is generally an effective reducing agent for hydrazones and azines,[149,206-208] but under forcing conditions it may cleave the N—N bond.[204] It often reduces the acyl groups of acylhydrazones at the same time[209] (but t-butoxycarbonyl groups survive well[210]). Some quaternary hydrazonium salts have been subjected to combined reduction and ring closure to prepare aziridines[211] (Eq. 2-91); yields are better than from the corresponding oximes.

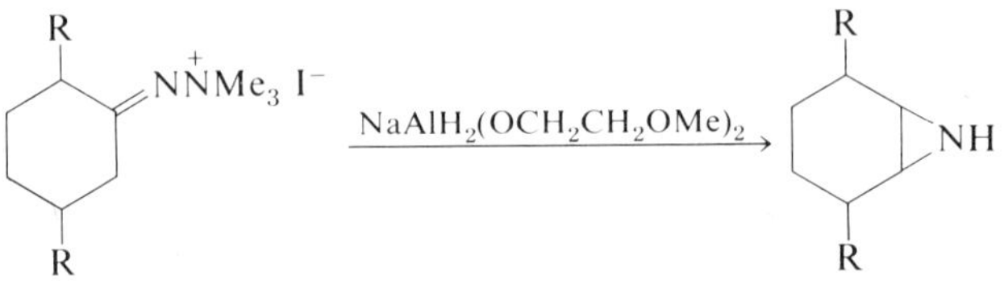

$$(2\text{-}91)$$

46–86%

Zinc and alcoholic HCl,[212] acetic acid, or calcium chloride[203] have been used to reduce hydrazones to amines. The Fischer indole cyclization may be a serious interference unless strong acid is avoided. Aluminum amalgam reduces azines to amines[213,214] (Eq. 2-92).

$$R_2C=NN=CR_2 \xrightarrow{AlHg} R_2CHNH_2 \; (+ \text{ some secondary amine}) \qquad (2\text{-}92)$$

Sodium amalgam in the presence of acid reduces hydrazones to amines; in base, only to the hydrazine stage.[215-219] It will also reduce aromatic azines, but aliphatic azines appear to be resistant. Partial reduction of aldazines to hydrazones also has been reported[215] (Eq. 2-93). (Further information on reduction of hydrazones may be found in Chap. 1 under Preparative Methods).

$$RCH{=}NN{=}CHR \xrightarrow{\text{NaHg}_x} RCH_2NHN{=}CHR \qquad (2\text{-}93)$$

Sulfonylhydrazones undergo reactions similar to the foregoing ones but have other features peculiar to the sulfonyl group. These are discussed in Chap. 3.

PREPARATIVE METHODS

Hydrazones are nearly always prepared from ketones or aldehydes and the appropriate hydrazine (Eq. 2-94), for it is a simple process that uses materials that are for

$$R_2C{=}O + RNHNH_2 \longrightarrow R_2C{=}NNHR + H_2O \qquad (2\text{-}94)$$

the most part easily available. It usually suffices to mix the reactants in aqueous or alcoholic solution, with or without addition of a catalytic amount of acid (acetic or stronger). Reaction often takes place readily at room temperature and in reactive cases may even require some cooling. When dinitrophenylhydrazine is a reactant, fairly strong solutions of sulfuric, hydrochloric, or phosphoric acid are customarily used. Because of the great usefulness of dinitrophenylhydrazones as derivatives for characterization of aldehydes and ketones, many recipes have been published for their preparation,[220] each with its own claims for advantage. Solutions of 2,4-dinitrophenylhydrazine in sulfuric acid for this purpose are sometimes known as *Brady's reagent.*

Formation of aryl and acyl hydrazones has a very favorable position of equilibrium,[221] and it is not necessary to take pains to remove the water formed in the reaction, or even to avoid using water as a solvent. Alkylhydrazines appear to react less completely with ketones, and the preparation of alkylhydrazones is usually carried out in the absence of water, often with provision to remove the water formed, such as a Dean-Stark trap.[7,32,98] High yields have been obtained by heating the reactants in absolute ethanol and then distilling the solvent.[222,223] The formation of unsubstituted hydrazones is experimentally similar to that of alkylhydrazones, except that a large excess of hydrazine is desirable to minimize formation of azine.[83] Benzaldehyde alkylhydrazones form readily in the presence of water, however.

Some alkylarylhydrazones have been found to form inconveniently slowly, perhaps as a result of steric hindrance. Higher overall yields have been obtained in a shorter time by alkylating the sodium salts of the corresponding arylhydrazones[89] (see Eq. 2-40). In other cases, where formation of a hydrazone is difficult, it may be advantageous to use instead the corresponding *gem*-dichloro compound[224] (Eq. 2-95).

$$R_2CCl_2 + R_2'NNH_2 \longrightarrow R_2C{=}NNR_2' \; ({+}\; R_2'NNH_2{\cdot}HCl) \qquad (2\text{-}95)$$

Unsubstituted hydrazones are often difficult to prepare directly, even with a large excess of hydrazine, owing to the predominance of azine formation.[72,83] In such cases it is sometimes convenient to prepare the azine deliberately and then to heat it with excess anhydrous hydrazine to displace the equilibrium toward the hydrazone[225,226] (Eq. 2-96). Alternatively, the dimethylhydrazone can be prepared first

$$2R_2CO + N_2H_4 \longrightarrow R_2C{=}NN{=}CR_2 \xrightarrow{N_2H_4} 2R_2C{=}NNH_2 \qquad (2\text{-}96)$$
$$(R = Me,\ 77\text{--}88\%)$$

and then be heated with anhydrous hydrazine (yields greater than 80%).[222,223] In all these processes, the important point is to avoid the presence of water, which promotes disproportionation of the hydrazone into azine and hydrazine (see Eq. 2-14).

There are two alternative routes to unsubstituted hydrazones, both reductive. If the corresponding imine is available (perhaps from reaction of a nitrile with a Grignard reagent), it can be nitrosated and the nitrosoimine can then be reduced with zinc and acetic acid[227] (Eq. 2-97). If the diazoalkane is more readily available

$$R_2C{=}NH \xrightarrow[\text{AcOH}]{NaNO_2} R_2C{=}NNO \xrightarrow[\text{AcOH}]{Zn} R_2C{=}NNH_2 \qquad (2\text{-}97)$$

than the hydrazone, as is the case with α-diazo ketones, reduction using a phosphine[97] or ammonium sulfide[228] may be advantageous (Eq. 2-98).

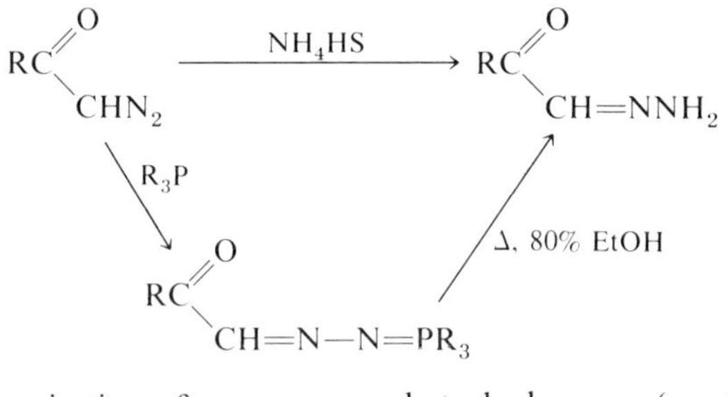

$$(2\text{-}98)$$

Isomerization of azo compounds to hydrazones (see Chap. 4) has preparative usefulness principally for the formation of arylhydrazones from diazonium salts in the Japp-Klingemann reaction[229,230] (Eq. 2-99). Since this reaction requires an ac-

$$A_2CH_2 + ArN_2{}^+ \longrightarrow A_2CH{-}N{=}NAr \longrightarrow A_2C{=}NNHAr \qquad (2\text{-}99)$$

tive methylene group in the starting material, it is suited to the preparation of mono(arylhydrazones) from β-dicarbonyl and related compounds.

Oxidation of *sym*-dialkyl hydrazines, which gives azo compounds, also can give rise to hydrazones if conditions that favor the necessary tautomerization are chosen (see Chaps. 1 and 4). Trialkylhydrazines can be oxidized directly to hydrazones, as shown by the preparation of formaldehyde dimethylhydrazone by oxidation of trimethylhydrazine with bromine (see Eq. 1-50).

Azines have potential use as sources of hydrazones in two ways. Partial reduction may give a monosubstituted hydrazone directly[215] (see Eqs. 2-89 and 2-93). Addition of Grignard reagents to azines can be a useful route to hydrazones bearing a tertiary substituent in special cases[114] (see Eq. 2-34). The addition of Grignard reagents to diazo compounds can also be managed so as to give useful yields of t-alkylhydrazones[132,231] (Eq. 2-100). Primary alkyllithium reagents form mono-alkylhydrazones by reaction with nitrous oxide[232] (Eq. 2-101).

$$Ph_2C{=}N_2 \xrightarrow[\text{(2) } H_2O]{\text{(1) } t\text{-BuMgCl}} Ph_2C{=}NNH\text{-}t\text{-Bu} \tag{2-100}$$

$$2BuLi + N_2O \longrightarrow BuNHN{=}CHPr \tag{2-101}$$

Azines are generally prepared[198,233] smoothly by treating an aldehyde or ketone with a limited amount of hydrazine; the reaction is often spontaneous, and the products, which commonly are only sparingly soluble, crystallize readily (Eq. 2-102). A direct synthesis of azines from nitriles has been reported, in which hydrazine acts as a reducing agent as well[37] (Eq. 2-103).

$$2R_2C{=}O + N_2H_4 \longrightarrow R_2C{=}N{-}N{=}CR_2 + 2H_2O \tag{2-102}$$

$$RCN + N_2H_4 \xrightarrow[\text{EtOH, 50-55}^\circ C]{\text{Raney Ni}} RCH{=}N{-}N{=}CHR \tag{2-103}$$

Azomethine imides can be prepared from aldehydes and hydrazines bearing an acid-strengthening substituent on the N' position; water is removed by codistillation with xylene and a Dean-Stark trap[126] (Eq. 2-104). In an instance where this method

$$RNHNHCOR' + R''CHO \longrightarrow R\overset{+}{N}\underset{CHR''}{\overset{NCOR'}{<}} \tag{2-104}$$

failed, benzaldehyde diethyl acetal was used with a trace of toluenesulfonic acid as catalyst and gave high yields; aliphatic acetals gave only poor yields, however.[234]

ANALYTICAL METHODS

There does not appear to be very much interest in determining hydrazones for their own sake; most of the analytical attention to them has resulted from the use of hydrazones in determining aldehydes and ketones. Dinitrophenylhydrazones, for example, which can be prepared quantitatively from them, can be estimated with a high degree of sensitivity colorimetrically, making use of the intense color developed when they are converted to their anions by base.[235] Dimethylhydrazones, and presumably other alkylhydrazones, can be distinguished from the corresponding hydrazines by potentiometric titration, making use of the fact that the hydrazones are markedly weaker bases.[236]

Arylhydrazones can be detected colorimetrically by the sequence hydrolysis, diazotization, and coupling to α-naphthylamine.[237] This method is specially suitable as a spot test.

2,4-Dinitrophenylhydrazine can be used as a reagent for more soluble hydrazones, especially alkylhydrazones, by carrying out an exchange reaction in much the same way that one would use the reagent with aldehydes and ketones.

In general, the methods for detection and determination of hydrazines (Chap. 1) can be adapted to hydrazones by incorporating a hydrolysis step.

REFERENCES

1. For example, W. Shorianetz and E. Kovats, *Helv. Chim. Acta* **53**, 251 (1970).
2. E. H. Cordes and W. P. Jencks, *J. Am. Chem. Soc.* **84**, 832 (1962).
3. M. C. B. Hotz and A. H. Spong, *J. Chem. Soc.*, 4283 (1962).
4. H. F. Harnsberger, E. L. Cochran, and H. H. Szmant, *J. Am. Chem. Soc.* **77**, 5048 (1955).
5. H. Gösl and A. Meuwsen, *Chem. Ber.* **92**, 2521 (1959).
6. R. L. Hinman, *J. Org. Chem.* **25**, 1775 (1960).
7. E. J. Corey and D. Enders, *Chem. Ber.* **111**, 1337 (1978).
8. K. G. Davenport, M. Newcomb, and D. E. Bergbreiter, *J. Org. Chem.* **46**, 3143 (1981).
9. E. Hyde, *Ber.* **32**, 1810 (1889).
10. L. A. Jones and N. L. Mueller, *J. Org. Chem.* **27**, 2356 (1962).
11. E. C. C. Baly and W. B. Tuck, *J. Chem. Soc.* **89**, 982 (1906).
12. G. J. Karabatsos, B. L. Shapiro, F. M. Vane, J. S. Fleming, and J. S. Ratka, *J. Am. Chem. Soc.* **85**, 2784 (1963).
13. G. J. Karabatsos and R. A. Taller, *J. Am. Chem. Soc.* **85**, 3624 (1963).
14. H. C. Yao and P. Resnick, *J. Org. Chem.* **30**, 2832 (1965).
15. Yu. P. Kitaev, B. I. Buzykin, and T. V. Toepol'skaya, *Usp. Khim.* **39**, 961 (1970) (English trans. p. 441): "The Structure of Hydrazones."
16. H. Ahlbrecht and H. Henk, *Chem. Ber.* **106**, 1659 (1975).
17. N. D. Heindel, P. D. Kennewell, and M. Pfau, *J. Org. Chem.* **35**, 80 (1970).
18. G. Zinner, W. Kliegel, W. Ritter, and H. Böhlke, *Chem. Ber.* **99**, 1678 (1966).
19. A. Grieder and P. Schiess, *Chimia* **24**, 25 (1970).
20. Yu. P. Kitaev and T. V. Troepol'skaya, *Izv. Akad. Nauk S.S.S.R., Otd. Khim. Nauk*, 454, 465 (1963) (English trans. pp. 408, 418).
21. N. B. Lebedeva, T. N. Masalitinova, O. N. Mon'yakova, and T. P. Oleinikova, *Zh. Org. Khim.* **16**, 256 (1980) (English trans. p. 226).
22. B. V. Ioffe and V. S. Stopskii, *Tetrahedron Lett.*, 18 (1960).
23. C. A. Bunnell and P. L. Fuchs, *J. Org. Chem.* **42**, 2614 (1977).
24. W. H. Mills and A. M. Bain, *J. Chem. Soc.*, **105**, 64 (1914).
25. W. H. Mills and B. C. Saunders, *J. Chem. Soc.*, 537 (1931).
26. B. Overton, *Ber.* **26**, 18 (1893).
27. M. Rosenblum, V. Nayak, S. K. Das Gupta, and A. Longroy, *J. Am. Chem. Soc.* **85**, 3874 (1963).
28. I. Fleming and J. Harley-Mason, *J. Chem. Soc.*, 5560 (1961).
29. G. R. Newkome and N. S. Bhacca, *J. Org. Chem.* **36**, 1719 (1971).
30. V. J. Stenberg, P. A. Barks, D. Bays, D. D. Hammargren, and D. V. Rao, *J. Org. Chem.* **33**, 4402 (1968).
31. P. A. S. Smith and E. E. Most, Jr., *J. Org. Chem.* **22**, 358 (1957).
32. R. H. Wiley and G. Irick, *J. Org. Chem.* **24**, 1925 (1959).
33. G. H. Brown and W. G. Shaw, Jr., *J. Org. Chem.* **24**, 132 (1959).
34. J. P. Phillips, *J. Org. Chem.* **29**, 982 (1964).
35. P. Grammaticakis, *Spectres d'Absorption Ultraviolets de Composes Organiques Azotes et Correlations Spectrochimiques,* Fasc. 3, Technique et Documentation, Paris, 1980.

36. B. V. Ioffe and L. M. Gershtein, *Zh. Org. Khim.* **5**, 268 (1969) (English trans. p. 257).

37. W. W. Zazac, Jr., and R. H. Denk, *J. Org. Chem.* **27**, 3716 (1962).

38. R. O'Connor, *J. Org. Chem.* **26**, 4375 (1961).

39. D. H. Kenny, *J. Chem. Ed.* **57**, 462 (1980).

40. J. Sire, *Acta Chem. Scand.* **31**, 589 (1977).

41. D. Goldsmith and C. Djerassi, *J. Org. Chem.* **31**, 3661 (1966).

42. W. C. Hamilton and S. J. La Place, *Acta Cryst.* **B24**, 1147 (1968).

43. P. J. L. Galigne and J. Falguerettes, *Acta Cryst.* **B24**, 1523 (1968).

44. K. Bjamer, S. Furberg, and C. S. Petersen, *Acta Chem. Scand.* **18**, 587 (1964).

45. A. Mannschreck and U. Koelle, *Tetrahedron Lett.* 863 (1967).

45*A*. S. Fischer and C. Wentrup, *J. Chem. Soc., Chem. Comm.* 502 (1980).

46. R. B. Hanson, P. Foley, Jr., E. L. Anderson, and M. H. Aldridge, *J. Org. Chem.* **35**, 1753 (1970).

47. G. R. Newkome and D. L. Fishel, *J. Org. Chem.* **37**, 1329 (1972).

48. H. E. Zimmerman and S. Somesekhara, *J. Am. Chem. Soc.* **82**, 5865 (1960).

49. S. S. Hirsch, *J. Org. Chem.* **32**, 2433 (1967).

50. R. L. Stern and J. G. Krause, *J. Org. Chem.* **33**, 212 (1968).

51. S. D. Carson and H. M. Rosenberg, *J. Org. Chem.* **35**, 2734 (1970).

52. R. W. Binkley, *J. Org. Chem.* **35**, 2796 (1970).

53. R. W. Binkley, *J. Org. Chem.* **34**, 931 (1969).

54. D. G. Horne and R. G. W. Norrish, *Proc. Roy. Soc. Lond.* [*A*] **315**, 301 (1970).

55. R. W. Binkley, *J. Org. Chem.* **33**, 2311 (1968).

56. R. W. Binkley, *J. Org. Chem.* **34**, 2072 (1969).

57. J. Gorse III and R. W. Binkley, *J. Org. Chem.* **37**, 575 (1972).

58. J. E. Hodgkins and J. A. King, *J. Am. Chem. Soc.* **85**, 2679 (1963).

59. R. W. Binkley, *J. Org. Chem.* **34**, 3218 (1969).

60. R. W. Binkley, *Tetrahedron Lett.*, 2085 (1970).

61. D. Todd, *J. Am. Chem. Soc.* **71**, 1353 (1949).

62. R. C. Corley and M. J. Gibian, *J. Org. Chem.* **37**, 2910 (1972).

63. E. J. Corey and D. Enders, *Chem. Ber.* **111**, 1362 (1978).

64. W. P. Jencks, *Prog. Phys. Org. Chem.* **2**, 63 (1964).

65. T. Pino and E. H. Cordes, *J. Org. Chem.* **36**, 1668 (1971).

66. W. Ried and A. Mühle, *Liebig's Ann. Chem.* **656**, 119 (1962).

67. S. R. Maynez, L. Pelavin, and G. Erker, *J. Org. Chem.* **40**, 3302 (1975).

68. C. E. Sacks and P. L. Fuchs, *Synthesis*, 456 (1976).

69. A. Fish and M. Saeed, *Chem. Ind.*, 571 (1963).

70. E. J. Corey and S. Knapp, *Tetrahedron Lett.*, 3667, 4687 (1976).

71. J. F. W. McOmie, *Protective Groups in Organic Chemistry*, Plenum Press, New York, 1973, pp. 340–342.

72. H. H. Szmant and C. McGinnis, *J. Am. Chem. Soc.* **72**, 2890 (1950).

73. B. Robinson, *Chem. Rev.* **69**, 227 (1969).

74. F. M. Miller and W. N. Schinske, *J. Org. Chem.* **43**, 3384 (1978).

75. R. B. Carlin and M. S. Moores, *J. Am. Chem. Soc.* **84**, 4107 (1962).

76. R. Fusco and F. Sannicolo, *Gaz. Chim. Ital.* **106**, 85 (1976).

77. R. B. Carlin, *J. Am. Chem. Soc.* **74**, 1077 (1952).

78. H. Plieninger and I. Nogradi, *Chem. Ber.* **88**, 1964 (1955).

79. A. W. Douglas, *J. Am. Chem. Soc.* **100**, 6463 (1978).

80. F. P. Robinson and R. K. Brown, *Can. J. Chem.* **42**, 1940 (1964).

81. P. W. Neber, *Liebig's Ann. Chem.* **471**, 113 (1929).

82. F. Sparatore, *Gaz. Chim. Ital.* **92**, 596 (1962).

83. (*a*) B. Robinson, *Tetrahedron* **20**, 515 (1964); (*b*) *Chem. Rev.* **63**, 397 (1963).

84. I. I. Grandberg, A. N. Kost, and Yu. A. Naumov, *Dokl. Akad. Nauk S.S.S.R.* **149**, 838 (1963).

85. (*a*) K. W. Frey and R. Hofmann, *Monatsh. Chem.* **22**, 760 (1901); (*b*) O. Piloty, *Ber.* **43**, 489 (1910).

86. R. Fusco and F. Sannicolo, *J. Org. Chem.* **46**, 83 (1981).

87. R. Fusco and F. Sannicolo, *J. Org. Chem.* **46**, 90 (1981).

88. F. E. Condon and D. Farcasiu, *J. Am. Chem. Soc.* **92**, 6625 (1970).

89. W. G. Kenyon and C. R. Hauser, *J. Org. Chem.* **30**, 292 (1965).

90. C. F. Bearn, R. S. Foote, and C. R. Hauser, *J. Heterocycl. Chem.* **9**, 183 (1972).

91. H. Ahlbrecht, E. O. Dueber, D. Enders, H. Eichenauer, and P. Weuster, *Tetrahedron Lett.*, 3691 (1978).

92. K. G. Davenport, H. Eichenauer, D. Enders, M. Newcomb, and D. E. Bergbreiter, *J. Am. Chem. Soc.* **101**, 5654 (1979).

93. D. Todd, *Org. Reactions* **4**, 378 (1948).

94. H. H. Szmant, *Angew. Chem. Int. Ed.* **7**, 120 (1968).

95. D. J. Cram and M. R. V. Sahyun, *J. Am. Chem. Soc.* **84**, 1734 (1962).

96. I. M. Hunsberger, E. R. Shaw, J. Fugger, R. Ketcham, and D. Lednicer, *J. Org. Chem.* **21**, 394, 2262 (1956).

97. H. J. Bestmann, H. Buckschewski, and H. Leube, *Chem. Ber.* **92**, 1345 (1959).

98. D. Todd, *J. Am. Chem. Soc.* **71**, 1356 (1949).

99. H. H. Szmant and C. M. Harmuth, *J. Am. Chem. Soc.* **86**, 2909 (1964).

100. H. H. Szmant, H. F. Harnsberger, T. J. Butler, and W. P. Barie, *J. Am. Chem. Soc.* **74**, 2724 (1952).

101. H. H. Szmant and C. E. Alciaturi, *J. Org. Chem.* **42**, 1081 (1977).

102. S. Robev, *Compt. Rend. Bulgar. Acad. Sci.* **12**, 141, 207 (1959); *Chem. Abstr.* **54**, 22463e, 4480d (1960).

103. M. F. Grundon and M. D. Scott, *J. Chem. Soc.*, 5674 (1964).

104. D. Y. Curtin and G. S. Russell, *J. Am. Chem. Soc.* **73**, 5450 (1951).

105. Z.-I. Yoshida, T. Harada, and Y. Tamaru, *Tetrahedron Lett.*, 3823 (1976).

106. R. F. Parcell, *Chem. Ind.*, 1396 (1963).

107. S. Sato, *Bull. Chem. Soc. Japan* **41**, 1440 (1968).

108. A. Kakehi, S. Ito, T. Manabe, T. Maeda, and K. Imai, *J. Org. Chem.* **42**, 2514 (1977).

109. R. F. Smith and A. C. Bates, *J. Chem. Ed.* **46**, 174 (1969).

110. R. F. Smith and L. E. Walker, *J. Org. Chem.* **27**, 4372 (1962).

111. R. F. Parcell and J. P. Sanchez, *J. Org. Chem.* **46**, 5229 (1981).

112. B. V. Ioffe and L. E. Poroshin, *Zh. Obshch. Khim.* **29**, 3154 (1959).

113. O. Westphal and M. Eucken, *Ber.* **76**, 1137 (1943).

114. C. G. Overberger and A. V. Di Giulio, *J. Am. Chem. Soc.* **80**, 6562 (1958).

115. P. A. S. Smith and H. H. Tan, *J. Org. Chem.* **32**, 2586 (1967).

116. F. E. Ziegler and P. A. Wender, *J. Org. Chem.* **42**, 2001 (1977).

117. W. Knobloch and G. Subert, *J. Prakt. Chem.* **[4]36**, 29 (1967).

118. I. Ugi, *Angew. Chem.* **74**, 9 (1962).

119. P. Foley, Jr., E. Anderson, and F. Dewey, *J. Chem. Eng. Data* **14**, 272 (1969).

120. B. V. Ioffe and N. L. Zelenina, *Zh. Org. Khim.* **4**, 1558 (1968) (English trans. p. 1496).

121. A. N. Kost and R. S. Sagitullin, *J. Gen. Chem. U.S.S.R.* **27**, 3338 (1957).

122. R. Ciusa and G. Rastelli, *Gaz. Chim. Ital.* **52II**, 121 (1922).

123.' D. Enders and H. Eichenauer, *Tetrahedron Lett.*, 191 (1977).

124. H.-U. Blaser and D. Reinehr, *Helv. Chim. Acta* **61**, 1118 (1978).

125. S. F. Nelsen, *J. Org. Chem.* **34**, 2248 (1969).

126. W. Oppolzer, *Angew. Chem. Int. Ed.* **16**, 10 (1977).

127. T. J. Curphey and K. S. Prasad, *J. Org. Chem.* **37**, 2259 (1972).

128. B. V. Ioffe, V. S. Stopskii, and N. B. Burmanova, *Khim. Geterotsikl. Soedin.*, 1066 (1969); *Chem. Abstr.* **72**, 132607 (1970).

129. J.-P. Chapelle, J. Elguero, R. Jacquier, and G. Tarrago, *Bull. Soc. Chim. France*, 4464 (1969).

130. J. W. Walker, *J. Chem. Soc.* **69**, 1280 (1896).

131. K. Bodendorf, *Liebig's Ann. Chem.* **623**, 109 (1959).

132. P. A. S. Smith, J. M. Clegg, and J. Lakritz, *J. Org. Chem.* **23**, 1595 (1958).

133. B. T. Gillis and K. F. Schimmel, *J. Org. Chem.* **27**, 413 (1962).

134. C. Friedel and A. Combes, *Bull. Soc. Chim. France* **[3]11**, 115 (1894).

135. C. Larsen and D. N. Harpp, *J. Org. Chem.* **46**, 2465 (1981).

136. S. Goldschmidt and W. L. C. Veer, *Rec. Trav. Chim. Pays Bas* **65**, 796 (1946).

137. D. E. Pearson, K. N. Carter, and C. M. Greer, *J. Am. Chem. Soc.* **75**, 5905 (1953).

138. L. G. Donaruma, U.S. Patent Nos. 2,777,841, 2,763,644 (1956); *Chem. Abstr.* **51**, 5822, 10565 (1957).

139. T. Taguchi, T. Matsuo, and M. Kojima, *J. Org. Chem.* **29**, 1104 (1964).

140. E. Bamberger and W. Pemsel, *Ber.* **36**, 57 (1903).

141. K. C. Kalia and A. Chakravorty, *J. Org. Chem.* **35**, 2231 (1970).

142. H. Feuer and L. F. Spinicelli, *J. Org. Chem.* **41**, 2981 (1976).

143. A. F. Hegarty and F. L. Scott, *J. Org. Chem.* **32**, 1957 (1967).

144. H. Hauptmann and A. C. de M. Perisse, *Experientia* **10**, 60 (1954).

145. O. Dimroth and S. Merzbacher, *Ber.* **43**, 2899 (1910).

146. F. D. Chattaway and G. O. Parkes, *J. Chem. Soc.*, 113 (1926).

147. J. L. Augert, A. F. Hurst. S. A. Lowry, and R. G. Landolt, *J. Org. Chem.* **46**, 168 (1981).

148. B. A. DellaColletta, J. G. Frye, T. L. Youngless. J. P. Zeigler, and R. G. Landolt, *J. Org. Chem.* **42**, 3057 (1977).

149. P. L. Grizzle, D. W. Miller, and S. E. Scheppele, *J. Org. Chem.* **40**, 1902 (1975).

150. M. W. Moon, *J. Org. Chem.* **37**, 383 (1972).

151. A. F. Hegarty and F. L. Scott, *J. Org. Chem.* **33**, 753 (1968).

152. T. Bacchetti, *Gaz. Chim. Ital.* **91**, 866 (1961).

153. H. Gehlen and K. Mockel, *Liebig's Ann. Chem.* **651**, 133 (1962).

154. F. D. Chattaway and A. B. Adamson *J. Chem. Soc.*, 157 (1930).

155. E. Klingsberg, *J. Org. Chem.* **25**, 572 (1960).

156. V. R. Mattox and E. C. Kendall, *J. Am. Chem. Soc.* **72**, 2290 (1950).

157. R. Baltzly, N. B. Mehta, P. B. Russell, R. E. Brooks, E. M. Grivsky, and A. M. Steinberg, *J. Org. Chem.* **26**, 3669 (1961).

158. W. Reusch, M. W. Di Carlo, and L. Traynor, *J. Org. Chem.* **26**, 1711 (1961).

159. W. Fischer and J.-P. Anselme, *J. Am. Chem. Soc.* **89**, 5312 (1967).

160. R. J. Theis and R. E. Dessy, *J. Org. Chem.* **31**, 624 (1966).

161. D. B. Mobbs and H. Suschitzky, *Tetrahedron Lett.*, 361 (1971).

162. A. J. Fry and J. N. Cawse, *J. Org. Chem.* **32**, 1677 (1967).

163. H. Biltz and O. Amme, *Liebig's Ann. Chem.* **321**, 1 (1902).

164. J. Buckingham, *Quart. Rev.* **23**, 37 (1969).

165. T. Milligan and B. Minor, *J. Org. Chem.* **27**, 4663 (1962).

166. C. Wintner and J. Wiecko, *Tetrahedron Lett.*, 1595 (1969).

167. I. Bhatnagar and M. V. George, *J. Org. Chem.* **32**, 2252 (1967).

168. W. Theilacker and H. J. Tomuschat, *Chem. Ber.* **88**, 1086 (1955).

169. R. Criegee and G. Lohaus, *Chem. Ber.* **84**, 219 (1951).

170. K. H. Pausacker, *J. Chem. Soc.*, 3478 (1950).

171. W. F. Taylor, H. A. Weiss, and T. J. Wallace, *J. Org. Chem.* **34**, 1759 (1969).

172. A. J. Bellamy and R. D. Guthrie, *J. Chem. Soc.*, 3528 (1965).

173. A. J. Bellamy and R. D. Guthrie, *J. Chem. Soc.*, 2788 (1965).

174. M. Busch and H. Kunder, *Ber.* **49**, 2347 (1916).

175. B. T. Gillis and F. A. Daniher, *J. Org. Chem.* **27**, 4001 (1962).

176. R. B. Woodward and C. Wintner, *Tetrahedron Lett.*, 2693 (1969).

177. B. T. Gillis and K. F. Schimmel, *J. Org. Chem.* **32**, 2865 (1962).

178. J. P. Freeman, *J. Org. Chem.* **28**, 2508 (1963).

179. B. T. Gillis and M. P. La Montagne, *J. Org. Chem.* **32**, 3318 (1967).

180. L. Horner and H. Ferkeness, *Chem. Ber.* **94**, 712 (1961).

181. W. M. Williams and W. R. Dolbier, Jr., *J. Org. Chem.* **34**, 155 (1969).

182. L. Horner, W. Kirmse, and H. Ferkeness, *Chem. Ber.* **94**, 279 (1961).

183. R. F. Smith, J. A. Albright, and A. M. Waring, *J. Org. Chem.* **31**, 4100 (1966).

184. E. Friedrich, W. Lutz, H. Eichenauer, and D. Endres, *Synthesis*, 893 (1977).

185. J. Warkentin, *Synthesis*, 279 (1970).

186. B. T. Gillis and M. P. La Montagne, *J. Org. Chem.* **33**, 762 (1968).

187. D. C. Iffland, R. Salisbury, and W. R. Schafer, *J. Am. Chem. Soc.* **83**, 747 (1961).

188. D. C. Iffland and T. M. Davies, *J. Am. Chem. Soc.* **85**, 2182 (1963).

189. (a) R. E. Erickson, A. H. Riebel, A. M. Reader, and P. S. Bailey, *Liebig's Ann. Chem.* **653**, 129 (1962); (b) R. E. Erickson, P. J. Andrulis, Jr., J. C. Collins, M. L. Lungle, and G. D. Mercer, *J. Org. Chem.* **34**, 2961 (1969).

190. G. A. Olah, J. Welsh, and M. Henninger, *Synthesis*, 308 (1977).

191. R. M. Hann and C. S. Hudson, *J. Am. Chem. Soc.* **66**, 735 (1944).

192. D. Y. Curtin and N. E. Alexandron, *Tetrahedron* **19**, 1697 (1963).

193. H. von Pechmann, *Liebig's Ann. Chem.* **262**, 265 (1891).

194. N. E. Alexandron and E. D. Micromastoras, *J. Org. Chem.* **37**, 2345 (1972).

195. E. J. Poth and J. R. Bailey, *J. Am. Chem. Soc.* **45**, 3001 (1923).

196. M. C. Chaco and N. Rabjohn, *J. Org. Chem.* **27**, 2765 (1962).

197. M. C. Chaco, P. R. Stapp, J. A. Ross, and N. Rabjohn, *J. Org. Chem.* **27**, 3371 (1962).

198. W. Sucrow in *Methodicum Chemicum*, Vol. 6, F. Zymalkowski (ed.), Academic Press, New York, 1975, pp. 109–110.

199. H. Bretschneider, A. de Jonge-Bretschneider, and N. Ajtai, *Ber.* **74**, 571 (1941).

200. Z. Földi and G. von Fodor, *Ber.* **74**, 589 (1941).

201. C. F. Winans, *J. Am. Chem. Soc.* **55**, 2051 (1933).

202. L. F. Audrieth and W. L. Jolly, *J. Phys. Coll. Chem.* **55,** 524 (1951).

203. N. H. Khan and A. R. Kidwai, *J. Org. Chem.* **38,** 822 (1973).

204. J. A. Blair and R. J. Gardner, *J. Chem. Soc. [C],* 1714 (1970).

205. G. N. Walker, M. A. Moore, and B. N. Weaver, *J. Org. Chem.* **26,** 2740 (1961).

206. R. Renaud and L. C. Leitch, *Can. J. Chem.* **32,** 545 (1954).

207. J. B. Class, J. G. Aston, and T. S. Oakwood, *J. Am. Chem. Soc.* **75,** 2937 (1953).

208. A. Ebnöther, E. Jucker, A. Lindenmann, E. Rissi, R. Steiner, R. Süess, and A. Vogel, *Helv. Chim. Acta* **42,** 533 (1959).

209. L. Spialter, D. H. O'Brien, G. L. Untereiner, and W. A. Rush, *J. Org. Chem.* **30,** 3278 (1965).

210. L. A. Carpino, A. A. Santilli, and R. W. Murray, *J. Am. Chem. Soc.* **82,** 2728 (1960).

211. Y. Girault, M. Decouzon, and M. Azzaro, *Tetrahedron Lett.,* 1175 (1976).

212. V. V. Feofilaktov, *Zh. Obshch. Khim.* **17,** 993 (1947); *Chem. Abstr.* **42,** 4537i (1948).

213. R. J. Block, *Chem. Rev.* **38,** 534, 546 (1946).

214. K. Kratzl and K. P. Berger, *Monatsh. Chem.* **89,** 83 (1958).

215. A. Wohl and C. Oesterlin, *Ber.* **33,** 2736 (1900).

216. B. W. Langley, B. Lythgoe, and L. S. Rayner, *J. Chem. Soc.,* 4191 (1952).

217. H. Franzen and F. Krafft, *J. Prakt. Chem.* **[2]84,** 122 (1911).

218. F. P. Jahn, *J. Am. Chem. Soc.* **59,** 1761 (1937).

219. B. Gisin and M. Brenner, *Helv. Chim. Acta* **53,** 1030 (1970).

220. For example, R. L. Shriner, R. C. Fuson, D. Y. Curtin, and T. C. Morrill, *The Systematic Identification of Organic Compounds,* Wiley, New York, 1980, pp. 162–179.

221. E. H. Cordes and W. P. Jencks, *J. Am. Chem. Soc.* **84,** 825 (1962).

222. G. R. Newkome and D. L. Fishel, *J. Org. Chem.* **31,** 677 (1966).

223. G. R. Newkome and D. L. Fishel, *Org. Syntheses* **50,** 102 (1970).

224. W. Theilacker and O. R. Leichtle, *Liebig's Ann. Chem.* **572,** 121 (1951).

225. H. Staudinger and A. Gaule, *Ber.* **49,** 1897 (1916).

226. A. R. Day and M. C. Whiting, *Org. Syntheses* **50,** 3 (1970).

227. R. A. Bartsch, S. Hünig, and H. Quast, *J. Org. Chem.* **37,** 3604 (1972).

228. M. L. Wolfrom and J. B. Miller, *J. Am. Chem. Soc.* **80,** 1678 (1958).

229. R. P. Linstead and A. B. Wang, *J. Chem. Soc.,* 807 (1937).

230. B. Eistert and M. Regitz, *Chem. Ber.* **96,** 2290 (1963).

231. G. H. Coleman, H. Gilman, C. E. Adams, and P. E. Pratt, *J. Org. Chem.* **3,** 99 (1938).

232. F. M. Beringer, J. A. Farr, and S. Sands, *J. Am. Chem. Soc.* **75,** 3984 (1953).

233. D. Kolbah and D. Korunčev, in *Houben-Weyl Methoden der Organischen Chemie,* Vol. X/2, E. Müller (ed.), Georg-Thieme-Verlag, Stuttgart, 1967.

234. P. A. Jacobi, A. Brownstein, M. Martinelli, and K. Grozinger, *J. Am. Chem. Soc.* **103,** 239 (1981).

235. S. Siggia and J. G. Hanna, *Quantitative Organic Analysis via Functional Groups,* 4th Ed., Wiley, New York, 1979, pp. 148–152.

236. *Ibid.,* pp. 109–113.

237. F. Feigl and V. Demant, *Mikrochim. Acta* **1,** 134 (1937).

3.

HYDRAZIDES AND RELATED COMPOUNDS
(Amidrazones, Hydrazidines, Acylhydrazones, etc.)

NOMENCLATURE

Hydrazides are acyl hydrazines, compounds derived from an acid by formal replacement of a hydroxyl group by a hydrazino group, which may itself bear other substituents. Although hydrazides exist for many types of acids, only hydrazides of carboxylic, sulfonic, sulfuric, and sulfurous acids will be given more than passing mention here.

In the IUPAC recommendations,[1] hydrazides are named by replacing the *-oic* or *-ic* ending of the name of the parent acid with *-ohydrazide,* as in benzohydrazide; the *o* is often omitted in common practice. If the name of the acid ends in *carboxylic,* it is converted to "carbohydrazide". The acylated nitrogen is designated by the locant N or $1'$, the other by N' or $2'$, as in N-methyl-N'-ethylacetohydrazide:

$$CH_3CON \begin{array}{c} \diagup Me \\ \diagdown NHEt \end{array}$$

If a substituent on a nitrogen carries a function of higher priority, the hydrazide portion is named as an acylhydrazino substituent, and the nitrogen atom attached to the parent structure is N or 1, as in N'-benzoyl-N-methylhydrazinoacetic acid:

$$PhCONHN \begin{array}{c} \diagup Me \\ \diagdown CH_2COOH \end{array}$$

Chemical Abstracts[2] indexes hydrazides under the name of the parent acid, followed by the word *hydrazide* with unprimed locants for the substituents, as in acetic acid, 2-methylhydrazide:

$$CH_3CONHNHCH_3$$

As a substituent, the group NH_2NHCO- can be named *hydrazinocarbonyl* or *carbazyl* (obsolescent).

Diacyl hydrazines (nameable as such) have sometimes been called "secondary hydrazides," but this is not a precise type of name. The IUPAC recommendations do not deal with these compounds, but by extrapolation one might reasonably expect them to be named as N-(N')-hydrazides, as in N-acetylacetohydrazide, Ac_2NNH_2, which *Chemical Abstracts* would index as "acetic acid, 1-acetylhydrazide." In earlier indexes, the diacyl hydrazine names were used in inverted form under "hydrazine."

Hydrazides of sulfonic acids are named in a fashion strictly analogous to those of carboxylic acids, as seen in N'-methylbenzenesulfonohydrazide or benzenesulfonic acid, 2-methylhydrazide for $PhSO_2NHNHMe$.

The enol tautomers of hydrazides are named *hydrazonic acids* according to the IUPAC recommendations, and derivatives can be named on that basis, as in methyl octanohydrazonate

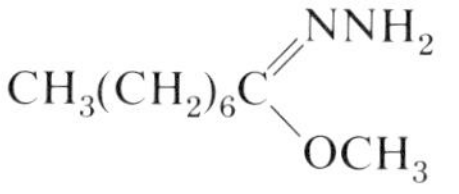

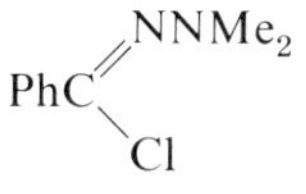

$$CH_3(CH_2)_6C\overset{NNH_2}{\underset{OCH_3}{\diagup}} \qquad PhC\overset{NNMe_2}{\underset{Cl}{\diagup}}$$

and N,N-dimethylbenzohydrazonoyl chloride. *Chemical Abstracts* uses essentially the same system in the *Ninth Collective Index*, but the reader should beware of the fact that although the hydrazides are indexed under "acetic" and "benzoic" acids, the hydrazonic derivatives are indexed under "ethanehydrazonic" and "benzene-carbohydrazonic" acid, respectively. Derivatives of the hydrazonic acid tautomers of *sym*-diacyl hydrazines have been named as azines in the few instances where there has been occasion to refer to them.

Amidine analogs of hydrazides are known generically in two types, amidrazones and hydrazidines (IUPAC):

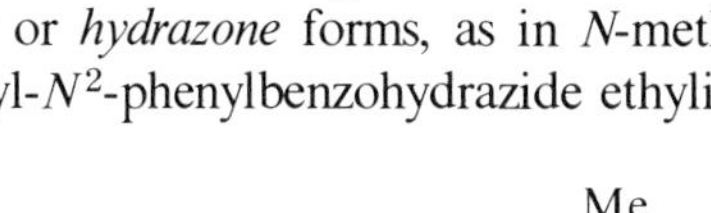

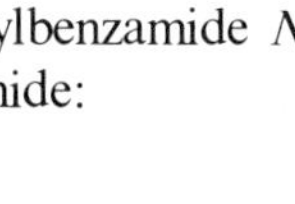

$$RC\overset{NH_2}{\underset{NNH_2}{\diagup}} \rightleftharpoons RC\overset{NH}{\underset{NHNH_2}{\diagup}} \qquad RC\overset{NNH_2}{\underset{NHNH_2}{\diagup}}$$

The generic names are used for individual compounds only when the particular tautomer cannot be specified (for hydrazidines this applies when there is a substituent at a terminal nitrogen). Otherwise, the IUPAC recommendations specify use of *imide* or *hydrazone* forms, as in N-methylbenzamide N-ethylhydrazone and N^1-methyl-N^2-phenylbenzohydrazide ethylimide:

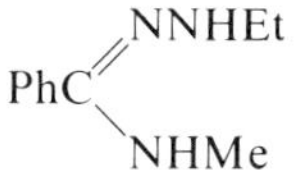

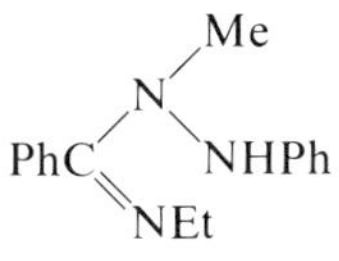

$$PhC\overset{NNHEt}{\underset{NHMe}{\diagup}} \qquad PhC\overset{\overset{Me}{|}}{\overset{N}{\underset{NEt}{\diagdown}}}NHPh$$

The problem of locants for amidrazones in which the double bond cannot be specifically located owing to tautomerism is met in the IUPAC recommendations by use of primes in the pattern $N\text{-}N'\text{-}CR\text{-}N''$. Thus the substance that might be N-benzimidoyl-N'-phenylhydrazine or its tautomer would be named N-phenyl-benzamidrazone.

For hydrazidines, with four nitrogen atoms to be accounted for, IUPAC recommends a different set of locants, as shown in N^1-methyl-N^2-ethyl-N^4-butylaceto-hydrazide hydrazone:

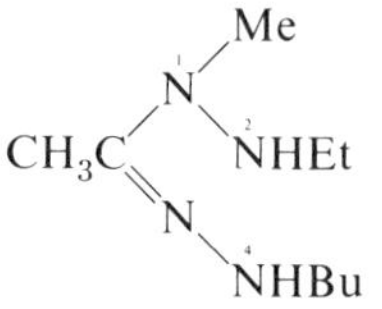

$$CH_3C \underset{N}{\overset{N}{<}} \begin{matrix} Me \\ NHEt \\ NHBu \end{matrix}$$

Chemical Abstracts does not cite the term *amidrazone* in the index to the 1976 Introduction to the *Index Guide,* but the 1977 *Index Guide* as well as the *Sixth Collective Index* contains the entries "Amidrazones, see Hydrazidines" (!). While it is true that in the late ninteenth century, the word *hydrazidine* was used to mean amidrazone, since then, virtually all chemical literature, including books and major reviews, has used the terms in the sense defined by IUPAC, and only *Chemical Abstracts* persists with the anachronism. Even the *Chemical Abstracts* abstractors are inclined to use *amidrazone* and *hydrazidine* in the IUPAC sense in the body of abstracts. In more recent *Chemical Abstracts* indexes, individual amidrazones are listed in the form "pentanohydrazonamide" or "pentanamidic acid, hydrazide," according to the tautomeric form, whereas the *General Subject Index* puts the generic class of amidrazones under "Hydrazidines." There is no general subject heading for true hydrazidines, but specific compounds are indexed under such names as "benzenecarbohydrazonic acid, hydrazide."

Alternative ways of naming or of locating substituents exist and have a significant following. According to a 1950 proposal,[3] adopted in an important review in 1970,[4] *N*-methylbenzamide *N*-ethylhydrazone is named N^1-ethyl-N^3-methylbenzamidrazone, according to the pattern N^1-N^2-CR-N^3. Hydrazidines have been named "dihydroformazans," after the azo analogs, $R_2N-N=CRN=NR$, in a recent major review, but that term must be rejected as ambiguous, for the site of the added hydrogen is not indicated. In fact, the same word has been used to refer to α-hydrazino azo compounds (the former use requires locants 1,2-, whereas the latter use requires 3,4-).

Ylides derived from quaternized hydrazides have been named variously as amine acylimides or *N*-ammonio amidates (*J. Chem. Soc.*):

$$R\overset{-}{CON}-\overset{+}{N}R_3$$

The compounds derived by formal loss of water from hydrazides should, by analogy with nitrile oxides, be named *nitrile imides,* but they are often called *nitrilimines.*

R_2N-NC	$RC\equiv\overset{+}{N}-\overset{-}{N}R$	$R_2N-N=C=O$
Isocyanoamines	Nitrile imides	Aminyl isocyanates

PROPERTIES[5]

Most unsubstituted hydrazides are solids. The smaller ones are very soluble in water and difficultly soluble in ether and nonpolar solvents. Substituted hydrazides and diacyl hydrazines are also mostly solids, although a few are viscous oils (Table 3-1). They are generally quite stable to storage in air.

Table 3-1 Physical Properties of Some Hydrazides

Structure	Melting Point, °C	Boiling Point, °C(mm)
$HCONHNH_2$	54	
HCONHNHMe		95(15)
$HCONHNMe_2$	61	
$AcNHNH_2$	67	129(18)
$AcNMeNH_2$	Liquid	
AcNMeNHMe		70(1.5)
$AcNHNMe_2$	24–25	95(1.5)
$AcNMeNMe_2$		60–65(15)
$PhCONHNH_2$	112.5	
$PhCONMeNH_2$	48–50	
PhCONHNHMe	86–88	
$PhCONMeNMe_2$	89–90	
HCONHNHCHO	159–160	
HCONHNHAc	96	
AcNHNHAc	140*	209(15)
Ac_2NNHAc		180–183(15)
Ac_2NNAc_2	86	141(15)
$O{=}C(NHNH_2)_2$	152	
$CH_3CSNHNH_2$	59	
$CH_3CSNMeNH_2$	79–80	
$HCSNHNMe_2$	84–86	
$HCSNMeNH_2$	Liquid	
$PhSO_2NHNH_2$	104–106	
$p\text{-}CH_3C_6H_4SO_2NHNH_2$	112	
$PhSO_2NHNHSO_2Ph$	228	
$PhNHN{=}SO$	105	

*Monohydrate, mp 80–100°C.

Benzamidrazone[6] (mp 78–79°C) appears to be the only unsubstituted nonfluorinated amidrazone to have been isolated. Acetamidrazone is known as its hydrochloride; the free base is extremely soluble in water. N-Phenylbenzamidrazone[7] (mp 86–87°C) is insoluble in water. Tetrasubstituted amidrazones are mostly distillable oils,[3] but N,N',N'-trimethylbenzhydrazide methylimide is a white solid (mp 71–72°C), whereas its isomer, N,N-dimethylbenzamide dimethylhydrazone, is a yellow oil (bp 124–128°C/20mm).[8] Many other unsubstituted[8] and alkyl-substituted amidrazones are known as salts, and aryl-substituted ones are known as free bases.

Unsubstituted hydrazidines appear not to have been isolated in the free state. Acethydrazidine is known as its hydrochloride; the free base is apparently extremely soluble in water.[10] N'-Phenylacethydrazide N-phenylhydrazone appears to be insoluble, but it was not obtained pure owing to rapid decomposition in the presence of air.[7] A number of other aryl-substituted hydrazidines have been found to be reasonably stable solids, but autoxidation occurs more or less readily with most free hydrazidines.[11]

All monoacyl hydrazines are weakly basic,[12] and most of them will dissolve in dilute mineral acid. A pK_a value of 3.24 has been reported for the conjugate acid of acethydrazide,[13] 2.97 for benzhydrazide, and only 1.6 for its trimethyl derivative (see hydrazine, $pK_a = 8$).[14] It is obvious that the electronic effect of the acyl group is transmitted strongly across the N—N bond. However, the drop in basicity with increasing alkylation parallels that of the parent hydrazines, and the explanation is probably to be found in solvation effects. Sulfonhydrazides are amphoteric, dissolve in both aqueous acid and base, and form isolable sodium salts and hydrochlorides.[15]

Amidrazones and hydrazidines appear in general to be slightly more basic than hydrazides, but not so strong as amidines. Like hydrazides, they are monoacid bases in water solution. Acetamide dimethylhydrazone (or tautomer) has a pK_a value of 10.0 for the conjugate acid.[16] Its trifluoro analog has a pK_a value of 10.2, and several other perfluoro amidrazones have similar values.[17] The information is less precise on hydrazidines, but acethydrazidine[10] is a strong enough base to precipitate calcium hydroxide, whereas triethylamine is reported to be strong enough to free hydrazidines from their salts,[11] and sodium carbonate has been observed to precipitate a diphenylhydrazidine from a solution of its hydrochloride.[7]

The hydrogens on the acylated nitrogen atom of hydrazides are weakly acidic; N'-phenylbenzhydrazide, for example, is soluble in aqueous alkali, as are sym-diacyl hydrazines and thiohydrazides. Unsubstituted thiohydrazides[18] have pK_a values of 10.2 to 11.2. Sym-diformylhydrazine forms both a mono- and a disodium salt. N'-Isopropylidenemethanesulfonhydrazide (acetone mesylhydrazone)[19] has a pK_a value of 8.5. Quaternized hydrazides are distinctly acidic in water solution:[20] N-acetyl-N',N',N'-trimethylhydrazinium ion has a pK_a value of 5.39 at 25°C, and the N-benzoyl analog has a pK_a value of 4.26.

Tautomerism in N-unsubstituted hydrazides is possible just as in amides, but the solids that have been examined are in the carbonyl form. In dodecanohydrazide and nicotinohydrazide, the O=C—N—N system is essentially planar, as would be expected from extended π-orbital overlap.[21] The C—N distances (1.332 Å) and the N—N distances (1.39–1.41 Å) are appreciably shorter than the normal single-bond distances. N,N'-Diformyl- and N,N'-diacetylhydrazine crystallize in a planar configuration having a center of symmetry (that is, oxygens opposed).[22,23] They also have C—N and N—N distances shorter than normal. However, sym-dibenzoylhydrazine appears to exist in a staggered conformation.[24]

Alkyl substitution influences conformation appreciably, as shown by NMR studies of sym-diacyl hydrazines bearing one or two N-benzyl groups.[25] Rotation about the N—N bond is considerably hindered, and enthalpies of activation as high as 23.4 kcal/mol have been reported. The conformations are therefore chiral, and the benzylic CH_2 hydrogens are nonequivalent. The preferred conformation is apparently far from planar.

The infrared spectra of hydrazides are characterized by a strong band in the 1630 to 1675-cm^{-1} region owing to carbonyl stretching; with incompletely substituted hydrazides, additional bands, analogous to amide II and III bands, may be present as well.[26] Trimethylbenzhydrazide has but a single band in the amide region, at 1650 cm^{-1}, and trimethyloctanohydrazide has one at 1659 cm^{-1}.[14] In quaternized hydrazides, the carbonyl-stretching frequency has been reported[27] in a wider range, 1550 to 1725 cm^{-1}; in amine acylimides, the frequency falls to 1555 to 1590 cm^{-1}. Stretching of N—H appears in the region 3050 to 3450 cm^{-1}, depending on solvent and dilution, in such a way as to imply the occurrence of both inter- and intramolecular hydrogen bonding.[28,29]

In amidrazones, the amide hydrazone structure is generally preferred, according to indications of infrared and NMR spectroscopy.[30] There may be as many as four bands in the N—H stretching region, and intramolecular hydrogen bonding is possible between the two remote nitrogens.[31] The C=N stretching frequency and N—H or NH$_2$ bending frequencies are generally seen in the 1590 to 1690-cm^{-1} range as a pair of bands. Although the complexity of the spectra makes firm interpretations difficult, there is general agreement among independent investigators, and it appears that there is restricted rotation about the C—N bonds.[32] N''-Aryl amidrazones appear to be different in structure from others and are believed to exist as hydrazide imides, presumably owing to conjugation between the aryl group and the C=N bond.[8] Some tetraalkyl amidrazones known to have the amide hydrazone structure are believed to have a configuration about the C=N bond with the two dialkylamino groups *anti* to each other.[33] This group of amidrazones has C=N stretching in the 1600 to 1645 cm^{-1} range and ultraviolet absorption owing to an $n \rightarrow \pi^*$ transition at 273 to 279 nm, which disappears on acidification.

Diacyl hydrazines can exist as *sym* and *asym* positional isomers, but the latter are known only with both positions on the terminal nitrogen substituted except for cyclic examples such as N-aminosuccinimide.[34] An early report[35] of isolation of N,N-diacetylhydrazine (mp 132°C) is in error, and the substance was an impure sample of the *sym* isomer (mp 140°C) obtained by direct acetylation of hydrazine, the structure of which has since been unequivocally established.[23] Indirect evidence suggests that *asym*-diacyl hydrazines may rearrange easily to the *sym* isomers when attempts are made to prepare them, unless they are stabilized by a ring structure.[36]

Tetraacyl hydrazines present some interesting structural problems, owing to the possibility of isomerism with the acid anhydride azine structure and restricted rotation about the N—N bond. The latter has been detected by NMR and was at first thought to be a result of steric repulsion between the four acyl groups, but now it is believed to be due to repulsion between the lone electron pairs on the two nitrogens (the analogous compounds (RCO$_2$N—CH(COR)$_2$ show free rotation, whereas their carbanions do not).[37] With cyclic tetraacyl hydrazides, the situation is especially confusing; the two dimaleylhydrazines and their isomer, maleic anhydride azine, have now been differentiated.[38] N,N,N',N'-Disuccinoylhydrazine (N,N'-bisuccinimide) is distinctly nonplanar, the two five-membered rings having an angle of 65° to one another, presumably in order to relieve repulsion between the lone pairs on the nitrogen atoms.[39]

Restricted rotation about the carbonyl-nitrogen single bond exists in hydrazides as in amides and leads to two geometric configurations, in which the amino nitrogen and the carbonyl oxygen are cis (Z) or trans (E) to each other. The subject has not been extensively studied, and it is undoubtedly complicated by hydrogen-bonding effects, but N,N',N'-trimethylbenzhydrazide appears to be (E), as does trimethylacethydrazide, whereas ab initio MO calculations imply that unsubstituted hydrazides should exist in the (Z) configuration.[14] The (Z) configuration has been found in two hydrazonoyl bromides that have been investigated by x-ray crystallography and NMR,[40] but this is a case of normal geometric isomerism about a double bond.

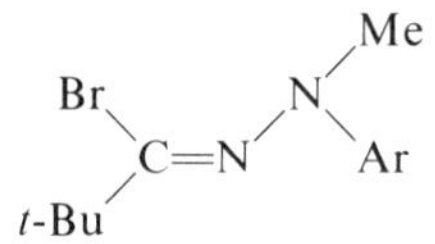

The NMR spectra of hydrazides commonly show two sets of signals close to each other and of unequal intensity, almost certainly owing to (E) and (Z) forms resulting from restricted rotation about the C—N bond.[41] In formhydrazide, the CH signals in dimethyl sulfoxide (chloroform does not dissolve it) are at δ 8.03 and 7.92, those of NH at δ 8.69 and 9.05, and NH_2 at 4.25 ppm for the (E) and (Z) isomers, respectively. Coupling of CH and NH has a $J = 10.5$. For acethydrazide, the CH_3 signals appear at 2.16 and 1.96 (in $CDCl_3$). For most hydrazides, the ratio E/Z is less than 1, but for formhydrazide it is greater than 1 and is quite sensitive to the solvent. In methylated hydrazides, the N-methyl signal is 0.5 to 1.5 ppm downfield of the N'-methyl resonance.[14]

The mass spectra of some N'-substituted benzhydrazides have been found to show strong molecular ion peaks. N,N'-Dibenzoylphenylhydrazine undergoes loss of a fragment of 122 mass, presumably benzoic acid, which could only arise through rearrangement.[42]

Thiohydrazides[43] are crystalline solids that are quite colorless if aliphatic and yellowish if derived from aromatic acids. They exist in two forms, which in many instances are separately isolable.[44] One form has the (E) configuration and normal structure, and the other has the (Z) configuration and a controversial structure, generally believed to be zwitterionic with chelate hydrogen bonding. The N'-Alkyl

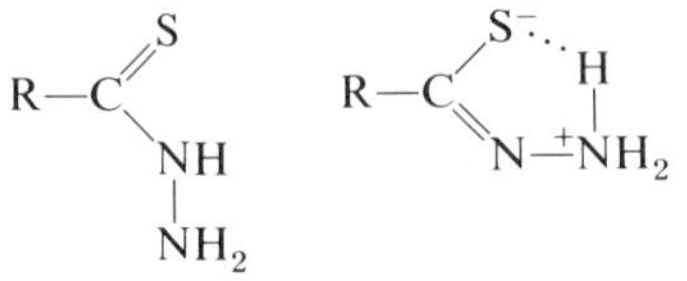

derivatives melt at markedly higher temperatures than the N-alkyl isomers. Thiohydrazides are 2 to 4 powers of 10 more strongly basic than ordinary hydrazides.[18]

The conjugate acids of unsubstituted examples have pK_a values of 5 to 6; N'-alkyl groups increase the basicity to pK_a values of 6.5 to 6.7, whereas N-alkyl groups increase pK_a values still further, to 7.2 to 7.5. Those bearing hydrogen at N^1 will dissolve in aqueous alkali.

Hydrazonic esters, the analogs of imidates, are known as stable liquids or solids, such as phenyl N-phenylbenzohydrazonate[45] (mp 151–154°C) and methyl N,N-dimethylbenzohydrazonate.[46] Geometric isomers are generally quite stable enough to be separately isolable, although they can be equilibrated by moderately strong heating.

Carbohydrazonoyl halides, especially with N-aryl substituents, are known in some variety and are quite stable. N-Phenylbenzenecarbohydrazonyl chloride has a melting point of 129.5–130.5°C, and N,N-disubstituted benzenecarbohydrazonyl chlorides are distillable oils.[47] The structure of N-methyl-N-(2,4-dinitrophenyl)-trimethylacetohydrazonyl bromide has been examined by x-ray crystallography; it exists in the (Z) configuration.[40]

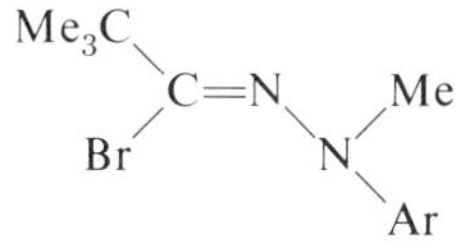

(Ar = 2,4-dinitrophenyl)

Acyl halide azines (1,1'-dihalo aldazines) are known, such as the azine of benzoyl chloride.[48] N-Arylcarbonohydrazonyl chlorides are sufficiently unreactive that they can be recrystallized from methanol.[49]

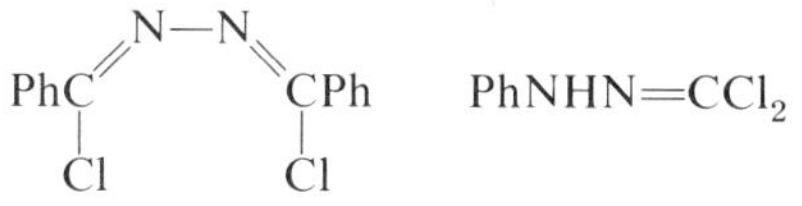

REACTIONS

Carbohydrazides

Heat and light. Hydrazides are commonly stable to moderate heating and, indeed, can even be prepared by controlled heating of hydrazine salts. At higher temperatures, however, usually beginning about 180°C, transacylation may come into play, giving diacyl hydrazines (Eq. 3-1). These are also labile to heating and are converted to 1,2,4-triazoles (mostly), 1,3,4-oxadiazoles, and small amounts of N—N cleavage products. Aminotriazoles can be prepared in some cases in synthetically useful yields by such reactions. A competing path of thermolysis leads to dihydrotetrazines (Eq. 3-2); it appears to be less important with benzhydrazide than with acethydrazide.[50,51] Heating acethydrazide in acetic acid causes disproportionation to diacetylhydrazine and hydrazine, however.[52] Stronger heating of diacetylhydrazine forms N-monoacetyldimethyldihydrotetrazine. Tetraacetyl hydrazines are surprisingly stable to heat, and the N—N bond does not break in spite of the accumulation

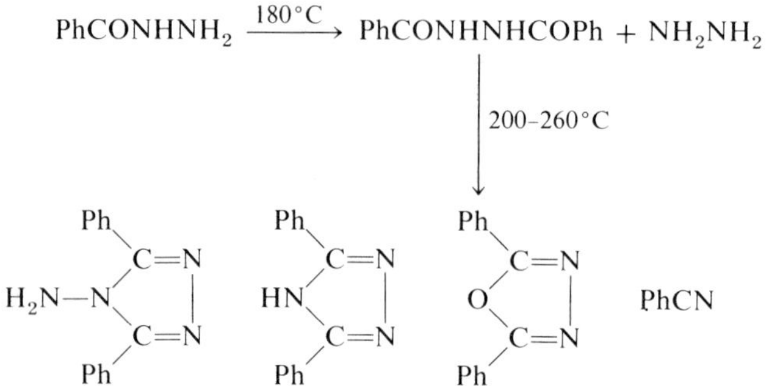

$$\text{PhCONHNH}_2 \xrightarrow{180°\text{C}} \text{PhCONHNHCOPh} + \text{NH}_2\text{NH}_2$$

$$\Big\downarrow \, 200\text{-}260°\text{C}$$

(3-1)

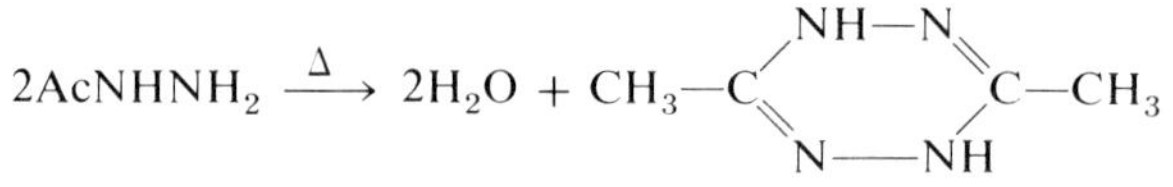

$$2\text{AcNHNH}_2 \xrightarrow{\Delta} 2\text{H}_2\text{O} + \text{CH}_3-\text{C}\underset{\text{N}\!-\!\text{NH}}{\overset{\text{NH}-\text{N}}{\diagdown}}\text{C}-\text{CH}_3 \tag{3-2}$$

of electron-withdrawing substituents. Instead, elimination of an anhydride takes place and an oxadiazole is formed[53,54] (Eq. 3-3).

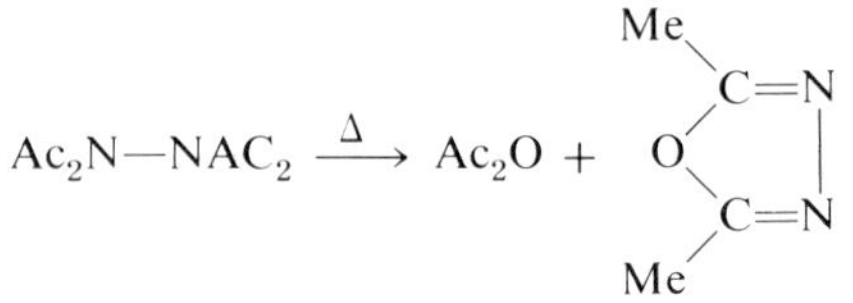

$$\text{Ac}_2\text{N}-\text{NAC}_2 \xrightarrow{\Delta} \text{Ac}_2\text{O} + \underset{\text{Me}}{\overset{\text{Me}}{\diagdown}}\text{O}\underset{\text{C}=\text{N}}{\overset{\text{C}=\text{N}}{\diagup}} \tag{3-3}$$

Photolysis of hydrazides in solution principally brings about reductive cleavage of the N—N bond[55] (Eq. 3-4); substituents that interfere with conjugation between the

$$\text{RCONHNPh}_2 \xrightarrow{h\nu} \text{RCONH}_2 + \text{Ph}_2\text{NH} \tag{3-4}$$

carbonyl group and the attached nitrogen favor an alternative cleavage leading to aldehydes (Eq. 3-5).

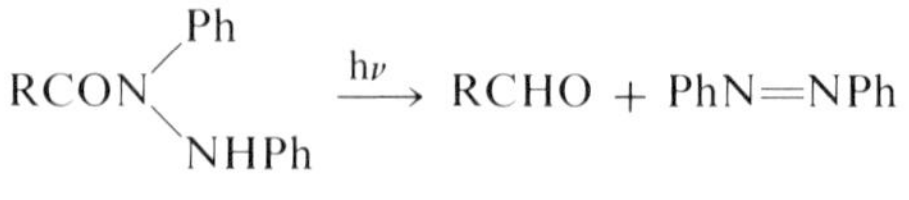

$$\text{RCON}\underset{\text{NHPh}}{\overset{\text{Ph}}{\diagup}} \xrightarrow{h\nu} \text{RCHO} + \text{PhN}{=}\text{NPh} \tag{3-5}$$

Amine acylimides undergo an analog of the Hofmann and Curtius rearrangements when heated[56,57] (Eq. 3-6). However, if one of the N'-substituents is benzylic,

$$\text{RCO}\overset{-}{\text{N}}-\overset{+}{\text{N}}\text{Me}_3 \xrightarrow{\Delta} \text{RNCO} + \text{Me}_3\text{N} \tag{3-6}$$

allylic, or propargylic, it migrates to the anionic nitrogen in an analog of the Stevens rearrangement[58] (Eq. 3-7). Studies of this rearrangement with amine carbamyli-

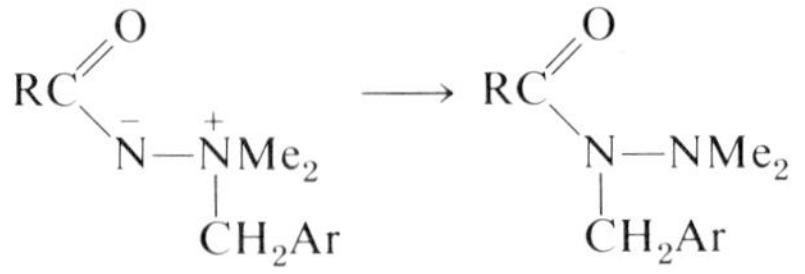

$$\text{RC}\overset{\text{O}}{\underset{\overset{|}{\underset{\text{CH}_2\text{Ar}}{\text{N}-\overset{+}{\text{N}}\text{Me}_2}}}{\diagup}} \longrightarrow \text{RC}\overset{\text{O}}{\underset{\overset{|}{\underset{\text{CH}_2\text{Ar}}{\text{N}-\text{NMe}_2}}}{\diagup}} \tag{3-7}$$

mides have shown that the allylic structure of substituted allylic groups is retained rather than inverted, and it has therefore been inferred that it takes place through a radical-pair process.[59]

Thiohydrazides readily lose hydrogen sulfide in a bimolecular reaction if unsubstituted and form dihydrotetrazines,[60,61] analogous to Equation 3-2. This reaction occurs far more easily than with ordinary hydrazides, and with unbranched aliphatic derivatives it follows so fast upon formation of a thiohydrazide even at −30°C that such thiohydrazides may not be isolable. However, the *tert*-butyl group of pivalothiohydrazide retards the reaction sufficiently to allow it to be isolated. A competing path of decomposition results from heating benzothiohydrazides; thiadiazoles are formed,[62] apparently through initial self-acylation, to generate an intermediate that may be a *sym*-bis (thioacyl)hydrazine (Eq. 3-8), or

$$2ArCSNHNH_2 \xrightarrow{\Delta} \underset{S}{Ar-C} \overset{N-N}{\underset{\|}{}} C-Ar + H_2S \,(+\, N_2H_4) \qquad (3\text{-}8)$$

oxidation may be involved. Substituted thiohydrazides cannot undergo these reactions and are more stable to heat. All thiohydrazides seem to be more or less sensitive to direct light, rapidly turning yellow.[18]

Acids, bases, and solvolysis.   Hydrazides are not hydrolyzed at an appreciable rate by water, and even acid- or base-catalyzed hydrolysis is slow. Salts can safely be prepared in water solution by reaction with mineral acids or alkalies. Hot aqueous acid slowly hydrolyzes hydrazides to carboxylic acids and hydrazine.[63,64] Although hydrolysis is quite successful with unsubstituted hydrazides, substitution may greatly retard it, and N,N'-dimethyl-N'-phenylacethydrazide, for example,[65] is reported to have resisted all attempts at hydrolysis. Hot alkalies also bring about slow hydrolysis,[66,67] but it is sometimes diverted partly or entirely to oxidation-reduction reactions.[53] Enzymic hydrolysis has also been observed, as in the action of α-chymotrypsin on tyrosine hydrazide.[68] Diacyl hydrazines can be hydrolyzed stepwise.[69] Alkylidene hydrazides (acyl hydrazones) can also be hydrolyzed in stages; the hydrazone function is generally hydrolyzed by aqueous acid far more readily then the hydrazide function.[70,71]

Ammonolysis is very difficult and apparently has an unfavorable position of equilibrium, but it has been proposed as a step in an industrial process for producing hydrazine[72] (Eq. 3-9).

$$RCONHNH_2 + NH_3 \xrightarrow[\text{200 psi}]{150°C} RCONH_2 + N_2H_4 \qquad (3\text{-}9)$$

Acid may catalyze certain reactions of hydrazides, including rearrangement in special instances,[73,74] such as that of disuccinoylhydrazine[75] (Eq. 3-10).

A strong base in the absence of water may catalyze condensation reactions. Self-condensation of phenylhydrazides of aliphatic acids leads to oxindoles, in a process

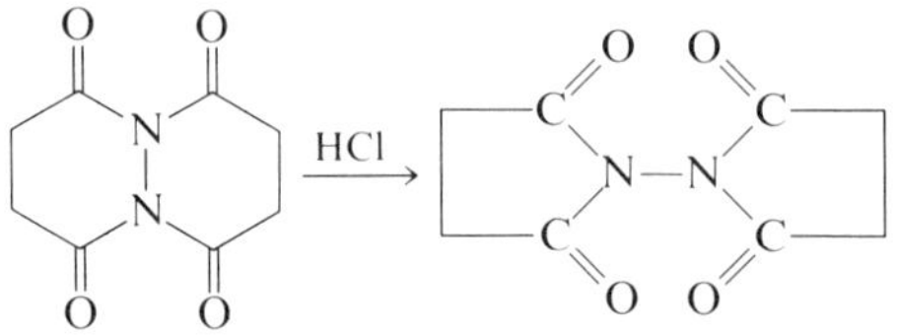

$$ \text{(3-10)} $$

reminiscent of the Fischer indole synthesis (Eq. 3-11). Self-condensation has been indicted as the cause of low yields when diacethylhydrazine is treated with strong

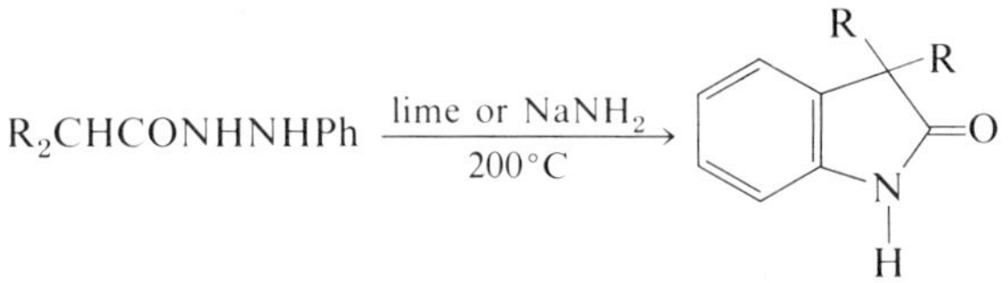

$$ R_2CHCONHNHPh \xrightarrow[200^\circ C]{\text{lime or NaNH}_2} \text{(3-11)} $$

base to generate its anion preparatory to alkylation (the more hindered diiso-butyrylhydrazine appears not to be so afflicted). A reaction closely related to the McFadyen-Stevens aldehyde synthesis takes place when benzhydrazides are exposed to prolonged contact with dilute sodium hydroxide solution at room temperature; benzaldehyde benzoylhydrazone is produced in yields up to 48%[76] (Eq. 3-12).

$$ ArCONHNH_2 \xrightarrow[\substack{\text{EtOH or H}_2\text{O,}\\ \text{many days}}]{\text{NaOH}} ArCH{=}NNHCOAr \qquad \text{(3-12)} $$

The stoichiometry has not been satisfactorily elucidated, but hydrolysis is a competing reaction, becoming more important at higher temperatures, and glucose appears to promote the reaction.

Alkylation.   Hydrazides are only sluggishly reactive toward alkylating agents in the neutral state. Benzhydrazide, when treated with a large excess of methyl iodide in ethanol, gave N',N'-dimethylbenzhydrazide, but in only 7 to 8% yield, and much benzhydrazide was recovered unreacted.[77] N'-Phenylformhydrazide did not react with ethyl iodide even after 3 hours at 100°C.[78] In the presence of strong base, alkylation may take place at either nitrogen, but most commonly[79] at N^2. On the basis of a rather limited number of examples, the generalization has been drawn that alkylation of the sodium derivatives of hydrazides takes place at the amide nitrogen in nonpolar solvents and at the terminal nitrogen in polar ones. The origin of this generalization appears to be an 1896 report of a very careful investigation of just one instance: methylation of N'-phenylformhydrazide[78] (Eq. 3-13), although benzhydrazide behaves similarly (only a low yield of N-methylbenzhydrazide was obtained from reaction in benzene, however).[77] The difference in behavior may not be due to polarity of the medium, for there are two other variables: the reaction takes place in one phase in ethanol and in two phases in benzene or ether, and the two nonpolar solvents are aprotic, whereas the only polar solvent used is protic.

The limits of the alkylation behavior are shown by the fact that N-ethyl-N'-phenylformhydrazide reacts with ethyl iodide in ethanol containing sodium

$$\text{HCONHNHPh} \xrightarrow{\text{NaOEt}}$$

$$\begin{array}{c}
\text{Na}^+ \\
\text{HCONNHPh}
\end{array}
\begin{array}{c}
\xrightarrow[\text{Et}_2\text{O or C}_6\text{H}_6]{\text{MeI}} \quad \text{HCON}\begin{array}{l}\diagup \text{Me}\\ \diagdown \text{NHPh}\end{array} \\[2em]
\xrightarrow[\text{MeI, EtOH}]{} \quad \text{HCONHN}\begin{array}{l}\diagup \text{Me}\\ \diagdown \text{Ph}\end{array}
\end{array}
\qquad (3\text{-}13)$$

ethoxide to form N,N'-diethyl-N'-phenylformhydrazide, whereas the isomer N'-ethyl-N'-phenylformhydrazide is inert,[78] and the fact that under similar conditions, benzhydrazide reacts with propyl bromide only to the stage of N',N'-dipropylbenzhydrazide, whereas with methyl iodide, three methyl groups are attached to form trimethylamine benzoylimide[57,77] (Eq. 3-14). N,N',N'-Trimethyl-

$$\text{PhCONHNH}_2 \xrightarrow{\text{NaOEt}} \left[\begin{array}{c}
\xrightarrow{\text{PrBr}} \text{PhCONHNPr}_2 \\[1.5em]
\xrightarrow{\text{MeI}} \text{PhCON}^- \text{---} \overset{+}{\text{N}}\text{Me}_3
\end{array}\right.
\qquad (3\text{-}14)$$

benzhydrazide, as well as trimethyloctanohydrazide, could not be alkylated by methyl iodide under any conditions.[14] This behavior seems to be a subtle steric effect, for the cyclic trialkyl hydrazide 1-benzoyl-2-methylpyrazolidine is methylated smoothly at N^2. There is a small difference in rotational conformation between the two compounds, affecting the proximity of the lone pairs on the two nitrogens. Earlier reports of alkylation of a hydrazide at oxygen to produce an alkyl carbohydrazonate are apparently erroneous.[77]

In addition to alkyl halides, acrylonitrile will alkylate hydrazides, producing N'-cyanoethyl derivatives.[64]

The kinetics of methylation of some N'-arylhydrazides have been studied as a function of substituents on the aryl group; not surprisingly, the kinetics were second-order, and the results were correlated nicely by the Hammett equation.[80]

Sym-diacyl hydrazines as such are not reactive toward alkylating agents, but they are easily converted to salts with strong bases, and the anions can be alkylated with primary alkyl halides. The most important use of this reaction is in the preparation of sym-dialkyl hydrazines,[63,81] obtained by subsequent hydrolysis (see Chap. 1). Alkylation can be held to the monoalkyl stage by use of monosodium salts[82] (Eq. 3-15). A second alkyl group of a different kind can then be attached by repeating the process[83] (Eq. 3-16). Yields are generally low with any group larger than ethyl, and

$$\text{PhCONHNHCOPh} + \text{RX} \xrightarrow{\text{NaOH}} \text{PhCONHN}\begin{array}{l}\diagup \text{R}\\ \diagdown \text{COPh}\end{array}
\qquad (3\text{-}15)$$

$$\text{PhCONHN}\begin{array}{l}\diagup \text{Et}\\ \diagdown \text{COPh}\end{array} + \text{Me}_2\text{SO}_4 \xrightarrow{\text{NaOH}} \begin{array}{l}\text{Me}\diagdown\\ \text{PhCO}\diagup\end{array}\text{N---N}\begin{array}{l}\diagup \text{Et}\\ \diagdown \text{COPh}\end{array}
\qquad (3\text{-}16)$$

it becomes extremely difficult to introduce a second alkyl group. Alkylation of *sym*-dibenzoylhydrazine with an excess of propyl bromide, for example, stopped at the monoalkyl stage even at 140°C; when the temperature was held at 160°C for 1 day, the main product was apparently N',N'-dipropylbenzhydrazide[82] (it was reported to be the O,N'-dipropyl isomer, almost certainly erroneously; the melting point, 100°C, is now known to be the same as that of the N',N'-dipropyl isomer). 1,3- And 1,4-dibromoalkanes, however, give fair yields of cyclic dialkylation products.[63,84] Alkylidene hydrazides can be alkylated at N^1 as their salts.[70,85]

Thiohydrazides undergo alkylation exclusively at sulfur, regardless of the presence or absence of strong base or of substituents on either nitrogen[86] (Eq. 3-17),

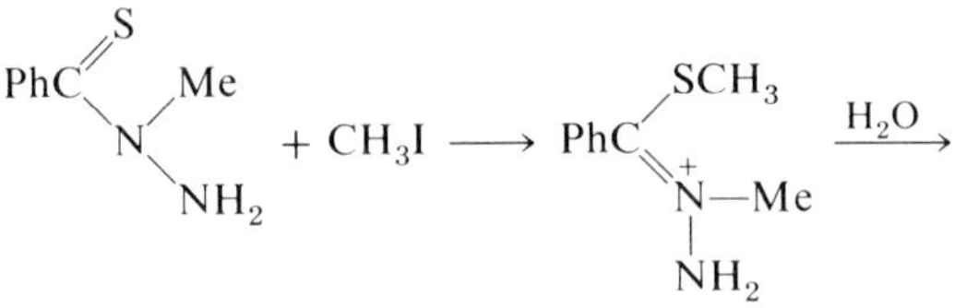

$$\text{PhC}\!\!\begin{array}{c}S\\\diagdown\end{array}\!\!\begin{array}{c}\text{Me}\\N\\\diagdown\\NH_2\end{array} + CH_3I \longrightarrow \text{PhC}\!\!\begin{array}{c}SCH_3\\\diagup\\\overset{+}{N}-Me\\|\\NH_2\end{array} \xrightarrow{H_2O}$$

$$PhCOSCH_3 + MeNHNH_2 \cdot HI \qquad (3\text{-}17)$$

except in the case of certain conjugated heterocyclic compounds that are formally cyclic thioacylhydrazones. The *S*-methyl derivatives of unsubstituted thiohydrazides are too reactive to be isolated, and they react with themselves to form dihydrotetrazines or thiadiazoles during the alkylation procedure (perfluoro derivatives are in exception[86b]). N'-Phenylformothiohydrazide reacts with methyl iodide to form an *S*-methyl derivative, apparently in only one geometric configuration,[87] whereas the individual configurational isomers of N',N'-diethylformothiohydrazide give distinct *S*-methyl derivatives with retention of configuration[88] (Eq. 3-18).

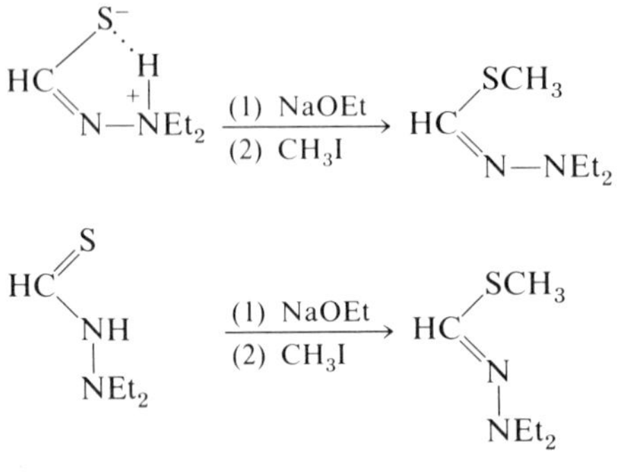

$$\text{HC}\!\!\begin{array}{c}S^-\\\diagdown\, \cdot\cdot\,H\\\overset{+}{N}-NEt_2\end{array} \xrightarrow[(2)\ CH_3I]{(1)\ NaOEt} \text{HC}\!\!\begin{array}{c}SCH_3\\\diagup\\N-NEt_2\end{array}$$

$$\text{HC}\!\!\begin{array}{c}S\\\diagdown\\NH\\|\\NEt_2\end{array} \xrightarrow[(2)\ CH_3I]{(1)\ NaOEt} \text{HC}\!\!\begin{array}{c}SCH_3\\\diagup\\N\\|\\NEt_2\end{array}$$

$$(3\text{-}18)$$

Aldehydes and ketones. Hydrazides having an unsubstituted amino group react more or less readily with all but highly hindered aldehydes and ketones to form acylhydrazones. This reaction has seen enormous use in the case of semicarbazide (carbamohydrazide) for making crystalline derivatives for identification, characterization, and isolation purposes (Eq. 3-19). The reaction is commonly carried out in

$$H_2NCONHNH_2 + R_2C{=}O \longrightarrow H_2NCONHN{-}CR_2 + H_2O \qquad (3\text{-}19)$$

weakly acid solution (aqueous or alcoholic) and leads quickly to the product, often spontaneously. Although the reaction is operationally simple, it is mechanistically complex and involves several intermediates and equilibria (see Chap. 1 and Eq. 11-12), which have been investigated with an uncommon degree of thoroughness and sophistication.[89-92]

There are two special applications of the formation of hydrazones from hydrazides. Hydrazides bearing a quaternary ammonium substituent were introduced in 1936 as reagents for converting aldehydes and ketones to water-soluble salts of value for extracting such compounds from complex mixtures.[93,94] Such hydrazides are usually called *Girard's reagents;* the two most common ones are shown:

$$Me_3\overset{+}{N}CH_2CONHNH_2 \ Cl^- \qquad \qquad \underset{\text{Girard's reagent P}}{\boxed{}\overset{+}{N}-CH_2CONHNH_2 \ Cl^-}$$

Girard's reagent T   Girard's reagent P

Hydrazides bearing a chiral center have been used in resolving racemic aldehydes and ketones into their enantiomers. Examples are 5-(1-phenylethyl)semioxamazide, $PhCH(CH_3)NHCOCONHNH_2$, and the hydrazide of N-(L-menthyl)glycine.[95,96]

It is ordinarily assumed, with some justification, that hydrazides substituted at the N' position will not react with aldehydes and ketones, since hydrazone formation is then not structurally possible. One might expect, however, that reactive aldehydes might, under suitable conditions, form adducts analogous to carbinolamines. In one interesting case,[97] such an adduct apparently undergoes dehydration (Eq. 3-20) even

$$ArCHO + \begin{matrix} HN \\ | \\ HN \end{matrix}\hspace{-0.3em}\rangle\hspace{-0.3em}\underset{O}{} \longrightarrow \left[\begin{matrix} ArCH \overset{OH}{\diagup} \\ N \\ HN \end{matrix}\hspace{-0.3em}\rangle\hspace{-0.3em}\underset{O}{} \right] \longrightarrow ArCH\hspace{-0.2em}=\hspace{-0.2em}\overset{+}{N}\hspace{-0.3em}\rangle\hspace{-0.3em}\underset{O^-}{N} \qquad (3\text{-}20)$$

though a conventional hydrazone cannot form; the generality of this behavior has not been explored. Examples of a reaction of N,N'-diacylhydrazines with benzaldehydes have been reported[98]; 1:1 adducts were formed with loss of water. Although the products were asserted to be diaziridines, the claim does not invite belief, for no evidence in support of such a structure was provided and more probable alternative structures were not considered (Eq. 3-21). It is significant that both

$$\underset{Y}{\boxed{}}\hspace{-0.3em}\begin{matrix} CONHNHCOAr \\ \\ OH \end{matrix} \xrightarrow{Ar'CHO}$$

$$\underset{Y}{\boxed{}}\begin{matrix} \overset{O}{\overset{\|}{C}} \\ \diagdown N-NHCOAr \\ | \\ O^{\diagup CH-Ar'} \end{matrix} \quad ? \quad \left[\underset{Y}{\boxed{}}\begin{matrix} \overset{O}{\overset{\|}{C}}-N\!\!-\!\!N-COAr \\ \underset{CH}{|} \\ OH \quad Ar' \end{matrix} \right] \quad ??? \qquad (3\text{-}21)$$

substrates were derivatives of salicylic acid and that the analogous reaction with dibenzoylhydrazine was not reported. An oxadiazoline structure, analogous to that obtained from N'-substituted thiohydrazides and benzaldehydes[85] (Eq. 3-22), can-

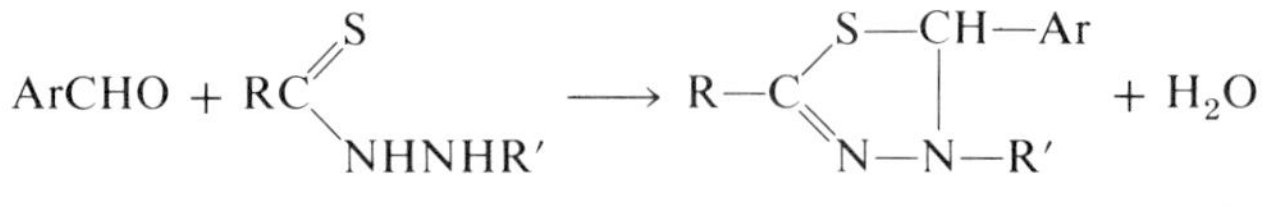

$$ArCHO + RC\overset{S}{\underset{NHNHR'}{\diagup}} \longrightarrow R-C\overset{S-CH-Ar}{\underset{N-N-R'}{\diagdown |}} + H_2O \qquad (3\text{-}22)$$

not be ruled out. Thiohydrazides having no substituent at the N' position form normal hydrazones.[85]

In the special case of trichloroacetaldehyde (chloral), unexpected difficulty was encountered during an attempt to prepare the corresponding hydrazone by reaction with a hydrazide in ethanol solution; an ester was formed, with loss of chloral hydrazone[99] (Eq. 3-23). It was subsequently determined that the hydrazones could

$$RCONHNH_2 + Cl_3CCHO \longrightarrow RCONHN{=}\underset{H}{\overset{|}{C}}CCl_3 \xrightarrow{\ EtOH\ }$$

$$RCO_2Et + H_2NN{=}\underset{H}{\overset{|}{C}}CCl_3 \qquad (3\text{-}23)$$

be prepared in the absence of nucleophiles (*caution:* a very vigorous reaction, especially when the reactants are mixed cold and allowed to warm up) and could be used as mild acylating agents in subsequent reaction with alcohols and amines. Since hydrazides are easily made from unhindered esters, the potential of this reaction in the synthesis of amides is obvious. The mechanism has not been established, but it may be that the azo tautomer of the hydrazone, $RCON{=}NCH_2CCl_3$, is the active acylating agent (see the section Oxidation).

Acylating and dehydrating agents. Hydrazides with at least one hydrogen on the terminal nitrogen are readily acylated at that site by acid chlorides and anhydrides, but only difficultly by esters (Eq. 3-24). Further acylation is more difficult, but has

$$RCONHNH_2 + R'COCl \longrightarrow RCONHNHCOR' \qquad (3\text{-}24)$$

been carried even to the tetraacyl stage by heating with anhydrides.[54] The presence of inorganic base[74] or tertiary amine[100,101] facilitates acylation with acid chlorides (Eq. 3-25).

$$PhCON\overline{H}COPh\ Na^+ + PhCOCl \longrightarrow$$

$$PhCONH(COPh)_2 + NaCl \qquad (3\text{-}25)$$

Acylation obeys second-order kinetics, and the effect of substituents on the rates conforms in a regular manner to expectations based on their known reactivity constants: electron-withdrawing influences on the acylating agent accelerate the reaction, but they decelerate the reaction when they are present in the hydrazide.[102,103] Acylation with anhydrides, which is usually carried out without added base, is auto-catalytic, because the rate is increased by the acid concurrently produced.[104] Such acylation may be promoted by added acid, which is effective in proportion to its strength.

An unusual case of acylation is the action of formic acid on 1,4-diphenylsemi-carbazide; the presumed intermediate formyl derivative is apparently dehydrated to form a mesoionic ring system[105,106] (Eq. 3-26). Closely related is the reaction of N-alkylbenzhydrazides with phosgene, which forms isosydnones[73] (Eq. 3-27).

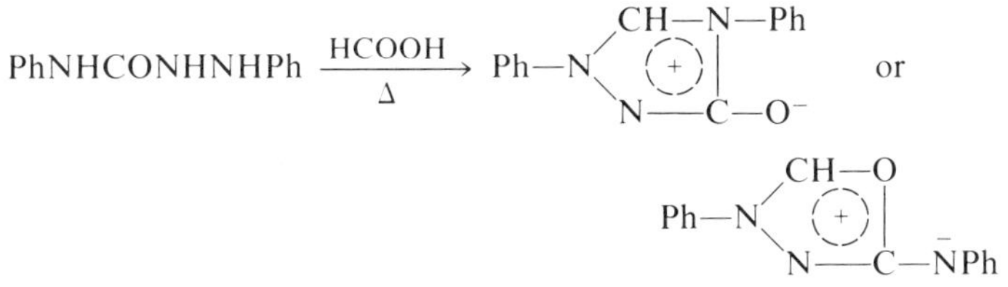

$$\text{PhNHCONHNHPh} \xrightarrow[\Delta]{\text{HCOOH}} \text{Ph—N} \underset{\text{N—C—O}^-}{\overset{\text{CH—N—Ph}}{\diagdown \diagup \overset{(+)}{}\mid}} \quad \text{or}$$

$$\text{Ph—N} \underset{\text{N—C—NPh}}{\overset{\text{CH—O}}{\diagdown \diagup \overset{(+)}{}\mid}} \tag{3-26}$$

$$\text{PhCON} \overset{R}{\underset{NH_2}{\diagdown}} + \text{COCl}_2 \longrightarrow \underset{R—N \overset{(+)}{\diagdown} C—O^-}{\overset{Ph—C—O}{\mid \diagup}} \tag{3-27}$$

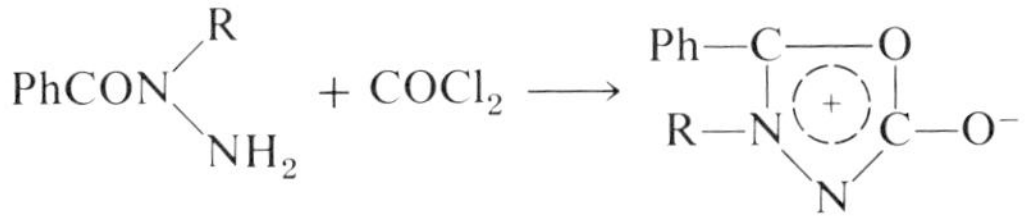

Acylation with orthoesters[107] or imino esters[108] leads to alkyl carbohydrazonates (hydrazono esters), which usually cyclize spontaneously to 1,3,4-oxadiazoles (Eq. 3-28).

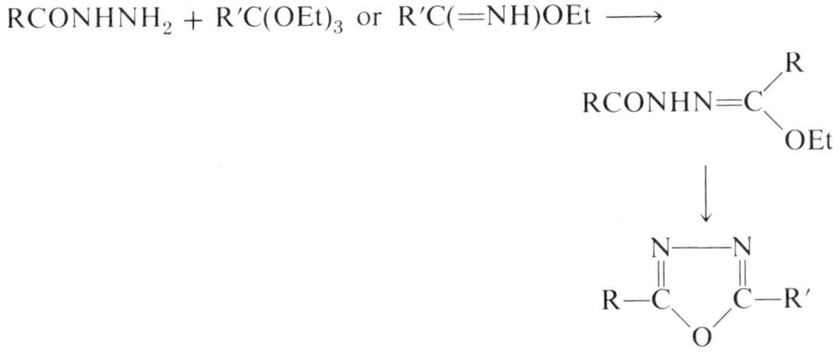

$$\text{RCONHNH}_2 + \text{R'C(OEt)}_3 \text{ or } \text{R'C(=NH)OEt} \longrightarrow$$

$$\text{RCONHN=C} \overset{R}{\underset{OEt}{\diagdown}}$$

$$\downarrow \tag{3-28}$$

$$\underset{O}{\overset{N——N}{R—C \diagdown \diagup C—R'}}$$

Hydrazides are also acylated by isocyanates to form 1-acylsemicarbazides[109,110] (Eq. 3-29), and by isothiocyanates[109,111] to form 1-acylthiosemicarbazides; the kinetics of the former reaction have been studied.[112]

$$\text{RCONHNH}_2 + \text{ArNCO} \longrightarrow \text{RCONHNHCONHAr} \tag{3-29}$$

Thiohydrazides can be acylated at the terminal nitrogen analogously to acylation of hydrazides; the products easily lose water on heating to form 1,3,4-thiadiazoles.[113]

In the foregoing discussion, the acylation of hydrazides of various kinds takes place at a previously unsubstituted position. There are, however, scattered reports in the literature of displacement of one acyl group by another, usually when rather vigorous conditions are used. As one example,[36] when acetone semicarbazone is boiled with acetic anhydride, the carbamyl group is lost, and acetone diacetyl-hydrazone is formed (Eq. 3-30). This phenomenon may have preparative advan-

$$\text{Me}_2\text{C}=\text{NNHCONH}_2 \xrightarrow[\text{boil}]{\text{Ac}_2\text{O}} \text{Me}_2\text{C}=\text{NNAc}_2 \qquad (3\text{-}30)$$

tages, for it has been reported that when acetone acetylhydrazone is heated with acetic anhydride, a quite different product, an oxadiazoline, results[114] (Eq. 3-31).

$$\text{Me}_2\text{C}=\text{NNHAc} \xrightarrow[\Delta]{\text{Ac}_2\text{O}} \underset{\substack{\text{Me} \quad \text{O}}}{\overset{\substack{\text{N}\text{---}\text{N}\text{---}\text{Ac}}}{\text{C} \qquad \text{CMe}_2}} \qquad (3\text{-}31)$$

Sulfonyl chlorides acylate hydrazides at the terminal nitrogen in much the same way as other acyl halides (Eq. 3-32). This reaction is important in connection with

$$\text{RCONHNH}_2 + \text{R}'\text{SO}_2\text{Cl} \xrightarrow{\text{C}_5\text{H}_5\text{N}} \text{RCONHNHSO}_2\text{R}' \qquad (3\text{-}32)$$

the McFadyen-Stevens aldehyde synthesis (see below).[115,116] Sulfinyl chlorides[117] also acylate hydrazides, and thionyl chloride produces N-acyl-N'-thionylhydrazines[118] (Eq. 3-33).

$$\text{RCONHNH}_2 + \text{SOCl}_2 \longrightarrow$$

$$\text{RCONHN}=\text{SO} \left\langle \begin{array}{l} \xrightarrow{\Delta} \text{RCOOH} + \text{N}_2 + \text{S} \\ \xrightarrow{\text{SOCl}_2} \text{RCOCl} \end{array} \right. \qquad (3\text{-}33)$$

Hydrazides of various types are susceptible to nitrosation by nitrous acid, oxides of nitrogen, and nitrosyl chloride. The most commonly encountered example is the conversion of unsubstituted hydrazides into acid azides for use in the Curtius rearrangement or as acylating agents in peptide synthesis.[119] It is usually carried out in a dilute solution of a strong mineral acid near 0°C, and under these conditions, it can be expected to give high yields (Eq. 3-34). Two side reactions may make incur-

$$\text{RCONHNH}_2 \xrightarrow{\text{HNO}_2, \text{H}^+}$$

$$[\text{RCONHNHNO}] \left\langle \begin{array}{l} \xrightarrow[0°C]{\text{Weak acid}} \text{RCONH}_2 + \text{N}_2\text{O} \\ \xrightarrow[0°C]{\text{Strong acid}} \text{RCON}_3 + \text{H}_2\text{O} \end{array} \right. \qquad (3\text{-}34)$$

sions if conditions allow. One of them leads to amide and nitrous oxide, presumably by fragmentation of the intermediate N'-nitroso hydrazide before it can eliminate water[120] (Eq. 3-34). The other is an acylation reaction between acid azide and yet unreacted hydrazide, forming a sym-diacylhydrazine (Eq. 3-35). It is favored by low acidity and slow addition of nitrous acid.[119]

$$\text{RCONHNH}_2 + \text{RCON}_3 \longrightarrow \text{RCONHNHCOR} + \text{HN}_3 \qquad (3\text{-}35)$$

Monosubstituted hydrazides can also be nitrosated, and the nitroso derivatives can often be isolated, although they are rather labile if there is a hydrogen on the adjacent nitrogen; nitroxyl is eliminated and an azo compound is formed[121] (Eq.

3-36). 1-Alkyl semicarbazides, however, consume two equivalents of nitrous acid and become converted to mesoionic oxatriazoles[122] (Eq. 3-37), whereas 2-alkyl semicarbazides and N-alkyl-N-ethoxycarbonylhydrazines are converted to alkyl azides[123] (Eq. 3-38).

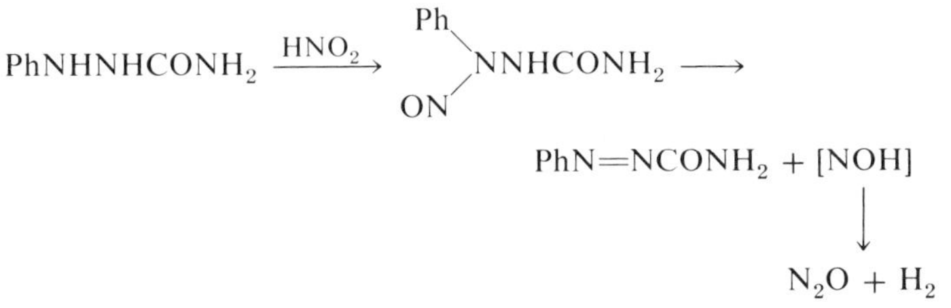

$$\text{PhNHNHCONH}_2 \xrightarrow{\text{HNO}_2} \begin{array}{c}\text{Ph}\\ \diagdown \\ \diagup \\ \text{ON}\end{array}\text{NNHCONH}_2 \longrightarrow$$

$$\text{PhN}\!=\!\text{NCONH}_2 + [\text{NOH}]$$

$$\downarrow \qquad\qquad (3\text{-}36)$$

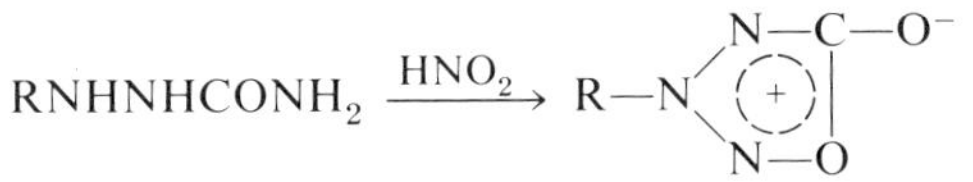

$$\text{N}_2\text{O} + \text{H}_2$$

$$\text{RNHNHCONH}_2 \xrightarrow{\text{HNO}_2} \text{R}\!-\!\text{N}\begin{array}{c}\diagup\text{N}\!-\!\text{C}\!-\!\text{O}^-\\ \big(\!+\!\big) \, \big|\\ \diagdown \text{N}\!-\!\text{O}\end{array} \qquad (3\text{-}37)$$

$$\begin{array}{c}\text{R}\\ \diagdown\\ \diagup\\ \text{H}_2\text{N}\end{array}\!\!\text{NCONH}_2 \xrightarrow{\text{HNO}_2} \text{RN}_3 + \text{CO}_2 + \text{N}_2 \qquad (3\text{-}38)$$

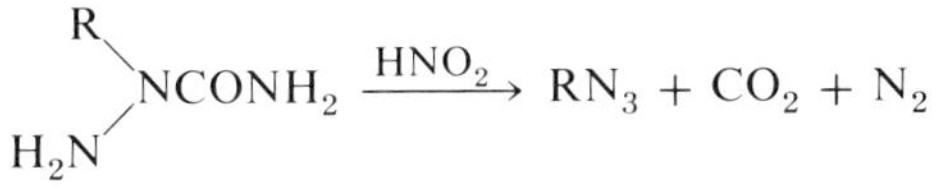

N',N'-Dialkyl hydrazides react with nitrous acid at the terminal nitrogen, as do other hydrazides, notwithstanding the fact that it is a tertiary site. Dealkylation takes place as it does with simple tertiary amines[124] (Eq. 3-39). In contrast, alkylidene

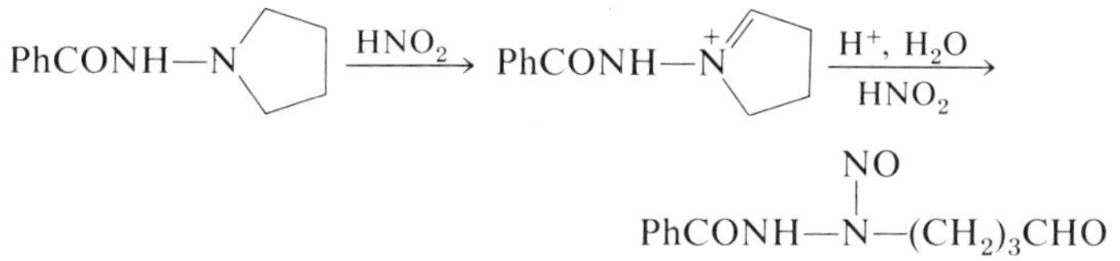

$$\text{PhCONH}\!-\!\text{N}\langle\!\!\bigcirc \xrightarrow{\text{HNO}_2} \text{PhCONH}\!-\!\overset{+}{\text{N}}\langle\!\!\bigcirc \xrightarrow[\text{HNO}_2]{\text{H}^+,\ \text{H}_2\text{O}}$$

$$\begin{array}{c}\text{NO}\\ |\end{array}$$
$$\text{PhCONH}\!-\!\text{N}\!-\!(\text{CH}_2)_3\text{CHO} \qquad (3\text{-}39)$$

hydrazides appear to undergo nitrosation at the amide nitrogen, the product of which then undergoes the general nitrosoamide rearrangement and finally appears as an α-acyloxy azide[125] (Eq. 3-40). Triacyl hydrazines give stable, crystalline N-nitroso derivatives.[126]

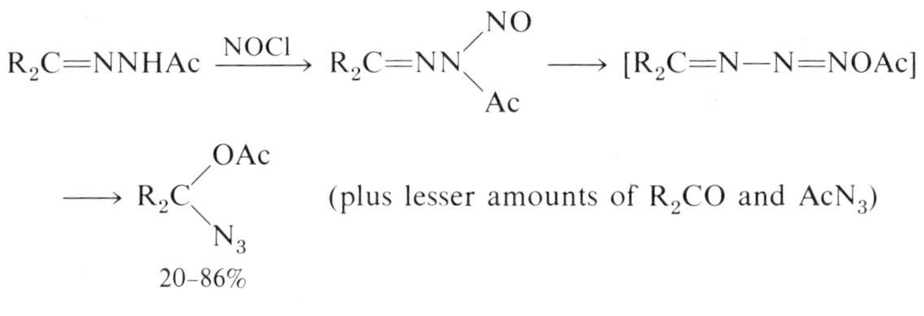

$$\text{R}_2\text{C}\!=\!\text{NNHAc} \xrightarrow{\text{NOCl}} \text{R}_2\text{C}\!=\!\text{NN}\begin{array}{c}\diagup\text{NO}\\ \diagdown\\ \text{Ac}\end{array} \longrightarrow [\text{R}_2\text{C}\!=\!\text{N}\!-\!\text{N}\!=\!\text{NOAc}]$$

$$\longrightarrow \text{R}_2\text{C}\begin{array}{c}\diagup\text{OAc}\\ \diagdown\\ \text{N}_3\end{array} \qquad \text{(plus lesser amounts of R}_2\text{CO and AcN}_3) \qquad (3\text{-}40)$$
$$20\text{-}86\%$$

Phosphorus pentachloride acts on *sym*-diacyl hydrazines as it does on amides, attacking at oxygen, which is removed as phosphoryl chloride, and forming α,α'-dichloro aldazines[48] (Eq. 3-41). If phosphoryl chloride is used instead, the result is cyclodehydration to form an oxadiazole, except in the case of cyclic examples,

$$\text{ArCONHNHCOAr} + 2\text{PCl}_5 \xrightarrow{140°C} \underset{\text{Cl} \quad \text{Cl}}{\text{Ar}-\overset{\displaystyle N-N}{\text{C}} \diagdown \diagup \text{C}-\text{Ar}} + 2\text{POCl}_3 + 2\text{HCl}$$

$$\downarrow \text{POCl}_3$$

$$\underset{O}{\text{Ar}-\overset{\displaystyle N-N}{\text{C}} \diagdown \diagup \text{C}-\text{Ar}}$$

(3-41)

which form α,α'-dichloro derivatives of cyclic azines,[127] as with the pentachloride. N'-Aryl hydrazides react analogously to form carbohydrazonoyl chlorides with phosphorus pentachloride[48,128] (Eq. 3-42), but a more or less stable intermediate,

$$\text{PhCONHNHPh} \xrightarrow[\text{(2) PhOH}]{\text{(1) PCl}_5} \text{PhC} \diagup{}^{\displaystyle \text{NNHPh}}_{\diagdown \text{Cl}}$$

(3-42)

presumably containing phosphorus, must be solvolyzed by treatment with phenol in ether to release the hydrazonoyl chloride. Unsubstituted benzhydrazides react with phosphorus pentachloride in large excess in a complex reaction leading to triazoles,[129] benzylidene chloride, and other products[130] (Eq. 3-43). In one instance,

$$\text{PhCONHNH}_2 \xrightarrow[25°C]{\text{PCl}_5} \underset{(?) \ 95\%}{\text{Ph}-\overset{\displaystyle O-\text{PCl}_3}{\underset{\displaystyle N-NH}{\text{C}}}} \xrightarrow[\Delta]{\text{PCl}_5}$$

$$\underset{65\%}{\text{PhCHCl}_2 + N_2 + \text{PCl}_3 + \text{POCl}_3} \qquad (3\text{-}43)$$

reaction at 25°C with a smaller amount of phosphorus pentachloride gave a high yield of a crystalline substance that reacted with more phosphorus pentachloride when heated to give the benzylidene chloride. Unfortunately, this intermediate, for which an N'-trichlorophosphoranylidenehydrazide structure was proposed rather than the cyclic structure proposed here, was inadequately characterized (analysis for Cl only, no spectra).

Acylation of hydrazides with chlorides of organic phosphorus acids parallels the reactions of carboxylic chlorides.[131]

Benzhydrazides can be converted to carbohydrazonoyl chlorides by another phosphorus reagent, triphenylphosphine, and carbon tetrahalide[47] (Eq. 3-44). Phos-

$$\text{PhCONHNR}^1\text{R}^2 \xrightarrow{\text{CX}_4/\text{Ph}_3\text{P}} \text{PhC} \diagup{}^{\displaystyle \text{NNR}^1\text{R}^2}_{\diagdown \text{X}}$$

$$39\text{-}77\%$$

(3-44)

$$(\text{R}^1\text{R}^2 = \text{HAr, MeAr, Me}_2; \ \text{X} = \text{Cl, Br})$$

phorus pentasulfide, P_4S_{10}, converts hydrazides to thiohydrazides (unsubstituted ones give very little).[132,133] Diacylhydrazines are converted to 1,3,4-thiadiazoles.[113]

Diazonium coupling. Hydrazides couple with diazonium salts at the terminal nitrogen to form tetrazenes[134] (Eq. 3-45). If there is no substituent on the terminal

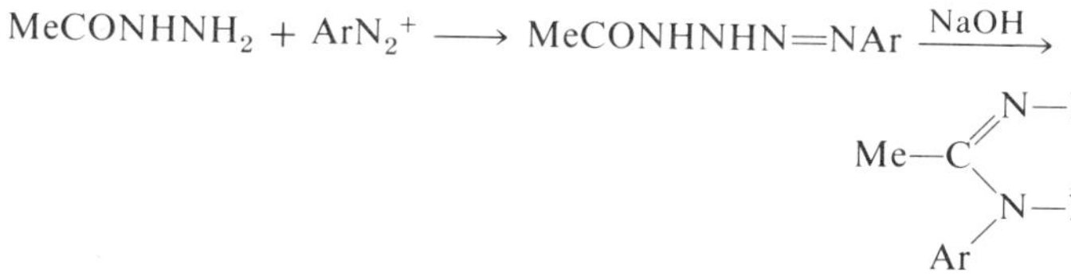

$$\text{MeCONHNH}_2 + \text{ArN}_2^+ \longrightarrow \text{MeCONHNHN}{=}\text{NAr} \xrightarrow{\text{NaOH}}$$

(3-45)

hydrazide nitrogen, treatment of the resulting tetrazene with alkali brings about cyclodehydration to a tetrazole. Diacyl hydrazines behave similarly; ring closure to a tetrazole takes place with the resulting diacyl tetrazene with expulsion of an acyl group. With an excess of diazonium salt, coupling occurs twice to form a diacyl hexazadiene[135] (Eq. 3-46).

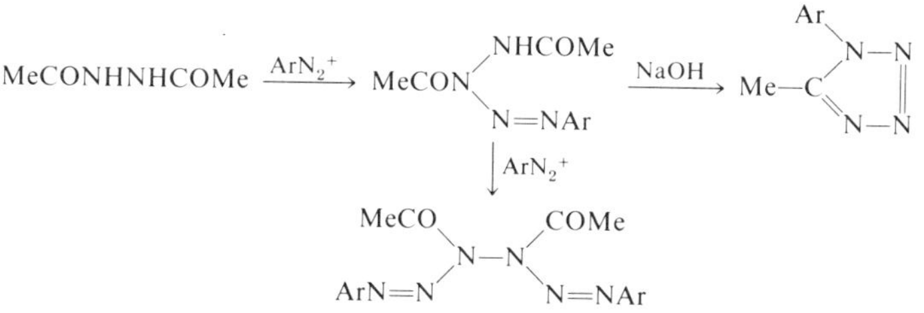

(3-46)

Organometallic reagents. Owing to the acidity of incompletely substituted hydrazides, reactions with organometallic reagents would in general be expected to consist of unproductive acid-base reactions, as has been reported for the example of triethylaluminum, which releases ethane and forms mono- and bis-(diethylalumino) derivatives.[136] Trisubstituted hydrazides, however, react at the carbonyl group with alkyllithium reagents, forming ketones after hydrolytic workup[14] (Eq. 3-47). Fully substituted acrylohydrazides undergo 1,4 addition with alkyllithium reagents.[137]

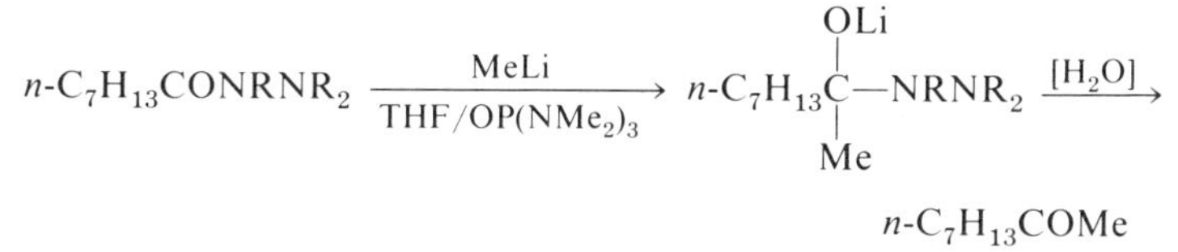

$$n\text{-}C_7H_{13}\text{CONRNR}_2 \xrightarrow[\text{THF/OP(NMe}_2)_3]{\text{MeLi}} n\text{-}C_7H_{13}\overset{\overset{\displaystyle \text{OLi}}{|}}{\underset{\underset{\displaystyle \text{Me}}{|}}{\text{C}}}\text{—NRNR}_2 \xrightarrow{[\text{H}_2\text{O}]}$$

$$n\text{-}C_7H_{13}\text{COMe} \qquad (3\text{-}47)$$

Oxidation. The behavior of hydrazides toward oxidizing agents can be roughly summed up by the statement that they are converted to azo compounds or their transformation products if there is a hydrogen on each nitrogen and to tetrazanes or tetrazenes if only one nitrogen bears a hydrogen. A wide variety of oxidizing agents can oxidize hydrazides: halogens, cupric salts, mercuric oxide, manganese dioxide, N-bromosuccinimide, potassium ferricyanide, etc. Fehling's and Tollens' reagents frequently, but not always, oxidize hydrazides. Acylhydrazyl radicals may be intermediates in some of these reactions, but they do not appear to be stable enough to be isolated, in contrast to triaryl hydrazyls (see Chap. 1).

The oxidation of *sym*-diacyl hydrazines has been the most thoroughly studied.[138]

They are converted cleanly to diacyl diazenes (α,α'-dicarbonyl azo compounds) (Eq. 3-48). Oxidation with oxygen in alkaline solution is accompanied by emission of

$$\text{RCONHNHCOR} \xrightarrow{[O]} \text{RCON}=\text{NCOR} \qquad (3\text{-}48)$$

light (chemiluminescence), which is due to decay of an excited state of the carboxylate formed in a subsequent step.[139] Simple hydrazides, such as benzhydrazide, are not chemiluminescent, but some that bear *ortho*-hydroxy or amino groups are weakly so.[140] N-Aryl *sym*-diacyl hydrazines lose an acyl group when oxidized and form acyl aryl diazenes.[141]

N'-Aryl hydrazides are oxidized to acyl aryl diazenes, which are rapidly attacked by even weak nucleophiles in the medium and may thereby be converted into carboxylic acids, esters, or amides[142,143] (Eq. 3-49). This reaction is of special impor-

$$\text{RCONHNHPh} \xrightarrow{\text{NBS, MnO}_2, \text{ or Cu}^{2+}} \text{RCON}=\text{NPh} \begin{cases} \xrightarrow[80\text{-}90\%]{\text{H}_2\text{O}} \text{RCOOH} + \text{N}_2 + \text{C}_6\text{H}_6 \\ \xrightarrow{\text{R'OH}} \text{RCOOR'} \\ \xrightarrow{\text{R'NH}_2} \text{RCONHR'} \end{cases} \qquad (3\text{-}49)$$

tance in peptide chemistry, where the phenylhydrazide group may be used to block a carboxyl group and may later either be oxidized to regenerate the carboxyl group or converted to a new peptide link by reaction with an amino acid.[144] It is probable that phenyldiazene is the species initially cleaved from the acyl aryl diazene; if it does not first fragment to nitrogen and hydrocarbon, it may be oxidized to a diazonium ion.[142]

N',N'-Dimethylhydrazides have been oxidized with S_2Cl_2 to tetrazanes[145] (Eq. 3-50); yields were good. With lead tetraacetate, however, oxidation takes an entirely

$$\text{RCONHNMe}_2 \xrightarrow[\text{C}_6\text{H}_6, \Delta]{\text{S}_2\text{Cl}_2} \begin{array}{c} \text{RCO} \\ | \\ \text{Me}_2\text{N}-\text{N}-\text{N}-\text{NMe}_2 \\ | \\ \text{RCO} \end{array} \qquad (3\text{-}50)$$

different course, as exemplified by N',N'-dimethylbenzhydrazide, which gives high yields of benzoic acid when the oxidation is carried out in the presence of aqueous acetic acid but under anhydrous conditions is converted largely to a mixture of oxadiazolones[146] (Eq. 3-51). The mechanism is apparently analogous to the oxida-

$$\text{PhCOOH} \xleftarrow[\text{anhydrous}]{\text{Pb(OAc)}_4} \text{PhCONHNMe}_2 \xrightarrow[\text{AcOH/H}_2\text{O}]{\text{Pb(OAc)}_4} \begin{array}{c} \text{N}-\!\!-\!\!\text{N}-\text{CH}_2\text{OAc} \\ \| \quad\quad | \\ \text{C} \quad\quad \text{C}=\text{O} \\ \diagdown \;\; \diagup \\ \text{Ph} \quad \text{O} \end{array} \qquad (3\text{-}51)$$

tion of tertiary amines, and the lost methyl group is believed to be first converted to a methylene-iminium function, which is cleaved by hydrolysis to formaldehyde. N',N'-Dibenzylbenzhydrazide behaves analogously, except that under anhydrous conditions the product is 2,5-diphenyl-1,3,4-oxadiazole.

Trialkylbenzhydrazides, of interest as a means of protecting carboxyl groups, can be converted to benzoic acid in high yields by a variety of oxidizing agents, including lead tetraacetate, periodic acid, chromic acid, and selenium dioxide, nitrous acid, and active manganese dioxide. One N' substituent is eliminated as an aldehyde, and the remainder of the hydrazine portion is cleaved off as an azo compound[146] (Eq. 3-52). N,N'-Dialkylhydrazides behave similarly, except that no alkyl

$$PhCONRN\begin{smallmatrix}Ar\\\\CH_2Ar'\end{smallmatrix} \xrightarrow{[O]} PhCOOH + RN{=}NAr + Ar'CHO \qquad (3\text{-}52)$$

group is cleaved from nitrogen. N,N'-Diisopropyl hydrazides have become desirable protecting groups because of the mildness and simplicity of this method for regenerating the carboxylic acid, and it is especially well suited to the chemistry of penicillins and cephalosporins, where gentle methods are essential. Yields of acid close to 100% can be obtained at room temperature in solutions close to neutral.[147] Alkylidenehydrazides have also been subjected to oxidative cleavage with good results through an apparently analogous mechanism.[146]

Several N-methylhydrazides have been oxidized to tetrazenes in good yield by permanganate, bromine, or hypobromite[70,85] (Eq. 3-53); demethylation is a com-

$$AcN\begin{smallmatrix}NH_2\\\\Me\end{smallmatrix} \xrightarrow[0°C]{KMnO_4 \text{ or } BrO^-} \begin{smallmatrix}Ac\\\\Me\end{smallmatrix}N{-}N{=}N{-}N\begin{smallmatrix}Ac\\\\Me\end{smallmatrix} \qquad (3\text{-}53)$$

peting reaction, forming methyl bromide, and may occur to the exclusion of tetrazene formation, but the circumstances determining the competition are obscure. However, the fact that tetrazenes have, in some instances, been obtained from oxidations in acidic solution shows that azamines, $RC\overset{+}{O}N(Me){=}N^-$, are not intermediates, for they do not dimerize in acid solution (see Chap. 5). Presumably, then, a tetrazane is first formed and oxidized further.

Unsym-diacyl hydrazides, such as N-aminophthalimide, react with oxidizing agents such as lead tetraacetate to form products that are best accounted for by the hypothesis that they arise from an intermediate azamine (diacylaminonitrene), which, for example, can be trapped by an alkene if one is present during the oxidation[148] (Eq. 3-54).

$$C_6H_4(CO)_2N{-}NH_2 \xrightarrow{Pb(OAc)_4} [C_6H_4(CO)_2NN] \xrightarrow{RCH{=}CHR}$$

$$C_6H_4(CO)_2N{-}N\begin{smallmatrix}CHR\\\\|\\\\CHR\end{smallmatrix} \qquad (3\text{-}54)$$

Oxidation of unsubstituted hydrazides may give several quite different types of product, depending on the oxidizing agent and conditions. They can be correlated by the hypothesis that the initial product is an acyldiazene, which has the potential for acting as a moderately good acylating agent if it is presented with a suitable nucleophile, may be halogenated, or may fragment to nitrogen and an aldehyde (Eq. 3-55).

$$
\begin{array}{l}
\text{RCONHN}{=}\text{NNHCOR} \xrightarrow{\ -\text{N}_2\ } \\[4pt]
\quad \uparrow{?} \qquad\qquad\qquad \left\{
\begin{array}{l}
\xrightarrow{\text{RCONHNH}_2} \text{RCONHNHCOR} \\[4pt]
\xrightarrow{\text{R}'\text{NH}_2} \text{RCONHR}' \\[4pt]
\xrightarrow{} \text{RCHO} + \text{N}_2 \\[4pt]
\xrightarrow{\text{Cl}_2 \text{ or } \text{S}_2\text{Cl}_2} \text{RCOCl} + \text{N}_2
\end{array}\right.
\end{array}
\qquad (3\text{-}55)
$$

$$\text{RCONHNH}_2 \xrightarrow{\ [\text{O}]\ } [\text{RCON}{=}\text{NH}]$$

Oxidation under mild, slow, and not too acidic conditions, such as by iodine, peroxy acids, dimethyl sulfoxide/dicyclohexylcarbodimide, phenylseleninic acid, or periodate,[149-153] favors formation of *sym*-diacyl hydrazine (the H_2N_2 presumably released is either oxidized or it disproportionates). If an amine is present during the oxidation, it may capture the acyldiazene and be converted to an amide; the process can be quite efficient and has been applied to synthesis of peptides.[152] Oxidation with alkaline ferricyanide gives rise to aldehydes in poor to moderate yields, which are generally lower than those obtained by the closely related McFadyen-Stevens reaction (see below); it has been referred to as the *Kalb-Gross reaction*.[154,155] Oxidation with halogen or S_2Cl_2 in nonaqueous medium produces acyl halides, often in high yields, perhaps through an *N*-acyl-*N'*-halodiazene.[156,157] A possible alternative oxidation, leading to a tetrazene, which loses nitrogen to become a diacyl hydrazine (Eq. 3-55), has been proposed.[138]

Thiohydrazides are oxidized at sulfur and give rise to carbohydrazonyl disulfides[86,158] (Eq. 3-56).

$$
\text{PhCSNHNHPh} \xrightarrow{\ I_2\ }
\underset{\substack{\text{PhNHN} \qquad\qquad \text{NNHPh}}}{\overset{\substack{\text{Ph} \qquad\qquad\quad \text{Ph}}}{\diagdown\text{C}{-}\text{S}{-}\text{S}{-}\text{C}\diagup}}
\qquad (3\text{-}56)
$$

(Orange-red, mp 149°C)

Reduction. Hydrazides may be reduced either by N—N cleavage or by conversion of the carbonyl group to a methylene. Catalytic hydrogenation (Pd or Ni) cleaves the N—N bond and forms amides; Raney nickel often gives poor yields, and reduction of phenylhydrazides has been proposed as a better route for preparing amides[159] (Eq. 3-57). However, hydrogenating acylhydrazones over platinum oxide

$$\text{RCONHNHPh} \xrightarrow[\text{Pd/C}]{\ H_2\ } \text{RCONH}_2 + \text{PhNH}_2 \qquad (3\text{-}57)$$

reduces only the C=N double bond.[160] Sodium amalgam and acetic acid[74] and lithium in liquid ammonia[161] also reduce hydrazides to amides, often in high yield.

Lithium aluminum hydride reduces hydrazides only very slowly if the nitrogens are not fully substituted, and yields of alkyl hydrazine are low.[162] This circumstance allows other functions, such as cyano or hydrazono, to be reduced selectively in the presence of a hydrazide function[64] (Eq. 3-58). When the nitrogens are fully substituted, reduction to alkyl hydrazines takes place readily in fair to moderate yields[162] (Eq. 3-59). Some failures,[163] notably benzhydrazide and its N',N'-dimethyl deriva-

$$RCONHNHCH_2CH_2CN \xrightarrow{\text{LiAlH}_4}$$
$$RCONHNHCH_2CH_2CH_2NH_2 \qquad (3\text{-}58)$$

$$\underset{\underset{R'}{|}}{\overset{\overset{R'}{|}}{RCON}}-NCOR \xrightarrow{\text{LiAlH}_4} \underset{\underset{R'}{|}}{\overset{\overset{R'}{|}}{RCH_2N}}-NCH_2R \qquad (3\text{-}59)$$

tive, have been attributed to interference from conversion of the amide NH to an anion, but the successful reduction of N',N'-dimethylformhydrazide[164] and of some N'-phenylhydrazides[165] casts doubt on this explanation; perhaps it is a question of the solubility of the metal derivatives first formed. Diborane, which would be free of such a problem, reduces both substituted and unsubstituted diacylhydrazines to alkyl hydrazines in moderate to good yields in hot tetrahydrofuran[166a] (Eq. 3-60).

$$EtCONHNHCOEt \xrightarrow[\text{THF, 65°C}]{\text{B}_2\text{H}_6} \underset{65\%}{EtCH_2NHNHCH_2Et} \qquad (3\text{-}60)$$

Some products of N—N cleavage are occasionally observed from hydride reductions; *sym*-dibenzoylhydrazine is inert to diborane, but it is reduced by lithium aluminum hydride to benzaldehyde benzoylhydrazone and benzylamine.[166b]

Amine acylimides undergo reductive cleavage when hydrogenated over Raney nickel.[57] This catalyst is also a very effective desulfurizing agent, and benzaldehyde thiosemicarbazone has been desulfurized with it to the formamidrazone stage [167] (Eq. 3-61).

$$PhCH{=}NNHCSNH_2 \xrightarrow[\text{Raney Ni}]{\text{H}_2} PhCH{=}NN{=}CHNH_2 \qquad (3\text{-}61)$$

Hydrazonoyl Derivatives: Halides and Esters

The chemistry of carbohydrazonoyl halides is characterized by displacement of halide by nucleophiles, often with subsequent cyclization[129,168,169] (Eq. 3-62), and

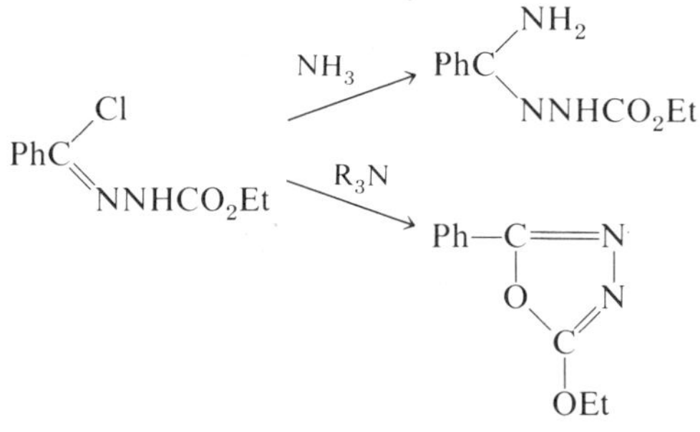

$$(3\text{-}62)$$

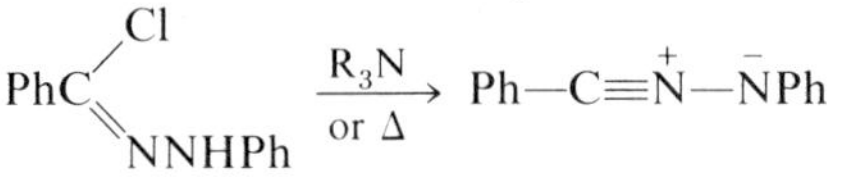

$$PhC\underset{NNHPh}{\overset{Cl}{\big<}} \xrightarrow[\text{or } \Delta]{R_3N} Ph-C\equiv\overset{+}{N}-\overset{-}{N}Ph \qquad (3\text{-}63)$$

by elimination of hydrogen halide to form a nitrile imide[170,171] (Eq. 3-63), which is usually trapped in situ by a cycloaddition reaction[172,173] by addition of a nucleophile, such as an alcohol. N-Phenylbenzohydrazonoyl bromide, for example, reacts with sodium phenoxides to form phenyl benzohydrazonates, either by addition followed by elimination or vice versa.[45] Base-catalyzed alcoholysis of hydrazonoyl halides is geometrically stereospecific, however, the (Z) bromide giving rise exclusively to (Z) ester, even through the (E) ester may be more stable.[40]

Carbohydrazonoyl chlorides are reduced by stannous chloride in ether (see the Sonn-Müller reduction of imidoyl chlorides); the N—N bond is cleaved, and aldimine chlorostannates are formed[174] (Eq. 3-64).

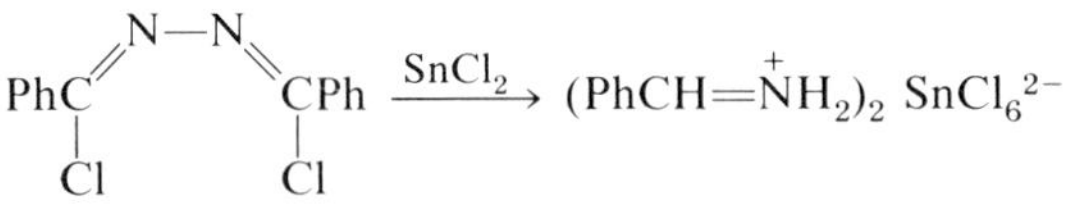

$$PhC\underset{Cl}{\overset{N-N}{\big<}}CPh\underset{Cl}{} \xrightarrow{SnCl_2} (PhCH=\overset{+}{N}H_2)_2\ SnCl_6^{2-} \qquad (3\text{-}64)$$

Alkyl carbohydrazonates are stable to heat, except for configurational equilibration,[40] which takes place readily at about 130°C. They do not undergo rearrangement with O-to-N migration.[45] Aryl carbohydrazonates, however, rearrange on heating (boiling a toluene solution is sufficient); although the aryl group migrates from O to N, it is not an analog of the Chapman rearrangement of aryl imidates, for the aryl group migrates to the terminal, not the amide, nitrogen[45,175] (Eq. 3-65).

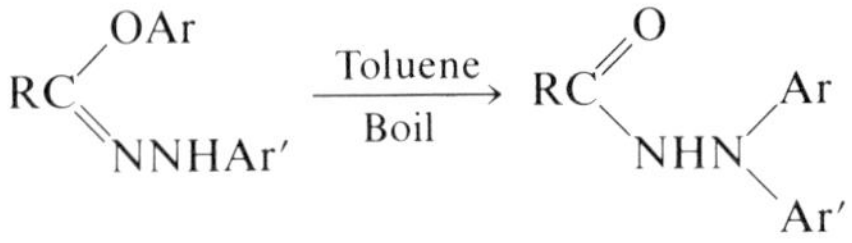

$$RC\underset{NNHAr'}{\overset{OAr}{\big<}} \xrightarrow[\text{Boil}]{\text{Toluene}} RC\underset{NHN\diagdown Ar'}{\overset{O}{\big<}}Ar \qquad (3\text{-}65)$$

Rearrangement is first-order and intramolecular, and the dependence of the rate on substituents on the migrating group obeys the Hammett relationship with a value for the reaction constant ρ of -0.4 to -0.52. This effect is the opposite of that observed in the Chapman rearrangement. Added dibenzoyl peroxide accelerates the rearrangement, which is believed to take place by a free-radical process through a hydrazyl intermediate.

A quaternary hydrazonic ester has been made; when treated with amines, the alkoxy group is not displaced, but instead, an alkyl group is displaced from nitrogen[46,176] (Eq. 3-66).

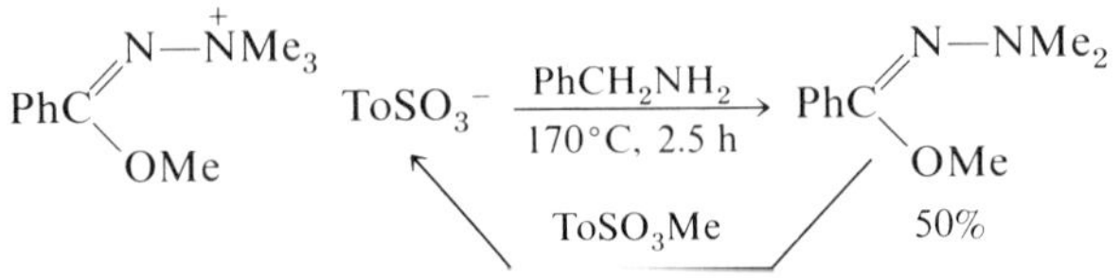

$$PhC\underset{OMe}{\overset{N-\overset{+}{N}Me_3}{\big<}} ToSO_3^{-} \underset{\diagdown\ ToSO_3Me}{\overset{\xrightarrow[\text{170°C, 2.5 h}]{PhCH_2NH_2}}{}} PhC\underset{OMe\quad 50\%}{\overset{N-NMe_2}{\big<}} \qquad (3\text{-}66)$$

Amidrazones.[4,177]

Amidrazones seem to be reasonably stable to heat, especially if they bear aryl groups, but unsubstituted ones easily condense to dihydrotetrazines in solution. An attempt to achieve an analog of the Chapman rearrangement of aryl imidates by heating N-methyl-N-phenylbenzamide phenylhydrazone in boiling dioxane, alone or in the presence of radical initiators, failed.[45] An attempt to achieve a Stevens rearrangement of the ylide derived from a quaternized amidrazone gave only products of N—N cleavage, which are most readily explained by the presumption that a benzyl group migrated from the cationic to the anionic nitrogen[178] (Eq. 3-67).

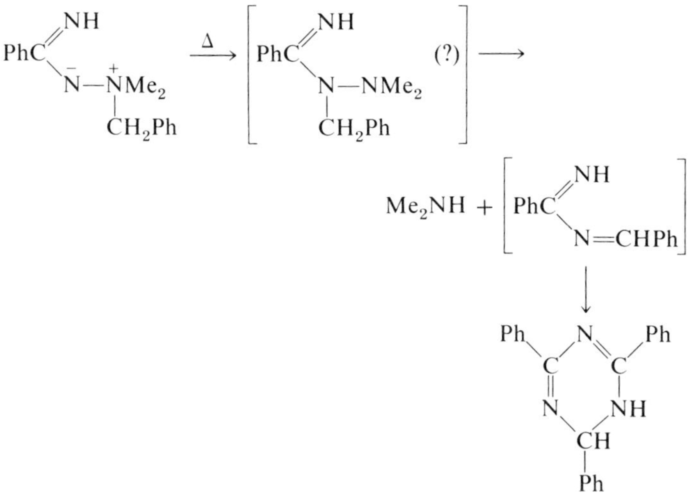

$$
PhC\begin{array}{c}\nearrow NH \\ \searrow \underset{\underset{CH_2Ph}{|}}{N-\overset{+}{N}Me_2}\end{array} \xrightarrow{\Delta} \left[PhC\begin{array}{c}\nearrow NH \\ \searrow \underset{\underset{CH_2Ph}{|}}{N-NMe_2}\end{array} \right] (?) \longrightarrow
$$

$$
Me_2NH + \left[PhC\begin{array}{c}\nearrow NH \\ \searrow N=CHPh\end{array} \right] \qquad (3\text{-}67)
$$

Amidrazones can be hydrolyzed all the way to the corresponding carboxylic acid with catalysis by acid or base, but the process is slow; the first stage is loss of an amine to form a hydrazide, which has been isolated in some instances[8,179] (Eq. 3-68). Solvolysis with hydrogen sulfide has been investigated with amidrazones de-

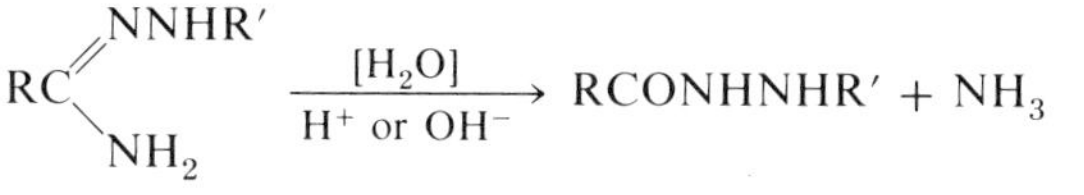

$$
RC\begin{array}{c}\nearrow NNHR' \\ \searrow NH_2\end{array} \xrightarrow[H^+ \text{ or } OH^-]{[H_2O]} RCONHNHR' + NH_3 \qquad (3\text{-}68)
$$

rived from perfluoro acids; it is analogous to hydrolysis, eliminating the amido group to form thiohydrazides.[86]

Amidrazones undergo aminolysis with primary or secondary amines so as to exchange the amino group and generate a new amidrazone, leaving the hydrazine portion intact[180] (Eq. 3-69). Hydroxylamine accomplishes an analogous change,

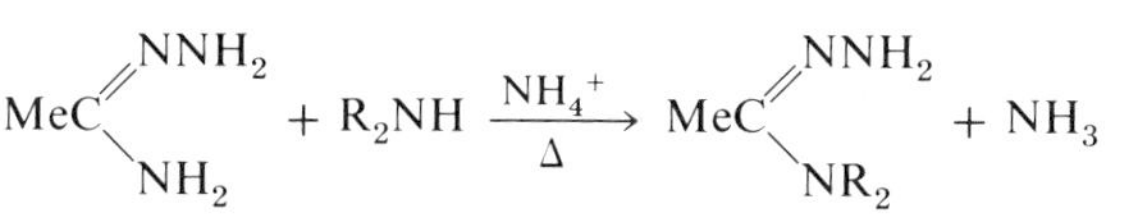

$$
MeC\begin{array}{c}\nearrow NNH_2 \\ \searrow NH_2\end{array} + R_2NH \xrightarrow[\Delta]{NH_4^+} MeC\begin{array}{c}\nearrow NNH_2 \\ \searrow NR_2\end{array} + NH_3 \qquad (3\text{-}69)
$$

giving rise to hydrazide oximes or their tautomers,[181,182] and hydrazines displace the amino group to form hydrazidines[4,183,184] (Eq. 3-70). However, amidrazones are

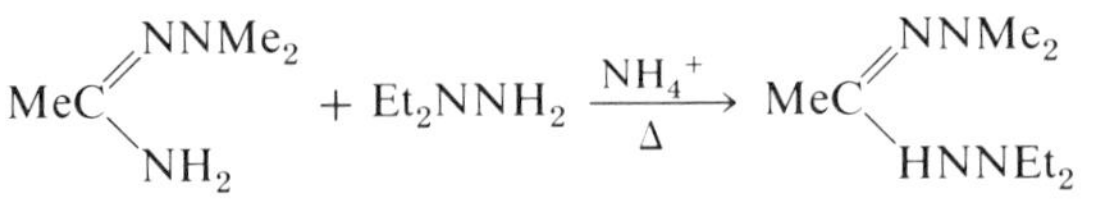

$$\underset{\substack{\diagdown \\ NH_2}}{\overset{\substack{NNMe_2 \\ \diagup}}{MeC}} + Et_2NNH_2 \xrightarrow[\Delta]{NH_4^+} \underset{\substack{\diagdown \\ HNNEt_2}}{\overset{\substack{NNMe_2 \\ \diagup}}{MeC}} \qquad (3\text{-}70)$$

reported to act on themselves contrary to this pattern and eliminate hydrazine to form amide azines (Eq. 3-71).

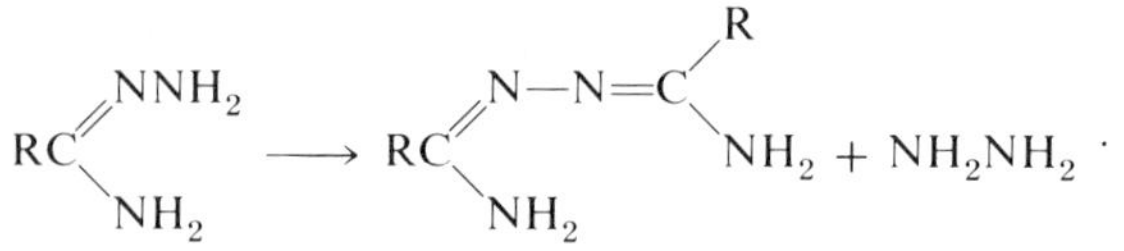

$$\underset{\substack{\diagdown \\ NH_2}}{\overset{\substack{NNH_2 \\ \diagup}}{RC}} \longrightarrow \underset{\substack{\diagdown \\ NH_2}}{\overset{\substack{N-N=C \\ \diagup}}{RC}} \overset{R}{\underset{NH_2}{\diagup}} + NH_2NH_2 \cdot \qquad (3\text{-}71)$$

Each of the nitrogens of amidrazones is a nucleophilic site, and examples of alkylation at each have been observed. Amide N',N'-dimethylhydrazones react with methyl iodide at the outer nitrogen[8] (Eq. 3-72). The corresponding methylphenyl-

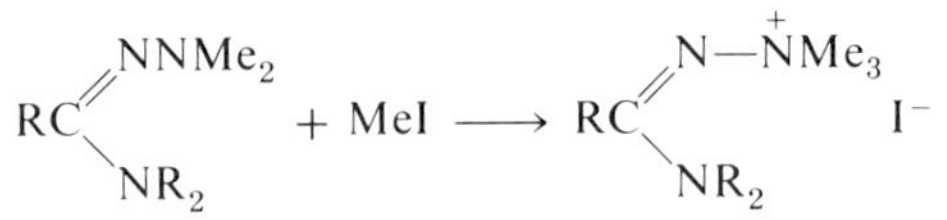

$$\underset{\substack{\diagdown \\ NR_2}}{\overset{\substack{NNMe_2 \\ \diagup}}{RC}} + MeI \longrightarrow \underset{\substack{\diagdown \\ NR_2}}{\overset{\substack{N-\overset{+}{N}Me_3 \\ \diagup}}{RC}} \quad I^- \qquad (3\text{-}72)$$

hydrazones, however, are methylated at the double-bonded nitrogen to form an amidinium system (Eq. 3-73), whereas benzanilide dimethylhydrazone gives a mixture of two amidinium derivatives[178] (Eq. 3-74), and trimethylbenzhydrazide imines are methylated at the imine nitrogen (Eq. 3-75). The amidrazone ylide trimethylamine acetimidylimide undergoes methylation at the imidoyl nitrogen (Eq. 3-76), but

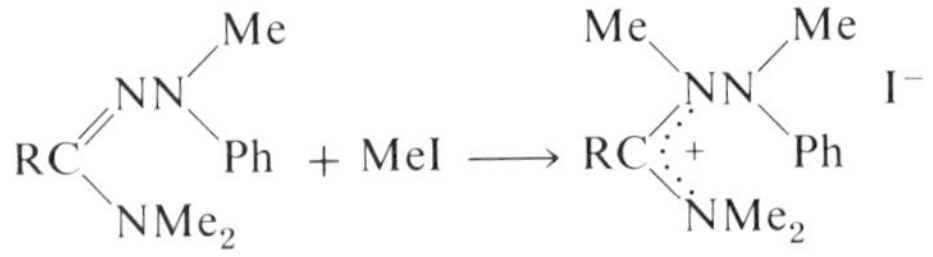

$$\underset{\substack{\diagdown \\ NMe_2}}{\overset{\substack{Me \\ NN \diagdown \\ \diagup}}{RC}} \overset{Me}{\underset{Ph}{}} + MeI \longrightarrow \underset{\substack{\diagdown \\ NMe_2}}{\overset{\substack{Me \diagdown \diagup Me \\ NN \\ \diagup}}{RC\colon+}} \overset{Me}{\underset{Ph}{}} \quad I^- \qquad (3\text{-}73)$$

$$\underset{\substack{\diagdown \\ NHPh}}{\overset{\substack{NNMe_2 \\ \diagup}}{PhC}} \xrightarrow{MeI} \underset{\substack{\diagdown \\ NHPh}}{\overset{\substack{NMeNMe_2 \\ \diagup}}{PhC\colon+}} \quad I^- \quad \text{and}$$

Major

$$\underset{\substack{\diagdown \\ N Ph \\ | \\ Me}}{\overset{\substack{NMeNMe_2 \\ \diagup}}{PhC\colon+}} \quad I^- \qquad (3\text{-}74)$$

Minor

$$\underset{\substack{\diagdown \\ NR}}{\overset{\substack{NMeNMe_2 \\ \diagup}}{PhC}} \xrightarrow{MeI} \underset{\substack{\diagdown \\ NR \\ | \\ Me}}{\overset{\substack{NMeNMe_2 \\ \diagup}}{PhC\colon+}} \qquad (3\text{-}75)$$

$$\underset{\substack{\diagdown \\ \overset{+}{N}-\overset{+}{N}Me_3}}{\overset{\substack{NH \\ \diagup}}{MeC\colon-}} + MeI \longrightarrow \underset{\substack{\diagdown \\ \overset{+}{N}-\overset{+}{N}Me_3}}{\overset{\substack{NMe \\ \diagup}}{MeC\colon-}} \qquad (3\text{-}76)$$

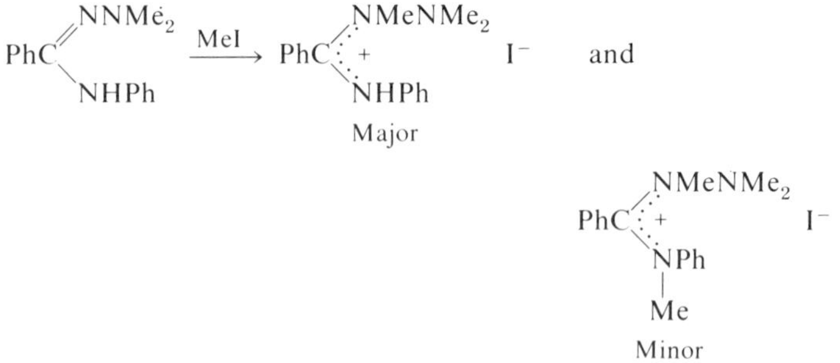

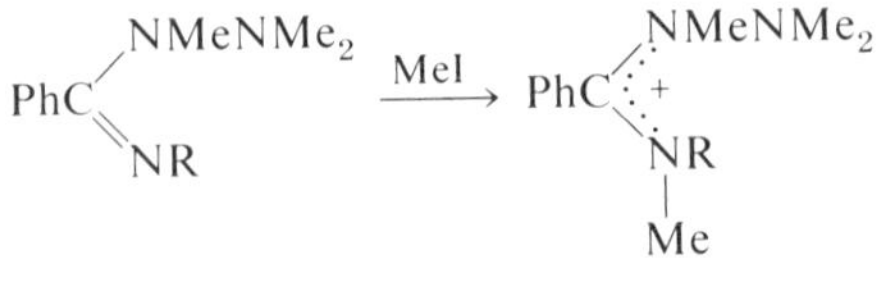

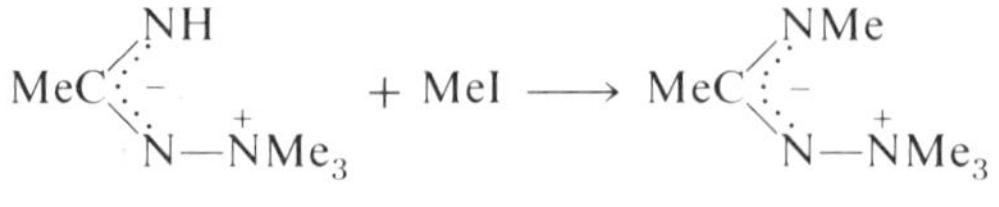

in some other instances, such compounds gave intractable mixtures.[46] Presumably all these reactions are kinetically controlled, and if the material balances were more nearly quantitative, small amounts of products alkylated at alternative sites would be found. The observed results probably reflect differences of only a few kilocalories per mole in the activation energies.

Amidrazones undergo acylation at the terminal hydrazone nitrogen[185] unless that site is disubstituted, in which case acylation takes place at the amine nitrogen[186,187] (Eqs. 3-77, 3-78). With acid chlorides or anhydrides, the acyl derivatives can be

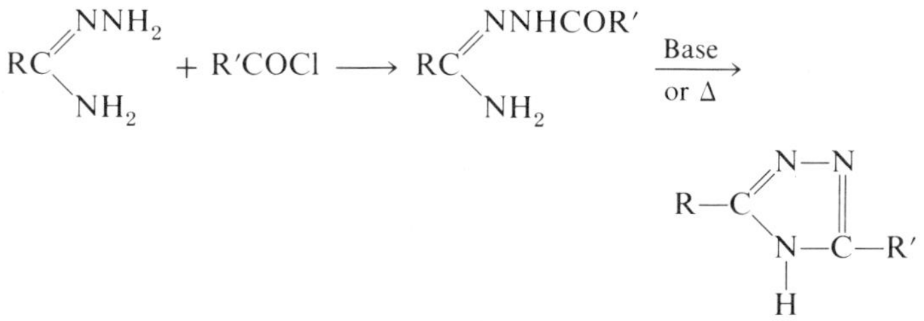

$$
RC\begin{array}{c} \diagup NNH_2 \\ \diagdown NH_2 \end{array} + R'COCl \longrightarrow RC\begin{array}{c} \diagup NNHCOR' \\ \diagdown NH_2 \end{array} \xrightarrow[\text{or } \Delta]{\text{Base}}
$$

$$
R-C\begin{array}{c} N-N \\ \diagdown N-C-R' \\ H \end{array} \tag{3-77}
$$

$$
RC\begin{array}{c} \diagup NNMePh \\ \diagdown NH_2 \end{array} + R'COCl \longrightarrow R-C\begin{array}{c} \diagup NNMePh \\ \diagdown NHCOR' \end{array} \tag{3-78}
$$

isolated, but they undergo cyclodehydration with base or heat to form triazoles. With carboxylic acids, esters, orthoesters, and imidic esters, which generally require heating to react, triazoles are the isolated products.[188,189] Benzanilide phenylhydrazone is converted to a triazolium salt by successive treatment with an acid chloride and perchloric acid[190] (Eq. 3-79).

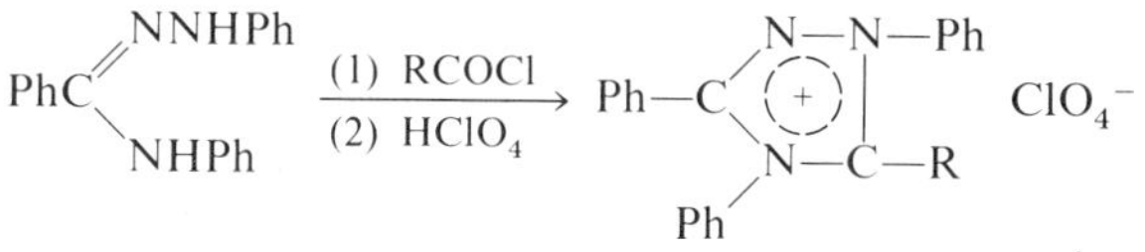

$$
PhC\begin{array}{c} \diagup NNHPh \\ \diagdown NHPh \end{array} \xrightarrow[(2)\ HClO_4]{(1)\ RCOCl} Ph-C\begin{array}{c} N-N-Ph \\ (+) \\ N-C-R \\ Ph \end{array} ClO_4^- \tag{3-79}
$$

Nitrous acid or alkyl nitrite converts amidrazones to tetrazoles[191,192] (Eq. 3-80), except that imides of N-substituted hydrazides are deaminated to the corresponding amidine[106] (Eq. 3-81).

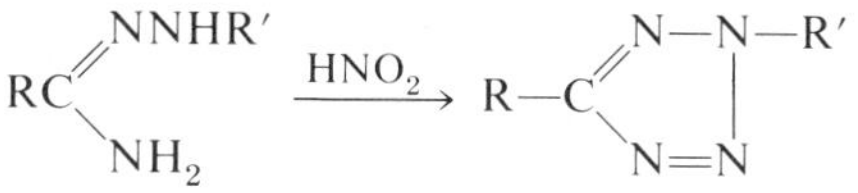

$$
RC\begin{array}{c} \diagup NNHR' \\ \diagdown NH_2 \end{array} \xrightarrow{HNO_2} R-C\begin{array}{c} N-N-R' \\ N=N \end{array} \tag{3-80}
$$

$$
\begin{array}{c} Ph \\ \diagdown \\ PhN \end{array} C-N\begin{array}{c} \diagup NH_2 \\ \diagdown Ph \end{array} \xrightarrow{HNO_2} PhC\begin{array}{c} \diagup NHPh \\ \diagdown NPh \end{array} + N_2O \tag{3-81}
$$

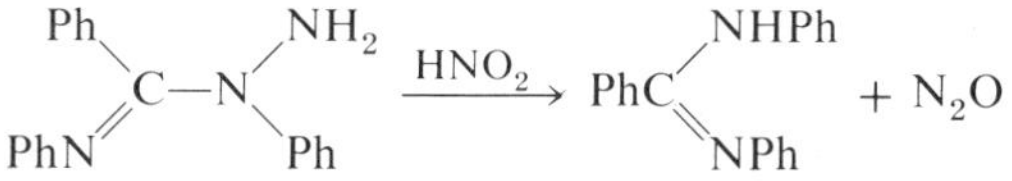

Amidrazones unsubstituted on the hydrazone part react with aldehydes and ketones to form alkylidene derivatives (carboximidylhydrazones).[106,193,194] Substitution on the inner nitrogen of the hydrazine does not interfere (Eq. 3-82). Monosub-

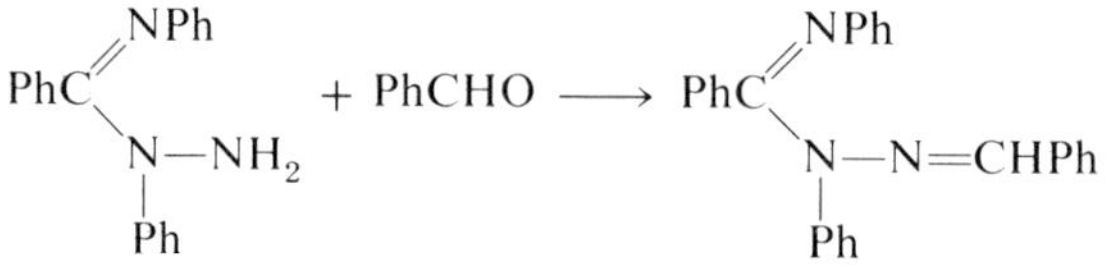

$$\text{PhC}\overset{\displaystyle NPh}{\underset{\displaystyle \underset{\displaystyle Ph}{|}}{N-NH_2}} + \text{PhCHO} \longrightarrow \text{PhC}\overset{\displaystyle NPh}{\underset{\displaystyle \underset{\displaystyle Ph}{|}}{N-N=CHPh}} \tag{3-82}$$

stitution at the terminal hydrazone nitrogen leads to cyclization to triazolines[4] (Eq. 3-83).

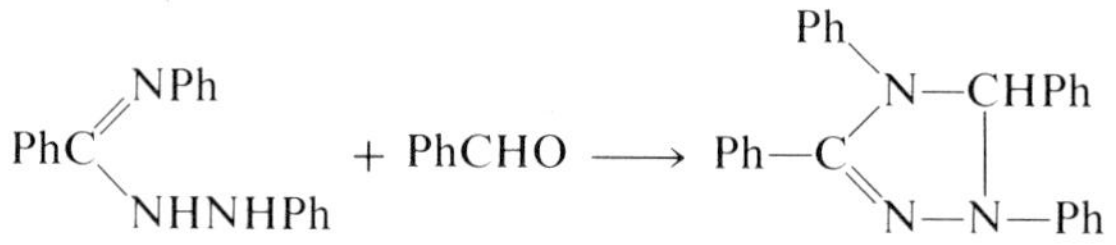

$$\text{PhC}\overset{\displaystyle NPh}{\underset{\displaystyle NHNHPh}{}} + \text{PhCHO} \longrightarrow \text{Ph}-\text{C}\overset{\displaystyle \overset{\displaystyle Ph}{\underset{\displaystyle |}{}}N-CHPh}{\underset{\displaystyle N-N-Ph}{|}} \tag{3-83}$$

Oxidation of amidrazones takes place readily if they are capable of assuming the N'-monosubstituted hydrazide imide structure; the products are the corresponding azo compounds[106] (Eq. 3-84). Imides of N-substituted hydrazides are inert, however.

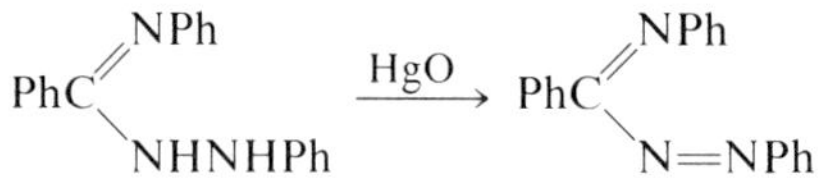

$$\text{PhC}\overset{\displaystyle NPh}{\underset{\displaystyle NHNHPh}{}} \xrightarrow{\text{HgO}} \text{PhC}\overset{\displaystyle NPh}{\underset{\displaystyle N=NPh}{}} \tag{3-84}$$

Unsubstituted amidrazones of perfluoro acids are converted to their hydrolysis products, the perfluoro acids, by hydrogen peroxide, apparently by an oxidative process.[17] However, the same reagent converts pyridine-1-carboxamidrazone to a mixture of the simple amide and the pyridylmethylene derivative of the amidrazone[195] (Eq. 3-85).

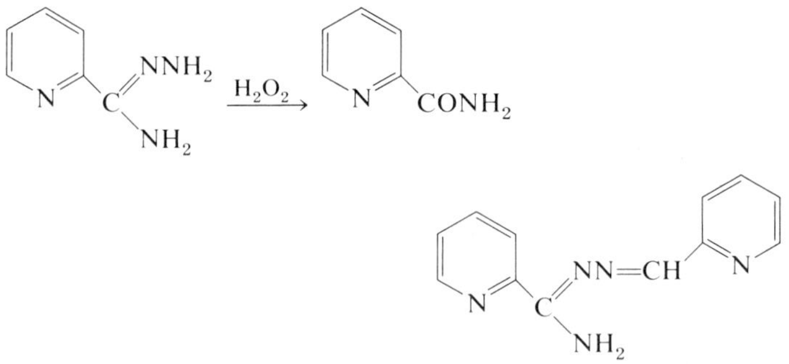

$$\tag{3-85}$$

Certain amidrazones of the amide azine type can be oxidized by a one-electron step to cation radicals, the salts of which are intensely colored (usually violet) and can be isolated[196,197] (Eq. 3-86).

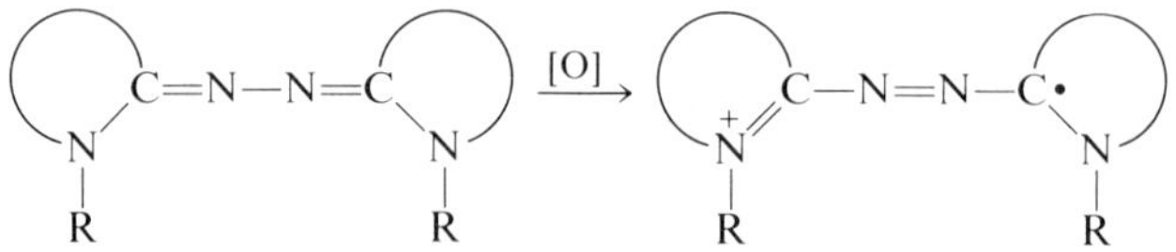

$$\tag{3-86}$$

Aminoguanidine as well as cyclic amidrazones having an unsubstituted hydrazono group can be made to undergo oxidative coupling with phenols, aromatic amines, and active methylene compounds to form azines or azo compounds having dyestuff properties,[198] analogous to the oxidative coupling of *unsym*-disubstituted hydrazines. Oxidation in the absence of a coupling partner has led to pentazadienes related to the cyanine dyes (Eq. 3-87). These reactions presumably take place through azo or azaminium intermediates.

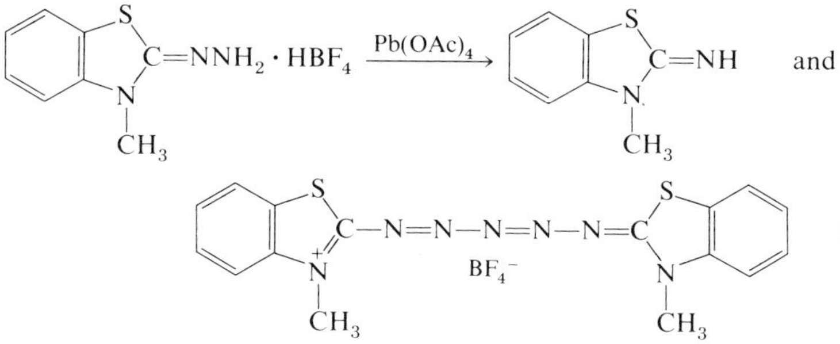

$$(3\text{-}87)$$

Hydrazidines

The chemistry of hydrazidines has been less completely studied than that of amidrazones. On warming in acidic solution, N'-phenylbenzhydrazide phenylhydrazone undergoes a redox disproportionation in which part of it suffers reductive cleavage of an N—N bond[199] (Eq. 3-88). Oxidation in general takes place easily,

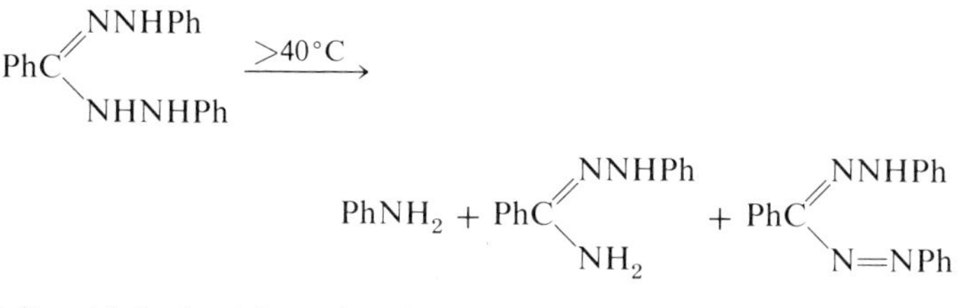

$$(3\text{-}88)$$

especially with hydrazidines that do not bear aryl substituents, which may be rapidly attacked by air as the free bases.[7,11] Certain tetraaryl hydrazidines have been oxidized to blue-green free radicals, which are isolable, but disproportionate in contact with acids into the parent hydrazidine and a formazan derivative[200] (Eq. 3-89).

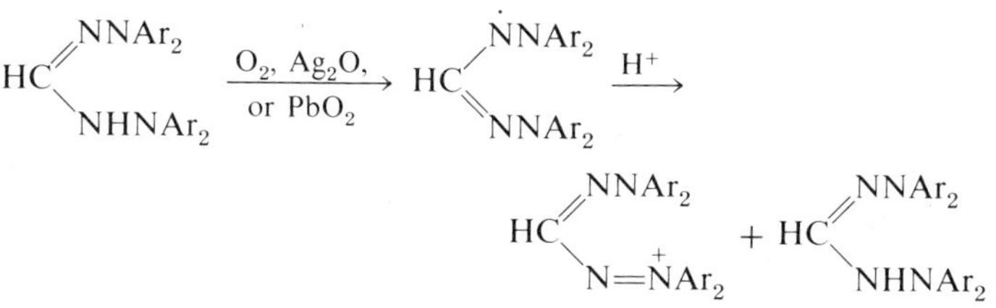

$$(3\text{-}89)$$

Some aryl-substituted hydrazidines have been reported to form adducts with methanol readily, but insufficient detail was given to enable one to judge whether the hydrazidine carbon became bonded to the oxygen or the product was only a hydrogen-bonded complex.[11] Elimination of hydrazine can take place, as exemplified by the behavior of acethydrazidine hydrochloride on heating; a 4-amino-1,2,4-triazole is formed[10] (Eq. 3-90).

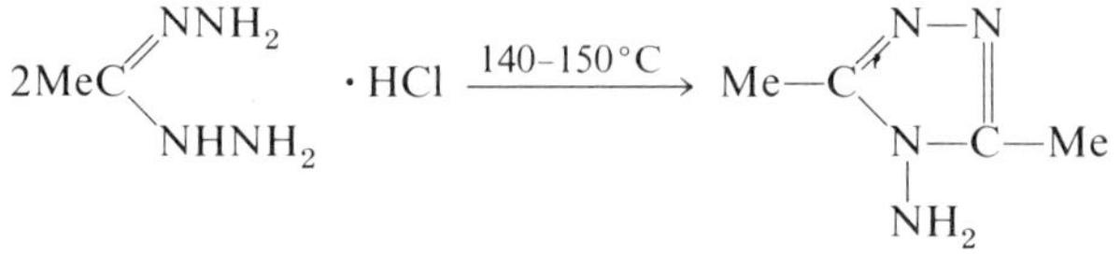

$$2MeC\overset{NNH_2}{\underset{NHNH_2}{\diagup}} \cdot HCl \xrightarrow{140-150\,^{\circ}C} Me-C\overset{N-N}{\underset{N-C-Me}{\diagup\diagdown}}\quad (3\text{-}90)$$
$$\underset{NH_2}{|}$$

Benzaldehyde forms hydrazones with hydrazidines having a free NH_2 group.[11] Reduction by such reagents as sodium dithionite or hydrogen and a catalyst forms amidrazones[7,201] (Eq. 3-91).

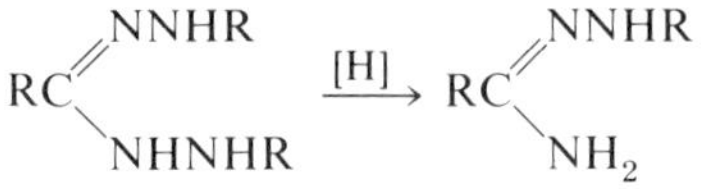

$$RC\overset{NNHR}{\underset{NHNHR}{\diagup}} \xrightarrow{[H]} RC\overset{NNHR}{\underset{NH_2}{\diagup}} \quad (3\text{-}91)$$

Some acyl hydrazidines have been cyclized to 4-amino-1,2,4-triazole derivatives by heating with acid or base.[202]

Sulfonhydrazides

The reaction behavior of sulfonhydrazides differs considerably from that of carbohydrazides owing to the lower reactivity of the bonds to oxygen and the ease with which the sulfur assumes a lower oxidation state.

Heat and light. *p*-Toluenesulfonhydrazide, a reasonably representative example, decomposes when strongly heated; the products clearly arise from more than one path, and the most likely initial step is elimination of toluenesulfinic acid with generation of unstable diazene[203] (Eq. 3-92). 2,4,6-Trialkylbenzenesulfonhydrazides

$$ArSO_2NHNH_2 \xrightarrow{\Delta} ArSO_2H\ (+\ HN{=}NH)\qquad ArSO_2NH_2$$
$$ArSO_2SAr \qquad ArSO_2^-NH_4^+ \qquad N_2 \quad (3\text{-}92)$$

decompose slowly even at room temperature.[204] N',N'-Disubstituted sulfonhydrazides appear to dissociate reversibly to azaminium and sulfinate ions[205] (Eq. 3-93)

$$R_2NNHSO_2To \rightleftharpoons R_2\overset{+}{N}{=}NH\ ToSO_2^- \quad (3\text{-}93)$$

(see the section Bases). N'-Monosubstituted sulfonhydrazides decompose when heated to form hydrocarbon and sulfinic acid, presumably through an analogous cleavage into an alkyldiazene; the reaction is more satisfactory with the help of base. Pyrolysis of alkylidene sulfonhydrazides (i.e., sulfonylhydrazones) produces a complex mixture of aldol condensation products, pyrazolines, sulfones, etc., all in low yield, accompanied by dark-colored material.[206] Photolysis or thermolysis of salts may give products from carbenes or carbocations.[207,208] The methylnorbornanone tosylhydrazones, when photolyzed in dilute sodium hydroxide, are converted to a mixture of methylnorbornanols corresponding to complete Wagner-Meerwein quilibration.[209]

Acids. Dilute sulfuric acid catalyzes disproportionation of p-toluenesulfonhydrazide into *sym*-disulfonylhydrazine and hydrazine[203] (Eq. 3-94); this may interfere

$$ArSO_2NHNH_2 \xrightarrow{H_2SO_4} ArSO_2NHNHSO_2Ar + N_2H_4 \qquad (3\text{-}94)$$

when sulfonhydrazones are being prepared, for the released hydrazine gives rise to azines. N',N'-Disubstituted sulfonhydrazides behave differently, although the reaction has not been extensively studied. A mixture of products is produced and can be rationalized by the hypothesis that the initial step is an elimination to form an azaminium salt[210] (Eq. 3-95). If heating is avoided, however, simple salt formation

$$p\text{-}ToSO_2NHNMe_2 \xrightarrow[100°C]{Conc.\ HCl} Me_2NNH_2 \qquad Me_2NNHMe \qquad Me_2NH$$

$$MeNH_2 \qquad MeCl \qquad p\text{-}ToSS\text{-}p\text{-}To \qquad p\text{-}ToSSO_2\text{-}p\text{-}To$$

$$NH_3 \qquad N_2 \qquad (3\text{-}95)$$

takes place, and further reactions are slow. N',N'-Dibenzyl-p-toluenesulfonhydrazide reacts similarly to the dimethyl compound with hot acid, but a benzyl group is converted to benzaldehyde. N'-Methyl-p-toluenesulfonhydrazide produces a slightly less complex mixture, in which methylhydrazine is the dominant nitrogenous product.

Bases. Sulfonhydrazides without a substituent on the amide nitrogen are acidic enough to form salts with aqueous alkalies, although it may be more advantageous to prepare them from sodium alkoxides or alkyllithium reagents. The salts are generally isolable, but they decompose easily with ejection of sulfinate ion. The fate of the remainder of the molecule depends on the kind of substituents, and the possibilities include the synthetically important McFadyen-Stevens and Bamford-Stevens reactions.

Unsubstituted sulfonhydrazides decompose through their anions into sulfinate and diazene in the primary step (Eq. 3-96). The unstable diazene is an active reduc-

$$RSO_2\bar{N}\text{—}NH_2 \longrightarrow RSO_2^- + (H_2\overset{+}{N}{=}N^- \rightleftharpoons HN{=}NH) \qquad (3\text{-}96)$$

ing agent and may either disproportionate into ammonia and nitrogen or act as a hydrogen donor to any reducible species to which it may be exposed[211] (Eq. 3-97).

$$(HN{=}NH) + CH_2{=}CHCH_2OH \longrightarrow N_2 + CH_3CH_2CH_2OH \qquad (3\text{-}97)$$

This type of decomposition occurs so readily that it may preclude simple base-catalyzed hydrolysis, but nevertheless, hydrolysis of salts of the parent sulfonic acids has been reported by heating toluene-p-, benzene-, and naphthalene-β-sulfonhydrazides with 5% sodium hydroxide solution.[212] p-Amino- and p-acetamido-benzenesulfonhydrazides gave only the sulfinates under these conditions.

N'-Alkyl or aryl sulfonhydrazides decompose readily when warmed with base to form sulfinate and hydrocarbon, which presumably arises from an intermediate alkyl diazene or azamine by extrusion of nitrogen[213-215] (Eq. 3-98). In an example in

$$\underset{\overset{|}{\text{Et}}}{\overset{\overset{|}{\text{Me}}}{\text{Ph}-\text{C}-\text{NHNHSO}_2\text{Ar}}} \xrightarrow[t\text{-BuOH}]{\text{OH}^-} \underset{\overset{|}{\text{Et}}}{\overset{\overset{|}{\text{Me}}}{\text{Ph}-\text{C}-\text{H}}} + \text{N}_2 + \text{ArSO}_2^- \qquad (3\text{-}98)$$

which the alkyl group is attached at a chiral center, an alkane was obtained with 80% retention of configuration. This stereospecificity has been interpreted to result from frontal electrophilic attack by H^+ on the carbon of the alkyldiazene or its anion,[216] but it might be an intramolecular reaction of the isomeric azamine in which geminal R and H combine synchronously with the departure of N_2. Formation of alkanes from N'-alkyl sulfonhydrazides is more commonly encountered as the final stage in the reduction of tosylhydrazones with complex hydrides.[217]

The behavior of N',N'-disubstituted sulfonhydrazides varies with the structure of the substituent and the conditions, but the initial step in all cases appears to be the same, formation of a disubstituted azamine. The sodium salt of N',N'-diethylbenzenesulfonhydrazide, for example, when decomposed in the aprotic solvent tetraethylene glycol dimethyl ether, produces tetraethyltetrazene[205] (Eq. 3-99),

$$\text{PhSO}_2\bar{\text{N}}\text{NEt}_2\ \text{Na}^+ \quad \begin{array}{l} \nearrow \overset{\substack{110\text{-}120°C \\ \text{Aprotic}}}{\xrightarrow{\hspace{1.2cm}}} \underset{80\%}{\text{Et}_2\text{N}-\text{N}=\text{N}-\text{NEt}_2} \\[1em] \searrow \underset{\text{Diethylene glycol}}{\overset{25°C,\ 2\ h}{\xrightarrow{\hspace{1.2cm}}}} \underset{70\%}{\text{EtNHN}=\text{CHCH}_3} \end{array} \qquad (3\text{-}99)$$

which might arise by dimerization of the azamine or attack by azamine on undecomposed sulfonhydrazide (see Chap. 5). Decomposition at a lower temperature and in a protic solvent gives a quite different result: acetaldehyde ethylhydrazone, an isomer of the azamine, is formed. Two reasonable routes to this product, rearrangement to an intermediate azoalkane or diaziridine, were excluded when it was determined that these compounds do not isomerize to the hydrazone under the experimental conditions. It is therefore concluded that the conversion begins with tautomerization of the azamine to an azomethine imide (Eq. 3-100).

$$\text{Et}_2\overset{+}{\text{N}}=\text{N}^- \longrightarrow \underset{\overset{|}{\text{Et}}}{\overset{\text{CH}_3\text{CH}}{\diagdown}}\text{N}^{\pm}-\text{NH}^- \longrightarrow \text{CH}_3\text{CH}=\text{NNHEt} \qquad (3\text{-}100)$$

N',N'-Diphenylbenzenesulfonhydrazide, the structure of which prevents tautomerism at any stage, decomposes to form diphenylamine and not azobenzene; tetraphenyltetrazene is believed to be an intermediate, which fragments into diphenylamino radicals, which abstract hydrogen from the medium[205] (Eq. 3-101). N',N'-Dibenzylbenzenesulfonhydrazide exemplifies still another mode of decomposition and forms bibenzyl[218,219] (Eq. 3-102), a type of reaction associated with substituents that can form radicals easily. Another dialkylsulfonhydrazide,

$$PhSO_2\bar{N}NPh_2 \ Na^+ \xrightarrow[\text{diethylene glycol}]{240\,°C} [Ph_2\overset{+}{N}=N^-] \longrightarrow$$

$$[Ph_2N-N=N-NPh_2] \longrightarrow N_2 + Ph_2N\cdot \xrightarrow{[H]} Ph_2NH \qquad (3\text{-}101)$$

$$PhSO_2\bar{N}N(CH_2Ph)_2 \ Na^+ \longrightarrow PhSO_2^-Na^+ + (PhCH_2)_2\overset{+}{N}-N^- \longrightarrow$$
$$N_2 + PhCH_2CH_2Ph \qquad (3\text{-}102)$$

N-(benzenesulfonamido)-pyrrolidine, exhibits yet a different behavior and fragments to two equivalents of ethylene, rather than cyclobutane or tetrahydropyridazine, when heated with base.

In general, there is a parallel between base-catalyzed fragmentation of sulfonhydrazides and oxidation of *unsym*-disubstituted hydrazines, a reaction that also involves azamines (see Chap. 1).

The most important base-catalyzed reactions of sulfonhydrazides are those of the alkylidene derivatives (sulfonylhydrazones) that have no substituent on the amide nitrogen, the Bamford-Stevens reactions[220] and variants thereof. The primary process is apparently the same as that in the foregoing reactions, elimination of a sulfinate ion to leave an azamine, but an alkylidene azamine is simply a diazoalkane. If the temperature and the proton-donating character of the medium are not too high, diazoalkanes can indeed be produced in preparative yield[221,222] (see Chap. 5) (Eq. 3-103). Heating the lithium salt under vacuum in the complete absence of

$$\underset{R}{\overset{Ar}{>}}C=NNHSO_2To \xrightarrow[\text{pyridine}]{NaOCH_3,\ 55\text{-}80\,°C} \underset{R}{\overset{Ar}{>}}\underset{20\text{-}70\%}{C=\overset{+}{N}=N^-} \qquad (3\text{-}103)$$

protic substances[223,224] is advantageous in converting tosylhydrazones to diazoalkanes that are sensitive to decomposition by even very feeble proton donors at the required temperatures and do not survive the conditions shown in Equation 3-103.

When conditions are not mild enough to favor survival of diazoalkanes, the products are usually alkenes, cyclopropanes, or a mixture of them. They arise from alkanediazonium ions, carbenium ions, and carbenes, and the alkenes may have a rearranged carbon skeleton as a consequence of Wagner-Meerwein rearrangement. The sometimes enigmatic dependence of the products on experimental conditions has recently been reviewed[225] with thoroughness and will only be outlined here.

The original conditions used by Bamford and Stevens were rather drastic: treatment with the sodium salt of ethylene glycol in boiling glycol. Under these and analogous conditions, alkenes are produced,[19,226] and Wagner-Meerwein rearrangement can be expected if structure favors it. Pinacolone tosylhydrazone, for example, is converted to tetramethylethylene rather than *t*-butylethylene[220] (Eq. 3-104), but

$$(CH_3)_3CC\underset{CH_3}{\overset{NNHSO_2To}{<}}\Big\langle \begin{array}{l} \xrightarrow[\text{boiling glycol}]{NaOCH_2CH_2OH} (CH_3)_2C=C(CH_3)_2 \\[2mm] \xrightarrow[\text{diethylcarbitol}]{NaOMe} (CH_3)_3CCH=CH_2 + (CH_3)_2C\underset{CH_2}{\overset{CHCH_3}{<}} \end{array} \qquad (3\text{-}104)$$

when the conditions are essentially aprotic, as with use of sodium methoxide in an ether type of solvent, the alkene with unrearranged skeleton is produced, along with a cyclopropane.[227] Although not all sulfonylhydrazones give different products under protic and aprotic conditions, aprotic (or very feebly protic) conditions generally·give alkenes with unrearranged skeletons, perhaps derived directly from an alkanediazonium ion, accompanied by products derived from a carbene, usually cyclopropanes (Eq. 3-105). Suitably active reagents may intercept any of these spe-

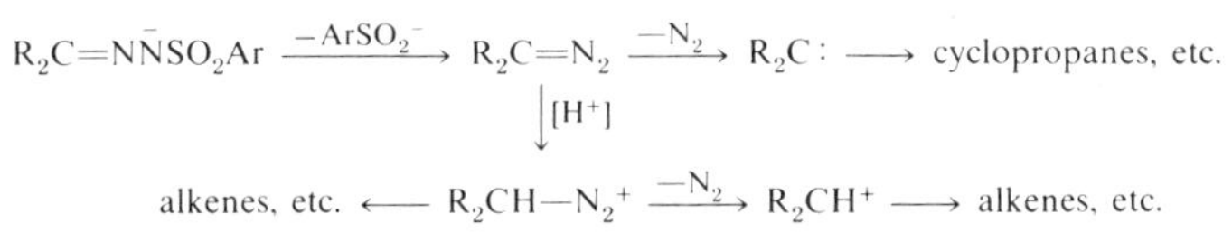

$$R_2C\!\!=\!\!N\bar{N}SO_2Ar \xrightarrow{-ArSO_2^-} R_2C\!\!=\!\!N_2 \xrightarrow{-N_2} R_2C\!: \longrightarrow \text{cyclopropanes, etc.}$$

$$\Big\downarrow [H^+]$$

(3-105)

$$\text{alkenes, etc.} \longleftarrow R_2CH\!\!-\!\!N_2{}^+ \xrightarrow{-N_2} R_2CH^+ \longrightarrow \text{alkenes, etc.}$$

cies to form other types of products. Furthermore, some sets of conditions may give rise to products from multiple competing paths. Pinacolone tosylhydrazone, for example, forms both tetramethylethylene and *t*-butylethylene, along with 2,3-dimethylbutene-1 and 1,1,2-trimethylcyclopropane, when heated with sodium methoxide in diethylene glycol.[227]

Extremely strong bases, especially alkyllithium reagents, but to some extent also sodium amide and sodium hydride, can convert arenesulfonylhydrazones having an α-hydrogen to a dianion, which is usually stable at $-78\,^\circ$C but at room temperature smoothly fragments to the lithium salt of a vinyldiazene and then, by loss of nitrogen, to a vinyllithium[224, 228] (Eq. 3-106). The most significant feature of this reac-

$$\begin{array}{c} i\text{-Pr} \\ \diagdown \\ \diagup \\ \text{Me} \end{array} C\!\!=\!\!NNHSO_2Ph \xrightarrow[\substack{\text{Decalin} \\ (2)\ H_2O}]{(1)\ BuLi,\ 70\text{--}110\,^\circ C} i\text{-PrCH}\!\!=\!\!CH_2 \text{ only}$$

(3-106)

tion is that the alkene that results when the reaction mixture is hydrolyzed is consistently found to be the least substituted alkene of unrearranged skeleton, as shown. The product may thus be different from those of either the protic or aprotic Bamford-Stevens reaction. Furthermore. the fragmentation takes place much more easily than that of the monoanions.

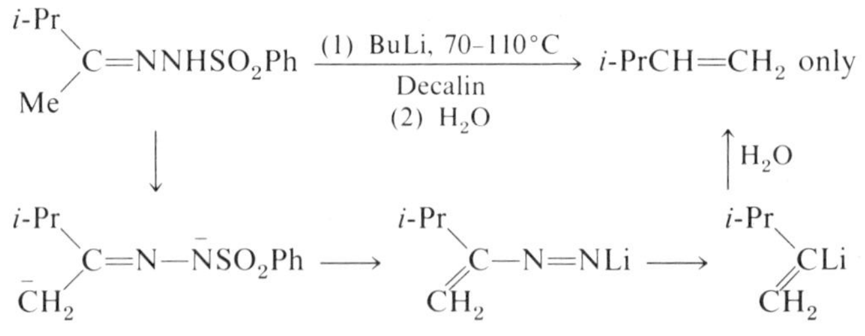

There is a competing reaction when sulfonylhydrazones are treated with excess alkyllithium, in which the hydrazono group is replaced by H and R; indanone tosylhydrazone, for example, is converted by methyllithium largely to indene, but also to some 1-methylindane[228] (Eq. 3-107). With diaryl ketone derivatives, which

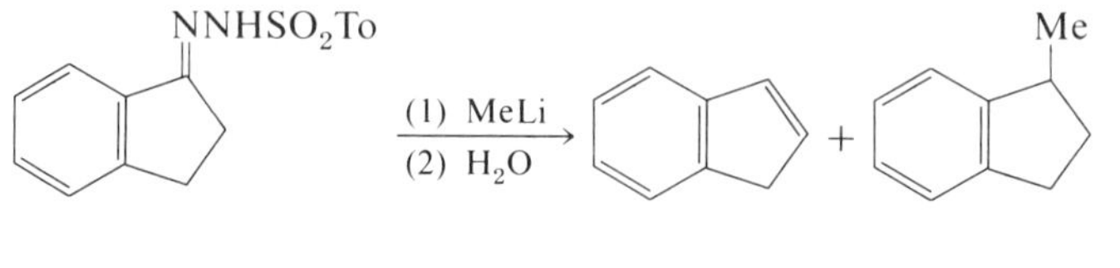

(3-107)

cannot form an alkene, the alkylated product is formed exclusively.[229] The reaction is believed to take place via a diazo anion; the hydrocarbon is released upon subsequent protonation (Eq. 3-108). Primary, secondary, and tertiary alkyllithium rea-

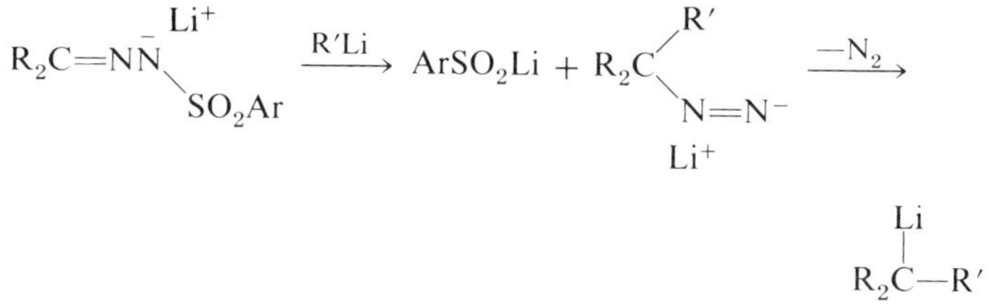

$$R_2C=N\overset{-}{N}\overset{Li^+}{\underset{SO_2Ar}{\diagdown}} \xrightarrow{R'Li} ArSO_2Li + R_2C\overset{R'}{\underset{\underset{Li^+}{N=N^-}}{\diagup}} \xrightarrow{-N_2}$$

$$R_2\overset{\overset{Li}{|}}{C}-R' \qquad (3\text{-}108)$$

gents have been used with success, but aryllithium reagents appear to be unreactive.[230] Yields of alkylated hydrocarbon may become high when a very large excess (e.g., eightfold) of alkyllithium is used.

This reaction has been adapted to an olefin synthesis that provides an alternative to the Wittig reaction. Alkyllithium reagents having an anionic leaving group on the α-carbon react with the lithium salts of aldehyde tosylhydrazones and the lithium salt of the leaving group is eliminated[231] (Eq. 3-109).

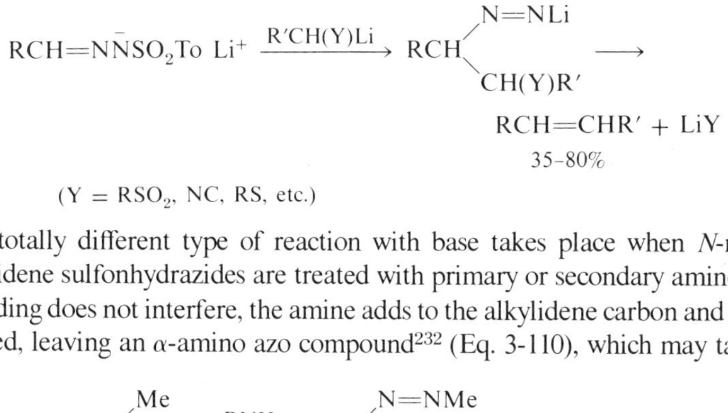

$$RCH=\overset{-}{N}NSO_2To\ Li^+ \xrightarrow{R'CH(Y)Li} RCH\overset{N=NLi}{\underset{CH(Y)R'}{\diagup}} \longrightarrow$$

$$RCH=CHR' + LiY \qquad (3\text{-}109)$$
$$35\text{-}80\%$$

$$(Y = RSO_2,\ NC,\ RS,\ etc.)$$

A totally different type of reaction with base takes place when N-methyl-N'-alkylidene sulfonhydrazides are treated with primary or secondary amines. If steric crowding does not interfere, the amine adds to the alkylidene carbon and sulfinate is ejected, leaving an α-amino azo compound[232] (Eq. 3-110), which may tautomerize

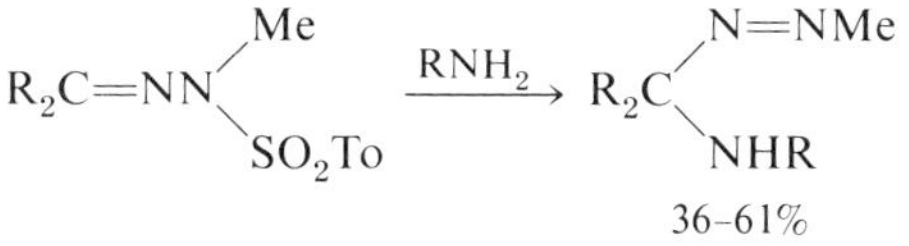

$$R_2C=NN\overset{Me}{\underset{SO_2To}{\diagdown}} \xrightarrow{RNH_2} R_2C\overset{N=NMe}{\underset{NHR}{\diagup}} \qquad (3\text{-}110)$$
$$36\text{-}61\%$$

to an amidrazone if the hydrazone is derived from an aldehyde. Little is known about the reaction of N-alkyl-N'-alkylidene sulfonhydrazides with stronger bases, but N-methyl-N'-benzylidenetoluenesulfonhydrazide cleaves to benzonitrile and N-methyltoluenesulfonamide when treated with t-butoxide[233] (Eq. 3-111).

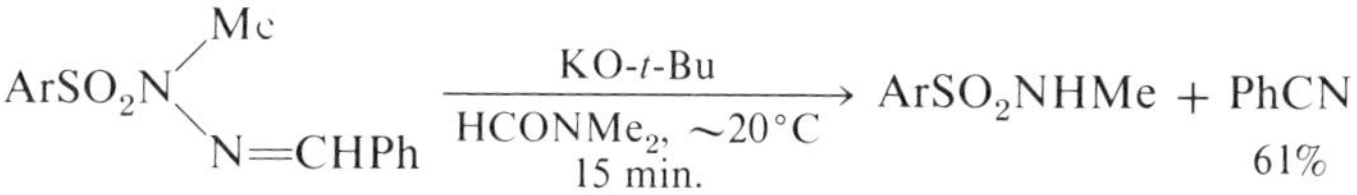

$$ArSO_2N\overset{Me}{\underset{N=CHPh}{\diagdown}} \xrightarrow[\underset{15\ min.}{HCONMe_2,\ \sim 20°C}]{KO\text{-}t\text{-}Bu} ArSO_2NHMe + PhCN \qquad (3\text{-}111)$$
$$61\%$$

N'-Acyl sulfonhydrazides are decomposed by bases to form aldehydes in moderate yields[215] (Eq. 3-112). This is the McFadyen-Stevens aldehyde synthesis.[115] Al-

$$RCONHNHSO_2Ar \xrightarrow[\substack{\text{Glycol} \\ 150-160\,°C}]{K_2CO_3} ArSO_2^- + [RCON{=}NH] \longrightarrow$$

$$RCHO + N_2 \qquad (3\text{-}112)$$

though it is rarely the method of choice, it has the valuable feature of providing a convenient route to 1-deuterio aldehydes if the reaction is carried out in D_2O. The reaction kinetics[234] are first-order in the acylsulfonhydrazide anion, and the rates for substituted benzoyl derivatives follow the Hammett relation with a reaction constant ρ of -1.38. Whereas the reaction succeeds with derivatives of benzoic and aromatic heterocyclic acids, it fails with aliphatic examples (although cyclopropane-carbaldehyde has been obtained in 16% yield[235]).

N-Allyl-N'-benzoylbenzenesulfonhydrazide is a curious case, for the overall effect of treating it with base is rearrangement to a sulfone[236] (Eq. 3-113). It is be-

$$PhCONHN{\underset{\displaystyle CH_2CH{=}CH_2}{\overset{\displaystyle SO_2Ph}{\diagup}}} \xrightarrow[\text{Glycol, 2 min.}]{Na_2CO_3,\ 130\,°C} PhCON{=}NCH_2CH{=}CH_2 + PhSO_2^-$$

$$(3\text{-}113)$$

$$\downarrow$$

$$PhCONHN{=}CHCH_2CH_2SO_2Ph \xleftarrow{PhSO_2^-} PhCONHN{=}CHCH{=}CH_2$$

lieved to begin with the usual elimination of sulfinate, followed by tautomerization to acrolein benzoylhydrazone, to which the eliminated benzenesulfinate re-adds.

Alkylation.   There has been relatively little interest in alkylation of sulfonhydrazides, probably because the expected products can be prepared easily from the alkylated hydrazine and sulfonyl chloride. N',N'-Dimethyltoluenesulfonhydrazide has been quaternized at N' by several primary alkyl halides (60–80%).[237,238] Benzyl bromide alkylates the free hydrazide or its lithium or sodium salt alike only at N'. Benzenesulfonhydrazide becomes picrylated at the N' position by picryl chloride.[215]

Tosylhydrazones have been alkylated at the amide nitrogen in high yields by alkyl halides and alkali under-phase-transfer conditions[239] (Eq. 3-114). This process has

$$R_2C{=}NNHSO_2To + R'X \xrightarrow[CH_2Cl_2]{15\%\ aq.\ NaOH} R_2C{=}NN{\underset{\displaystyle SO_2To}{\overset{\displaystyle R'}{\diagup}}} \qquad (3\text{-}114)$$

preparative value, for the alternative route to the products, reaction of an N-alkyl sulfonhydrazide with a carbonyl compound, can be beset with competitive decomposition of the sulfonhydrazide when unreactive ketones are involved. Similar alkylation has been reported as a side reaction in the Bamford-Stevens reaction, presumably as a result of capture of an intermediate alkylium or alkanediazonium ion by sulfonylhydrazone anion.[240] Some N'-acyl sulfonhydrazides have been alkylated at the sulfonamide nitrogen with concomitant ejection of sulfinate; the azo compound presumably produced tautomerizes to an acylhydrazone[241] (Eq. 3-115).

$$t\text{-BuOCONHNHSO}_2\text{CF}_3 \xrightarrow[\text{MeCN}]{\text{K}_2\text{CO}_3, \text{R}_2\text{CHX}} [t\text{-BuOCONHN} \overset{\text{CHR}_2}{\underset{\text{SO}_2\text{CF}_3}{\diagdown}}] \longrightarrow$$

$$[t\text{-BuOCON}{=}\text{NCHR}_2] \longrightarrow t\text{-BuOCONHN}{=}\text{CR}_2 \xrightarrow{[\text{H}_2\text{O}]}$$

$$\text{R}_2\text{C}{=}\text{O} \qquad (3\text{-}115)$$

When this sequence is combined with hydrolysis of the product, it accomplishes the conversion of an alkyl halide to an aldehyde (or ketone) in good yield (primary and activated secondary halides).

Ketones and aldehydes. In the absence of severe steric hindrance, ketones and aldehydes react readily with sulfonhydrazides to form sulfonylhydrazones[226] (Eq. 3-116). Reaction may take place without a catalyst, but warming in the presence of

$$\text{RSO}_2\text{NHNH}_2 + \text{R}_2'\text{C}{=}\text{O} \longrightarrow \text{RSO}_2\text{NHN}{=}\text{CR}_2' + \text{H}_2\text{O} \qquad (3\text{-}116)$$

acid (acetic, hydrochloric) is often used. Many examples are to be found in papers dealing with the Bamford-Stevens reaction.[225]

Although sulfonylhydrazones are normally very stable, certain α-diketones do not form isolable tosylhydrazones, for the monohydrazones fragment to α-diazo ketone as fast as they are formed.[242,243] This has been observed, for example, with phenanthrene-9,10-quinone.

Tosylhydrazones undergo acid-catalyzed exchange with carbonyl compounds, a reaction that has been adapted to regeneration of carbonyl compounds from their tosylhydrazones[244] (Eq. 3-117).

$$\text{R}_2\text{C}{=}\text{NNHSO}_2\text{To} \xrightarrow[\text{Et}_2\text{O} \cdot \text{BF}_3, \text{6-18 h}]{\text{aq. acetone}}$$

$$\text{R}_2\text{C}{=}\text{O} + \text{Me}_2\text{C}{=}\text{NNHSO}_2\text{To} \qquad (3\text{-}117)$$
$$80\text{-}97\%$$

Acylation. Sulfonhydrazides are readily acylated at the N' position by acid chlorides and anhydrides. Benzenesulfonhydrazide, for example, is converted to its N'-acetyl derivative by acetic anhydride,[15] to N,N'-dibenzenesulfonylhydrazine by benzenesulfonyl chloride,[15] and to N-benzenesulfonyl-N'-benzoylhydrazine by treatment with benzoyl chloride and pyridine[215] (Eq. 3-118). Acylation at the amide

$$\text{PhSO}_2\text{NHNH}_2 \xrightarrow[\text{C}_5\text{H}_5\text{N}]{\text{PhCOCl}}$$

$$\text{PhSO}_2\text{NHNHCOPh} \; (+ \; \text{PhSO}_2\text{NHN(COPh)}_2) \qquad (3\text{-}118)$$

nitrogen is apparently more difficult, for N',N'-dimethyl-p-toluenesulfonhydrazide gave rise to its N-benzoyl derivative in only 10.5% yield[55] (Eq. 3-119). The sodium salts of some tosylhydrazones, however, have been converted in better yields to ditosylhydrazones by treatment with tosyl chloride in dimethylformamide.[245]

$$ToSO_2NHNMe_2 + PhCOCl \xrightarrow[Et_2O]{Et_3N} \begin{array}{c} PhCO \\ \diagdown \\ \diagup \quad NNMe_2 \\ ToSO_2 \end{array} \qquad (3\text{-}119)$$

10.5%

Nitrosation of sulfonhydrazides takes place readily with nitrous acid in dilute acid solution; the reaction is an excellent method for preparing sulfonyl azides[246] (Eq. 3-120).

$$ToSO_2NHNH_2 \xrightarrow[\text{aq. AcOH}]{NaNO_2} ToSO_2N_3 \qquad (3\text{-}120)$$

$>95\%$

Oxidation. Oxidation of unsubstituted sulfonhydrazides has not been systematically studied, but it appears that the initial step is conversion to a sulfonyldiazene. If the oxidation is accomplished with bromine, the final product is the sulfonyl bromide, in good yield[247] (Eq. 3-121). Sulfur monochloride oxidizes sulfonhydrazides

$$ArSO_2NHNH_2 \xrightarrow[\text{aq. HCl}]{Br_2} ArSO_2Br + N_2 \qquad (3\text{-}121)$$

to sulfonyl chlorides. Ferric salts are said to produce sulfinates (as their iron salts) and nitrogen.[248] Iodine causes liberation of nitrogen quantitatively, but the other product is mostly diaryl disulfide, and the role of the iodine may be more catalytic than oxidative.[15]

N'-Aryl sulfonhydrazides are oxidized by ferricyanide to isolable but explosive azo compounds, which can be decomposed to sulfones[249] (Eq. 3-122). N'-Alkyl

$$ArSO_2NHNHAr' \xrightarrow{K_3Fe(CN)_6} ArSO_2N{=}NAr' \xrightarrow[\text{water}]{Boiling}$$

$$ArSO_2Ar' \qquad (3\text{-}122)$$

sulfonhydrazides are apparently oxidized to similar azo compounds by peroxide, but the products actually obtained are alkyl hydroperoxides, in excellent yields[250] (Eq. 3-123). N,N'-Dibenzenesulfonylhydrazine has been oxidized to the corresponding azodisulfone with bromine.[251]

$$ArSO_2NHNHR \xrightarrow[\text{Tetrahydrofuran}]{30\% \ H_2O_2, \ Na_2O_2} ROOH \qquad (3\text{-}123)$$

87–95%

Somewhat more attention has been paid to oxidation of sulfonylhydrazones, which have been oxidized to the parent ketones (or aldehydes) by N-bromosuccinimide (up to 74% yield),[252] sodium hypochlorite (60–85%),[253] hydrogen peroxide[254] (Eq. 3-124), and cobalt trifluoride (17–52%).[255] Oxidation with N-bromosuccinimide

$$RR'C{=}NNHSO_2To \xrightarrow[\text{MeOH or dioxane, 0°C}]{27.5\% \ H_2O_2, \ K_2CO_3} RR'C{=}O \qquad (3\text{-}124)$$

76–96%

(R′ = alkyl or H)

in methanol and acetone produces dimethyl ketals rather than the ketones, which are quickly generated when the product is chromatographed on silica gel. The toluenesulfonyl group appears as the acid bromide or methyl ester.[252]

Reduction. Interest in reduction of sulfonhydrazides has been concerned mostly with their hydrazones. Sodium borohydride,[256-258] cyanoborohydride,[259,260] and diborane[217] reduce the hydrazones to N'-alkyl sulfonhydrazides; under suitable conditions, these may be cleaved in situ to alkane and sulfinate[257,261] (Eq. 3-125),

$$R_2C\text{=}NNHAr \xrightarrow{NaBH_3CN} R_2CHNHNHSO_2Ar \xrightarrow[80°C]{Base}$$

$$R_2CH_2 + ArSO_2^- \qquad (3\text{-}125)$$

thus accomplishing an alternative to the Wolff-Kishner reduction. The alkyl hydrazides can be isolated in particularly high yields (87–96%) by reduction of the mercuric salts of the hydrazones.[262] When lithium aluminum deuteride, necessarily used in aprotic medium, is used, followed by hydrolysis with ordinary water, the monodeuterio hydrocarbon R_2CHD is obtained.[263] Reduction with $LiAlH_4$ is generally not so satisfactory as with borohydride reagents, because of a competing reaction caused by the strongly basic conditions, leading to alkene (the least-substituted isomer, analogous to Eq. 3-106).[225]

Reduction of tosylhydrazones of α,β-unsaturated aldehydes and ketones is always accompanied by migration of the C=C bond toward the hydrazone carbon, even when a thermodynamically less stable alkene is thereby produced[258] (Eq. 3-126). It has been proposed that this circumstance is a result of intramolecular delivery of hydrogen to the β position by an intermediate diazene.

$$PhCH\text{=}CH\text{—}CH\text{=}NNHSO_2To \xrightarrow[gl.\ AcOH]{NaBH_4} PhCH_2CH\text{=}CH_2 \qquad (3\text{-}126)$$
$$42\%$$

Raney nickel cleaves the N—N bond of sulfonylhydrazones to form sulfonamides and primary amines[264] (Eq. 3-127). Unsubstituted sulfonhydrazides are also reduced to sulfonamides by this means.

$$ArSO_2NHN\text{=}CR_2 \xrightarrow{H_2,\ Raney\ Ni} ArSO_2NH_2 + R_2CHNH_2 \qquad (3\text{-}127)$$

Thionyl Hydrazines

As a class, thionyl hydrazines have been studied relatively little.[265,266] The acidic hydrogen of thionylphenylhydrazine allows sodium or magnesium salts to be prepared. Treatment of such salts with benzoyl chloride brings about deep-seated decomposition. Alkylation probably takes place at sulfur, for treatment with benzyl chloride produces phenyl benzyl sulfoxide, a reasonable product of decomposition of a phenylazo sulfoxide (Eq. 3-128) (thionyl-N-phenyl-N-benzylhydrazine, prepared in another way, is stable). Grignard reagents merely form salts with monosubstituted thionyl hydrazines, whereas disubstituted ones are unreactive. Oxidation

$$PhNHN{=}SO \xrightarrow{RMgX} (PhNNSO)_2Mg \xrightarrow{PhCH_2Cl}$$

$$[PhN{=}NS\overset{\displaystyle O}{\underset{\displaystyle CH_2Ph}{\diagup}}] \longrightarrow N_2 + PhS\overset{\displaystyle O}{\underset{\displaystyle CH_2Ph}{\diagup}} \qquad (3\text{-}128)$$

of thionylphenylhydrazine gives benzene and diphenyl sulfone, probably by initial attack at sulfur to form an azosulfinic acid (or tautomer) (Eq. 3-129). Reduction

$$PhNHN{=}SO + H_2O_2 \longrightarrow PhSO_2Ph + PhH \qquad (3\text{-}129)$$

with sodium under warm toluene follows a somewhat analogous path, thiophenols being formed, presumably through an azothiol (or tautomer)[267] (Eq. 3-130).

$$ArNHN{=}SO \xrightarrow{Na} [ArN{=}NSH] \longrightarrow ArSH + N_2 \qquad (3\text{-}130)$$

PREPARATIVE METHODS

Carbohydrazides and Derivatives[5]

The most widely used method to prepare unsubstituted hydrazides is the reaction of esters with hydrazine (or its hydrate)[119,268,269] (Eq. 3-131). Anhydrous hydrazine

$$RCOOEt + NH_2NH_2 \longrightarrow RCONHNH_2 + EtOH \qquad (3\text{-}131)$$

reacts much faster than its hydrate and requires less heating[270] (simple, unhindered esters may react at room temperature, and in some cases the reaction may be spontaneous and exothermic even with the hydrate). Esters seldom produce significant amounts of diacylhydrazine, but less reactive esters may require inconveniently long reaction times or undesirably severe conditions. Ethyl cinnamate, for example, reacts with hydrazine only when heated sufficiently strongly that cyclization to a pyrazolidone ensues; the hydrazide can be obtained by the use of a more active acylating agent[271] (Eq. 3-132). The reaction of esters can be accelerated with sodium methoxide.[272] Active esters, such as cyanomethyl[273,274] and p-nitrobenzyl,[275,276] can be used in difficult cases.

$$PhCH{=}CHCOOCOOEt + NH_2NH_2 \longrightarrow$$

$$PhCH{=}CHCONHNH_2 \qquad (3\text{-}132)$$

Anhydrides, acid chlorides,[277] acid azides, and acyl imidazoles[278] all react with hydrazine to give hydrazides faster then esters do. However, acid chlorides are so reactive that it is difficult to stop the reaction short of diacylation. In general, this group of acylating agents is a second choice, to be resorted to when the ester is unavailable or unreactive.

Amides, including peptides, can be converted to hydrazides by reaction with hydrazine,[279-281] but the reaction is somewhat slow.[282] Sodium hydrazide reacts more readily to form hydrazides with N,N-disubstituted amides, but other amides simply form the amide anion.[283]

It is also possible to prepare hydrazides directly from carboxylic acids by heating their hydrazine salts.[75,284] Yields may be high with unhindered acids free of other reactive centers. The most effective procedure seems to be to use a butanol solvent and activated alumina as a catalyst and to distill the water out[285] (Eq. 3-133). Alter-

$$\text{Me}_2\text{CHCO}_2\text{H} + \text{NH}_2\text{NH}_2 \cdot \text{H}_2\text{O} \xrightarrow[\text{BuOH, boil}]{\text{Al}_2\text{O}_3}$$

$$\text{Me}_2\text{CHCONHNH}_2 \qquad (3\text{-}133)$$
$$82\%$$

natively, dicyclohexylcarbodiimide can be used with a carboxylic acid (yields up to 67%).[272]

In all the foregoing methods, consideration should be given to the fact that hydrazine frequently is contaminated with small amounts of ammonia, which may compete for the acylating agent and generate amide impurities.[286]

Diacyl, triacyl, and tetraacyl hydrazines are prepared by further acylation of hydrazides, usually with acid chlorides or anhydrides[13] (Eqs. 3-134, 3-135), although

$$\text{HCONHNH}_2 + \text{Ac}_2\text{O} \longrightarrow \text{HCONHNHAc} \qquad (3\text{-}134)$$

$$\text{AcNHNH}_2 + \text{Ac}_2\text{O} \text{ (excess)} \longrightarrow \text{Ac}_2\text{N}-\text{NAc}_2 \qquad (3\text{-}135)$$

in the especially favorable case of succinic acid, a cyclic tetraacyl hydrazine has been obtained by heating the hydrazine salt.[75] A route to unsym-diacyl hydrazines makes use of a protecting group, by treating t-butoxycarbonylhydrazine with succinic anhydride and then removing the protecting group with dry hydrogen chloride, leaving N,N-succinoylhydrazine (N-aminosuccinimide)[287] (Eq. 3-136). This

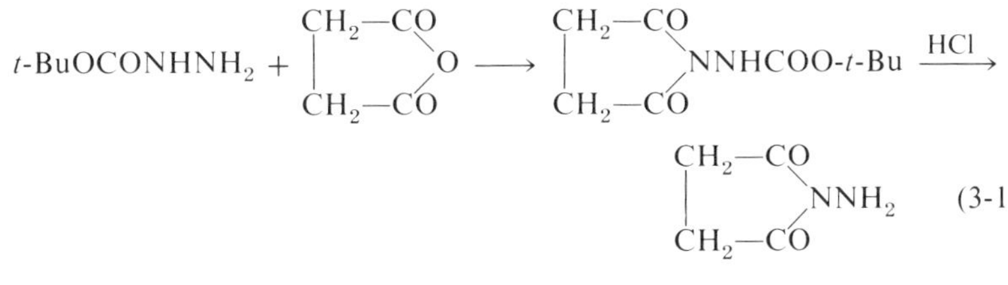

$$(3\text{-}136)$$

method does not appear to have been applied to noncyclic examples, with which difficulties can be anticipated, since there is reason to suspect that the desired products might rearrange spontaneously to the sym isomers during the process.

Substituted hydrazides are generally prepared by acylating the corresponding substituted hydrazine. The characteristics of acylation reactions of hydrazines were discussed in Chapter 1 and will not be repeated here. When there are two possible sites for acylation, as in the case of monosubstituted hydrazines, steric effects appear to dominate acylation by esters, whereas electronic effects appear to be more important with acid chlorides, but mixtures of the two possible isomers can be expected. Acylation can be directed exclusively to the substituted nitrogen by use of an alkylidene-blocking agent; acetone or benzaldehyde hydrazones are most commonly used. The block can be removed after acylation by treatment with hydrochloric acid in ethanol[71,288,289] (Eq. 3-137). This procedure is especially useful when the N-substituent is highly branched (for example, t-Bu) or aryl.

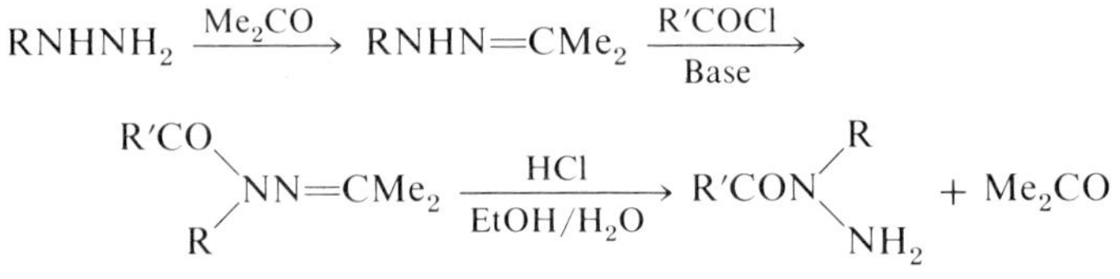

$$RNHNH_2 \xrightarrow{Me_2CO} RNHN=CMe_2 \xrightarrow[\text{Base}]{R'COCl}$$

$$\underset{R}{\overset{R'CO}{>}}NN=CMe_2 \xrightarrow[EtOH/H_2O]{HCl} R'CON\underset{NH_2}{\overset{R}{<}} + Me_2CO \qquad (3\text{-}137)$$

Attempts to direct acylation to the N position of phenylhydrazine using its sodium salt, Na^+ $PhNNH_2$, have been fruitless; the product is nearly or entirely the N,N'-diacyl derivative.[290]

Substituted hydrazides can also be prepared by alkylation of simpler hydrazides or their metal salts (see Eqs. 3-13 through 3-16).

There are a number of routes to hydrazides involving ring-opening of heterocyclic compounds,[5] but for most of them, their utility has yet to pass the test of time. One that has the advantage of producing N-aryl hydrazides is the hydrolysis of sydnones, which are readily prepared from N-substituted glycines[105] (Eq. 3-138). Another example is the hydrolysis of acylated diaziridines.[291]

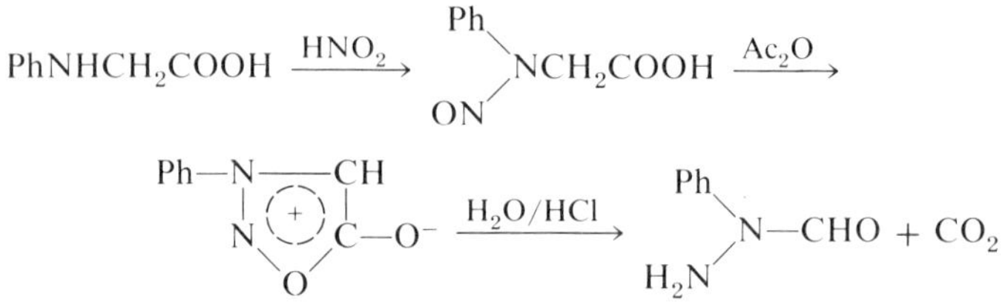

$$PhNHCH_2COOH \xrightarrow{HNO_2} \underset{ON}{\overset{Ph}{>}}NCH_2COOH \xrightarrow{Ac_2O}$$

$$\begin{matrix} Ph-N{-\!-\!-}CH \\ | \quad (+) \quad | \\ N \quad \underset{O}{\smile} \quad C{-}O^- \end{matrix} \xrightarrow{H_2O/HCl} \underset{H_2N}{\overset{Ph}{>}}N{-}CHO + CO_2 \qquad (3\text{-}138)$$

Reductive methods appear to be significant only with N-nitroso and N-nitro amides,[72] particularly ureas, which are of considerable importance in the preparation of semicarbazides (carbamohydrazides)[124,292,293] (Eq. 3-139). Reduction of acylazo

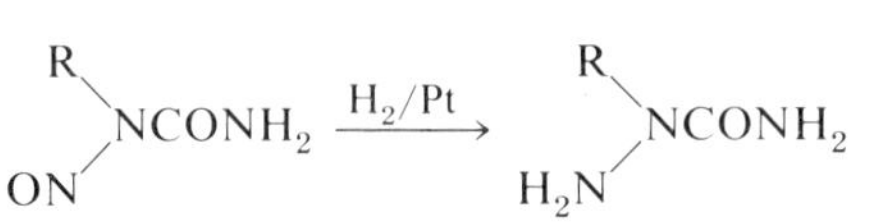

$$\underset{ON}{\overset{R}{>}}NCONH_2 \xrightarrow{H_2/Pt} \underset{H_2N}{\overset{R}{>}}NCONH_2 \qquad (3\text{-}139)$$

compounds also gives hydrazides, but since the azo compounds must usually be prepared from the hydrazides in the first place, the process is not of preparative value.

Oxidative methods are not of preparative importance, although hydrazides can be made from aldehyde hydrazones by reaction with oxygen,[294] halogen,[168,169] or peroxy acids,[295] followed by hydrolysis if required (Eq. 3-140).

Some trialkyl hydrazides have been made by a carbonylation reaction, in which formyl, α-hydroxy, or α-keto acylhydrazines may be formed[161] (Eq. 3-141).

A variety of ester-hydrazides of carbonic acid can be made through electrophilic substitution by azoformic esters on alkenes or arenes.[296-298] The products are N,N'-di(alkoxycarbonyl)-hydrazines bearing an allylic or aryl substituent, but in some

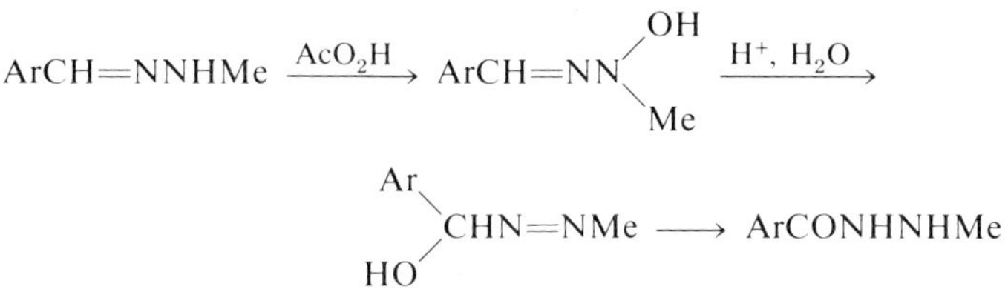

$$ArCH=NNHMe \xrightarrow{AcO_2H} ArCH=NN\begin{smallmatrix}OH\\ \diagup \\ \diagdown \\ Me\end{smallmatrix} \xrightarrow{H^+, H_2O}$$

$$\begin{smallmatrix}Ar\\ \diagdown\\ \quad CHN=NMe\\ \diagup\\ HO\end{smallmatrix} \longrightarrow ArCONHNHMe \qquad (3\text{-}140)$$

$$Pr_2NNPr\ Li^+ \xrightarrow[-75°C]{CO} Pr_2NNPr \atop \underset{OCLi}{|} \begin{cases} \xrightarrow{H_2O} Pr_2NN\!\!\begin{smallmatrix}Pr\\ \diagup\\ \diagdown\\ CHO\end{smallmatrix} \\[2em] \xrightarrow[(2)\ H_2O]{(1)\ R'CHO} Pr_2N\!\!\begin{smallmatrix}Pr\\ \diagup\\ \diagdown\\ COCHOHR'\end{smallmatrix} \\ \qquad\qquad\qquad 60\text{-}85\% \\[2em] \xrightarrow{RCOOEt} Pr_2NN\!\!\begin{smallmatrix}Pr\\ \diagup\\ \diagdown\\ C-CR\\ \|\ \ \|\\ O\ \ O\end{smallmatrix} \\ \qquad\qquad ca.\ 35\% \end{cases} \qquad (3\text{-}141)$$

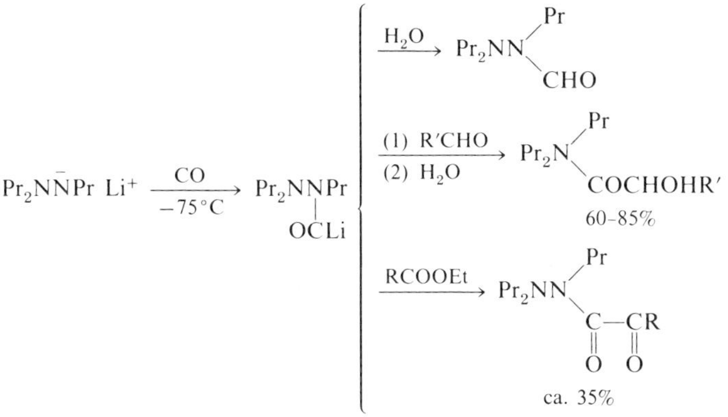

instances, one of the ester groups can be reduced to a methyl group (see Eqs. 4-103 through 4-109). Active-methylene compounds, such as malonic ester and many ketones, also add to azoformic esters to produce ester-hydrazides[299] (see Eq. 4-103).

There is a group of methods in which the N—N bond is formed in the final step. Although these reactions are rarely used for preparative purposes for hydrazides, the yields are satisfactory to excellent, and in special situations they may be useful (for example, in structure proof), especially where it is important to avoid a mixture of positional isomers such as may result from direct acylation of a monosubstituted hydrazine.

The oldest of these methods is the reaction of *N*-chloro amides with sodamide[300] (Eq. 3-142). It can be modified so as to produce *N,N'*-diacyl hydrazines by using the sodium salt of an amide instead of sodamide.

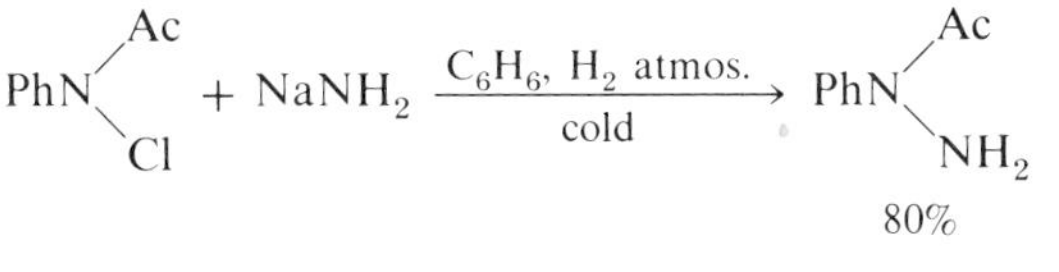

$$PhN\!\!\begin{smallmatrix}Ac\\ \diagup\\ \diagdown\\ Cl\end{smallmatrix} + NaNH_2 \xrightarrow[cold]{C_6H_6,\ H_2\ atmos.} PhN\!\!\begin{smallmatrix}Ac\\ \diagup\\ \diagdown\\ NH_2\end{smallmatrix} \qquad (3\text{-}142)$$
$$80\%$$

Some *unsym*-diacyl hydrazines have been prepared by treating a salt of the corresponding imide, such as potassium phthalimide, with an *O*-aryl hydroxylamine[301,302] (Eq. 3-143); yields as high as 88% have been obtained. *O*-Acyl hydroxylamines can also be used.[303] 3-Phenyloxaziridines bearing various *N*-acyl groups are readily obtainable and undergo ring cleavage when treated with amines to give hydrazides in high yield[304,305] (Eq. 3-144).

Ester-hydrazides of carbonic acid can be prepared from *N*-chloro ureas in moder-

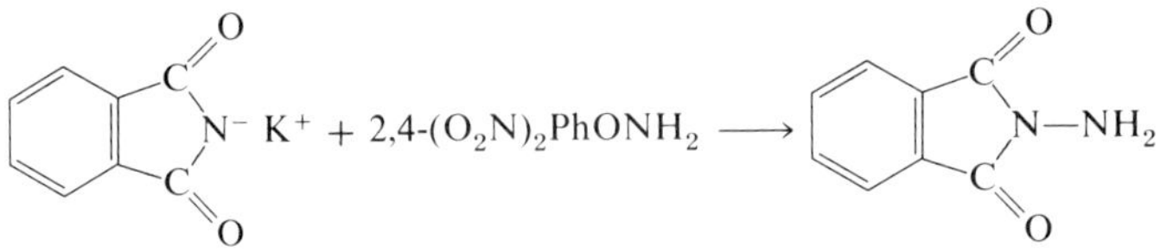

$$\text{(phthalimide-N}^-\text{)} \; K^+ + 2,4\text{-}(O_2N)_2PhONH_2 \longrightarrow \text{(phthalimide-N—NH}_2\text{)} \qquad (3\text{-}143)$$

$$PhCH\!-\!NCOR + R_2'NH \longrightarrow RCONHNR_2' + PhCHO \qquad (3\text{-}144)$$
$$\underset{O}{\diagup}$$

ate to high yields according to a patent claim; the reaction presumably proceeds through a diaziridinone[306] (Eq. 3-145).

$$RNHCONHCl \xrightarrow{\text{NaOH}} [RN\underset{NH}{\overset{}{\diagdown}}C\!=\!O] \xrightarrow{\text{R'OH}}$$

$$RNHNHCO_2R' \qquad (3\text{-}145)$$

The most generally satisfactory method for preparing thiohydrazides is acylation of hydrazine or substituted hydrazines with carboxymethyl dithioates[60,132] (Eq. 3-146). Methyl and ethyl esters can also be used,[307] but dithio acids, which also give

$$RC\overset{S}{\underset{SCH_2CO_2H}{\diagup}} + NH_2NH_2 \longrightarrow$$

$$RCSNHNH_2 + HSCH_2CO_2^- \; N_2H_5^+ \qquad (3\text{-}146)$$

rise to hydrazides, are inclined to give complex mixtures with reduction products and heterocyclic compounds.[308] Monosubstituted hydrazines give mixtures of positional isomers with all these reagents, the proportions depending considerably on steric requirements; aryl and *t*-alkyl hydrazines, however, give only *N'*-substituted thiohydrazides.

Less widely used methods include treating hydrazides with phosphorus penta-sulfide[133] (Eq. 3-147); yields are only moderate. Solvolysis of isosydnones with hy-

$$RCONHNH_2 + P_4S_{10} \longrightarrow RCSNHNH_2 \qquad (3\text{-}147)$$

drogen sulfide has the valuable feature of producing *N*-aryl thiohydrazides, which are not obtainable by direct acylation[309] (Eq. 3-148). Solvolysis of hydrazonoyl

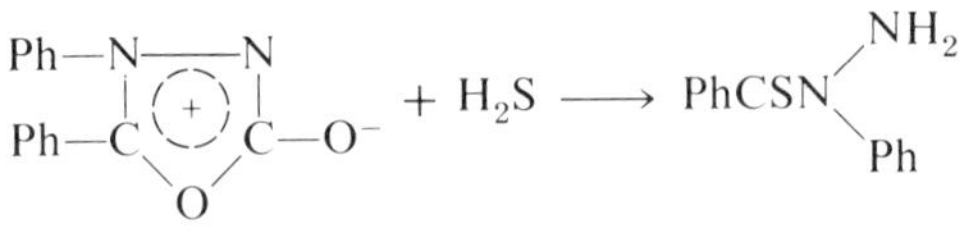

$$\begin{array}{c} Ph\!-\!N\!\!-\!\!N \\ \underset{Ph-C}{|} \overset{(+)}{\underset{O}{\diagdown}} \underset{C-O^-}{|} \end{array} + H_2S \longrightarrow PhCSN\overset{NH_2}{\underset{Ph}{\diagdown}} \qquad (3\text{-}148)$$

halides, obtainable from aldehyde hydrazones, with hydrogen sulfide gives *N'*-substituted thiohydrazides[310] (Eq. 3-149); the method is essentially confined to benzoic acid derivatives owing to accessibility of the hydrazonoyl halides.

$$ArCH\!=\!NNR_2 \xrightarrow{X_2} ArC\overset{X}{\underset{NNR_2}{\diagup}} \xrightarrow{H_2S} ArC\overset{S}{\underset{NHNR_2}{\diagup}} \qquad (3\text{-}149)$$

Hydrazonoyl Derivatives: Halides and Esters

Hydrazonoyl halides are most commonly prepared by direct halogenation of alde-hyde hydrazones, as described in Chapter 2, under Reactions (see also Eq. 3-149). They have also been prepared by the action of phosphorus pentachloride on hydrazides, as described earlier in this chapter (Eqs. 3-41, 3-42).

Carbonohydrazonoyl halides (phosgene hydrazones), $ArNHN=CX_2$, have been .prepared in moderate yield by treating formaldehyde hydrazones (or the corre-sponding hydrazonoyl halides) with chlorine or bromine, but it is more efficient to halogenate α-hydrazono β-diketones, $(MeCO)_2C=NNHAr$, which are readily ob-tained from acetylacetone and diazonium salts. The acyl groups are then easily removed by solvolysis (overall yields ~70%). Arylhydrazones of glyoxalic acid also give carbonohydrazonoyl halides upon halogenation (the carboxyl group is lost as CO_2).[49] N,N-Bis(alkoxycarbonyl)carbohydrazonoyl chlorides, $(RO_2C)_2NN=CCl_2$, have been prepared by the reaction of azoformate esters with dichloromethylene (generated by fragmentation of sodium trichloroacetate).[311]

Hydrazonic esters can be prepared by alcoholysis or phenolysis of hydrazonoyl halides (or the nitrile imides derived from them), as described earlier in this chapter in the section Reactions). Alkyl thiohydrazonates, however, are made by direct reaction of thiohydrazides with alkyl halides (see Eq. 3-17).

Amidrazones[4,177]

Most of the preparative methods for amidrazones are solvolytic. The most widely used methods fall into one class, hydrazinolysis of imidic acid derivatives, as gener-alized in Equation 3-150. The reverse procedure, ammonolysis (or aminolysis) of hydrazonic acid derivatives (Eq. 3-151), also leads to amidrazones, but it is less

$$RC\overset{Y}{\underset{NR'}{\big\langle}} + R_2'NNH_2 \longrightarrow RC\overset{NHNR_2'}{\underset{NR'}{\big\langle}} \rightleftharpoons RC\overset{NNR_2'}{\underset{NHR'}{\big\langle}} \qquad (3\text{-}150)$$

$$RC\overset{Y}{\underset{NNR_2}{\big\langle}} + R_2'NH \longrightarrow RC\overset{NR_2'}{\underset{NNR_2}{\big\langle}} \qquad (3\text{-}151)$$

commonly used, because the starting materials are not so readily obtainable. Closely related to these methods is the addition of hydrazines to nitriles. Less im-portant are certain reductive methods. Addition of hydrazines to ketenimines is a rarely used method, but addition to carbodiimides is an important route to amino-guanidines.

Imidic esters, used as their salts with acids or as free bases, give amidrazones readily if the reaction is carried out cold ($<0°C$).[197,312] Otherwise, if hydrazine is being used, further hydrazinolysis may take place, giving rise to hydrazidines or diamino azines (see Eqs. 3-70, 3-71). Substituted hydrazines give less trouble in this respect[184,313] (Eq. 3-152), but even they can give rise to hydrazidines by further

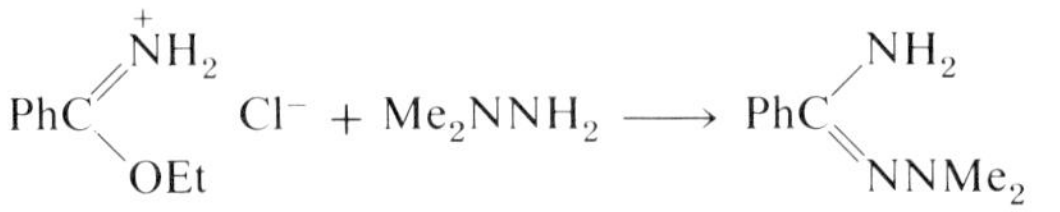

$$PhC\overset{\overset{+}{N}H_2}{\underset{OEt}{\diagdown}}Cl^- + Me_2NNH_2 \longrightarrow PhC\overset{NH_2}{\underset{NNMe_2}{\diagdown}} \qquad (3\text{-}152)$$

reaction with the hydrazine when the temperature is too warm. N-Substituted imidic esters can also be used with good results[31,314] (Eq. 3-153). In general, increasing

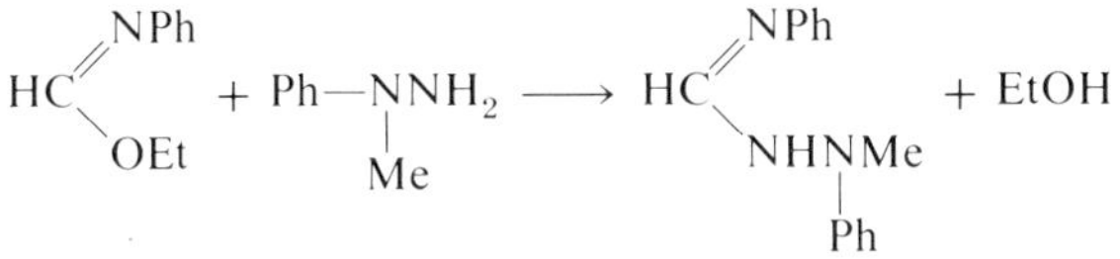

$$HC\overset{NPh}{\underset{OEt}{\diagdown}} + Ph\underset{Me}{\overset{|}{-}}NNH_2 \longrightarrow HC\overset{NPh}{\underset{NHNMe}{\diagdown}} + EtOH \qquad (3\text{-}153)$$

$$\underset{Ph}{|}$$

84%

bulk on either the imidic or the hydrazine nitrogen retards secondary reactions and leads to better yields of amidrazones. Hydrazinolysis of imidic esters with mono-alkyl hydrazines, for example, gives preparative yields when the alkyl group is secondary or tertiary, but with primary alkyl groups, reaction with two equivalents of imidic ester with subsequent ring closure to a 1,2,4-triazole[315] (Eq. 3-154) seri-

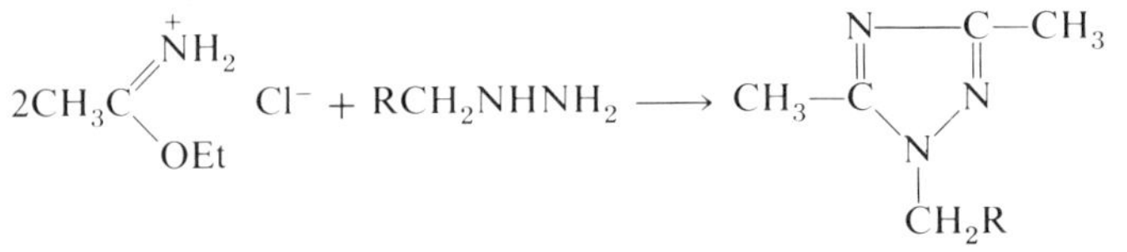

$$2CH_3C\overset{\overset{+}{N}H_2}{\underset{OEt}{\diagdown}}Cl^- + RCH_2NHNH_2 \longrightarrow CH_3-C\overset{N\text{------}C-CH_3}{\underset{\underset{CH_2R}{\overset{|}{N}}}{\diagdown\diagup}}N \qquad (3\text{-}154)$$

ously lowers the yields. Nevertheless, completely unsubstituted acetamidrazone has been prepared from ethyl acetimidate hydrochloride and hydrazine in ethanol at $-10°C$ in 50% yield.[10]

Imidoyl chlorides, prepared from N-substituted amides, react with hydrazines much like imidic esters and give amidrazones in preparative yields[8,106] (Eq. 3-155).

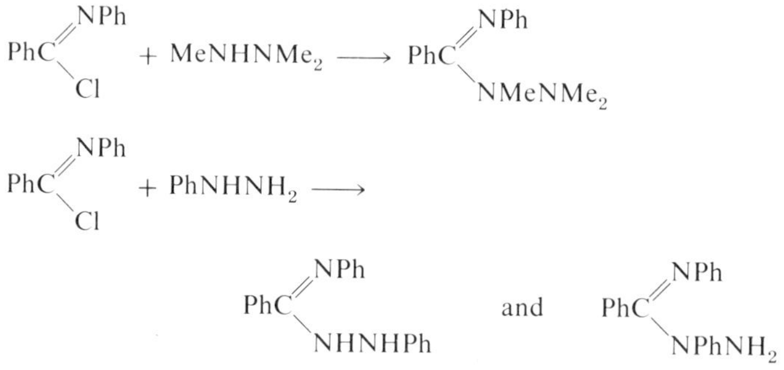

$$PhC\overset{NPh}{\underset{Cl}{\diagdown}} + MeNHNMe_2 \longrightarrow PhC\overset{NPh}{\underset{NMeNMe_2}{\diagdown}}$$

$$PhC\overset{NPh}{\underset{Cl}{\diagdown}} + PhNHNH_2 \longrightarrow \qquad\qquad (3\text{-}155)$$

$$PhC\overset{NPh}{\underset{NHNHPh}{\diagdown}} \qquad and \qquad PhC\overset{NPh}{\underset{NPhNH_2}{\diagdown}}$$

(ratio 4:1)

In reaction with monosubstituted hydrazines, there is the possibility of attack at either nitrogen, resulting in isomeric amidrazones.[106]

Amidines have seen less use as sources of amidrazones, probably because of the extra effort to obtain them.[17,316] In some instances, better yields have been claimed than from the corresponding imidic ester.[315] (Eq. 3-156).

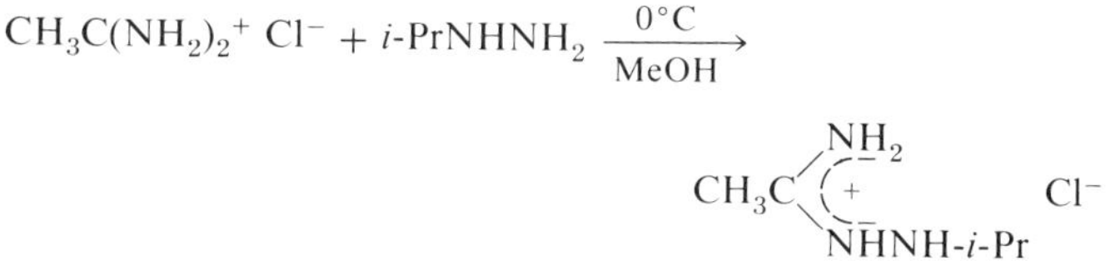

$$CH_3C(NH_2)_2{}^+ \ Cl^- + \textit{i-}PrNHNH_2 \ \xrightarrow[\text{MeOH}]{0\,^\circ C}$$

$$CH_3C\!\!\begin{array}{c}\text{NH}_2\\ \diagup \\ \overset{+}{\diagdown}\\ \text{NHNH-}\textit{i-}\text{Pr}\end{array}\qquad Cl^- \qquad (3\text{-}156)$$

72%

(40% from ethyl acetimidate)

Although amides themselves do not react with hydrazines to form amidrazones, their O-acyl derivatives, prepared by treatment with phosphorus oxychloride[3] or arenesulfonyl chloride,[317] usually give amidrazones in good yield (Eq. 3-157).

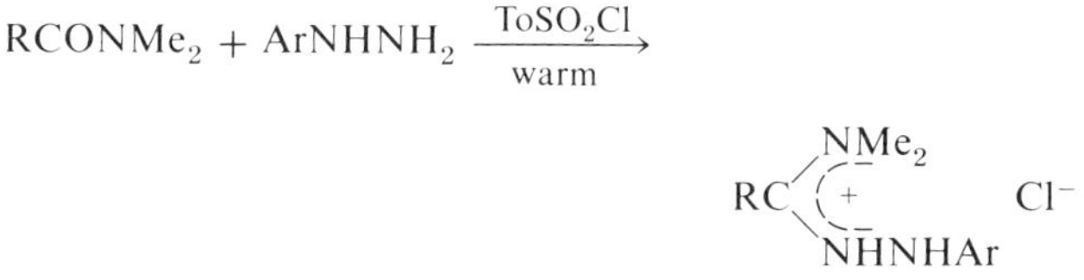

$$RCONMe_2 + ArNHNH_2 \ \xrightarrow[\text{warm}]{ToSO_2Cl}$$

$$RC\!\!\begin{array}{c}\text{NMe}_2\\ \diagup \\ \overset{+}{\diagdown}\\ \text{NHNHAr}\end{array}\qquad Cl^- \qquad (3\text{-}157)$$

68%

(R = H, Ar = 2,4-dinitrophenyl)

Thioamides react with hydrazines to eliminate hydrogen sulfide and form amidrazones[6,318] (Eq. 3-158). At higher temperatures, dihydrotetrazines become

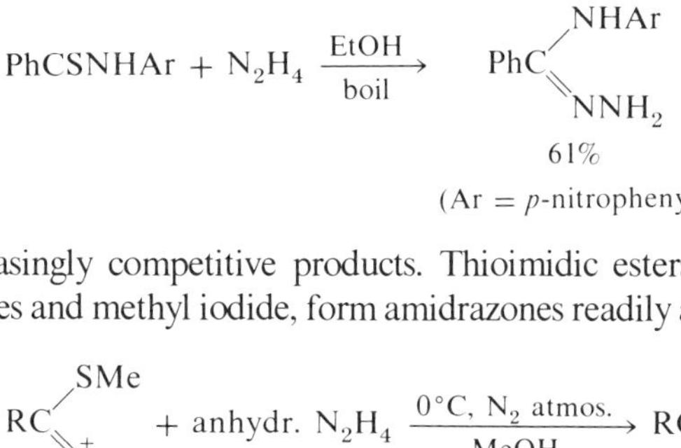

$$PhCSNHAr + N_2H_4 \ \xrightarrow[\text{boil}]{EtOH} \ PhC\!\!\begin{array}{c}\text{NHAr}\\ \diagup \\ \diagdown\\ \text{NNH}_2\end{array} \qquad (3\text{-}158)$$

61%

(Ar = p-nitrophenyl)

increasingly competitive products. Thioimidic esters, easily prepared from thio-amides and methyl iodide, form amidrazones readily and in good yield[8] (Eq. 3-159).

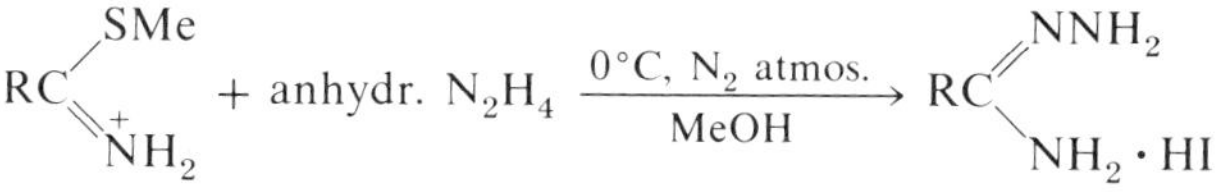

$$RC\!\!\begin{array}{c}\text{SMe}\\ \diagup \\ \overset{+}{\diagdown}\\ \text{NH}_2\end{array} + \text{anhydr. } N_2H_4 \ \xrightarrow[\text{MeOH}]{0\,^\circ C,\ N_2\ \text{atmos.}} \ RC\!\!\begin{array}{c}\text{NNH}_2\\ \diagup \\ \diagdown\\ \text{NH}_2\cdot HI\end{array} \qquad (3\text{-}159)$$

78–85%

Nitriles will add hydrazines with varying ease, and for good results, an electron-poor nitrile seems to be required[17,319-321] (Eq. 3-160). Heating is undesirable, for it

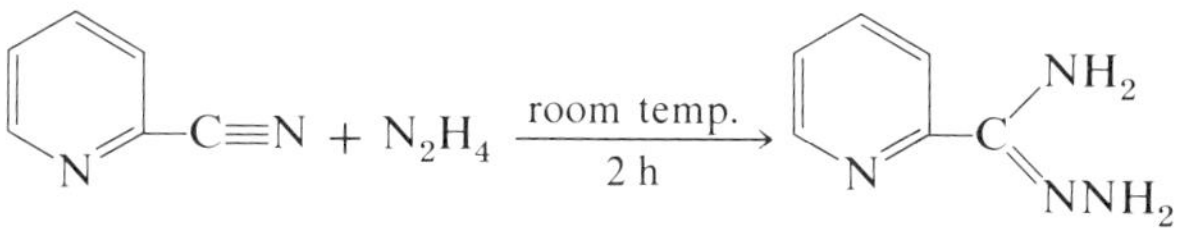

$$(3\text{-}160)$$

95%   67%

usually leads to α,α'-diamino azines. Use of sodium hydrazide ($NaNHNH_2$) expands the utility of the reaction to most nitriles, other than those with very acidic α-hydrogens.[322] Lithium hydrazides may also be used.[323] The sodium hydrazides may be prepared in situ by adding metallic sodium to a solution of the hydrazine and nitrile in benzene[324] (Eq. 3-161). The method is more successful with *unsym-*

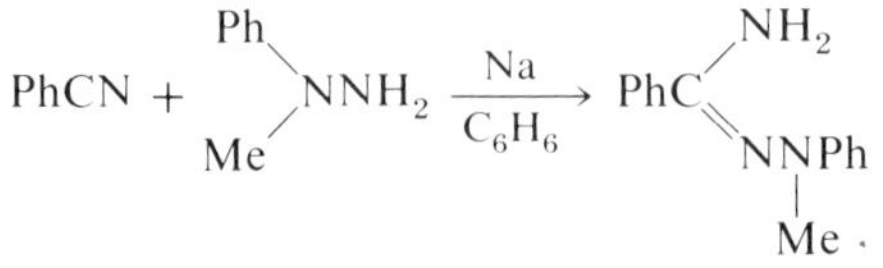

$$\text{PhCN} + \begin{array}{c} \text{Ph} \\ \diagdown \\ \diagup \\ \text{Me} \end{array}\!\!\text{NNH}_2 \xrightarrow[\text{C}_6\text{H}_6]{\text{Na}} \text{PhC}\!\!\begin{array}{c} \diagup\text{NH}_2 \\ \diagdown \\ \text{NNPh} \\ | \\ \text{Me} \cdot \end{array} \tag{3-161}$$

disubstituted hydrazines, for when it was applied to phenylhydrazine and benzonitrile, the principal product was 3,5-diphenyl-1,2,4-triazole, although methylhydrazine has given a 68% yield of amidrazone by reaction with 2-cyanopyridine.[321]

The reaction of hydrazines with nitriles can also be catalyzed by Lewis acids, which coordinate at the nitrile nitrogen and increase the electrophilicity of the carbon. Decaborane[325] and aluminum chloride[31] have been used (Eq. 3-162).

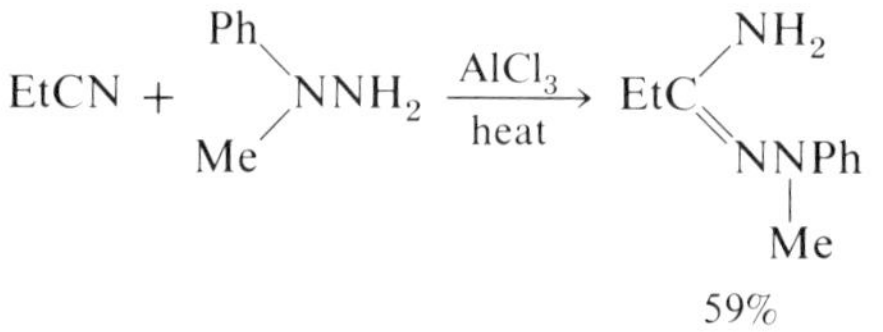

$$\text{EtCN} + \begin{array}{c} \text{Ph} \\ \diagdown \\ \diagup \\ \text{Me} \end{array}\!\!\text{NNH}_2 \xrightarrow[\text{heat}]{\text{AlCl}_3} \text{EtC}\!\!\begin{array}{c} \diagup\text{NH}_2 \\ \diagdown \\ \text{NNPh} \\ | \\ \text{Me} \end{array} \tag{3-162}$$
$$59\%$$

Aminolysis of hydrazonic esters and halides is largely confined to derivatives of benzoic and other aromatic carboxylic acids, for the corresponding hydrazonic esters and halides are the most readily obtainable[31,190] (Eq. 3-163). Thiohydrazonic

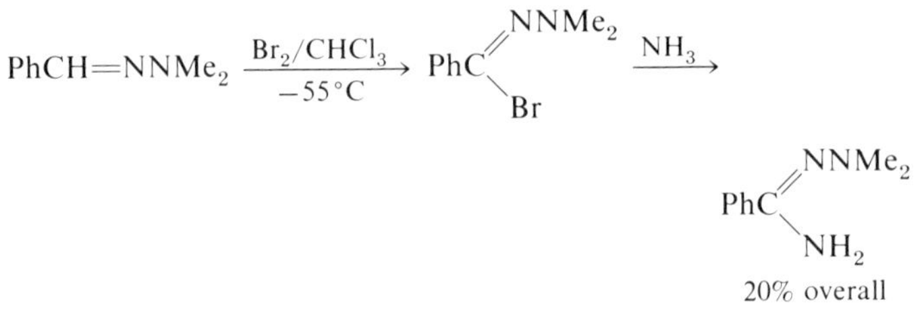

$$\text{PhCH}{=}\text{NNMe}_2 \xrightarrow[-55\,^\circ\text{C}]{\text{Br}_2/\text{CHCl}_3} \text{PhC}\!\!\begin{array}{c} {\diagup\!\!\!=}\text{NNMe}_2 \\ \diagdown \\ \text{Br} \end{array} \xrightarrow{\text{NH}_3}$$

$$\text{PhC}\!\!\begin{array}{c} {\diagup\!\!\!=}\text{NNMe}_2 \\ \diagdown \\ \text{NH}_2 \end{array} \tag{3-163}$$
$$20\% \text{ overall}$$

esters can also be aminolyzed. Nitrazones (hydrazones of 1-nitro aldehydes) undergo aminolysis with displacement of nitrite ion, giving amidrazones[326] (Eq. 3-164). This reaction has a potentially wide generality, for nitrazones can be pre-

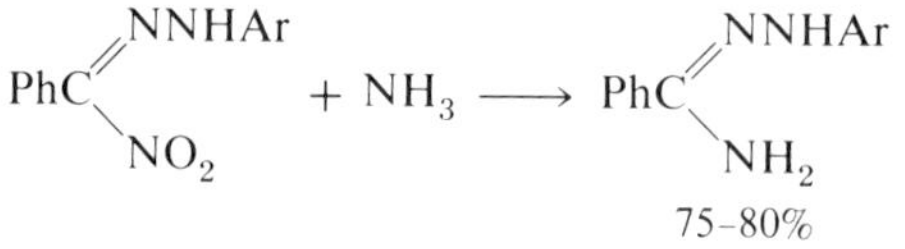

$$\text{PhC}\!\!\begin{array}{c} {\diagup\!\!\!=}\text{NNHAr} \\ \diagdown \\ \text{NO}_2 \end{array} + \text{NH}_3 \longrightarrow \text{PhC}\!\!\begin{array}{c} {\diagup\!\!\!=}\text{NNHAr} \\ \diagdown \\ \text{NH}_2 \end{array} \tag{3-164}$$
$$75\text{--}80\%$$

pared from nitroalkanes and diazonium salts. Nitrazones can also be reduced to amidrazones[327] (Eq. 3-165).

Tetrazolium salts, and the formazans[328] to which they are easily reduced, can be reduced all the way to amidrazones, presumably through hydrazidines, by catalytic hydrogenation or sodium dithionite[201] (Eq. 3-166). Reduction has also been used to prepare some formamidrazones through desulfurization of thiosemicarbazides.

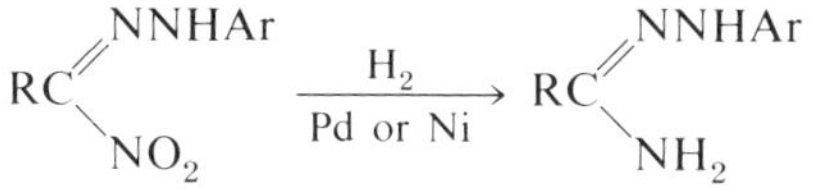

$$RC\overset{NNHAr}{\underset{NO_2}{\diagdown}} \xrightarrow[\text{Pd or Ni}]{H_2} RC\overset{NNHAr}{\underset{NH_2}{\diagdown}} \qquad (3\text{-}165)$$

94% (R = Ph, Ar = p-tolyl)
65% (R = Me, Ar = Ph)

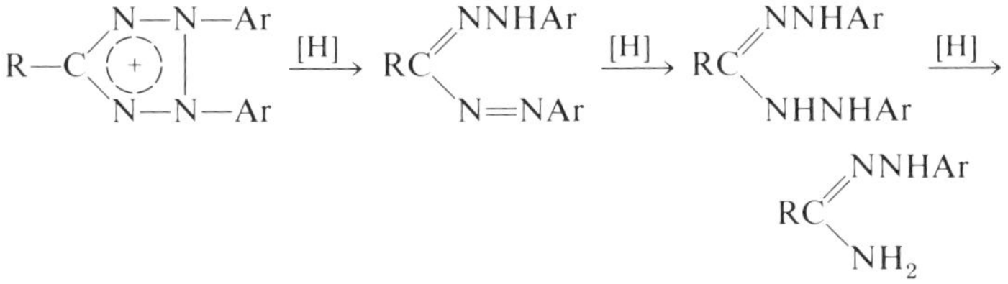

$$R-C\overset{N-N-Ar}{\underset{N-N-Ar}{\diagup}} \xrightarrow{[H]} RC\overset{NNHAr}{\underset{N=NAr}{\diagdown}} \xrightarrow{[H]} RC\overset{NNHAr}{\underset{NHNHAr}{\diagdown}} \xrightarrow{[H]}$$

$$RC\overset{NNHAr}{\underset{NH_2}{\diagdown}} \qquad (3\text{-}166)$$

The only oxidative method for preparing amidrazones appears to be the coupling of amidines by permanganate[329] (Eq. 3-167).

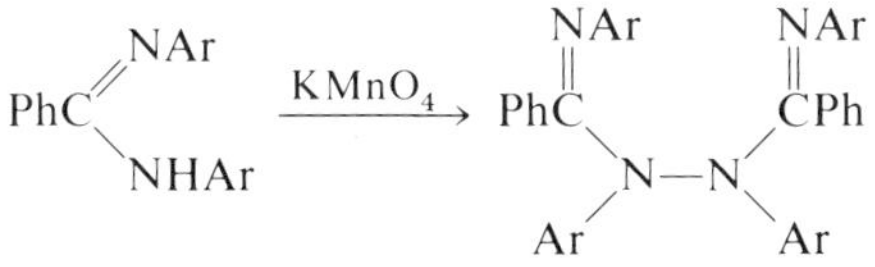

$$PhC\overset{NAr}{\underset{NHAr}{\diagdown}} \xrightarrow{KMnO_4} PhC\overset{NAr}{\underset{N-N}{\Vert}}\underset{Ar}{\underset{\diagdown}{\,}}\overset{NAr}{\underset{\Vert}{C}}Ph \qquad (3\text{-}167)$$

Semicarbazides, as their structure is ordinarily written, are tautomers of an amidrazone structure, and their synthesis should therefore be mentioned here. Among the many methods are reaction of hydrazines with isocyanates and reduction of nitro or nitroso ureas, as well as methods analogous to those used for ordinary amidrazones. They have been reviewed.[292] Alkoxycarbonamidrazones, in which the amidrazone tautomer is fixed by the presence of an alkyl group, have been prepared by the addition of alkyl cyanates and hydrazine[330] (Eq. 3-168). Simi-

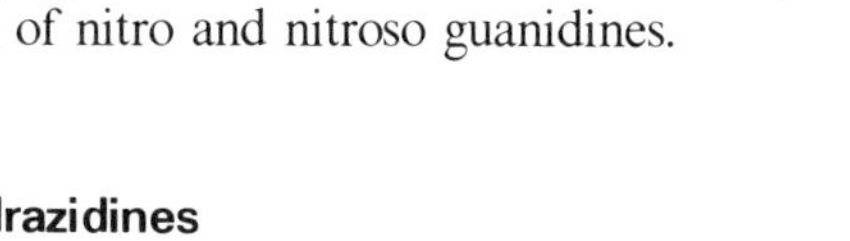

$$ROC\equiv N + N_2H_4 \longrightarrow ROC\overset{NH_2}{\underset{NNH_2}{\diagdown}} \qquad (3\text{-}168)$$

larly, aminoguanidines may be considered as amidrazones of carbamic acid; a large body of chemistry is devoted to their preparation and reactions. Prominent among the preparative methods are addition of hydrazines to carbodiimides[331] and reduction of nitro and nitroso guanidines.

Hydrazidines

The general methods for preparing hydrazidines make use of hydrazinolysis of hydrazonoyl chlorides[128,332] (Eq. 3-169), S-Methyl thiohydrazonates[11] (Eq. 3-170), or amidrazones (Eq. 3-70). The last method is a reaction frequently encountered as

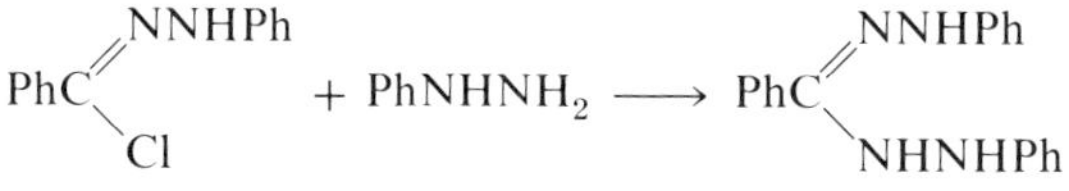

$$PhC\overset{NNHPh}{\underset{Cl}{\diagdown}} + PhNHNH_2 \longrightarrow PhC\overset{NNHPh}{\underset{NHNHPh}{\diagdown}} \qquad (3\text{-}169)$$

$$RC\overset{NNR_2}{\underset{SMe}{\diagup}} + R_2'NNH_2 \xrightarrow{EtOH} RC\overset{NNHR_2}{\underset{NHNR_2'}{\diagup}} \tag{3-170}$$
$$40–97\%$$

a secondary reaction in the preparation of amidrazones by hydrazinolysis procedures.[183,184] It is generally slow, however, and has not seen much use, but acethydrazidine has been prepared from ethyl acetimidate hydrochloride and excess hydrazine, through intermediate acetamidrazone, for example.[10]

The thiohydrazonate reaction is sometimes accompanied by formation of amidrazones derived from decomposition of certain arylhydrazines when they are used. Since hydrazine itself is easily attacked at both nitrogens, it is best avoided; instead, t-butoxycarbonylhydrazine can be used. The protecting group can be removed without destroying the hydrazidine group.[11]

By careful control, formazans can be reduced to hydrazidines[7] (Eq. 3-166). Hydrazines bearing ethoxycarbonyl substituents have been obtained by the reaction of aldehyde hydrazones with azodiformic ester (see Eq. 4-104).

Sulfonhydrazides and Thionyl Hydrazines

The general method for preparing sulfonhydrazides is acylation of hydrazine with sulfonyl chlorides, using either sodium bicarbonate or an excess of hydrazine to take up the HCl[333] (Eq. 3-171). Esters cannot be used, because alkyl sulfonates behave as alkylating agents.

$$ToSO_2Cl + 2N_2H_4 \longrightarrow ToSO_2NHNH_2 + N_2H_5^+ \ Cl^- \tag{3-171}$$
$$91–94\%$$

No systematic study of sulfonylation of substituted hydrazines has been reported, but the behavior appears to be parallel to acylation with chlorides of carboxylic acids, and N-methyltoluenesulfonhydrazide, for example, has been prepared satisfactorily in this way[233] (Eq. 3-172).

$$ToSO_2Cl + 2MeNHNH_2 \xrightarrow[0–60\,°C]{toluene}$$

$$ToSO_2N\overset{Me}{\underset{NH_2}{\diagup}} + MeNHNH_2 \cdot HCl \tag{3-172}$$
$$82\%$$

Sulfonylation of arylhydrazines provides N'-aryl sulfonhydrazides in good yield (Eq. 3-173). Examples in which there are like aryl groups on N and S are easily available from the reaction of diazonium salts with sulfur dioxide[214,334] (Eq. 3-174);

$$ArSO_2Cl + Ar'NHNH_2 \longrightarrow ArSO_2NHNHAr' \tag{3-173}$$

$$2ArNH_2 \xrightarrow[SO_2]{HNO_2} ArSO_2NHNHAr \tag{3-174}$$

the synthesis from an aniline is customarily carried out as one step, although the actual steps are diazotization, formation of sulfinic acid, coupling to form a diazosulfone, and reduction.

There are numerous examples of preparation of N',N'-disubstituted sulfonhydrazides by reaction of sulfonyl chlorides with *unsym*-disubstituted hydrazines.[335] An attempt to prepare N,N',N'-trimethyl-p-toluenesulfonhydrazide by direct acylation failed, however, perhaps because the expected product could not withstand the conditions (see Eq. 1-26).

Some N-alkyl sulfonhydrazides have been prepared by amination of N-alkyl sulfonamides with mesityloxyamine[336] (Eq. 3-175); presumably other aminating agents could accomplish the same result.

$$PhSO_2NHCH_2Ph + 2,4,6\text{-}Me_3C_6H_2ONH_2 \xrightarrow[\text{or NaH}]{\text{NaOMe}}$$

$$PhSO_2N\begin{array}{c} \diagup CH_2Ph \\ \diagdown NH_2 \end{array} \qquad (3\text{-}175)$$
$$57\%$$

N'-Alkyl sulfonhydrazides can be prepared fairly generally by reducing sulfonylhydrazones with sodium borohydride,[256,258] sodium cyanoborohydride,[259,260] or diborane (see Eq. 3-125) (N'-cyclohexyl-p-toluenesulfonhydrazide, 55–75%).[217]

Thionyl hydrazines are prepared by the action of thionyl chloride on substituted hydrazines[118,337] (Eq. 3-176).

$$RNHNH_2 + SOCl_2 \longrightarrow RNHN{=}SO \qquad (3\text{-}176)$$

ANALYTICAL METHODS[338]

Unsubstituted hydrazides can be detected and also quantitatively determined by first hydrolyzing them to hydrazine, which reacts with p-dimethylaminobenzaldehyde (Ehrlich's reagent) to form an intensely red azine. The color can be used to identify hydrazide spots in paper chromatography in some instances where enough spontaneous hydrolysis occurs,[339] or it can be adapted to colorimetric determination in solution.[340] Another qualitative test makes use of 2-diphenylacetylindandione, which reacts at the side-chain carbonyl group to form an acylhydrazone (or tautomer) that fluoresces brilliantly.[341] Some heavy-metal ions form strongly colored complexes that have been adapted to colorimetric detection and determination.[342,343]

N'-Phenylhydrazides give the Bülow reaction, in which a solution of the substrate in concentrated sulfuric acid is treated with a small amount of an oxidizing agent (sodium nitrite or chromate, but usually ferric chloride), whereupon a violet color develops.[344] Although no systematic exploration of its scope appears to have been made, the reaction seems to be quite general.

Oxidative titration of hydrazides has been described, using nitrite,[345,346] iodate,[347] N-bromosuccinimide,[348] etc.[349] A potentiometric method has been developed in which hydrazides can be determined in the presence of the corresponding hydrazine salts in order to follow the progress of the formation of hydrazides by heating carboxylic acids with hydrazine.[285] Hydrazides can also be titrated as acids, using pyridine solutions and tetrabutylammonium hydroxide, after first converting the hydrazide to its hydrazone with 2,4-dichlorobenzaldehyde.[350] *Sym*-diacyl hydrazines can be titrated in a similar manner. Voltammetric analysis has been applied to various hydrazides, especially maleic hydrazide (an important plant-growth inhibitor)[351] and the hydrazides of the phthalic acids.[352]

Amidrazones form colored complexes with many transition-metal ions (cobalt, for example), and it has been suggested that they might be adaptable for analytical purposes.[313] For certain heterocyclic amidrazone derivatives, both acid-base and oxidimetric titration have been applied.[197]

Thiohydrazides can be detected by the Wuyts reaction if they are *N'*-monsubstituted. Treatment with benzaldehyde closes a thiadiazoline ring (see Eq. 3-22). When the products are dissolved in concentrated sulfuric acid and treated with hydrogen peroxide or sodium nitrite, a blue to green color develops.[310,353]

REFERENCES

1. *Nomenclature of Organic Chemistry (IUPAC)*, 1979 Edition, prepared by J. Rigaudy and S. Klesney, Pergamon Press, Oxford and New York, 1979.

2. *Naming and Indexing of Chemical Substances for Chemical Abstracts during the Ninth Collective Period (1972–1976)*, Chemical Abstracts Service, Columbus, Ohio, 1976.

3. H. Rapoport and R. M. Bonner, *J. Am. Chem. Soc.* **72**, 2783 (1950).

4. D. G. Neilson, R. Rogers, J. W. M. Heatlie, and L. R. Newlands, *Chem. Rev.* **70**, 151 (1970).

5. Reviewed by H. Paulsen and D. Stoye, The Chemistry of Hydrazides, in *The Chemistry of Amides*, J. Zabicky (ed.), Wiley, New York, 1970, pp. 518–525 (properties), 545–586 (reactions), 525–544 (preparation), 587–589 (analysis).

6. W. J. van der Burg, *Rec. Trav. Chim. Pays Bas* **74**, 257 (1955).

7. D. Jerchel and R. Kuhn, *Liebig's Ann. Chem.* **568**, 185 (1950).

8. R. F. Smith, D. S. Johnson, R. A. Abgott, and M. J. Madden, *J. Org. Chem.* **38**, 1344 (1973).

9. K. M. Doyle and F. Kurzer, *Synthesis*, 583 (1974).

10. W. Oberhummer, *Monatsh. Chem.* **63**, 285 (1933).

11. R. Grashey, M. Baumann and H. Bauer, *Chem.-Ztg.* **96**, 224, 225 (1972).

12. H.-H. Stroh and H. Tengler, *Chem. Ber.* **101**, 751 (1968).

13. C. R. Lindergren and C. Niemann, *J. Am. Chem. Soc.* **71**, 1504 (1949).

14. S. Knapp, B. H. Toby, M. Sebastian, K. Krogh-Jespersen, and J. A. Potenza, *J. Org. Chem.* **46**, 2490 (1981).

15. T. Curtius and F. Lorenzen, *J. Prakt. Chem.* **[2]58**, 160 (1898).

16. R. R. Tarasyants, G. M. Petrova, G. S. Gol'din, V. G. Poddubnyi, S. G. Fedorov, and A. A. Simonova, *J. Gen. Chem. U.S.S.R.* **43**, 495 (1973).

17. H. C. Brown and D. Pilipovich, *J. Am. Chem. Soc.* **82**, 4700 (1960).

18. K. Jensen, H. R. Baccaro, O. Ruchardt, G. E. Olsen, C. Pedersen, and J. Toft, *Acta Chem. Scand.* **15**, 1109 (1961).

19. J. W. Powell and M. C. Whiting, *Tetrahedron* **7**, 305 (1959).

20. W. H. Beck, M. Liler and D. G. Morris, *J. Chem. Soc., [Perkin II]*, 1876 (1977).

21. L. H. Jensen, *J. Am. Chem. Soc.* **76**, 4663 (1954); **78**, 3993 (1956).

22. Y. Tomiie, C. H. Koo, and S. Nitto, *Acta Cryst.* **11**, 774 (1958).

23. R. Shintani, *Acta Cryst.* **13**, 609 (1960).

24. H. Lumbroso and J. Barassin, *Bull. Soc. Chim. France*, 3190 (1964).

25. G. J. Bishop, B. J. Price, and I. O. Sutherland, *J. Chem. Soc. Chem. Commun.*, 672 (1967).

26. D. M. Wiles and T. Suprunchuk, *Can. J. Chem.* **46**, 701 (1968).

27. M. Liler and D. G. Morris, *J. Chem. Soc.* [*Perkin II*], 909 (1977).

28. J. B. Jensen, *Acta Chem. Scand.* **10**, 667 (1956).

29. D. Prevoŕsek, *Bull. Soc. Chim. France*, 795 (1958).

30. Reviewed by W. H. Prichard, in *The Chemistry of Amidines and Imidates*, S. Patai (ed.), Wiley, New York, 1975, pp. 170–172.

31. W. Walter and H. Weiss, *Liebig's Ann. Chem.* **758**, 162 (1972).

32. P. Jakobsen and S. Treppendahl, *Acta Chem. Scand.* **31**, 92 (1977).

33. K. N. Zelenin, V. A. Khrustalev, and V. P. Sergutina, *Zh. Org. Khim.* **16**, 276 (1980).

34. T. Curtius, K. Hochschwender, and H. Thiemann, *J. Prakt. Chem.* [2]**92**, 102 (1915).

35. K. A. Hofmann and E. C. Marburg, *Liebig's Ann. Chem.* **305**, 191 (1899).

36. R. A. Turner, *J. Am. Chem. Soc.* **69**, 875 (1947).

37. S. M. Verma and R. M. Singh, *J. Org. Chem.* **40**, 897 (1975).

38. E. Hedaya, R. L. Hinman and S. Theodoropoulos, *J. Org. Chem.* **31**, 1311 (1966).

39. G. S. D. King, *J. Chem. Soc.* [*B*], 1224 (1964).

40. A. F. Hegarty, M. T. McCormack, B. J. Hathaway, and L. Hulett, *J. Chem. Soc.* [*Perkin II*], 1136 (1977).

41. P. Bouchet, J. Elguero, R. Lacquièr, and J.-M. Pereillo, *Bull. Soc. Chim. France*, 2264 (1972).

42. T. W. Bentley, R. A. W. Johnston, and A. F. Neville, *J. Chem. Soc.* [*Perkin I*], 449 (1973).

43. W. Walter and K. J. Reubke, in *The Chemistry of Amides*, J. Zabicky (ed.), Wiley, New York, 1970, pp. 477–514.

44. W. Walter and K. J. Reubke, *Angew. Chem. Int. Ed.* **6**, 368 (1967).

45. A. J. Hegarty, J. A. Kearney, and F. L. Scott, *J. Chem. Soc.* [*Perkin II*], 1422 (1973).

46. R. F. Smith, L. L. Kinder, D. G. Walker, L. A. Buckley, and J. M. Hammond, *J. Org. Chem.* **42**, 1862 (1977).

47. P. Wolkoff, *Can. J. Chem.* **53**, 1333 (1975).

48. R. Stollé, *J. Prakt. Chem.* [2]**75**, 416 (1907).

49. M. W. Moon, *J. Org. Chem.* **37**, 2005 (1972).

50. R. Stollé, *J. Prakt. Chem.* [2]**69**, 145 (1904).

51. G. Pellizzari, *Gazz. Chim. Ital.* **39I**, 520 (1909).

52. H. M. Abdou and T. Medwick, *J. Org. Chem.* **43**, 15 (1978).

53. T. Curtius and G. Struve, *J. Prakt. Chem.* [2]**50**, 295 (1894).

54. J. A. Young, W. S. Durrell, and R. D. Dresdner, *J. Am. Chem. Soc.* **84**, 2105 (1962).

55. A. C. Watterson, Jr., and S. A. Shama, *J. Org. Chem.* **40**, 19 (1975).

56. M. S. Gibson and A. W. Murray, *J. Chem. Soc.*, 880 (1965).

57. S. Wawzonek and R. C. Gueldner, *J. Org. Chem.* **30**, 3031 (1965).

58. W. J. McKillip, E. A. Sedor, B. M. Culbertson, and S. Wawzonek, *Chem. Rev.* **73**, 255 (1973).

59. R. F. Smith, R. D. Blondell, R. A. Abgott, K. B. Lipkowitz, J. A. Richmond, and K. A. Fountain, *J. Org. Chem.* **39**, 2036 (1974).

60. K. A. Jensen and C. Pedersen, *Acta Chem. Scand.* **15**, 1124 (1961).

61. H. E. Wijers, C. H. D. Van Ginkel, L. Brandsma, and J. F. Arens, *Rect. Trav. Chim. Pays Bas* **86**, 907 (1961).

62. B. Holmberg, *Arkiv. Kem. Mineral. Geol.* **25A**, 18 (1947); **17A**, No. 23 (1944).

63. H. Stetter and H. Spangenberger, *Chem. Ber.* **91**, 1982 (1958).

64. A. Ebnöther, E. Jucker, A. Lindenmann, E. Rissi, R. Steiner, R. Süess, and A. Vogel, *Helv. Chim. Acta* **42**, 533 (1959).

65. E. Fischer, *Liebig's Ann. Chem.* **239**, 248 (1887).

66. Y. Nitta and K. Okui, Japanese Patent No. 13,972 (1962); *Chem. Abstr.* **59**, 9802f (1963).

67. H. Röhnert, *Z. Chem.*, 302 (1965).

68. R. Lutwack, H. F. Mower, and C. Niemann, *J. Am. Chem. Soc.* **79**, 2179 (1957).

69. O. Widman, *Ber.* **27**, 2964 (1894).

70. W. S. Wadsworth, Jr., *J. Org. Chem.* **34**, 2994 (1969).

71. G. Lockemann, *Ber.* **43**, 2223 (1910).

72. W. E. Hanford, U.S. Patent No. 2,717,200; *Chem. Abstr.* **50**, 2131a (1956).

73. C. Ainsworth, *Can. J. Chem.* **43**, 1607 (1965).

74. T. Taguchi, J. Ishibashi, T. Matsuo, and M. Kojima, *J. Org. Chem.* **29**, 1097 (1964).

75. E. Hedaya, R. L. Hinman, and S. Theodoropoulos, *J. Am. Chem. Soc.* **85**, 3052 (1963).

76. T. Curtius and H. Melsbach, *J. Prakt. Chem.* **[2]81**, 501 (1910).

77. R. L. Hinman and M. C. Flores, *J. Org. Chem.* **24**, 660 (1959).

78. P. C. Freer and P. L. Sherman, *Am. Chem. J.* **18**, 562 (1896).

79. I. S. Berdinskii, I. A. Ustyuzhaninov, and I. N. Gorelysheva, *Zh. Org. Khim.* **1**, 454 (1965).

80. I. S. Berdinskii, E. Yu Posyagina, G. S. Posyagin, and V. F. Ust'-Kachkintsev, *Zh. Org. Khim.* **4**, 91 (1968) (English trans. 87).

81. H. H. Hatt, *Org. Syntheses,* Coll. **II**, 208 (1943).

82. R. Stollé and A. Benrath, *J. Prakt. Chem.* **[2]70**, 263 (1904).

83. T. J. Curphey and K. S. Prasad, *J. Org. Chem.* **37**, 2259 (1972).

84. R. L. Hinman and R. L. Landborg, *J. Org. Chem.* **24**, 724 (1959).

85. K. Ronco, B. Prijs, and H. Erlenmeyer, *Helv. Chim. Acta* **39**, 1253 (1956).

86. (*a*) B. Holmberg, *Arkiv. Kem.* **9**, 47 (1956); (*b*) H. C. Brown and R. Pater, *J. Org. Chem.* **30**, 3739 (1965).

87. T. Sato and M. Ohta, *Bull. Chem. Soc. Japan* **27**, 624 (1954).

88. W. Walter and K. J. Reubke, *Chem. Ber.* **102**, 2117 (1969).

89. W. P. Jencks, *Progr. Phys. Org. Chem.* **2**, 63 (1964).

90. W. P. Jencks, *Catalysis in Chemistry and Enzymology*, McGraw-Hill, New York, 1969, Chapter 10.

91. E. H. Cordes and W. P. Jencks, *J. Am. Chem. Soc.* **84**, 4319 (1962).

92. T. Pino and E. H. Cordes, *J. Org. Chem.* **36**, 1668 (1971).

93. A. Girard and G. Sandulesco, *Helv. Chim. Acta* **19**, 1095 (1936).

94. H. H. Inhoffen et al., *Liebig's Ann. Chem.* **570**, 58 (1950).

95. N. J. Leonard and J. H. Boyer, *J. Org. Chem.* **15**, 42 (1950).

96. R. B. Woodward, T. P. Kohman, and G. C. Harris, *J. Am. Chem. Soc.* **63**, 120 (1941).

97. J. C. Howard, G. Gever and P. H.-L. Wei, *J. Org. Chem.* **28**, 868 (1963).

98. C. von Plessing, *Arch. Pharm.* (*Weinheim*) **297**, 240 (1964).

99. T. Kametani and O. Umezawa, *Chem. Pharm. Bull.* (*Tokyo*) **14**, 369 (1966).

100. E. Bellasio, A. Ripamonti, and E. Testa, *Gazz. Chim. Ital.* **98**, 3 (1968).

101. A. Winterstein, B. Hegedüs, B. Fust, E. Böhni, and A. Studer, *Helv. Chim. Acta* **39**, 229 (1956).

102. I. S. Berdinskii, G. F. Piskunova, G. S. Posyagin, E. S. Ponosova, and I. M. Shevaldina, *Zh. Org. Khim.* **3**, 1645 (1967) (English trans. 1602).

103. A. P. Grekov and M. S. Marakhova, *Stsintillyatory i Stsintillyats. Materialy* (*Kharkov*), 24 (1963); *Chem. Abstr.* **63**, 8144g (1965).

104. A. P. Grekov and V. K. Skripchenko, *Zh. Org. Khim.* **3**, 1251, 1287, 1844 (1967); **4**, 243 (1968) (English trans. p. 1212, 1248, 1800, 236).

105. W. Baker and P. D. Ollis, *Quart. Rev.* **11**, 15 (1957).

106. M. Busch and R. Ruppenthal, *Ber.* **43**, 3001 (1910).

107. C. Ainsworth, *J. Am. Chem. Soc.* **87**, 5800 (1965).

108. R. Kraft, H. Paul, and G. Hilgetag, *Chem. Ber.* **101**, 2028 (1968).

109. Ng Ph. Buu-Hoi, N. D. Xuong, and E. Lescot, *Bull. Soc. Chim. France,* 441 (1957).

110. H.-H. Stroh and H. Beitz, *Liebig's Ann. Chem.* **700**, 78 (1966).

111. G. J. Durant, *J. Chem. Soc.* [C], 92 (1967).

112. A. P. Grekov and V. V. Shevchenko, *Zh. Org. Khim.* **3**, 1294 (1967) (English trans. p. 1255).

113. H. Eilingsfeld, *Chem. Ber.* **98**, 1308 (1965).

114. R. S. Sagitullin and A. N. Kost, *Vestnik Moskovsk. Univ.* **14(4)**, 187, 190 (1959); *Chem. Abstr.* **54**, 17383h (1960).

115. E. Mosettig, *Org. Reactions* **8**, 232 (1954).

116. M. S. Newman and E. G. Caflisch, Jr., *J. Am. Chem. Soc.* **80**, 862 (1958).

117. H.-H. Stroh and G. Jöhnchen, *Z. Chem.* **8**, 24 (1968).

118. P. Hope and L. A. Wiles, *J. Chem. Soc.,* 5386 (1965).

119. P. A. S. Smith, *Org. Reactions* **3**, Chap. 9 (1946).

120. J. Honzl and J. Rudinger, *Coll. Czech. Chem. Commun.* **26**, 2333 (1961).

121. O. Widman, *Ber.* **28**, 1925 (1895).

122. J. H. Boyer and F. C. Canter, *J. Am. Chem. Soc.* **71**, 1280 (1955).

123. P. A. S. Smith, J. M. Clegg, and J. Lakritz, *J. Org. Chem.* **23**, 1595 (1958).

124. P. A. S. Smith and H. G. Pars, *J. Org. Chem.* **24**, 1325 (1959).

125. R. H. McGirk, C. R. Cyr, W. D. Ellis, and E. H. White, *J. Org. Chem.* **39**, 3851 (1974).

126. T. Koenig and L. Lam, *J. Org. Chem.* **34**, 956 (1969).

127. M. Robba, B. Roques and J. LeGuen, *Bull. Soc. Chim. France,* 4220 (1967).

128. H. von Pechmann and L. Seeberger, *Ber.* **27**, 2121 (1894).

129. R. Stollé, *J. Prakt. Chem.* [2]**73**, 277 (1906).

130. G. I. Matyushecheva, A. V. Norbut, G. I. Derkach, and M. M. Yagupol'skii, *Zh. Org. Khim* **3**, 2254 (1967) (English trans. p. 2205).

131. H. Tolkmith, *J. Am. Chem. Soc.* **84**, 2097 (1962).

132. K. A. Jensen and C. Pedersen, *Acta Chem. Scand.* **15**, 1097 (1961).

133. H. Bredereck, B. Föhlisch and K. Walz, *Liebig's Ann. Chem.* **688**, 93 (1965).

134. O. Dimroth and G. de Montmollin, *Ber.* **43**, 2904 (1910).

135. J. P. Horwitz and V. A. Grakauskas, *J. Am. Chem. Soc.* **79**, 1249 (1957).

136. T. Kauffmann, L. Bán, and D. Kuhlmann, *Angew. Chem. Int. Ed.* **6**, 256 (1967).

137. S. Knapp and J. Calienni, *Synth. Commun.* **10**, 837 (1980).

138. E. Fahr and H. Lund, *Angew. Chem. Int. Ed.* **5**, 372 (1966).

139. E. H. White and M. M. Bursey, *J. Am. Chem. Soc.* **86**, 941 (1964).

140. E. H. White, M. M. Bursey, D. F. Roswell, and J. H. M. Hill, *J. Org. Chem.* **32**, 1198 (1967).

141. W. A. F. Gladstone, J. B. Aylward, and R. O. C. Norman, *J. Chem. Soc.* [*C*], 2587 (1969).

142. R. B. Kelly, G. R. Umbreit, and W. F. Liggett, *J. Org. Chem.* **29**, 1273 (1964).

143. R. M. Hann and C. S. Hudson, *J. Am. Chem. Soc.* **56**, 957 (1934).

144. H. B. Milne and C. F. Most, Jr., *J. Org. Chem.* **33**, 169 (1968).

145. P. Hope and L. A. Wiles, *J. Chem. Soc.* [*C*], 2636 (1967).

146. M. J. V. de Oliveira Baptista, A. G. M. Barrett, D. H. R. Barton, M. Girijavallabhan, R. C. Jennings, J. Kelly, V. J. Papadimitriou, J. V. Turner, and N. A. Usher, *J. Chem. Soc.* [*Perkin I*], 1477 (1977).

147. D. H. R. Barton, M. Girijavallabhan, and P. G. Sammes, *J. Chem. Soc.* [*Perkin I*], 929 (1972).

148. R. S. Atkinson and J. R. Malpass, *J. Chem. Soc.* [*Perkin I*], 2242 (1977).

149. D. M. Lemal, T. W. Rave, and S. D. McGregor, *J. Am. Chem. Soc.* **85**, 1944 (1963).

150. T. G. Back, S. Collins, and R. G. Karr, *J. Org. Chem.* **46**, 1564 (1981).

151. U. Lerch and J. G. Moffatt, *J. Org. Chem.* **36**, 3861 (1971).

152. Y. Wolman, P. M. Gallop, A. Patchornik, and A. Beyer, *J. Am. Chem. Soc.* **84**, 1889 (1962).

153. D. H. Marrian, P. B. Russell, A. R. Todd, and W S. Waring, *J. Chem. Soc.*, 1365 (1947).

154. H. N. Wingfield, W. R. Harlan, and H. R. Hammer, *J. Am. Chem. Soc.* **74**, 5796 (1952).

155. L. Kalb and O. Gross, *Ber.* **59**, 727 (1926).

156. L. A. Carpino, *J. Am. Chem. Soc.* **79**, 19 (1957).

157. P. Hope and L. A. Wiles, *J. Chem. Soc.* [*Supp. 1*], 5837 (1964).

158. B. Holmberg, *Arkiv Kem.* **7**, 517 (1954).

159. M. J. Hearn and E. S. Chung, *Synth. Commun.* **10**, 3 (1980).

160. J. Szmuszkovic, U.S. Patent No. 3,022,345 (1962); *Chem. Abstr.* **57**, 3298a (1962).

161. V. Rautenstrauck and F. Delay, *Angew. Chem. Int. Ed.* **19**, 726 (1980).

162. R. L. Hinman, *J. Am. Chem. Soc.* **78**, 1645 (1956).

163. R. L. Hinman, *J. Am. Chem. Soc*, **78**, 2463 (1956).

164. R. T. Beltrami and E. R. Bissell, *J. Am. Chem. Soc.* **78**, 2467 (1956).

165. K. Kratzl and K. P. Berger, *Monatsh. Chem.* **89**, 83 (1958).

166a. H. Fever and F. Brown, Jr., *J. Org. Chem.* **35**, 1468 (1970).

166b. J. A. Blair and R. J. Gardner, *J. Chem. Soc.* [*C*], 1714 (1970).

167. E. Hoggarth, *J. Chem. Soc.*, 2202 (1951).

168. T. Bacchetti, *Gazz. Chim. Ital.* **91**, 866 (1961).

169. F. L. Scott and M. Holland, *Proc. Chem. Soc.*, 106 (1962).

170. F. D. Chattaway and A. J. Walker, *J. Chem. Soc.* **127**, 2407 (1925).

171. A. Padwa and S. Nahm, *J. Org. Chem.* **46**, 1402 (1981).

172. Reviewed by R. Huisgen, R. Grashey, and J. Sauer, in *The Chemistry of Alkenes*, S. Patai (ed.), Interscience Publishers, New York, 1964, pp. 806ff.

173. A. Padwa, *Angew. Chem. Int. Ed.* **15**, 123 (1976).

174. A. Sonn and W. Meyer, *Ber.* **58**, 1096 (1925).

175. A. S. Shawali, H. M. Hassaneen, and S. Almousawi, *Bull. Chem. Soc. Japan* **51**, 512 (1978).

176. R. F. Smith, T. A. Craig, T. C. Rosenthal, and R. F. Oot, *J. Org. Chem.* **41**, 1555 (1976).

177. K. M. Watson and D. G. Neilson, in *The Chemistry of Amidines and Imidates*, S. Patai (ed.), Wiley, New York, 1975, Chap. 10.

178. R. F. Smith, A. S. Craig, L. A. Buckley, and R. R. Soelch, *Tetrahedron Lett.*, 4193 (1979).

179. (*a*) R. F. Smith, D. S. Johnson, C. L. Hyde, T. C. Rosenthal, and A. C. Bates, *J. Org. Chem.* **36**, 1155 /1971); (*b*) A. S. Shawali, A. Osman, and H. H. Hassaneen, *Ind. J. Chem.* **10**, 965 (1972).

180. G. S. Gol'din, V. G. Poddubnyi, A. A. Simonova, E. V. Orlova, and G. S. Shor, *Zh. Org. Khim* **5**, 1411 (1969).

181. H. G. O. Becker, G. Goermar, and H.-J. Timpe, *J. Prakt. Chem.* **312**, 601 (1970).

182. H. Neunhoeffer, F. Weischedel, and V. Böhnisch, *Liebig's Ann. Chem.* **750**, 12 (1971).

183. G. S. Gol'din, V. G. Poddubnyi, E. V. Orlova, and A. V. Kisin, *Zh. Org. Khim.* **9**, 517 (1973).

184. G. S. Gol'din, V. G. Poddubnyi, A. A. Simonova, G. S. Shor, and E. A. Rybakov, *Zh. Org. Khim.* **5**, 1404 (1969).

185. H. Reimlinger, W. R. F. Lingier, and J. J. M. Vanderwalle, *Chem. Ber.* **104**, 839 (1971).

186. B.-G. Baccar and J. Barrans, *Compt. Rend. Acad. Sci. Paris* **263**, 743 (1966).

187. E. P. Nesynov, M. M. Besprozvannaya, and P. S. Pel'kis, *Zh. Org. Khim.* **6**, 805 (1970).

188. M. Yanai, T. Kinoshita, S. Takeda, M. Nishimura, and T. Kuraishi, *Chem. Pharm. Bull. (Tokyo)* **20**, 1617 (1972).

189. K. T. Potts and C. R. Surapaneni, *J. Heterocycl. Chem.* **7**, 1019 (1970).

190. R. Fusco and P. Dalla Croce, *Gazz. Chim. Ital.* **99**, 69 (1969).

191. T. Kuraishi and R. N. Castle, *J. Heterocycl. Chem.* **1**, 42 (1964).

192. D.-I. Shiko and S. Tagami, *J. Am. Chem. Soc.* **82**, 4044 (1960).

193. E. C. Taylor and S. F. Martin, *J. Org. Chem.* **37**, 3958 (1972).

194. F. H. Case, *J. Heterocycl. Chem.* **10**, 353 (1973).

195. S. Kubota, O. Kirino, Y. Koida, and K. Miyake, *J. Pharm. Soc. Japan* **92**, 275 (1972).

196. S. Hünig et al., *Angew. Chem. Int. Ed.* **7**, 335 (1968).

197. S. Hünig, H. Balli, H. Conrad, and A. Schott, *Liebig's Ann. Chem.* **676**, 36, 52 (1964).

198. S. Hünig, H. Balli, E. Breither, F. Brühne, H. Geigher, E. Grigat, F. Müller, and H. Quast, *Angew. Chem.* **74**, 818 (1962).

199. H. Lund, *Studier over Elektrodereaktioner i Organisk Polarografi og Voltammetri, Aarhus Stiftsbogtrykkerie, Aarhus*, 1961, quoted by H. Lund in *The Chemistry of Amidines and Imidates*, S. Patai (ed.), Wiley, New York, 1975. pp. 246, 249.

200. R. Kuhn, F. A. Neugebauer, and A. Trischmann, *Angew. Chem.* **76**, 230 (1964).

201. D. Jerchel and W. Woticky, *Liebig's Ann. Chem.* **605**, 191 (1957).

202. B. T. Gillis and F. A. Daniher, *J. Org. Chem.* **27**, 4001 (1962).

203. H. S. Hertz, B. Coxon, and A. R. Siedle, *J. Org. Chem.* **42**, 2508 (1977).

204. N. J. Cusack, C. B. Reese, A. C. Risins, and B. Roozpeikar, *Tetrahedron* **32**, 2157 (1976).

205. D. M. Lemal, F. Menger, and E. Coats, *J. Am. Chem. Soc.* **86**, 2395 (1964).

206. R. A. Henry and D. W. Moore, *J. Org. Chem.* **32**, 4145 (1967).

207. W. G. Dauben and F. G. Willey, *J. Am. Chem. Soc.* **84**, 1497 (1962).

208. G. L. Closs, L. E. Closs, and W. A. Bell, *J. Am. Chem. Soc.* **85**, 3796 (1963).

209. W. Kirmse, M. Hartmann, R. Siegfried, H.-J. Wroblowsky, B. Zang, and V. Zellmer, *Chem. Ber.* **114**, 1793 (1981).

210. S. Wawzonek and W. McKillip, *J. Org. Chem.* **27**, 3946 (1962).

211. R. S. Dewey and E. E. van Tamelen, *J. Am. Chem. Soc.* **83**, 3729 (1961).

212. V. M. Rodionov and A. M. Fedorova, *Trudy Moskov. Khim. Tekhnol. Inst. im D. I. Mendeleeva*, No. 23, 21 (1956); *Chem. Abstr.* **53**, 1267h (1959).

213. W. L. F. Armarego, *Proc. Chem. Soc.*, 459 (1961).

214. S. Coffey, *J. Chem. Soc.*, 637 (1926).

215. J. S. McFadyen and T. S. Stevens, *J. Chem. Soc.*, 584 (1936).

216. D. J. Cram and J. S. Bradshaw, *J. Am. Chem. Soc.* **85**, 1108 (1963).

217. S. Cacchi, L. Caglioti, and G. Paolucci, *Bull. Chem. Soc. Japan* **47**, 2323 (1974).

218. R. L. Hinman and K. L. Hamm, *J. Am. Chem. Soc.* **81**, 3294 (1959).

219. L. A. Carpino, *J. Am. Chem. Soc.* **79**, 4427 (1957).

220. W. R. Bamford and T. S. Stevens, *J. Chem. Soc.*, 4735 (1952).

221. D. G. Farnum, *J. Org. Chem.* **28**, 870 (1963).

222. Reviewed by M. Regitz, in *Methodicum Chemicum*, Vol. 6, F. Korte (ed.), Academic Press, New York, 1975, pp. 215–223.

223. G. M. Kaufman, J. A. Smith, G. G. Van der Strouwe, and H. Shechter, *J. Am. Chem. Soc.* **87**, 935 (1965).

224. G. Kaufman, F. Cook, H. Shechter, J. Bayless, and L. Friedman, *J. Am. Chem. Soc.* **89**, 5736 (1967).

225. R. H. Shapiro, *Org. Reactions* **23**, Chap. 3 (1976).

226. C. H. DePuy and D. H. Froemsdorf, *J. Am. Chem. Soc.* **82**, 634 (1960).

227. L. Friedman and H. Shechter, *J. Am. Chem. Soc.* **81**, 5512 (1959).

228. R. H. Shapiro and M. J. Heath, *J. Am. Chem. Soc.* **89**, 5734 (1967).

229. R. H. Shapiro and T. Gadek, *J. Org. Chem.* **39**, 3418 (1974).

230. J. E. Herz and C. V. Ortiz, *J. Chem. Soc.* [C], 2294 (1971).

231. E. Vedejs, J. M. Dolphin, and W. T. Stolle, *J. Am. Chem. Soc.* **101**, 249 (1979).

232. M. N. Makhova, A. N. Mikhajluk, G. A. Karpov, N. V. Protopopova, B. N. Khasapov, L. I. Khmelnitski, and S. S. Novikov, *Tetrahedron* **34**, 413 (1978).

233. A. Nürrenbach and H. Pommer, *Liebig's Ann. Chem.* **721**, 34 (1969).

234. S. B. Martin, J. Cymerman-Craig, and R. P. K. Chan, *J. Org. Chem.* **39**, 2285 (1974).

235. J. D. Roberts, *J. Am. Chem. Soc.* **74**, 2959 (1951).

236. M. S. Newman and I. Ungar, *J. Org. Chem.* **27**, 1238 (1962).

237. S. Wawzonek and D. Meyer, *J. Am. Chem. Soc.* **76**, 2918 (1954).

238. J. E. Baldwin and J. E. Brown, *J. Org. Chem.* **36**, 3642 (1971).

239. S. Cacchi, F. La Torre, and D. Misiti, *Synthesis*, 301 (1977).

240. D. M. Lemal and A. J. Fry, *J. Org. Chem.* **29**, 1673 (1964).

241. J. B. Hendrickson and D. D. Sternbach, *J. Org. Chem.* **40**, 3450 (1975).

242. O. Süs, H. Steppan, and R. Dietrich, *Liebig's Ann. Chem.* **617**, 20 (1958).

243. M. P. Cava, R. Litle, and D. R. Napier, *J. Am. Chem. Soc.* **80**, 2257 (1958).

244. C. E. Sacks and P. L. Fuchs, *Synthesis*, 456 (1976).

245. J. F. W. Keana, D. P. Dolata, and J. Ollerenshaw, *J. Org. Chem.* **38**, 3815 (1973).

246. T. Curtius and G. Kraemer, *J. Prakt. Chem.* [2]**125**, 323 (1930).

247. A. C. Poshkus, J. E. Herweh, and F. A. Magnotta, *J. Org. Chem.* **28**, 2766 (1963).

248. F. Lober, *Angew. Chem.* **64**, 71 (1952).

249. H. Meerwein, G. Dittmar, G. Kauffmann, and R. Rave, *Chem. Ber.* **90**, 853 (1957).

250. L. Caglioti, F. Gasparrini, D. Misiti, and G. Palmieri, *Tetrahedron* **34**, 185 (1978).

251. H. Kwart and A. A. Khan, *J. Org. Chem.* **33**, 1537 (1968).

252. G. Rosini, *J. Org. Chem.* **39**, 3504 (1974).

253. T.-L. Ho and C. M. Wong, *J. Org. Chem.* **39**, 3453 (1974).

254. J. Jiricny, D. M. Orere, and C. B. Reese, *Synthesis*, 919 (1978).

255. G. A. Olah, J. Welch, and M. Henninger, *Synthesis*, 308 (1977).

256. L. Caglioti, *Org. Syntheses* **52**, 122 (1972).

257. M. Fisher, Z. Pelah, D. H. Williams, and C. Djerassi, *Chem. Ber.* **98**, 3236 (1965).

258. R. O. Hutchins and N. R. Natale, *J. Org. Chem.* **43**, 2299 (1978).

259. R. O. Hutchins, B. E. Maryanoff, and C. A. Milewski, *J. Am. Chem. Soc.* **93**, 1793 (1971).

260. R. O. Hutchins, M. Kacher, and L. Run, *J. Org. Chem.* **40**, 923 (1975).

261. L. Caglioti, *Tetrahedron* **22**, 487 (1966).

262. G. Rosini and A. Medici, *Synthesis*, 530 (1976).

263. D. N. Kirk and M. P. Hartshorn, *Steroid Reaction Mechanisms*, Elsevier, New York and Amsterdam, 1968, pp. 339–342.

264. T. Ueda and T. Tsuji, *Chem. Pharm. Bull.* (*Tokyo*) **9**, 71 (1961).

265. A. Michaelis, *Liebig's Ann. Chem.* **270**, 117 (1892).

266. D. Klamann, U. Krämer, and P. Weyerstahl, *Chem. Ber.* **95**, 2694 (1962).

267. G. Leandri and D. Spinelli, *Ann. Chim.* (*Rome*) **49**, 1689 (1959).

268. P. A. S. Smith, *Org. Syntheses*, Coll. **IV**, 819 (1963).

269. Kinetics: T. C. Bruice, and S. J. Benkovic, *J. Am. Chem. Soc.* **86**. 418 (1964).

270. L. Spialter, D. H. O'Brien, G. L. Untereiner, and W. A. Rush, *J. Org. Chem.* **30**, 3278 (1965).

271. W. D. Godtfredsen and S. Vangedal, *Acta Chem. Scand.* **9**, 1498 (1955).

272. R. F. Smith, A. C. Bates, A. J. Battisti, P. G. Byrnes, C. T. Muroz, T. J. Smearing, and F. X. Albrecht, *J. Org. Chem.* **33**, 851 (1968).

273. T. Rinderspacher and B. Prijs, *Helv. Chim. Acta* **41**, 22 (1958).

274. P. Grudzinska, *Roczniki Chem.* **34**, 1687 (1960); *Chem. Abstr.* **56**, 5863c (1962).

275. J. Gloede, K. Poduška, H. Gross, and J. Rudinger, *Coll. Czech. Chem. Commun.* **33**, 1307 (1968).

276. H. Gross, *Chem. Ber.* **95**, 2270 (1962).

277. C. Naegeli and G. Stefanovich, *Helv. Chim. Acta* **11**, 609 (1928).

278. H. A. Staab, M. Licking, and F. H. Dürr, *Chem. Ber.* **95**, 1275 (1962).

279. R. D. Twelves, U.S. Patent No. 3,023,241 (1962); *Chem. Abstr.* **57**, 3297d (1962).

280. K. K. Moll, H. Seefluth, L. Brüsehaber, and G. Schrattenholz, *Pharmazie* **23**, 36 (1968).

281. C. I. Niu and H. Fraenkel-Conrat, *J. Am. Chem. Soc.* **77**, 5882 (1955).

282. H. Paulsen and D. Stoye, *Chem. Ber.* **99**, 908 (1966).

283. T. Kauffmann, *Angew. Chem. Int. Ed.* **3**, 342 (1964).

284. T. Curtius and H. Franzen, *Ber.* **35**, 3239 (1902).

285. T. Rabini and G. Vita, *J. Org. Chem.* **30**, 2486 (1965).

286. U. P. Basu and S. Dutta, *J. Org. Chem.* **30**, 3562 (1965).

287. J. G. Krause, S. Kwon, and B. George, *J. Org. Chem.* **37**, 2040 (1972).

288. L. M. Grivnak and B. G. Boldyrev, *Zh. Org. Khim.* **3**, 2160 (1967) (English trans. p. 2111).

289. T. Taguchi, J. Ishibashi, T. Matsuo, and M. Kojima, *J. Org. Chem.* **29**, 1097 (1963).

290. A. Michaelis and F. Schmidt, *Liebig's Ann. Chem.* **252**, 300 (1889).

291. E. Schmitz, *Angew. Chem.* **57**, 197 (1964).

292. L. Peyron, *Bull. Soc. Chim. France*, D12 (1954).

293. C. Weygand, *Organic Preparations*, Interscience Publishers, New York, 1945.

294. M. Busch and H. Kunder, *Ber.* **49**, 2345 (1916).

295. B. T. Gillis and K. F. Schimmel, *J. Org. Chem.* **27**, 413 (1962).

296. R. Huisgen, F. Jakob, W. Siegel, and A. Cadus, *Liebig's Ann. Chem.* **590**, 1 (1954).

297. B. T. Gillis and P. E. Beck, *J. Org. Chem.* **28**, 3177 (1963).

298. A. F. Hegarty, in *The Chemistry of the Hydrazo, Azo and Azoxy Groups*, S. Patai (ed.), Wiley, New York, 1975, pp. 707–711.

299. R. Huisgen and F. Jakob, *Liebig's Ann. Chem.* **590**, 37 (1954).

300. W. F. Short, *J. Chem. Soc.* **119**, 1445 (1921).

301. F. D. Greene, R. L. Camp, V. P. Abegg, and G. O. Pierson, *Tetrahedron Lett.*, 4091 (1973).

302. T. Sheradsky, *Tetrahedron Lett.*, 1909 (1968).

303. L. A. Carpino, *J. Org. Chem.* **29,** 2820 (1964).

304. E. Schmitz, S. Schramm, and R. Ohme, *J. Prakt. Chem.*, **[4]36,** 86 (1967).

305. E. Schmitz and S. Schramm, *Tetrahedron Lett.*, 1857 (1965).

306. C. S. Sheppard and L. E. Korczykowski, U.S. Patent No. 3,956,366 (1976); *Chem. Abstr.* **85,** 176881 (1976); Ger. Offen. No. 1,818,020 (1972); *Chem. Abstr.* **87,** 38848 (1977).

307. R. Gompper, R. R. Schmidt, and E. Kutter, *Liebig's Ann. Chem.* **684,** 37 (1965).

308. H. Wuyts, *Bull. Soc. Chim. Belges* **46,** 27 (1937).

309. A. R. McCarthy, W. D. Ollis, and G. A. Ramsden, *J. Chem. Soc. Chem. Commun.*, 499 (1968).

310. W. Walter and K. J. Reubke, *Tetrahedron Lett.*, 5973 (1968).

311. D. Seyferth and H.-M. Shih, *J. Org. Chem.* **39,** 2329 (1974).

312. H. Neunhoeffer and H. W. Frühauf, *Liebig's Ann. Chem.* **760,** 102 (1972).

313. M. R. Atkinson and J. B. Polya, *J. Chem. Soc.*, 3319 (1954).

314. F. A. Hussein and A. A. Kadir, *J. Ind. Chem. Soc.* **45,** 729 (1968).

315. V. A. Khrustalev, K. N. Zelenin, and V. P. Seragutina, *Zh. Org. Khim.* **15,** 2280 (1979) (English trans. p. 2066).

316. A. Lottermoser, *J. Prakt. Chem.* **[2]54,** 113 (1896).

317. W. Hoyle, *J. Chem. Soc. [C]*, 690 (1967).

318. A. Spassov, E. Golovinsky, and G. Demirov, *Chem. Ber.* **99,** 3734 (1966).

319. R. L. Hinman and D. Fulton, *J. Am. Chem. Soc.* **80,** 1895 (1958).

320. F. H. Case, *J. Heterocycl. Chem.* **5,** 223 (1968).

321. S. Kubota, Y. Koida, T. Kosaka, and O. Kirino, *Chem. Pharm. Bull.* (*Tokyo*) **18,** 1696 (1970).

322. T. Kauffmann, S. Spaude, and D. Wolf, *Chem. Ber.* **97,** 3436 (1964).

323. K. H. Linke, R. Taubert, K. Bister, W. Bornatsch, and B. J. Liem, *Z. Naturforsch.* **B26,** 296 (1971).

324. R. Engelhardt, *J. Prakt. Chem.* **[2]54,** 143 (1896).

325. M. M. Fein, J. Bobinski, J. E. Paustian, D. Grafstein, and M. S. Cohen, *Inorg. Chem.* **4,** 422 (1965).

326. G. Ponzio, *Gazz. Chim. Ital.* **40I,** 77, 312; **40II,** 153 (1910).

327. D. Jerchel and H. Fischer, *Liebig's Ann. Chem.* **574,** 85 (1951).

328. E. Bamberger and P. de Gruyter, *Ber.* **26,** 2783 (1893); E. Bamberger and H. Witter, *ibid.,* 2786 (1893).

329. S. P. Joshi, A. P. Khanolkar, and T. S. Wheeler, *J. Chem. Soc.*, 793 (1936).

330. G. Zinner and I. Holdt, *Arch. Pharm.* (*Weinheim*) **307,** 644 (1974).

331. F. Kurzer and D. R. Hanks, *J. Chem. Soc. [C]*, 1375 (1968).

332. R. Stollé and F. Helwerth, *Ber.* **47,** 1132 (1914).

333. L. Friedman, R. C. Litle, and V. R. Reichle, *Org. Syntheses*, Coll. **V,** 1055 (1973).

334. H. Limpricht, *Ber.* **20,** 1238 (1887).

335. B. V. Ioffe and L. A. Kartsova, *Zh. Org. Khim.* **9,** 1209 (1973); (English trans. p. 1237).

336. L. A. Carpino, *J. Org. Chem.* **30,** 321 (1965).

337. A. Michaelis and J. Ruhl, *Liebig's Ann. Chem.* **270,** 114 (1892).

338. A. F. Krivis, in *Analytical Chemistry, Nitrogen and its Compounds*, C. A. Streuli and P. R. Averell (eds.), Wiley, New York, 1970, pp. 313–317.

339. R. L. Hinman, *Analyt. Chim. Acta* **15,** 125 (1956).

340. P. R. Wood, *Analyt. Chem.* **25,** 1879 (1953).

341. W. A. Mosher, I. S. Bechara, and E. J. Pozomek, *Talanta* **15,** 482 (1968).

342. A. P. Grekov and S. A. Malyutenko, *Zh. Anal. Khim.* **23,** 639 (1968) (English trans. p. 549).

343. V. Scardi and V. Bonavita, *Clin. Chem. Acta* **4,** 161 (1959).

344. C. Bülow, *Liebig's Ann. Chem.* **236,** 194 (1886).

345. A. P. Grekov and V. A. Yamchevskii, *Metody Anal. Khim. Reaktivov Prep.* **12,** 103 (1966); *Chem. Abstr.* **67,** 17638v (1967).

346. J. Vulterin and J. Zyka, *Chem. Listy* **50,** 364 (1965).

347. A. P. Grekov and S. A. Sakhorukova, *Metody Anal. Khim. Reaktivov Prep.* **12,** 109 (1966); *Chem. Abstr.* **67,** 17660w (1967).

348. M. Z. Barakat and M. Shaker, *Analyst* **91,** 466 (1966).

349. N. K. Mathur and C. K. Narany, *The Determination of Organic Compounds with N-Bromosuccinimide and Allied Reagents,* Academic Press, New York, 1975, pp. 82-84.

350. A. A. Latour, E. J. Kuchar, and S. Siggia, *Anal. Chem.* **36,** 2479 (1964).

351. P. J. Elving and C. Teitelbaum, *J. Am. Chem. Soc.* **71,** 3916 (1949).

352. A. J. Krivis, E. S. Gazda, G. R. Supp, and P. Kippur, *Anal. Chem.* **35,** 1955 (1963).

353. H. Wuyts, *Bull. Sci. Acad. Roy. Belg.* **20,** 156 (1934); *Chem. Abstr.* **28,** 3407 (1934); *Bull. Soc. Chim. Belg.* **43,** 261 (1934).

4.

DIAZONIUM AND AZO COMPOUNDS
(Derivatives of Diazene)

NOMENCLATURE

Diazonium compounds are normally named by means of the suffix *-diazonium*. This term is a suffix and not a compound in itself; consequently, the only approved form (IUPAC, *Chemical Abstracts*) is to affix it to the name of the parent substance, usually a hydrocarbon, never to the name of a radical. Thus the species $C_6H_5N_2^+$ can be found in *Chemical Abstracts* indices only under the name "benzenediazonium". In the rare situation where it is necessary to designate the function by a prefix, the form is *diazonio-*. Derivatives that are believed to be covalent rather than ionic (for example, $Ar—N=N—OH$ rather than $ArN_2^+OH^-$) are named differently by interposing the form *diazo* between the name of the parent compound and the potential anion: benzenediazohydroxide (IUPAC). With the *Ninth Collective Index*, *Chemical Abstracts* has abandoned this method in favor of names based on the parent compound *diazene*, $HN=NH$ (formerly called "diimide"), as in 1-hydroxy-2-phenyldiazene. Salts of this example are known as *diazoates* (formerly "diazotates"), as in sodium benzenediazoate. The analogous anhydrides, $ArN=N—O—N=NAr$, are called *diazoanhydrides*. Diazo compounds of this type exhibit geometric isomerism, and when it is necessary to designate configuration, the terms *syn-* and *anti-* have customarily been used, rather than *cis-* or *trans-*. The preservation of this seemingly pointless distinction serves one useful purpose: so long as the stereochemistry of the diazo compounds is at all uncertain, it allows the isomers to be designated clearly with respect to historical precedent, without committing the user to a particular geometry that may have been firmly established later (in the latter situation, *E* and *Z* are becoming more widely used).

Azo compounds are customarily named by the prefix *azo-* attached to the name of the parent hydrocarbon derivative if the structure is symmetrical, as in azomethane, $CH_3N=NCH_3$ (IUPAC). This system was used by *Chemical Abstracts* until the *Ninth Collective Index*, in which compounds are indexed under "diazene" ("diazene, dimethyl"). *Azo* names may serve as stems to which substituents can be attached, as in 4-chloro-3'-ethylazobenzene. However, when the parent structures at each end are not the same, as in $C_6H_5N=NCH_3$, the system is modified thus: benzeneazomethane. When *Chemical Abstracts* is not using the diazene nomenclature, it differs from IUPAC trivially but confusingly: "phenylazomethane." There are some common compounds that would have somewhat awkward names by the rigid application of the foregoing methods, however. Informal names for them have become customary, and since they are logical, unambiguous, and easily comprehensible, there seems to be no objection. Examples are "azodicarboxylic ester" or "azoformic ester" for $EtO_2C—N=N—CO_2Et$, and "azodibenzoyl" for $PhCON=NCOPh$.

The foregoing discussion of azo compounds assumes the presence of two substituents, for until recently, only such examples were known. When only one substituent is present, the diazene nomenclature is used exclusively: $C_6H_5N{=}NH$, phenyldiazene. Similarly, cations (conjugate acids), having any number of substituents, are named *diazenium ions*. In this book, the dipolar isomers of azo compounds, $R_2\overset{+}{N}{=}N^-$, are referred to as *azamines* and are to be found in Chapter 5.

The structure $H\overset{①}{N}{=}\overset{②③}{N}CH{=}\overset{④}{N}{-}\overset{⑤}{N}H_2$ represents a special case; the existence of tautomerism when it is incompletely substituted makes it desirable to consider it as a functional group in its own right. Such compounds are called *formazans* and named as derivatives of the unsubstituted parent formazan (diazenecarbaldehyde hydrazone). *Chemical Abstracts* now uses the formazan root, but it has also used "azo hydrazono" names that in some cases verge on the incomprehensible. The somewhat analogous structures $RN{=}NCR{=}NOH$ and $RN{=}NCR{=}NO_2H$ are sometimes referred to as "nitrosazones" and "nitrazones," respectively.

There is a class of compound that can be named as *o*- or *p*-diazoniophenoxides or as 2- or 4-diazocyclohexadienones and has the properties of a hybrid of them. Such compounds are commonly called *benzenediazooxides* or, regrettably, *quinone diazides* (see Chap. 5).

PROPERTIES

Diazonium salts of the strong mineral acids are nearly colorless, water-soluble, ionic solids,[1] whose stability in the solid state varies from fleeting to moderate. They usually deflagrate before melting, and some, notably nitrates, detonate. Fluoroborates, as well as some other salts with complex anions, are sparingly soluble and more easily isolated. By far the most stable diazonium salts are those in which the diazonio group is attached to a tricoordinate carbon, which provides some stabilization by π-orbital delocalization. Although most of the known diazonium salts are derivatives of benzene or naphthalene, many other systems can sustain a diazonium ion, such as heterocyclic nuclei with aromatic character (pyridine, pyrazole, triazole), and even metal chelate systems, such as *tris*-(acetylacetonato)chromium(I):[2]

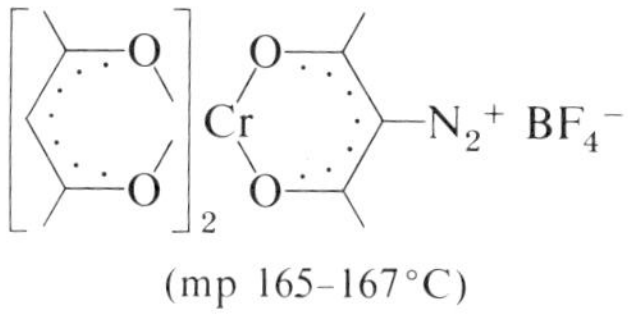

(mp 165–167°C)

I

Some alkenediazonium salts, $R_2C{=}CHN_2^+SbCl_6^-$, have been made,[3] and some diazonium salts with saturated substituents, such as $(CF_3)_2CHN_2^+$,[4] are known in solution.[5] Crown ethers both solubilize and stabilize arenediazonium salts.[6,7]

Alkali metal arenediazoates are known as water-soluble, crystalline solids,[8,9] and even the less stable alkanediazoates have been made (CH_3N_2OLi and CH_3N_2OK

have been obtained solid).[10,11] Diazohydroxides themselves appear not to have been isolated, but a few examples of their tautomers, the aryl nitrosamines, are known as very unstable solids, obtainable only at low temperatures,[8] except in the case of certain nitrosamino heterocycles.[12,13] Diazoanhydrides are explosive, yellow solids that decompose slowly at room temperature if undisturbed (bis-(p-chlorophenyl)diazoanhydride has a half-life of 402 minutes at 19.5°C).[14,15] There have been contentions that the diazoanhydrides were really nitrosotriazenes, but the unambiguous synthesis of the latter compounds, which were found to be quite different, has resolved the question.[15]

Monosubstituted diazenes are in general unstable substances that decompose at room temperature,[16] but some, notably p-nitrophenyldiazene, can be kept at room temperature for over 24 hours.[17] trans-Methyldiazene is a volatile, yellow substance.[18,19] The cis isomer, which is more difficult to obtain, is much less volatile.[20]

Azo compounds bearing two substituents are much more stable than the foregoing types of compounds, and most of them are more or less indefinitely stable at room temperature. The aliphatic members are volatile, light-yellow liquids or gases that have an unpleasant, sweetish odor, such as trans-azomethane (mp −78°C, bp 2°C) and trans-2,2'-azopropane (bp 88°C); cis-azomethane is less stable and boils nearly 100° higher than the trans isomer.[21] In general, azoalkanes boil at 10 ± 3°C lower than the isosteric alkenes. Azomethane is appreciably soluble in water. Azobenzene exists as a trans isomer, an orange-yellow solid (mp 68°C), and a less stable cis isomer (mp 71°C); the difference in melting points between geometric isomers is usually larger, however, as in p,p'-azotoluene (mp 70°C, trans; 37°C, cis).

The basicity of azo compounds is so low that protonation is not observed in aqueous solution except when other groups in conjugation markedly modify the nature of the system. The situation then becomes part of the involved subject of acid-base indicators, such as methyl orange. The base strengths of azoalkanes have not been measured, owing to rapid isomerization to hydrazones brought about by strong acids, but trans-azobenzene ($pK_a = -2.95$) and its cis isomer ($pK_a = -2.25$) are weaker bases than water or methanol.[22,23] Salt-like addition compounds with strong acids have nevertheless been isolated. The site of protonation has been a source of controversy, the question of debate being whether the proton is linked to a particular nitrogen or is bound symmetrically to the π-electrons of the double bond.[23,24] The existence of distinct conjugate acids from trans- and cis-azobenzenes is difficult to reconcile with π-bond protonation, however, and x-ray photoelectron spectroscopy has indicated that protonation is on nitrogen.[25]

Tertiary diazenium salts having the ion ArN=$\overset{+}{N}$Ar$_2$ are known, in which Ar is a 3-pyrazolyl group,[26] and aliphatic diazenium salts such as II have been prepared by oxidation of the corresponding hydrazines.[27]

$$(CH_3)_3C-N=\overset{+}{N}\underset{CH_2CH=CH_2}{\overset{C(CH_3)_3}{<}} \quad BF_4^-$$

II

The acid-base characteristics of diazonium compounds form an exceptionally involved topic.[28-30] Water solutions of diazonium salts of the strong mineral acids are not highly acidic, and the diazohydroxides from which they are formally derived must therefore be somewhat dissociated. Determination of the dissociation constant is not possible by direct measurement, owing to a multiplicity of equilibria. When a solution of a diazonium salt is made basic, diazohydroxides are not formed to an extent of more than a percent or so; instead, the diazonium salt consumes two equivalents of hydroxide ion and is converted to the diazoate ion (Eq. 4-1). The pH

$$ArN_2^+ + OH^- \xrightarrow[\text{fast}]{} ArN_2OH \underset{\text{fast}}{\xrightarrow{OH^-}} ArN_2O^- \qquad (4\text{-}1)$$

of solutions of diazoate salts, 7 to 12, shows that the parent acids are weak to fairly strong. We are dealing with the phenomenon of a dibasic acid, ArN_2^+, the first acidity constant of which is much smaller than the second.[31-33] This is the opposite of all common experience and was for many decades a cause of confusion in the interpretation of diazonium reactions.

The unusual characteristics of diazonium ions as dibasic acids result from the fact that the original cation is a Lewis acid, but the diazohydroxide is a Brønsted acid. When a diazonium salt is treated with one equivalent of base, only about half the salt reacts to form diazoate, and half remains (ca. 1% of diazohydroxide and some diazoanhydride are usually present as well). The potentiometric curve for addition of two equivalents of base shows only one inflection, and from it can be obtained only $pK_{app} = (pK_1 + pK_2)/2$. For unsubstituted benzenediazonium ion, $pK_{app} = 11.9$, for the p-nitro derivative, 9.44, and for p-methyl, 12.59.[31] The Hammett regression constant ρ is thus unusually large $(+6.3)$.

Investigation of acid-base relationships among diazonium derivatives is further complicated by isomerism in the diazoates and diazohydroxides, which is discussed later in connection with structure. The initially formed diazoate anions have traditionally been assigned the syn configuration; they isomerize more or less slowly to less reactive forms, commonly considered to be the anti isomers. The two forms have different acidity constants, and the isomerization may be fast enough to interfere with accurate determination of their values. However, by means of kinetic rather than equilibrium measurements, the constants for the various equilibria can be separately evaluated for the most part.[30] The magnitude of the problem is shown by Equation 4-2, which includes the possibility of tautomerism with the nitrosamine

$$ArN_2^+ + OH^- \rightleftharpoons syn\text{-}ArN{=}NOH \underset{}{\overset{OH^-}{\rightleftharpoons}} syn\text{-}ArN{=}NO^-$$

$$anti\text{-}ArN{=}NOH \overset{OH^-}{\rightleftharpoons} anti\text{-}ArN{=}NO^- \qquad (4\text{-}2)$$

structure, which is also capable of geometric isomerism owing to restricted rotation about the N—N bond (see below).

The *anti*-diazoates are favored in basic solution, and the pK values of *anti*-diazohydroxides are estimated to be about one pK unit smaller (i.e., stronger) than the *syn* forms.[34] The value of pK_a for *anti*-benzenediazohydroxide is reported to be 7.2×10^{-7} at 20°C,[35] and for *anti*-p-toluenediazohydroxide, a pK_a value of 7.40 is reported.[34] For several substituted examples, constants for the two neutralization stages of the diazonium ion have been reported; for 3,5-dichlorobenzenediazonium ion, $K = 6.3 \times 10^2$ for formation of ArN_2OH and 10^{-7} for its conversion to *syn*-ArN_2O^-.[36]

The infrared spectra of diazonium salts show absorption near 2260 to 2300 cm^{-1} owing to triple-bond stretching.[7,37,38] The value of this stretching frequency correlates very well with the Hammett σ^+ constants.[39] Such a band is not found in the convalent diazo compounds, such as benzenediazocyanide, which shows only nitrile absorption in the triple-bond region, and p-chlorobenzenediazoanhydride, which has no absorption between 1500 and 2800 cm^{-1}.[40] Benzenediazooxides, which can be considered to be conjugated diazo ketones or p-diazoniophenoxides, absorb at 2070 to 2080 cm^{-1} and thus resemble open-chain diazo ketones more than diazonium salts.[41] The *ortho*-diazooxides are similar, but their sulfur analogs are transparent in the 2000 cm^{-1} region and must be considered to be benzothiadiazoles.[42]

Infrared absorption assignable to N=N stretching is absent or very weak in most azo compounds, owing to electronic symmetry. Such absorption has been identified variously with bands between 1460 and 1550 cm^{-1}.[8,43-45] A study of a group of unsymmetrical azo compounds, however, revealed a weak, common band at 1562 ± 5 cm^{-1},[46] and methyldiazene absorbs at 1575 cm^{-1}, surely owing to N=N.[18]

The electronic spectra of diazonium salts have not received much attention, being overshadowed by the great importance of the spectra of azo compounds in connection with their widespread use as dyes and indicators. Benzenediazonium ion, which is colorless, shows two moderately strong maxima in the near UV (263 and 298 nm); *meta*-substituted examples are similar, but *para*-substituted examples are reported to show only the shorter wavelength absorption.[47] The extensive literature on the subject of color and constitution of azo dyes has been reviewed elsewhere[28,48] and will not be taken up here. Aliphatic azo compounds have a weak yellow color that results from end absorption of a low-intensity band in the near UV in the region 340 to 380 nm, believed to be due to an $n{\rightarrow}\pi^*$ transition;[44-46,49] this band may appear as an inflection or shoulder on a stronger band farther from the visible. For azomethane, the maximum comes at 343 to 357 nm, and for methyldiazene, at 361 nm.[16] With electronically active groups, the $n{\rightarrow}\pi^*$ transition may occur anywhere in the range 300 to 600 nm.[50] The absorption is more intense in aliphatic *cis*-azo compounds, and much more so in *cis*-azobenzene.

Conjugation with one or two aryl substituents introduces the possibility of strong bands owing to $\pi{\rightarrow}\pi^*$ transitions, but these do not appear in the visible range unless auxochromic groups, such as amino or hydroxy, are also present in the *ortho* or *para* positions. It is the resulting shift of this band into the visible region that is responsible for the deep color of the azo dyes (the color of azobenzene itself results from the weak $n{\rightarrow}\pi^*$ absorption at 343 nm; the $\pi{\rightarrow}\pi^*$ absorption is at 319 nm). Since the absorption maximum is sensitive to the electronic character of the substituents, it is easily affected by protonation or deprotonation of the common auxochromic groups; this phenomenon is responsible for the acid-base indicator properties of such compounds as methyl orange. With p-dimethylaminoazobenzene, a simple representative, the conjugate acid is of redder color. The evidence indicates that protonation occurs in part on the amino group and in part on the azo group, with the latter favored.[51-53] It is significant, however, that even with unsubstituted azobenzene, the $\pi{\rightarrow}\pi^*$ transition is shifted toward the visible, and the conjugate acid is deep red.[49] The photoelectron spectra of *cis* and *trans* azoalkanes have been recorded.[54]

The NMR spectra of diazonium salts demonstrate the strongly electron-withdrawing effect of the diazonio substituent.[7] p-Chlorobenzenediazonium fluoroborate, for example, shows signals at δ 9.06 and 8.50 ppm, considerably downfield from those of the corresponding potassium diazoate (δ 7.74 and 7.26 ppm for the *syn* isomer, δ 7.83 and 7.74 ppm for the *anti*)[55] and of azobenzene (δ 7.9 and 7.45 ppm for the *trans* isomer in chloroform). The azo group is markedly deshielding in azoalkanes, as shown by *trans*-azomethane, δ 3.68 ppm, *trans*-2,2′azopropane, δ 3.56 ppm (α—H), and *cis*-2,2′-azopropane, δ 4.05 ppm (α—H).[56,57] For ^{13}C NMR, the diazonio substituent exhibits the expected downfield shift.[7,58,59] Electron-donating *para*-substituents have an interesting but logical effect of shifting the ^{15}N NMR downfield for both nitrogens, presumably owing to the fact that conjugation increases the resemblance to the aliphatic diazoalkane structure.[59]

The substituent effect of the diazonio group manifests itself in chemical as well as spectroscopic consequences. p-Diazoniophenol, for example, has pK_a of 3.40, about 10^6 times that of phenol.[60,61] The Hammett substituent constant σ_p is estimated to be between 1.5 and 3, and σ_m, 1.7. The azo group, however, is apparently *ortho-para* directing toward electrophilic aromatic substitution, and only when it is protonated does it strongly activate a benzene ring toward nucleophilic attack.[62] p-Phenylazophenol ($pK_a = 8.38$ in 20% ethanol) is about 10 times stronger than phenol itself.[63a] The acid strength of substituted benzeneazophenols has been correlated[63b] with the substituent constant σ^-.

The structure of diazonium and related azo compounds was a matter of major controversy for decades, but most of the questions have been resolved as more powerful physical methods have become available. Most of the questions center about isomerism. Ionic diazonium salts have not been observed in isomeric forms, but diazoates and most azo compounds are known in two forms. As has been implied in the discussion of properties thus far, these forms are generally believed to be a consequence of geometric isomerism about the N=N bond. The historical development of the evidence has been reviewed in detail elsewhere[28] and will be only briefly surveyed here.

Diazoates, diazohydroxides, diazocyanides, etc., when first formed from diazonium salts, are usually in a less stable, reactive form that more or less slowly changes to a less reactive isomeric form. By analogy with oximes and olefins, the less stable forms were assigned the *syn-* (*cis*) configuration. The two forms are completely distinct in solution as well as in the solid state. Differences in electronic absorption spectra, dipole moments, and other physical properties are consistently parallel among a wide variety of diazoates,[64,65] diazocyanides,[40] diazosulfonates,[66,67] and especially significantly, azobenzenes. From the large amount of accumulated data, it seems inescapable that the existence of pairs of isomers is in all cases due to the same cause.

The existence of geometric isomers is on the firmest footing in the case of azobenzene, where both forms have been investigated by x-ray diffraction.[68,69] The stable form, definitely shown to be *trans*, is converted by irradiation into the *cis* form, which slowly reverts back in the dark.

trans, mp 68°C, dipole moment:[70] 0   *cis*, mp 71°C, dipole moment: 3.0 D.

The N—N distances are the same, but are longer than those determined in diazonium salts,[71-74] 1.10–1.11 Å. Some *trans*-arenediazocyanides have also been investigated by x-ray crystallography, with comparable results (Ar—N, 1.29–1.42 Å; N=N, 1.25–1.32 Å; N—C, 1.36–1.4 Å; C=N, 1.14–1.15 Å).[75,76] Unfortunately, the *cis* isomers were not sufficiently stable for structure determination, but the proposal that they might be isocyanides has been disposed of by synthesizing the latter, as well as *anti*-diazocyanides, in other ways.[77] Almost all other evidence is also against it.[78] The potassium salt of *cis*-methanediazoate has also been examined by x-ray crystallography.[79] It has the presumed structure, with C—N = 1.477 Å, N=N = 1.246 Å, and N—O = 1.306 Å. At least in this example, the results rule out the alternative suggestion that the presumed *syn*-diazoates might be structural isomers having oxygen on the substituted nitrogen, R—N(→O)N⁻.[80] The *trans* configuration of the stable form of azomethane has been confirmed by electron diffraction.[81]

Less is known about *cis*-azoalkanes, and investigation of the energy difference between isomers is hindered by the fact that the *cis* isomers lose nitrogen as readily as they isomerize, and the equilibrium cannot be directly determined in every case. However, whereas *cis*-2,2'-azopropane loses N_2 easily, 1,1'-azoadamantane and 1,1'-azonorbornane are more stable, and their isomerization to the *trans* forms can be followed.[82,83]

Ab-initio calculations of geometry and energy of nine isomers of H_2N_2O imply that the *cis*-diazohydroxides are the most stable, followed closely by the trans isomers and then the nitrosamines.[84]

Formazans[85] behave as members of a class distinct from either simple azo compounds or hydrazones, and the formazyl group may properly be considered as a unit functional group. Most of the known representatives bear aryl groups on nitrogen and are deeply colored. The formazan function is very weakly basic, and salts are hydrolyzed by dilution. The anion is symmetrical about the central carbon, similar to carboxylate anions (Eq. 4-3), and it is thus not surprising that formazans should be acidic enough to form salts with strong alkalies.

$$R-C \overset{N=N-Ar}{\underset{N-NH-Ar'}{<}} \quad \underset{H^+}{\overset{B:}{\rightleftharpoons}} \quad R-C \overset{N=N-Ar}{\underset{N-N-Ar'}{<}} \overset{\ominus}{\ominus} \tag{4-3}$$

Nitrosazones and nitrazones are in general crystalline solids with a distinct acidic character.[86,87] N-Phenylnitrosazone (phenylazoformaldoxime) is a yellow solid (mp 94°C); N-phenylnitrazone, $PhN=N-CH=NO_2H$, exists in two modifications (α, mp 75°C; β, mp 85°C), which may perhaps be geometric isomers. The pK_a values of N-arylnitrosazones have been correlated with σ_x, $\rho = -0.6$, for *para* substituents; N-phenylnitrosazone has a pK_a of 8.48. The UV spectra have a weak maximum near 430 nm ($n\rightarrow\pi^*$) and an intense one near 300 nm ($\pi\rightarrow\pi^*$); the NMR of the central CH falls at δ 8.5–8.8 ppm.[88]

The discovery of exotic and unstable structures in living things continues to yield surprises, and even a diazonium salt, [p-(hydroxymethyl)benzenediazonium], has been identified as a product of enzymatic oxidation of the corresponding hydrazine in the stem of the common and delectable mushroom *Agaricus bisporus*.[89]

With regard to toxicity, although some diazonium species have antiseptic properties, they should be assumed to be toxic, and possibly carcinogenic, in view of their close relation to anilines. Certain azo dyes have been found to have low-level carcinogenicity and have been banned as food colors. Azo compounds are not regarded as dangerous in the laboratory, but caution against prolonged contact with the skin is advisable.

REACTIONS

Diazonium compounds

The most clean-cut distinction among reactions of diazonium compounds is whether nitrogen is retained or not. Those in which it is retained consist of coupling reactions, in which the diazonium group joins to a nucleophilic site, reduction, oxidation, and substitution on the arene ring induced by a diazonio substituent. This group will be taken up first, in the order mentioned.

Coupling to oxygen.   Coupling of diazonium ions appears to require a fairly nucleophilic site, and oxygen sites qualify only marginally. Coupling to hydroxide ion to form diazohydroxides and diazoates has been described in the previous section. Coupling to alkoxide ions has been difficult to observe, owing to strong competition with dediazoniation processes. With CH_3O^-, equilibration with *syn*-diazoether occurs (Eq. 4-4); isomerization to the *anti* isomer takes place slowly, in competition with a free-radical reaction leading to loss of N_2 (to be discussed later). The rate of disappearance of ArN_2^+ is second-order. For the m-chlorophenyl example, the equilibrium constant[90] is 4.2×10^5. Very exceptionally, formation of a diazoether

$$\text{ArN}_2{}^+ + \text{CH}_3\text{O}^- \;\rightleftharpoons\; \underset{\text{N}=\text{N}}{\overset{\text{Ar} \qquad \text{OCH}_3}{\diagdown \qquad \diagup}} \xrightarrow{\text{slow}} \underset{\text{N}=\text{N}}{\overset{\text{Ar}}{\diagdown \qquad}} \underset{\text{OCH}_3}{\diagdown} \qquad (4\text{-}4)$$

by coupling with an alcohol has been reported,[91] as well as to the oxygen of p-nitrophenol[92] and naphthol.[93] Formation of diazo acetates from diazonium and acetate ions has been estimated to have a very unfavorable equilibrium constant (ca. 10^{-5}).[94] The formation of diazo anhydrides from diazonium ions and diazoate, mentioned in the previous section, is another example.

Coupling to sulfur.   Coupling to thiophenols occurs rapidly to form *syn*-diazosulfides,[90] which then may isomerize to the *anti* configuration, but they also decompose quite easily into diaryl sulfides, etc.[95,96] The equilibrium constant for formation of p-NCC$_6$H$_4$NNSPh is 1.8×10^{10} at $23\,°$C. Although displacement of N$_2$ has been thought to be the dominant reaction with aliphatic mercaptans, crystalline diazothio ethers have been isolated,[97] as shown in Equation 4-5.[98] With hydrogen

$$\text{ArN}_2{}^+\text{BF}_4{}^- + \text{RSH} \xrightarrow[\text{acetone, } 0\,°\text{C}]{\text{pH 5-6}} \text{ArN}=\text{NSR} \quad (syn \xrightarrow{\text{slow}} anti) \qquad (4\text{-}5)$$

sulfide, diazothioanhydrides (ArN=N—S—N=NAr) are formed;[99] they are dangerously explosive, even when wet.[100]

Sulfur in the $+4$ oxidation state couples readily. Sulfinates give rise to diazosulfones[101,102] (Eq. 4-6). Sulfite ion forms *syn*-diazosulfonates, which rather rapidly isomerize to the *anti* forms.[66,103] This isomerization can be reversed photochemically.[104] Reduction of the diazonium group is a competitive reaction that may become dominant in acidic solution.[105] Displacement of the diazonium group does not appear to be important, except in the case of sulfur dioxide (discussed later). The *syn*-diazosulfonates equilibrate with diazonium ion and can therefore take part in coupling reactions with aromatic rings to produce azo dyes. This fact makes photoisomerization of *anti* to *syn* important in the photographic industry, for the *anti*-diazosulfonates are stable for long periods in the dark.

$$\text{ArN}_2{}^+ + \text{Ar}'\text{SO}_2{}^- \;\rightleftharpoons\; \text{Ar}-\text{N}=\text{N}-\text{SO}_2\text{Ar}' \qquad (4\text{-}6)$$

Coupling to nitrogen.   Diazonium salts seem to be able to couple with most types of nitrogen compounds having an N-attached hydrogen. The initial products are often stable enough to isolate, especially when α-elimination cannot occur with formation of an azide (examples of such reactions are to be found in the section Preparation in Chap. 6). Ammonia reacts with two equivalents of diazonium ion to form pentazadienes[106] (Eq. 4-7), and only rarely have aryltriazenes been isolated.[107]

$$\text{ArN}_2{}^+ + \text{NH}_3 \longrightarrow \text{ArN}=\text{N}-\text{NH}_2 \xrightarrow{\text{ArN}_2{}^+}$$
$$\text{ArN}=\text{N}-\text{NH}-\text{N}=\text{NAr} \qquad (4\text{-}7)$$

Primary amines, aliphatic and aromatic, also couple easily with diazonium salts (Eq. 4-8). The triazenes produced, sometimes called *diazoamino compounds,* do not cou-

$$\text{ArN}_2{}^+ + \text{RNH}_2 \longrightarrow \text{ArN}{=}\text{N}{-}\text{NHR} \longrightarrow (\text{ArN}{=}\text{N}{-})_2\text{NR} \qquad (4\text{-}8)$$

ple further when derived from anilines. With aliphatic amines, such as methylamine, it is difficult to obtain only triazenes, unless the diazonium salt bears an electron-withdrawing substituent,[108] for diazonium salts readily couple twice to form pentazadienes, accompanied by some diaryltriazene, which is also formed when the diazonium salt is treated with acetamide, urea, or even sodium acetate, presumably as a result of reactions that generate some aniline first. Triazenes have been obtained more cleanly by treating dry diazonium fluoroborates with primary amines and solid Na_2CO_3 in dimethylformamide.[109] With secondary amines, coupling is limited to formation of triazenes. If an N-alkyl group also bears a suitable leaving group at a β position, internal alkylation to form a triazolinium salt follows.[110]

Coupling to aromatic amino groups must compete with coupling to a ring carbon. Since coupling to nitrogen is reversible, especially in acid solution, diaryltriazenes are more or less easily isomerized to aminoazobenzenes, the products of ring coupling (see Chap. 7). Furthermore, since triazenes with two different aryl groups equilibrate readily with their tautomers, another consequence of reversibility of coupling to nitrogen is diazonium-amine interchange (Eq. 4-9), which can occur in

$$\text{ArN}_2{}^+ + \text{Ar}'\text{NH}_2 \rightleftharpoons \text{ArNHN}{=}\text{NAr}' \rightleftharpoons \text{ArN}{=}\text{NNHAr}' \rightleftharpoons \text{ArNH}_2 + \text{Ar}'\text{N}_2{}^+ \qquad (4\text{-}9)$$
$$\downarrow$$
$$p\text{-ArN}{=}\text{N}{-}\text{Ar}'\text{NH}_2$$

acid solution. It has been demonstrated by isotopic labeling that the alternative possibility, reversible diazotization, is not involved.[111]

Hydrazine derivatives may couple with one or two equivalents of diazonium salt; in the latter case, coupling occurs preferentially at separate nitrogens[112] (Eq. 4-10).

$$\text{RNH}{-}\text{NHR} \xrightarrow{\text{ArN}_2{}^+} \text{RN} \overset{\text{NHR}}{\underset{\text{N}{=}\text{NAr}}{}} \xrightarrow{\text{ArN}_2{}^+} \underset{\underset{\text{ArN}}{\overset{\|}{\text{N}}} \quad \underset{\text{NAr}}{\overset{\|}{\text{N}}}}{\text{RN}{-}\text{NR}} \qquad (4\text{-}10)$$

Unless both nitrogens are already substituted, however, elimination to form an azide (Eq. 4-11) occurs so easily, as it does with the coupling product from hydra-

$$\text{ArN}_2{}^+ + \text{Ar}'\text{NHNH}_2 \rightleftharpoons \text{Ar}'\text{N} \overset{\text{NH}_2}{\underset{\text{N}{=}\text{NAr}}{}} \longrightarrow \text{Ar}'\text{NH}{-}\text{NH}{-}\text{N}{=}\text{NAr}$$
$$\longrightarrow \text{ArN}_3 \qquad \text{Ar}'\text{N}_3 \qquad \text{ArNH}_2 \qquad \text{Ar}'\text{NH}_2 \qquad (4\text{-}11)$$

zine itself, that the initial coupling products can seldom be isolated.[113] With aryl hydrazines, attack is apparently faster at the arylated nitrogen, but in acid solution this is reversible, allowing the isomeric 1,4-diaryltetrazene to form slowly and then fragment.[114] Most of the known examples of double coupling to form a hexazadiene involve reaction with a diacylhydrazine.[112]

Coupling of diazonium salts to hydrogen azide (or azide ion) occurs very rapidly and is usually complete in less than a minute at $0°C$. In the first stage, unstable substances of the formula $Ar—N_5$ are formed; in most instances, these decompose spontaneously below room temperature to an aryl azide and nitrogen (Eq. 4-12).

$$Ar—\overset{*}{N}≡\overset{+}{N} + N_3^-$$

$$Ar—N=\overset{*}{N}—\overset{+}{N}=N=N^- \longrightarrow Ar—N=\overset{*}{N}=\overset{+}{N}N^- + N_2$$
$$(\text{Major})$$

$$Ar—N\underset{N=N}{\overset{\overset{*}{N}=N}{\diagdown}} \longrightarrow \begin{cases} Ar—N=\overset{+}{N}=N^- + N\overset{*}{N} \\ Ar—N=\overset{*}{N}=\overset{+}{N}^- + NN \end{cases}$$
$$(\text{Minor})$$

$$(4\text{-}12)$$

Kinetic studies of the evolution of nitrogen reveal that two distinct ArN_5 species are present, each of which decomposes at its own rate in a first-order process.[115] The more stable ArN_5 has been isolated in several instances. Labeling with ^{15}N at appropriate sites in the reactants reveals that the more-stable ArN_5 contains only three chemically different nitrogen sites; this fact indicates that they are arylpentazoles, as shown in Equation 4-12. The less stable isomers, which have not been isolated and in which the separate identity of each of the five nitrogens is preserved, must therefore be diazoazides, having a chain of five nitrogen atoms. The rate-limiting step for the overall conversion of a diazonium salt to an aryl azide may be either the coupling step or the subsequent fragmentation, depending on structure and solvent.[90]

Hydroxylamine couples with diazonium salts presumably to give hydroxytriazenes, but very few have been isolated. Elimination of water to form an aryl azide follows very rapidly. When the coupling is accomplished in basic solution, however, the principal products are the aryl amine and nitrous oxide (Eq. 4-13), notwith-

$$ArN_2^+ + NH_2OH \begin{cases} \xrightarrow{\text{OH}^-} ArNH_2 + N_2O \\ \xrightarrow{pH \leq 7} ArN_3 + H_2O \end{cases}$$

$$(4\text{-}13)$$

standing the fact that coupling in acid or neutral solution followed rapidly by addition of base gives only azide.[116] Presumably, prototropy is involved, for the formation of amine and nitrous oxide corresponds to decomposition of the nitrosohydrazine tautomer. N-Substituted hydroxylamines[117-120] give rise to more-stable coupling products, as shown in Equation 4-14.

$$ArN=N—\underset{OH}{N}R \xleftarrow{\text{RNHOH}} ArN_2^+ \xrightarrow{R_2NOH} ArN=N—\overset{+}{N}\underset{O^-}{R_2}$$

$$(4\text{-}14)$$

For coupling with other nitrogen compounds, the chapters dealing with them should be consulted. The coupling products (triazenes, pentazadienes, etc.) are discussed in Chapter 7.

Coupling to carbon. The diazonium group is capable of electrophilic attack on a benzene or naphthalene ring providing that the ring is suitably activated by hydroxy or amino groups; azobenzenes are formed. This reaction has been extensively investigated because of its great importance in the dye industry. Diazonium salts bearing electron-withdrawing substituents, particularly nitro, are the most strongly electrophilic and couple the most readily.[121] Phenols react best as the anions,[122] in which the ring is more strongly activated, and aryl ethers react only sluggishly. Dialkylanilines couple in the form of the free base, and the ammonium ions are inactive. As a consequence, it is possible to control the site of coupling on an aminonaphthol by proper selection of the acidity of the medium (Eq. 4-15).

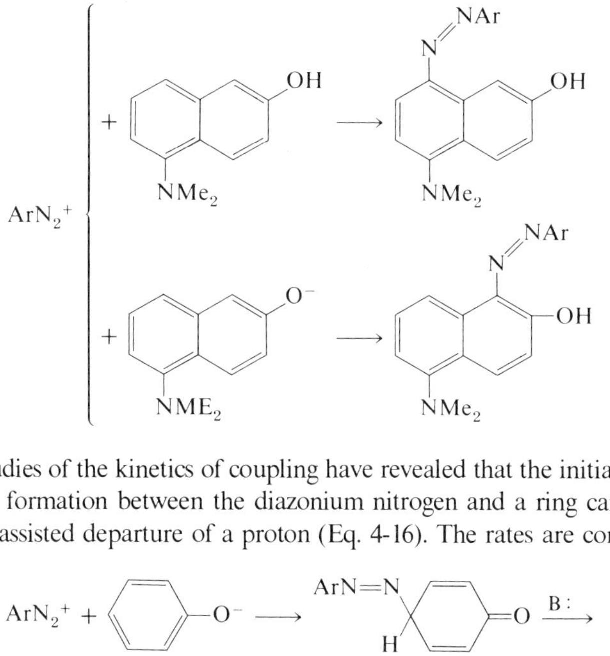

$$(4\text{-}15)$$

Studies of the kinetics of coupling have revealed that the initial step is reversible bond formation between the diazonium nitrogen and a ring carbon, followed by base-assisted departure of a proton (Eq. 4-16). The rates are correlated[39] with σ^+.

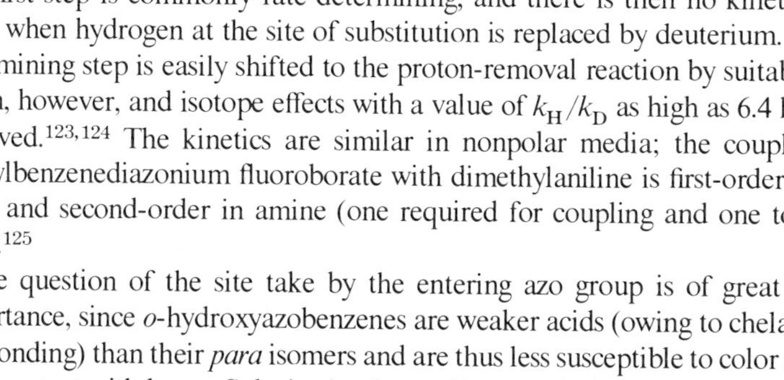

$$(4\text{-}16)$$

The first step is commonly rate-determining, and there is then no kinetic isotope effect when hydrogen at the site of substitution is replaced by deuterium. The rate-determining step is easily shifted to the proton-removal reaction by suitable substitution, however, and isotope effects with a value of k_H/k_D as high as 6.4 have been observed.[123,124] The kinetics are similar in nonpolar media; the coupling of p-t-butylbenzenediazonium fluoroborate with dimethylaniline is first-order in diazonium and second-order in amine (one required for coupling and one to act as a base).[125]

The question of the site take by the entering azo group is of great technical importance, since o-hydroxyazobenzenes are weaker acids (owing to chelate hydrogen bonding) than their *para* isomers and are thus less susceptible to color alteration from contact with bases. Substitution is usually overwhelmingly *para* with benzene derivatives, but coupling to the naphthalene ring is sensitive both to the nature of the diazonium compound and to the reaction conditions. It can be made to occur predominantly *ortho* to a hydroxy group situated at an α position. These effects have been attributed to the incursion of species other than ArN_2^+ as the attacking agent and to the steric requirements of the base required to remove a proton from the ring. The rate of mixing has also been found to influence the site of coupling.[126] Multiple coupling to produce bis- and tris-azo compounds has been observed.

The preference for *para* coupling with benzene derivatives is so strong that the entering diazonium group will even displace another substituent already present if it can be ejected as a cation of not too high energy. *p*-Dimethylaminobenzyl alcohols, for example, are cleaved in good yields by diazonium salts to aldehydes and *p*-dimethylaminoazobenzenes[127,128] (Eq. 4-17).

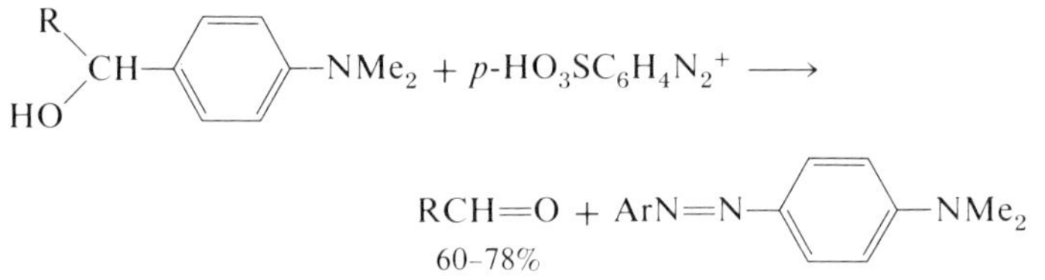

$$\begin{array}{c}R\\ \diagdown\\ CH \\ \diagup \\ HO\end{array}\!\!-\!\!\bigcirc\!\!-\!\!NMe_2 + p\text{-}HO_3SC_6H_4N_2^+ \longrightarrow$$

$$RCH{=}O + ArN{=}N\!\!-\!\!\bigcirc\!\!-\!\!NMe_2 \qquad (4\text{-}17)$$
$$60\text{-}78\%$$

Coupling to aryl alkyl ethers, which occurs with the more strongly electrophilic diazonium ions, is accompanied by replacement of the alkoxy group by hydroxy.[129] Experiments with ^{18}O labeling have shown that the entire alkoxy group is lost and the hydroxy group in the product comes from water, evidently by a nucleophilic displacement at the intermediate cationic stage[130] (Eq. 4-18).

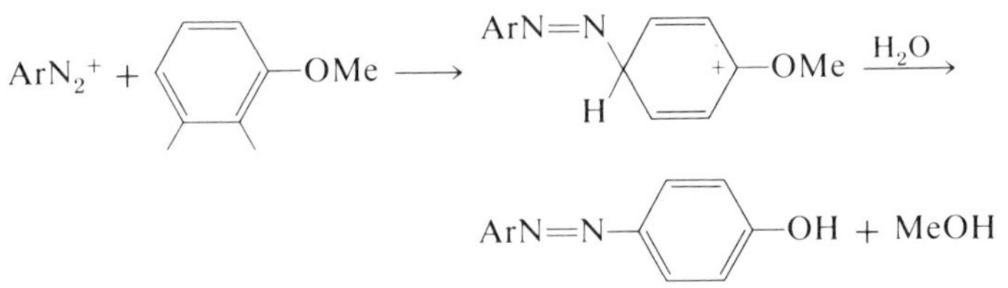

$$ArN_2^+ + \bigcirc\!\!-\!\!OMe \longrightarrow \overset{ArN{=}N}{\underset{H}{\diagup}}\!\!\bigcirc\!\!\overset{+}{-}\!OMe \xrightarrow{H_2O}$$

$$ArN{=}N\!\!-\!\!\bigcirc\!\!-\!\!OH + MeOH \qquad (4\text{-}18)$$

Alkylbenzenes undergo coupling only in exceptional circumstances; 2,4,6-trinitrobenzenediazonium ion will couple with mesitylene, 1,2,3,5-tetramethylbenzene, or pentamethylbenzene, but not durene.[131,132] Polynuclear aromatic hydrocarbons may react somewhat more readily, and benz[*a*]pyrene, for example, has been observed to couple with *p*-nitrobenzenediazonium ion.[133]

Coupling to nonaromatic carbon occurs readily if the site is sufficiently nucleophilic. The simplest case is probably cyanide ion, which first gives orange *syn*-diazocyanides, which then rearrange to the somewhat redder *anti* isomers.[64,102,134] The potential carbanions of organometallic compounds will engage in coupling if dry diazonium salts are used[135] (Eq. 4-19). The most important reactions, however,

$$ArN_2^+ + R_2Zn \xrightarrow{\text{DMF}} ArN{=}NR \qquad (4\text{-}19)$$
$$90\%$$

are those with compounds having activated CH, as found in β-dicarbonyl compounds.[136] Reaction occurs with the enolate and is thus facilitated by base. If the site is an active methylene group, the initially formed azo compound quickly isomerizes to an arylhydrazone[137,138] (Eq. 4-20). With β-keto acids, decarboxylation accompanies the process, and arylhydrazones of α-keto aldehydes are formed[139] (Eq. 4-21).

$$ArN_2^+ + RCOCH_2COR \longrightarrow ArN{=}N{-}CH(COR)_2 \longrightarrow$$

$$ArNH{-}N{=}C(COR)_2 \qquad (4\text{-}20)$$

$$ArN_2^+ + RCOCH_2CO_2H \longrightarrow RCOCH{=}N{-}NHAr + CO_2 \qquad (4\text{-}21)$$

However, double coupling is a general consequence if excess diazonium ion is present, and formazans are then the products. With β-keto esters, the acyl group is cleaved off[140] (Eq. 4-22). Malonic ester reacts with up to three equivalents of diazo-

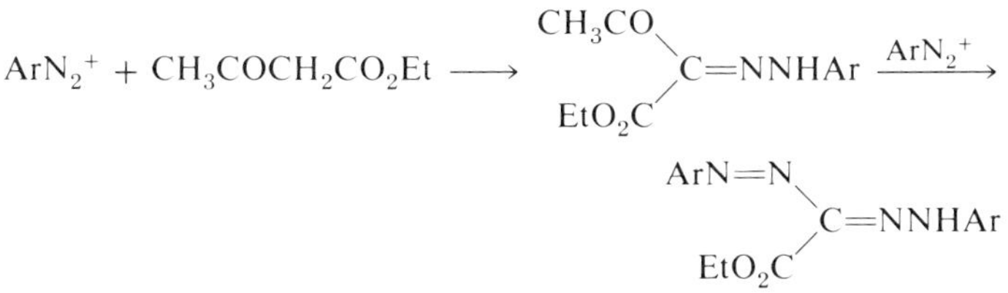

$$ArN_2^+ + CH_3COCH_2CO_2Et \longrightarrow \begin{array}{c} CH_3CO \\ \diagdown \\ \diagup \\ EtO_2C \end{array} C{=}NNHAr \xrightarrow{ArN_2^+}$$

$$\begin{array}{c} ArN{=}N \\ \diagdown \\ \diagup \\ EtO_2C \end{array} C{=}NNHAr \qquad (4\text{-}22)$$

nium salt, eventually losing both ester groups solvolytically to give rise to azo-formazans[141] (Eq. 4-23).

$$3ArN_2^+ + CH_2(CO_2Et)_2 \longrightarrow ArNH{-}N{=}C(N{=}NAr)_2 \qquad (4\text{-}23)$$

Compounds with only one activated CH, such as α-alkyl-β-keto esters, at first form isolable azo compounds, but the driving force for formation of the hydrazone structure is so great that a C-acyl group is easily ejected to permit it[138] (Eq. 4-24).

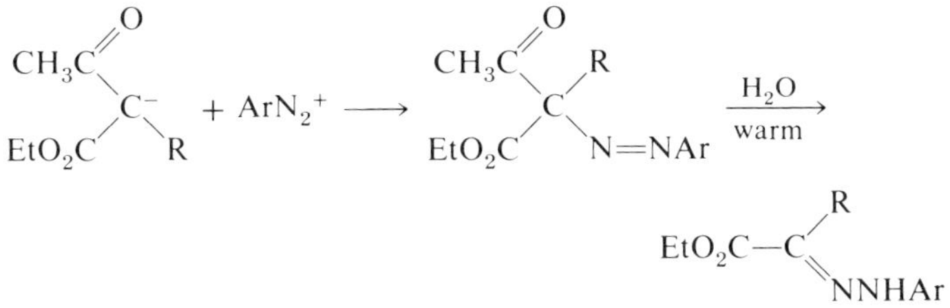

$$\begin{array}{c} O \\ \diagup\!\!\!\parallel \\ CH_3C \\ \diagdown \\ C^- \\ \diagup \diagdown \\ EtO_2C \quad R \end{array} + ArN_2^+ \longrightarrow \begin{array}{c} O \\ \diagup\!\!\!\parallel \\ CH_3C \quad R \\ \diagdown \diagup \\ C \\ \diagup \diagdown \\ EtO_2C \quad N{=}NAr \end{array} \xrightarrow[\text{warm}]{H_2O}$$

$$\begin{array}{c} R \\ \diagup \\ EtO_2C{-}C \\ \diagdown \\ NNHAr \end{array} \qquad (4\text{-}24)$$

The overall process is known as the Japp–Klingemann synthesis,[142] and it is a useful route to α-amino acids when followed by reduction of the hydrazone function to —NH$_2$.

Nitriles generally require an additional activating substituent for easy coupling, as do simple ketones. However, acetone can be made to react twice at the same site to form an acetylformazan,[143] and pyruvic acid reacts more readily in the same way, producing formazanoxalic acids.[144] A nitro group, however, is sufficient by itself to promote diazo coupling at the α position.[145] Nitromethane easily couples twice to form nitroformazans as the principal products[146,147] (Eq. 4-25).

$$2ArN_2^+ + CH_3NO_2 \longrightarrow ArN=N-C\overset{NO_2}{\underset{N-NHAr}{<}} \tag{4-25}$$

As the foregoing reactions imply, aldehyde hydrazones can undergo coupling at the aldehydic CH; this is, in fact, a convenient route to formazans.[84,85,148] Olefins, however, are not generally attacked by diazonium ions unless the double bond is unusually nucleophilic. Enamines are generally susceptible, as shown by vinyl-dimethylamine[149] and styryldialkylamines[150] (Eq. 4-26). Even 1-*p*-methoxyphenyl-

$$PhCH=CHNR_2 + ArN_2^+ \longrightarrow PhC\overset{N=NAr}{\underset{CHNR_2}{<}} \tag{4-26}$$

propene is attacked, but the result is more than simple coupling, for the propene chain is cleaved at the double bond[151] (Eq. 4-27). *o*-Diazoniostilbene reacts analogously with itself to form indazole and benzaldehyde.[152]

$$p\text{-}CH_3OC_6H_4CH=CHCH_3 + ArN_2^+ \longrightarrow$$
$$p\text{-}CH_3OC_6H_4CH=NNHAr + CH_3CH=O \tag{4-27}$$

Terminal acetylenes couple to form azoacetylenes if their silver salts are treated with dry diazonium salts[153] (Eq. 4-28), but in water solution, the triple bond is hydrated and the products are arylhydrazones of α-keto aldehydes.[154]

$$ArN_2^+X^- \begin{cases} + PhC{\equiv}CH \longrightarrow [Ph\overset{+}{C}=CHN_2Ar] \xrightarrow{H_2O} PhC\overset{O}{\underset{CH=NNHAr}{<}} \\[2mm] + ArC{\equiv}CAg \xrightarrow{EtOH/CHCl_3} ArC{\equiv}C-N=NAr \\[1mm] \qquad\qquad\qquad\qquad\quad 33\text{–}87\% \end{cases} \tag{4-28}$$

Intramolecular coupling.   If a site for coupling at C, N, or S is so situated that a five- or six-membered ring can be formed from an *ortho*-substituted diazonium salt, indazoles, benzotriazoles, or benzothiadiazoles may be formed. Benzothiadiazole is formed in 95% yield by diazotization of *o*-benzylthioaniline, with removal of the benzyl group[155] (Eq. 4-29). When the ring closure is to an *o*-methyl group, as in

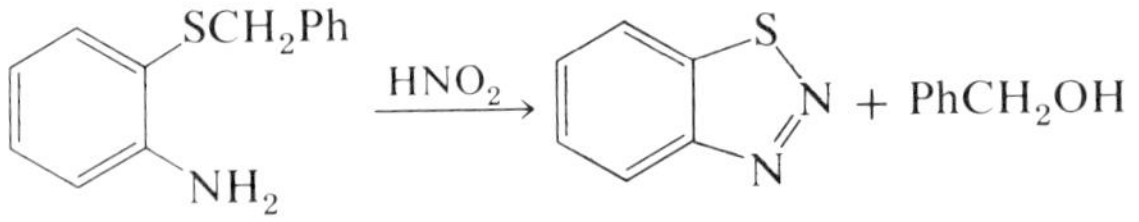

$$\underset{NH_2}{\overset{SCH_2Ph}{\bigcirc}} \xrightarrow{HNO_2} \underset{N}{\overset{S}{\bigcirc}}N + PhCH_2OH \tag{4-29}$$

o-toluenediazonium salts, the diazohydroxide or diazoacetate must be used, and results are inclined to be erratic. Removal of a proton from the methyl group to form a zwitterionic species (Eq. 4-30) is believed to be required.[156] *o*-Diazooxides, derived from *o*-aminophenols by diazotization, do not detectably cyclize to benzo-oxadiazoles. Diazotized *o*-aminostyrene cyclizes to cinnoline, and diazotized *o*-aminophenyl ketones close to 4-hydroxycinnolines.[157,158] These are known as the *Widman-Stoermer* and *Borsch cinnoline syntheses*, respectively. Diazotized

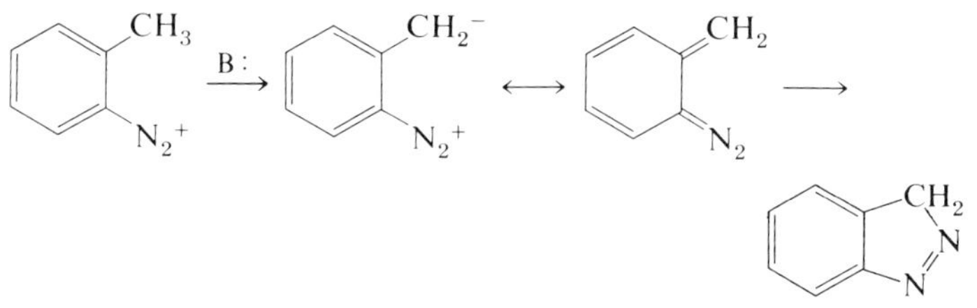

$$
\text{(structures)} \qquad (4\text{-}30)
$$

o-aminophenylacetylenes give the same products as *o*-aminophenyl ketones; this route is known as the *Richter synthesis*.

Coupling to a metal atom. There are many instances of coupling of a diazonium ion to a transition metal in complexed form. It thus becomes a ligand of the complex, with or without displacement of a ligand already present. A simple example is shown in Equation 4-31; elaboration of this subject may be found elsewhere.[159, 160]

$$
Pt(PPh_3)_3 + ArN_2^+X^- \longrightarrow ArN{=}NPt(PPh_3)_3{}^+ \ X^- \qquad (4\text{-}31)
$$

Nucleophilic substitution activated by the diazonium group. There is a group of reactions in which a substituent on a benzene ring *ortho* or *para* to a diazonium function is displaced unusually easily by nucleophiles, while leaving the diazonium group itself intact.[161] The powerful electron-withdrawing effect of the diazonium function is responsible; it is greater than either a nitro or a quaternary ammonium substituent and approximates that of two nitro groups.[162] When the effect of the diazonium function is reinforced by a nitro group also *ortho* or *para* to a displaceable group, displacement can occur in cold aqueous solution, even during the diazotization process[163] (Eq. 4-32). Displaceable groups include nitro (as NO_2^-),

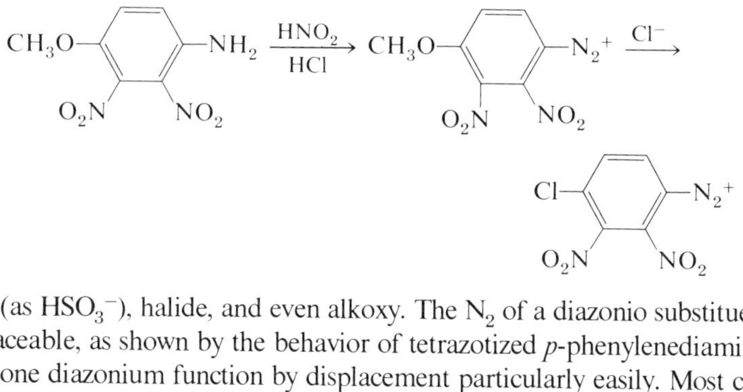

$$
\text{(structures)} \qquad (4\text{-}32)
$$

sulfo (as HSO_3^-), halide, and even alkoxy. The N_2 of a diazonio substituent is also displaceable, as shown by the behavior of tetrazotized *p*-phenylenediamine, which loses one diazonium function by displacement particularly easily. Most commonly the displacing nucleophile in all these reactions is water, but halide ions, nitrite, and thiocyanate have also been observed to react in this way.[163-165]

Reduction with retention of one or both nitrogens. Reduction of diazonium salts may proceed to aryldiazenes, arylhydrazines, arylamines, azobenzenes, arene hydrocarbons, or biaryls, depending on reagent, substrate, and conditions. Of these, the conversion to aryldiazenes was the last to be developed, largely because it was not understood how to handle the extremely sensitive products.[16] With deft experimentation, sodium borohydride will reduce diazonium salts to aryldiazenes (Eq. 4-33); cycloheptatriene can also act as the hydrogen donor.[17,166]

$$ArN_2^+ + NaBH_4 \longrightarrow ArN=NH \tag{4-33}$$

Reduction to hydrazines is an important general method for their synthesis. The two principal methods, formation and further reduction of diazosulfonates[167] and direct reduction with stannous chloride,[168] are fairly general and usually give good yields, although the paths are considerably different. syn-Diazosulfonates, formed from diazonium ions and bisulfite, are reduced in situ by excess bisulfite to arylhydrazinesulfonic acids, which are easily hydrolyzed in acid solution (Eq. 4-34).

$$
\begin{array}{c}
ArN_2^+ + HSO_3^- \longrightarrow \quad \underset{N=N}{\overset{Ar\diagdown \quad \diagup SO_3H}{}} \quad \xrightarrow{HSO_3^-} ArNHNHSO_3H \\
\downarrow \qquad\qquad\qquad\qquad \downarrow H^+/H_2 \\
\underset{HO_3S}{\overset{Ar\diagdown}{}}N-NHSO_3H \xleftarrow{HSO_3^-} \underset{SO_3H}{\overset{Ar\diagdown}{N=N}} \qquad ArNHNH_3^+ + SO_4^{2-}
\end{array}
\tag{4-34}
$$

The anti-diazosulfonates are also reduced, but the products are arylhydrazine-disulfonic acids.[169,170] Reduction by stannous chloride probably passes through the aryldiazene; the final products commonly crystallize as the arylhydrazinium chlorostannite or chlorostannate (Eq. 4-35).

$$ArN_2^+ Cl^- + SnCl_2 \longrightarrow (ArNHNH_3^+) SnCl_6^{2-} \tag{4-35}$$

Reduction of diazonium compounds to amines is of little or no practical importance, but it has been accomplished with sodium amalgam and acetic acid,[171] and in one case, benzene-1,2-bis-diazonium chloride, with stannous chloride.[172]

Reduction of diazonium compounds to azobenzenes involves coupling as well as reduction, and half the original nitrogen is lost. It is a side reaction sometimes encountered with the Sandmeyer reaction.[173-175] The same result can sometimes be accomplished with sulfurous acid[176] (Eq. 4-36), but this method has only quite

$$\alpha\text{-}C_{10}H_7N_2^+ \xrightarrow{H_2SO_3} \alpha\text{-}C_{10}H_7-N=N-\alpha\text{-}C_{10}H_7 \tag{4-36}$$

limited applicability. With α-naphthalenediazonium salts, no change in the orientation could be detected when specifically deuterated naphthylamine was used, and no binaphthyl was formed. These observations have led to the conclusion that the azo compounds are formed from two aryldiazenyl radicals, $ArN=N\cdot$, or from one of them and an ion. The cuprous ammonia complex has also been used for reductive coupling.[177] It works well with diazotized toluidines and m- and p-aminobenzoic acids, but with the nitroanilines and anthranilic acid, biaryls predominate (see Eq. 4-75). Copper or Cu[I] is especially effective in the absence of halides.[178] An attempt to obtain unsymmetrical azobenzenes by the reduction of a mixture of two diazonium salts was unsuccessful, but this may be a result of different rates of reduction of the examples chosen. The formation of unsymmetrical azobenzenes has indeed been accomplished by the interaction of a diazonium salt with a *syn*-diazosulfonate (Eq. 4-37); the evolved nitrogen, interestingly, was found to come from the diazonium ion.[179]

$$ArN_2^+ + syn\text{-}Ar'N{=}NSO_3H \longrightarrow Ar{-}N{=}N{-}Ar' + N_2 \qquad (4\text{-}37)$$

Oxidation. Diazonium salts have been oxidized in alkaline solution as diazoates, as have aliphatic diazoates. The products are nitramides (as their salts)[180,181] (Eq. 4-38). Their formation may be looked upon as the addition of a second oxygen to a

$$\begin{array}{c} ArN{=}NO^- \\ \updownarrow \qquad \xrightarrow{[O]} \quad Ar{-}\bar{N}{-}NO_2 \longleftrightarrow Ar{-}N{=}NO_2^- \qquad (4\text{-}38) \\ Ar\bar{N}{-}N{=}O \end{array}$$

nitrosamine, the anion of which is the same as the diazoate ion. The preferred oxidizing agent is ferricyanide, although permanganate, tribromide, and hydrogen peroxide have been used. The last can also be made to undergo coupling to form diazoperoxides.[182]

Alkylation and acylation. Alkylation of diazoates,[183,184] which can give either diazo ethers (from silver salts) or nitrosamines (from sodium salts), will be discussed in connection with nitrosamines in a companion volume. Acylation of diazoates, which gives *N*-nitroso carboxamides or diazocarboxylates, will also be taken up there.

Reactions of diazonium compounds with loss of both nitrogens are complicated by multiple paths, both ionic and free-radical.[185,186] The diazonium group may be replaced by almost any kind of group, from hydroxy to aryl or hydrogen. In many cases, the reaction is actually a combination of coupling to form an unstable azo derivative that subsequently loses nitrogen. The effect of solvents may be substantial,[187] and heterogeneous catalysis, particularly by copper compounds, is common. Even such a seemingly simple process as hydrolysis presents surprises and enigmas, but it is a good place to begin the subject.

Hydrolysis. Since they were first recognized over a century ago, diazonium salts have been known to be unstable, decomposing slowly even at room temperature and in water solution. The decomposition is not clean, but among the products of this classic reaction can be found phenols. They are generally considered to be the primary products, and their formation in this way provides a route for converting aromatic primary amines to phenols. Much of the difficulty arises from the fact that the phenols are easily attacked by undecomposed diazonium salt and are converted into azo compounds. For preparative purposes, the recommended procedure is to add a solution of the diazonium salt to a boiling solution of aqueous sulfuric acid, distilling the phenol with steam as it is formed[188,189] (Eq. 4-39). Even so, yields are

$$ArN_2^+ + aq.\ H_2SO_4 \xrightarrow{\text{steam distill}} ArOH + N_2 \qquad (4\text{-}39)$$

often indifferent to poor, and there are many contaminants. The experimental variations have been reviewed and compared, and alternatives have been suggested.[190]

The mechanism of hydrolysis of diazonium ions is generally believed to involve dissociation into aryl cations. This hypothesis has substantial experimental support, which, however, indicates that there is much more to the mechanism than simple dissociation. The kinetics in acidic aqueous solution are first-order, and the rate is quite independent of the presence of many anions.[191-193] The intermediates are evidently very reactive and unselective, for in water solution, formation of phenols predominates over capture of more nucleophilic species. The rate constants for reaction with water and with chloride ion have been determined to have the ratio 2.5:1 at 100°C by comparing the ratio ArOH/ArCl in the products with the composition of the medium.[194] It is evidently this lack of selectivity that defeats the attempt to replace the diazonium function by many of the common anions by uncatalyzed reaction in water. However, it is nevertheless desirable to avoid the presence of halide ions if the object is to prepare a phenol.

Some anions, particularly bromide and thiocyanate, accelerate decomposition of diazonium ions in water. The acceleration by bromide is accompanied by an increase in the proportion of ArBr in the products; a bimolecular reaction between bromide and ArN_2^+ or an intermediate derived from it is therefore indicated.[195] Against this stands the observation that the ratio ArOH/ArBr is independent of pressure when the decomposition of p-nitrobenzenediazonium ion in aqueous bromide solutions is studied up to very high pressures.[196] A most important observation comes from experiments with diazonium ions in which one nitrogen is labeled with ^{15}N enrichment. When it is allowed to decompose partially and the remaining diazonium salt is examined, it is found that a considerable amount of isotope scrambling has taken place, corresponding to a partial turnaround of the diazonium function.[197] A diazonium salt that has not been allowed to decompose in part does not show scrambling.

The foregoing observations and the fact that acceleration of decomposition by thiocyanate ion is greater than the concurrent isotope scrambling have led to the proposal that there are two discrete intermediates, at least, preceding formation of product and that each of these may revert to ArN_2^+.[185,197] Furthermore, exchange with molecular N_2 in the medium has been demonstrated.[198] It has been proposed that one of the intermediates might be a spirodiazirine, in which both nitrogens are equivalent, but the sum of the more recent evidence is that the steps are partial fragmentation to an "encounter complex" which then dissociates into completely separated Ar^+ and N_2, reversibly[199,200] (Eq. 4-40).

$$ArN_2^+ \rightleftarrows Ar^+ \cdot N_2 \rightleftarrows Ar^+ + N_2 \xrightarrow{H_2O} ArOH + H^+ \qquad (4\text{-}40)$$

$$\Big\downarrow X^- \qquad \Big\downarrow X^-$$

$$ArX$$

Support for the intermediacy of aryl cations is found in the kinetic effect of substituents and kinetic isotope effects.[191,201] These intermediates have apparently been intercepted by reaction with CO[198] (Eq. 4-41). Further evidence is found in the occurrence of hydride shifts when *ortho*-substituted diazonium salts are decomposed in pyridinium polyhydrogen-fluoride[202] (Eq. 4-42). In water solution, the

$$ArN_2^+ \longrightarrow Ar^+ \xrightarrow{CO} ArCO^+ \xrightarrow{F_3CCH_2OH} ArCO_2CH_2CF_3 \qquad (4\text{-}41)$$

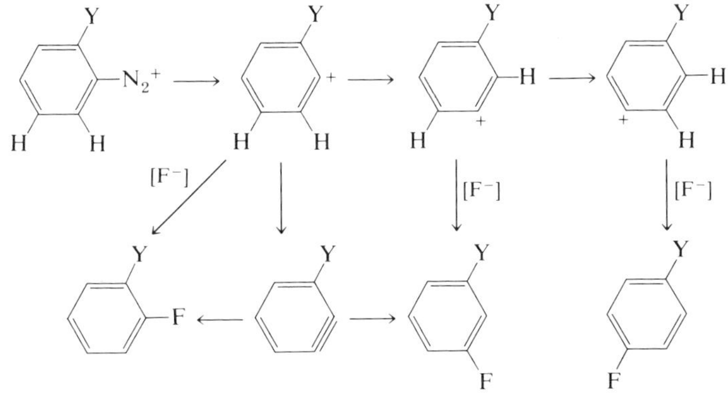

$$(4\text{-}42)$$

evidence is against significant reaction through an intermediate aryne, for there is almost no exchange of ring H with D_2O.[191]

Hydrolysis with the intent of obtaining a phenol is usually carried out in acidic solution in order to minimize coupling of diazonium salt to phenol. There is another reason as well, for in alkaline solution, electron exchange between phenoxide and diazonium ions can produce phenoxy and diazenyl radicals, which lead to other products, especially products of arylation. Such reactions are discussed later, but it is appropriate to include here a free-radical alternative for converting diazonium salts to phenols. By use of a large excess of cupric nitrate in the presence of cuprous oxide, a variety of diazonium fluoroborates have been converted to phenols in high yields (up to 95%).[190] The method seems to be especially advantageous in cases where cyclization normally interferes massively with hydrolysis, such as in the example of diazotized *o*-aminobenzophenone, which gives mostly fluorenone under ordinary conditions, but yields up to 85% of *o*-hydroxybenzophenone by the copper-catalyzed route.

Alcoholysis. Decomposition of diazonium salts in alcoholic solution can give good yields of aryl alkyl ethers, presumably by a mechanism analogous to that for hydrolysis. Most of the reported examples have been with methanol, for practical reasons. Heterolytic alcoholysis must compete with homolytic chain reactions that result in reduction to hydrocarbon (discussed later), and only when such reactions are suppressed are ethers formed cleanly. Oxygen is an effective inhibitor, and alcoholysis carried out in good contact with air usually gives good conversion to ether[203-205] (Eq. 4-43). Weakly electrophilic diazonium ions are the least susceptible to free-radical competition, and with them, formation of ethers may predominate even in the absence of oxygen; strongly electrophilic diazonium ions, such as those bearing nitro substituents, are especially prone to free-radical hydrodediazoniation. At-

$$\mathrm{ArN_2^+ + CH_3OH} \xrightarrow[\text{under N}_2]{\text{under O}_2} \begin{array}{l} \mathrm{ArOCH_3 + N_2} \\[1em] \mathrm{ArH + CH_2{=}O + N_2} \end{array} \qquad (4\text{-}43)$$

tempted solvolysis with phenols gives only poor yields of diaryl ethers, owing to the predominance of free-radical arylation.[206]

Replacement by halogen or pseudohalogens: The Sandmeyer and related reactions.

Reference has already been made, in connection with hydrolysis, to uncatalyzed replacement of the diazonium group by chloride or bromide, which is virtually useless for preparative purposes. For these two groups and the cyano group, the Sandmeyer reaction provides an effective method for substitution; it requires a copper catalyst. Replacement by iodine is efficient without a catalyst, and replacement by fluorine is apparently not subject to catalysis by copper, perhaps because of the rapid disproportionation of CuF. Consequently, special methods must be used.

Preparation of aryl fluorides from diazonium fluoroborates was introduced in 1927, long after the other replacements,[143] and is known as the *Schiemann reaction*.[207-210] It is best accomplished by gently heating the strictly dry, solid salt (Eq. 4-44); hexafluorophosphates may also be used. Use of crown ethers permits the

$$\mathrm{ArN_2^+BF_4^-} \longrightarrow \mathrm{ArF + BF_3 + N_2} \qquad (4\text{-}44)$$

reaction to be carried out in solution in hydrocarbons or their halogenated derivatives.[6] Mechanistic studies carried out in methylene chloride solution[211] demonstrate that the fluorine is transferred in an ionic process with BF_4^- acting as the nucleophile rather than F^-. The reaction has also been accomplished by photolyzing solid films of the fluoroborate.[212] Yields are in some cases better than with the thermolytic method, but the relative convenience may be a matter of taste.

Conversion of diazonium salts to aryl iodides proceeds spontaneously and usually smoothly at room temperature in aqueous solution containing hydriodic acid.[213] The reaction has not been so extensively studied as the copper-catalyzed replacements by chloride or bromide, but the facts have been interpreted on the basis of a free-radical process that occurs especially easily because of the low oxidation potential of iodide ion, which allows electron transfer to give rise to $ArN_2 \cdot$ and $I \cdot$.[214] Alternatively, the ease of the replacement might be attributed to the high nucleophilicity of iodide ion. However, cyanide ion, which is also a good nucleophile but is not such a strong reducing agent, does not convert diazonium ions to nitriles without a catalyst, and the free-radical path therefore seems more likely to be correct. Although yields are usually good, better yields have been reported[215] from photolysis of diazonium chlorostannates in solutions of I_2 (Eq. 4-45), but the process is not so simple to carry out.

$$\mathrm{[\textit{p}\text{-}ClC_6H_4N_2^+]_2SnCl_6^{2-} + I_2} \xrightarrow{h\nu} \textit{p}\text{-}ClC_6H_4I \qquad (4\text{-}45)$$
$$92\%$$

The reaction that has come to be known as the *Sandmeyer reaction* was originally distinguished into three processes, according to whether precipitated copper, cuprous compounds, or cupric compounds were used as catalysts, but it has long since been realized that it is all the same fundamental reaction. The most generally successful procedure is to treat a diazonium halide with the corresponding cuprous complex in water solution (Eq. 4-45). If X^- from the acid used in diazotization is not the same as that used to form the catalyst, a contaminated product may result. A distinct variant is the *Schwechten reaction*, in which the catalyst is a mercuric salt; high yields of aryl bromides may be obtained[216] (Eq. 4-46).

$$[ArN_2^+] \, HgBr_4^{2-} \xrightarrow{\Delta} ArBr \qquad\qquad (4\text{-}46)$$

The Sandmeyer procedures have been used to prepare chlorides,[217,218] bromides,[219] nitriles,[220] isocyanates,[221] and thiocyanates.[222] Although these reactions are usually carried out in water solution, a critical examination of them has led to the recommendation that use of diazonium ion and CuX_2^- in 2:1 proportion in acetonitrile solution is likely to give the best results (higher yields, fewer by-products).[223] Preparative yields commonly fall in the 60 to 80% range.

The Sandmeyer reaction is subject to several side reactions: formation of biaryls, azoarenes, and, if a solvent other than water is used, reduction to arene.[174,175] When $CuBr_2^-$ is used as the catalyst, *o*- or *p*-bromo products may be formed; in an extreme case, aniline has been reported to give rise to 57% bromobenzene and 40% *p*-dibromobenzene.[223] The rate of formation of biaryls and azoarenes depends on the square of the catalyst concentration, and optimum conditions for a pure product demand a low concentration.[175]

The mechanism of the Sandmeyer reaction has been a major enigma for many decades, but it is slowly yielding to investigation. There is little doubt that it is a free-radical chain reaction, closely related to the Meerwein arylation reaction (discussed later).[224-226] It is not clear how free the presumed intermediate aryl radicals are, but it is significant that the Sandmeyer reaction on optically active *o*-diazoniobiphenyls, in which the presence of the diazonio or halo substituent is necessary to prevent racemization, gives a product that retains 80 to 90% of the optical activity.[227] If the aryl radical is indeed free, it must have a lifetime shorter than that required for rotation about the aryl-aryl bond. Kinetic investigations of the formation of aryl chlorides[173,174] have shown that the rate is *inversely* proportional to the free chloride concentration. This fact almost certainly indicates the removal of the active catalytic species by its reaction with Cl^-. Such a reaction might be coordination with Cu, as in conversion of $CuCl_2^-$ to $CuCl_4^{3-}$, and suggests the possibility that the Sandmeyer reaction proceeds through coordination of the diazonium group in a coupling reaction with Cu^I (Eq. 4-47). Subsequent reaction of the aryl radical formed in the

$$ArN_2^+ + CuCl_2^- \longrightarrow [ArN{=}NCuCl_2] \longrightarrow Ar\cdot + N_2 + CuCl_2 \qquad (4\text{-}47)$$

decomposition of this adduct probably occurs by transfer of chlorine from $CuCl_2$ or other chlorine-containing Cu^{II} species. The aryl radical and/or arylcopper species could lead to the observed by-products by reasonable paths as well.[228]

An interesting variant[229] to the Sandmeyer reaction involves combined diazotization and replacement by treating an amine with a cupric halide–nitric oxide complex under an atmosphere of NO (Eq. 4-48); yields are lower than in the conventional method.

$$ArNH_2 + CuX_2 \cdot NO \longrightarrow ArX + N_2 + H_2O + CuX \tag{4-48}$$

Replacements with arylation at N or S.   Apart from the conversion of diazonium salts to thiocyanates and isocyanates mentioned earlier, there are other reactions that do not generally require copper catalysts and are not considered to be Sandmeyer reactions.

Replacement with the nitro group by reaction with nitrites[230] is of erratic applicability. It has been studied extensively for the purpose of developing a general preparative procedure, but although high yields can be obtained in favorable cases, no general method has emerged. The best method appears to be adding a diazonium solution rapidly to a warm solution of sodium nitrite buffered with sodium bicarbonate, with or without a cuprous oxide catalyst[231] (Eq. 4-49) or with a cupric car-

$$p\text{-}O_2NC_6H_4N_2{}^+HSO_4{}^- \xrightarrow[\text{NaHCO}_3,\ 60\,^\circ\text{C}]{\text{NaNO}_2} p\text{-}O_2NC_6H_4NO_2 \tag{4-49}$$
$$97\%$$

bonate catalyst,[232] but 4-nitro-1-naphthylamine has been converted to 1,4-dinitronaphthalene in good yield through reaction of the diazonium bisulfate with sodium nitrite in the presence of $CuSO_3$,[233] and cobalt nitrite complexes have been used as catalyst with some success, among other various examples.[234] The kinetics of the reaction of diazonium nitrites show a different order of dependence on the concentrations of diazonium and nitrite ions according to the experimental conditions.[230,235] Conversion to nitroarene is inhibited by oxygen, and m-nitrobenzenediazonium ion, for example, is converted in over 70% yield to m-dinitrobenzene under a nitrogen atmosphere at pH 6.5. In air, however, the yield falls to 20%. If the decomposing aqueous diazonium salt is stirred with benzene, the major product becomes m-nitrobiphenyl, presumably as a result of trapping m-nitrophenyl radicals, and if $BrCCl_3$ is present, m-nitrobromobenzene is formed in yields up to 86%. These facts point to a free-radical chain reaction in which the key step is the homolysis of a diazo nitrite into diazenyl radicals and nitrogen dioxide.

Replacement by nitrogen has also been accomplished by reaction of diazonium fluoroborates with nitriles.[236,237] The products are N-arylcarboximidoyl fluoroborates, $ArN{\equiv}CR\ BF_4{}^-$, and the strong resemblance to the Ritter reaction of nitriles suggests that this is a heterolytic reaction involving Ar^+.

Replacement by reduced sulfur has several variants; all probably take place by initial coupling to diazosulfide derivatives of varying stability, which decompose by a homolytic process, although nucleophilic displacement of the diazonium function cannot be totally ruled out as a competing path. Reaction with thiosulfate and a copper catalyst produces some diaryl sulfide, along with azo compounds,[238] and reaction with hydrogen sulfide or NH_4HS has been reported to give diphenyl sulfide accompanied by diphenyl disulfide,[239] but these reactions may not be general. With thiophenol anions, coupling takes place in the cold[95,240] (Eq. 4-5), and on warming, the diazosulfides evolve nitrogen more or less rapidly, sometimes alarmingly and dangerously, to form unsymmetrical diaryl sulfides. This transformation has been managed in a controlled manner by using copper powder as a catalyst, which allows it to take place at 5°C[241] (Eq. 4-50). Sodium disulfide converts diazonium salts to

$$\text{PhN}_2^+ + p\text{-CH}_3\text{OC}_6\text{H}_4\text{SH} \xrightarrow[\text{Cu}]{\text{OH}^-} \underset{66\%}{\text{Ph—S-}p\text{-C}_6\text{H}_4\text{OCH}_3} \tag{4-50}$$

diaryl disulfides, a rather unwanted species, but useful because they are easily reduced to thiophenols[242] (Eq. 4-51). Replacement by reaction with tetramethylthiourea

$$o\text{-H}_2\text{N—C}_6\text{H}_4\text{CO}_2\text{H} \longrightarrow \text{ArN}_2\text{Cl} \xrightarrow[5-20°C]{\text{Na}_2\text{S}_2} \text{ArS—SAr} \xrightarrow[\text{AcOH}]{\text{Zn}}$$

$$\underset{71-84\%}{o\text{-HSC}_6\text{H}_4\text{CO}_2\text{H}} \tag{4-51}$$

gives a thiuronium compound that is easily hydrolyzed to a thiophenol[91,243] (Eq. 4-52).

$$\text{ArN}_2^+ + \text{S}{=}\text{C(NMe}_2)_2 \longrightarrow \text{Ar—SC(NMe}_2)_2^+ \xrightarrow{[\text{H}_2\text{O}]} \text{ArSH} \tag{4-52}$$

The most general method for converting diazonium compounds to thiophenols appears to be by reaction with xanthate salts (the Leuckart procedure), which initially forms S-aryl O-alkyl dithiocarbonates. These are easily hydrolyzed to thiophenols, but they may also be converted to aryl alkyl thioethers by heat[244-246] (Eq. 4-53). m-Thiocresol has been prepared in 63 to 75% yield in this manner.[247] A

$$\text{ArN}_2^+ + \text{ROCS}_2^-\text{K}^+ \longrightarrow \text{ArSC}{\overset{S}{\underset{\text{OR}}{\diagdown}}} \quad \begin{matrix} \xrightarrow{[\text{H}_2\text{O}]} & \text{ArSH} \\ \xrightarrow{\Delta} & \text{ArSR} \end{matrix} \tag{4-53}$$

side reaction leading to S,S-diaryl dithiocarbonates has been encountered; it is minimized by carrying out the replacement at low pH (ca. 1).[248]

Replacement with oxidized sulfur can be accomplished by reaction with sulfur dioxide. Solvolysis of diazonium chlorides in liquid sulfur dioxide to form sulfonyl chlorides is an interesting although not very practical reaction.[249] The same transformation can be brought about more simply by adding an aqueous diazonium chloride solution to one of sulfur dioxide in glacial acetic acid in the presence of cupric or cuprous chloride[243,250,251] (Eq. 4-54). With a powdered copper catalyst, reductive displacement takes place, yielding sulfinic acids[252].

$$\text{ArN}_2^+\text{Cl}^- + \text{SO}_2 \xrightarrow[\text{AcOH}]{\text{CuCl}} \text{ArSO}_2\text{Cl}$$

$$\text{ArN}_2^+ + \text{SO}_2 \xrightarrow{\text{Cu}} \text{ArSO}_2\text{H} \tag{4-54}$$

Replacement with arylation at other elements. Diazonium salts react with sodium arsenite to form salts of arenearsonic acids; the reaction is catalyzed by copper compounds, and yields are low to moderate[253,254] (Eq. 4-55). Diaryl selenides are formed by direct reaction with polyselenide salts[255] (Eq. 4-56). Arylmercuric chlo-

$$\text{PhN}_2{}^+\text{Cl}^- + \text{Na}_3\text{AsO}_3 \xrightarrow[\text{Bart reaction}]{\text{Cu}^{\text{II}}} \text{PhAsO}_3{}^{2-} \text{ 2Na}^+ \qquad (4\text{-}55)$$

$$\text{PhN}_2{}^+\text{Cl}^- + \text{K}_2\text{Se}_x \longrightarrow \underset{79\text{-}86\%}{\text{Ph}_2\text{Se}} \qquad (4\text{-}56)$$

rides have been prepared in considerable variety by treating diazonium chloride-mercuric chloride double salts with copper, which is a reactant and not just a catalyst[256] (Eq. 4-57). Replacement with arylation of antimony, bismuth, tellurium, tin, platinum, etc. is possible by reactions analogous to the foregoing.[257]

$$\text{ArN}_2\text{Cl} \cdot \text{HgCl}_2 + 2\text{Cu} \longrightarrow \text{ArHgCl} + \text{N}_2 + 2\text{CuCl} \qquad (4\text{-}57)$$

Arylation at carbon. Replacement of the diazonium group with formation of a new carbon-carbon bond may occur by reaction with another arene ring (*Gomberg-Bachmann reaction*), with ring closure (*Pschorr cyclization*), with olefins (*Meerwein reaction*), with cyanide (*Sandmeyer reaction*, already discussed), with formaldoxime, or with nickel carbonyl.

By means of nickel carbonyl in glacial acetic acid, diazonium fluoroborates can be converted into carboxylic acids (Eq. 4-58), thereby providing an alternative to the

$$p\text{-CH}_3\text{OC}_6\text{H}_4\text{N}_2{}^+\text{BF}_4{}^- + \text{Ni(CO)}_4 \xrightarrow{\text{AcOH}} \underset{74\%}{p\text{-CH}_3\text{OC}_6\text{H}_4\text{CO}_2\text{H}} \qquad (4\text{-}58)$$

Sandmeyer reaction with cyanide.[258] Arylation of formaldoxime takes place at carbon in the presence of copper sulfate and sodium sulfite and provides a route to benzaldehydes[259] (Eq. 4-59).

$$\text{ArN}_2{}^+ + \text{CH}_2\text{=NOH} \xrightarrow[\text{Na}_2\text{SO}_3]{\text{CuSO}_4} \text{ArCH=O} \qquad (4\text{-}59)$$

The Meerwein arylation of olefins has usually been carried out by adding an aqueous solution of the diazonium salt to a solution of the olefin in acetone or acetonitrile, followed by addition of CuCl_2, at room temperature.[260,261] The product is either the aryl olefin or an aryl chloro alkane corresponding to addition of Ar and Cl to the double bond (Eq. 4-60). Side products always found to some extent are

$$\text{ArN}_2{}^+ + \underset{\text{H}}{\overset{\diagdown}{\text{C}}}\text{=}\overset{|}{\text{C}}\text{—A} \xrightarrow{\text{CuCl}_2} \text{Ar—}\overset{|}{\text{CH}}\text{—}\underset{\text{Cl}}{\overset{|}{\text{C}}}\text{—A} \quad \text{and/or}$$

$$\text{Ar—}\overset{|}{\text{C}}\text{=}\overset{|}{\text{C}}\text{—A} \qquad (4\text{-}60)$$

ArH and ArCl. An electron-withdrawing group, A, is generally required for satis-factory results, and conjugated carbonyl, cyano, or aryl olefins are therefore usually used; even so, yields are usually modest at best. If the aryl olefin is the desired product, any chloro aryl alkane formed can usually be dehydrohalogenated in a subsequent step. A modification that gives improved yields (of aryl chloro alkane) is to carry out the diazotization in situ by adding a solution of the aniline in acetoni-trile to a mixture of *tert*-butyl nitrite, cupric chloride, olefin, and acetonitrile at room temperature[262] (Eq. 4-61). A further variant is the use of zero-valent palla-

$$PhNH_2 + CH_2{=}CHCN \xrightarrow[\text{CuCl}_2]{t\text{-BuONO}} \underset{71\%}{PhCH_2{-}CHClCN} \tag{4-61}$$

dium compounds as catalysts, but reactivity is then different, electron-rich olefins giving good yields and ethyl acrylate, which reacts well in the conventional Meerwein procedure, giving poor results.[263]

The mechanism of the Meerwein reaction has been studied extensively, but in spite of substantial advances, it remains enigmatic.[264,265] There seems little doubt that the reaction is homolytic in nature, but the evidence for the formation of free aryl radicals is inconsistent; there may well be more than one path to the products. A study of product ratios with CuCl and CuCl$_2$ catalysts has been interpreted as being inconsistent with formation of an intermediate complex, but consistent with formation of aryl radicals, which may react in three ways: attacking an olefin, abstracting a hydrogen, or abstracting a chlorine.[266] However, a mechanism that generates aryl radicals before involvement of the olefin is inconsistent with the observation that evolution of nitrogen is markedly accelerated when an olefin is added to a mixture of diazonium salt and catalyst in certain cases,[267] although generally the kinetics are first-order in [ArN$_2^+$] and [CuCl$_2^-$] with only low sensitiv-ity to the concentration of olefin.[268] Furthermore, in the presence of air, olefins retard the decomposition of 2,4-dichlorobenzenediazonium ion, whereas in the ab-sence of air, acrylates and acrylonitrile approximately double the rate. Some of the apparent inconsistencies might be resolved if there are two ways in which the diazo-nium ion reacts with the catalyst (Eq. 4-62). Subsequent reactions are believed to

$$\text{Ar}N_2\cdot + \text{CuCl}_2 \xleftarrow{\text{CuCl}_2^-} \text{Ar}N_2^+ \xrightarrow[\text{RCH=CHA}]{\text{CuCl}_2^-} \text{ArN=N-CuCl}_2\cdot \underset{\text{CHA}}{\overset{\text{CHR}}{\|}}$$
$$\downarrow \qquad\qquad\qquad\qquad\qquad\qquad\qquad\qquad \downarrow \tag{4-62}$$
$$\text{Ar}\cdot + N_2 \qquad\qquad\qquad\qquad \text{Ar}\cdot + N_2 + \text{CuCl}_2 + \underset{\text{CHA}}{\overset{\text{CHR}}{\|}}$$

proceed as in Equation 4-63. The side products could arise by reaction of the aryl radicals with CuCl$_2$ to form ArCl or with acetone to form ArH; the former might also arise directly from the proposed initial complex. The general absence of polym-erized olefin and of biaryls is understandable if the rates of capture of the interme-diate radicals by the processes of Equation 4-63 are greater than the rates of the

$$\text{Ar} \cdot + \text{RCH=CHA} \longrightarrow \underset{\text{Ar}}{\overset{\text{R}}{\diagdown}}\text{CH}\!-\!\overset{\cdot}{\text{C}}\text{HA} \xrightarrow{\text{CuCl}_2}$$

$$\underset{\text{Ar}}{\overset{\text{R}}{\diagdown}}\text{CH}\!-\!\text{CH}\underset{\text{Cl}}{\overset{\text{A}}{\diagup}} + \text{CuCl}$$

$$\text{or} \quad \underset{\text{Ar}}{\overset{\text{R}}{\diagdown}}\text{C=CHA} + \text{CuCl} + \text{H}^+ + \text{Cl}^- \qquad (4\text{-}63)$$

unobserved competing reactions. The presence of a high concentration of cupric chloride, essential for the Meerwin reaction, would be expected to preempt free-radical polymerization.

Conjugated dienes react with diazonium salts in polar media to form pyridazines by 2 + 4 cycloaddition.[102]

Intramolecular arylation of an aromatic ring is known as the *Pschorr cyclization*, exemplified by the conversion of *o*-diazonio-*cis*-stilbene to phenanthrene[269] (Eq. 4-64). A copper catalyst is generally required, but examples of analogous ring clo-

$$\text{(structure)} \xrightarrow{\text{Cu}} \text{(phenanthrene)} + \text{N}_2 \qquad (4\text{-}64)$$

sures are known that proceed in acidic solution without a catalyst, as well as in basic solution. Closure may take place so as to form either six- or five-membered rings; in the latter case, the bridging vinylidene group of the example in Equation 4-64 must be replaced by a single carbon or hetero atom.

The catalyzed Pschorr cyclization (Cu powder or Cu$_2$O) must be a free-radical process analogous to the Meerwein arylation and the Sandmeyer reaction. The rate is faster and the yields are higher with a cuprous oxide catalyst, as shown by the example of *o*-diazoniobenzophenone, which goes almost entirely to fluorenone with a catalyst, whereas in the absence of a catalyst, formation of *o*-hydroxybenzophenone competes to the extent of 35%.[270] The uncatalyzed reaction in acidic solution is believed to be heterolytic, but in basic solution it is apparently homolytic, even without catalysis.[271] The consequences of the two mechanisms are seen not only in the formation of side products, but in the site selectivity when two sites for ring closure are present. Aryl cations, the presumed intermediates in acidic medium without a catalyst, are highly reactive and thus not very selective, whereas aryl radicals are considerably more selective. 2-(*o*-Diazoniobenzoyl)naphthalene provides an example.[272] Ring closure occurs preferentially at the 1 position under all conditions, but in acid solution, the ratio of 1-closure to 3-closure without a catalyst is 2.4:1, whereas in the presence of cuprous oxide, the ratio is about 9:1, and in basic solution without a catalyst, it is 9.5:1 (Eq. 4-65).

The Pschorr cyclization can also be carried out electrochemically by cathodic

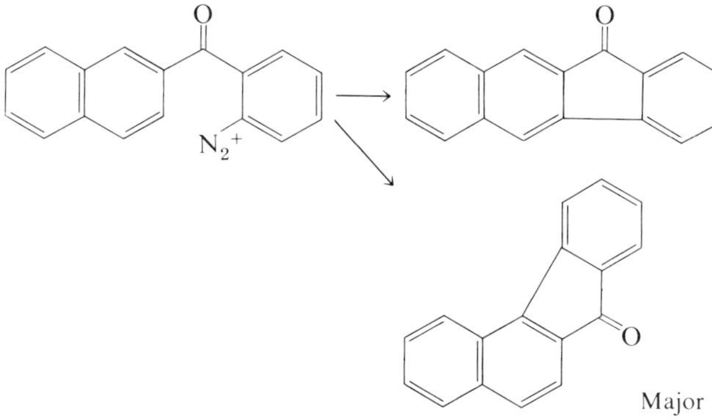

(4-65)

Major

oxidation of the diazonium fluoroborates. This is apparently a free-radical process.[273]

Other side reactions may be encountered besides those already mentioned. The cyclization of o-diazonio-cis-stilbene to indazole in acid solution without a catalyst was mentioned earlier.[152] Rearrangement, involving migration of an aryl group, has been encountered with benzhydrol derivatives, in a process reminiscent of the pinacol rearrangement[274] (Eq. 4-66).

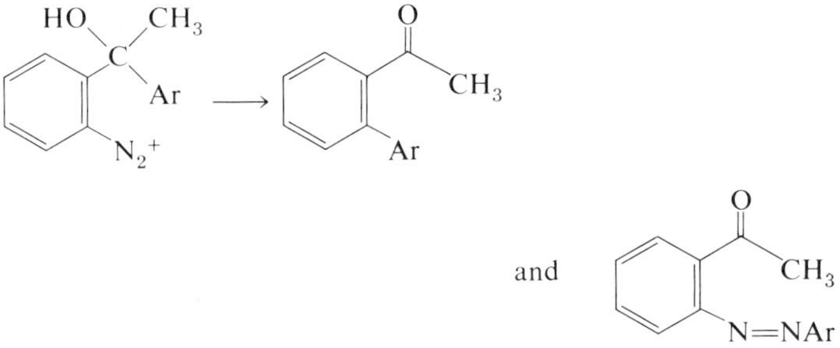

and

(4-66)

Arylation of aromatic rings intermolecularly by reaction with diazonium compounds in neutral to mildly basic medium is known as the *Gomberg-Bachmann reaction*.[275] It is a synthetically useful route to biaryls and may be carried out on heterocyclic rings such as pyridine. Benzene rings with strongly electron-withdrawing substituents, such as nitrobenzene, are arylated mostly at the *ortho* and *para* positions, and they react more readily than do substrates without such substituents. However, compounds such as toluene are arylated more slowly, and the site of attack is more or less indiscriminate. These characteristics are inconsistent with electrophilic attack and imply a free-radical process.

The reaction has usually been carried out by adjusting a diazonium solution to optimum pH and agitating it with excess aromatic substrate in a two-phase system. Better results are claimed for use of phase-transfer agents or of acetonitrile as a mutual solvent.[276] Another one-phase process has been reported[277] in which an arylamine and an alkyl nitrite are heated in solution in excess substrate (Eq. 4-67).

$$o\text{-}ClC_6H_4NH_2 + C_5H_{11}ONO \xrightarrow{\text{benzene}} o\text{-}ClC_6H_4\text{—}C_6H_5 \qquad (4\text{-}67)$$
$$40\%$$

Another variant is to substitute an *N*-nitrosoacetanilide for the diazonium salt or to convert the diazonium salt to a diazoacetate; these amount essentially to the same thing, for nitrosoacetanilides isomerize to diazoacetates. 1-Aryltriazenes, derived from coupling a diazonium salt with a secondary amine, have also been used. By any method, yields are moderate to low.

When the diazonium salt and base are used, the kinetics of nitrogen evolution are second-order, a fact that has been explained by invoking formation of a diazoanhydride as the active reagent.[278,279] Indeed, there appears to be an optimum pH for each diazonium compound, corresponding to that at which $[ArN_2^+]$ and $[ArN_2O^-]$ are equal and therefore to maximum concentration of the anhydride, ArN_2ON_2Ar. The anhydride then decomposes homolytically, setting up a free-radical chain reaction (Eq. 4-68). The diazonium acetate/nitrosoacetanilide variant of the reaction

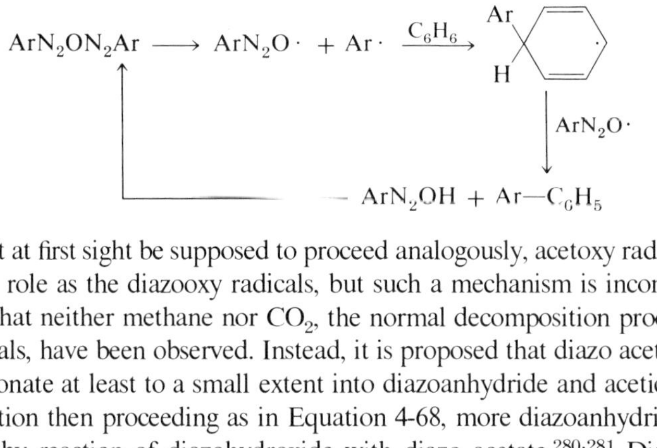

$$ (4\text{-}68) $$

might at first sight be supposed to proceed analogously, acetoxy radicals playing the same role as the diazooxy radicals, but such a mechanism is inconsistent with the fact that neither methane nor CO_2, the normal decomposition products of acetoxy radicals, have been observed. Instead, it is proposed that diazo acetates can disproportionate at least to a small extent into diazoanhydride and acetic anhydride, the arylation then proceeding as in Equation 4-68, more diazoanhydride being generated by reaction of diazohydroxide with diazo acetate.[280,281] Diazo nitrites also arylate arenes.[230]

Although the diazooxy radical has been detected by electron-spin resonance in Gomberg-Bachmann arylations in a variety of media,[282] it is not necessarily involved in the principal path to products. An alternative mechanism, which can proceed as a chain reaction once a small amount of aryl radical arises in one way or another to start the chain, makes use of electron transfer from the cyclohexadienyl radical to diazonium ion in the key step, a proton subsequently being removed by acetate ion (Eq. 4-69). Evidence for the intermediacy of aryl radicals is found in the

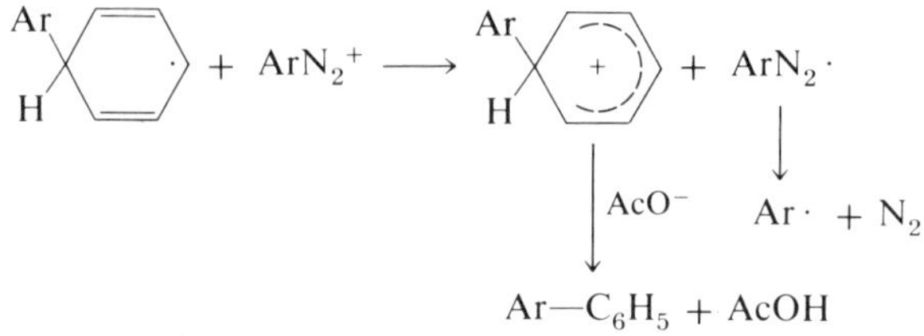

$$ (4\text{-}69) $$

orientation of attack on substituted benzenes and CIDNP,[283] but there are nevertheless some difficulties. Optically active 2-methyl-6-nitrobiphenyl-2'-diazonium salts undergo the Gomberg-Bachmann arylation reaction with nearly 80% retention of optical activity.[227] Since maintenance of chirality depends on interference with rotation about the aryl-aryl bond of the diazonium salt due to the bulk of the diazonio substituent, loss of that substituent to form either a free radical or an aryl cation should result in racemization, unless subsequent reaction is faster than the rate of rotation. However, aryl radicals must be stable enough to undergo numerous molecular collisions if they are to be invoked to explain the selectivity for site of substitution, and extensive racemization would be expected in the example at hand.

Arylation of benzenes can also be accomplished under acidic conditions with a copper catalyst, as in the Meerwein reaction. The product ratios for arylation of substituted benzenes are the same as for arylation with nitrosoacetanilides.[260,284,285] With $Cu^I \cdot$ amine complexes, free-radical arylation can be carried out in nearly neutral solution.[271]

An electrophilic path for arylation of aromatic rings also exists and may come into play when structures and conditions, including the polarity of the solvent, are favorable. The reaction of diazonium fluoroborates in polar aprotic solvents, such as acetonitrile, generally satisfies the requirements.[270] Warming dry diazonium chlorides with an arene in the presence of aluminum chloride has also been used, but formation of aryl chloride competes seriously.[286] The occurrence of electrophilic arylation is usually revealed by the site selectivity.

A reaction related to the foregoing examples is arylation through a benzyne. Benzyne is formed by fragmentation of o-diazoniobenzoate zwitterion. It reacts by a Diels-Alder cycloaddition with suitable aromatic partners, such as anthracene and furan[287] (Eq. 4-70).

$$(4\text{-}70)$$

Replacement by hydrogen (deamination or hydrodediazoniation). Deamination, the replacement of NH_2 by H by diazotization and reduction, is a widely used reaction in synthesis and structure proof.[288] The reaction has most often been carried out using hypophosphorous acid as the reducing agent, but many other hydrogen donors have been used with good results, such as hydroquinone, ethers (dioxan, tetrahydrofuran, dioxolane, diglyme),[289] dialkylamides (tetramethylurea,[290] dimethylformamide,[291,292] hexamethylphosphoramide[293]), etc. (Eq. 4-71). Tetra-

$$ArN_2^+ + AH \longrightarrow ArH + N_2 + [A^+] \qquad (4\text{-}71)$$

methylurea is reported to give good results when the diazonium salt bears electron-withdrawing substituents, but H_3PO_2 or alkaline formaldehyde give better results with electron-donating substituents. A comparison of different reagents by one research group, using a telescoped procedure (adding an aniline slowly at 65°C to a mixture of alkyl nitrite and hydrogen donor) showed dimethylformamide to give better results than six other reagents.[292] Alcohols, particularly methanol, can give high yields if oxygen is excluded.[294] Electrochemical reduction is possible, but the results do not seem to be so good as chemical methods.[295,296] Catalysis by cuprous oxide leads to particularly high yields with H_3PO_2 as the hydrogen donor, but it is then desirable to use the diazonium fluoroborate or hexafluorophosphate in order to avoid competition from the Sandmeyer reaction.[297] Phase-transfer agents have also been used to facilitate reduction,[298] using diazonium fluoroborates in methylene chloride in the presence of dicyclohexano-18-crown-6 ether and powdered copper (Eq. 4-72). Formation of fluoroarene competes in some instances. Rhodium com-

$$4\text{-Cl-3-}O_2NC_6H_3N_2^+BF_4^- \xrightarrow[\substack{\text{crown ether}\\40°C}]{CH_2Cl_2,\ Cu} o\text{-}ClC_6H_4NO_2 \qquad (4\text{-}72)$$
$$93\%$$

plexes have been used to catalyze hydrodediazoniation by dimethylformamide, but the yields are only low to moderate.[299] Sodium borohydride gives fair yields.[300]

The cumulative evidence seems quite conclusive that hydrodediazoniations take place by free-radical chain reactions,[185,301-303] as shown in Equation 4-73. With

$$\begin{aligned} &H_2\overset{O}{\overset{\|}{P}}OH \xrightarrow{[O]} HP\overset{O}{\underset{OH}{\diagup}} + H^+ \\ &H_2PO_2\cdot\ + ArN_2^+ \longrightarrow \qquad\qquad\qquad\qquad (4\text{-}73)\\ &\qquad\qquad Ar\cdot\ + N_2 + H_2PO_2^+\ (\xrightarrow{H_2O} H_3PO_3 + H^+)\\ &Ar\cdot\ + H_3PO_2 \longrightarrow ArH + H_2PO_2\cdot \end{aligned}$$

H_3PO_2, the kinetic isotope effect[304] in D_2O is $k_H/k_D = 2$ to 6, and the benzene formed in reductions with H_3PO_2 exhibits the CIDNP phenomenon.[305] However, it has been proposed that reductions by ethers occur through aryl cations.[289] Photolysis of diazonium salts in ethanol gives mostly hydrodediazoniation, but other free-radical products have been identified, such as 1- and 2-arylethanol and tetramethylene glycol.[215,306]

Reduction by dialkyl amides can take place intramolecularly, as shown by the behavior of diazotized anthranilic amides. The dibenzyl example gives both cyclic and debenzylated products (Eq. 4-74), but the product-determining steps have not been indentified.[307] With catalysis by Cu_2O, the dimethyl example goes more cleanly to N-methylbenzamide and formaldehyde.[308]

Reductive dediazoniation may be brought about by $Cu(NH_3)_2^+$ with a different consequence, coupling to form a biaryl[177] (Eq. 4-75). The occurrence of this reaction, which must compete with formation of an azobenzene, seems to depend more on structure than on experimental conditions, and its generality is sharply limited. In favorable cases, such as the formation of diphenic acid from anthranilic acid and of 2,2'-dinitrobiphenyl from o-nitroaniline, the reaction has distinct preparative value.

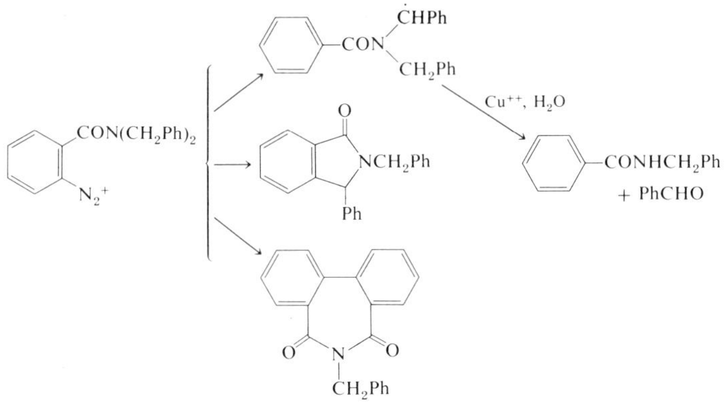

$$(4\text{-}74)$$

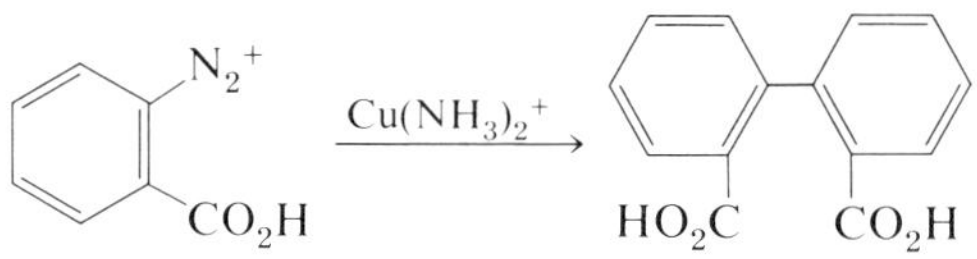

$$(4\text{-}75)$$

Azo Compounds

Aromatic and aliphatic azo compounds differ from each other considerably in their chemistry. The reactions of aromatic azo compounds (azobenzenes) are limited largely to oxidation, reduction, and cycloaddition. Aliphatic azo compounds also undergo such reactions, but their most prominent reaction is fragmentation with loss of nitrogen; tautomerism with hydrazones, which is not possible for azobenzenes, is also a prominent feature of their chemistry.

Thermolysis and photolysis. The simplest azo compounds, monosubstituted diazenes, decompose spontaneously at room temperature; the principal products are arene and diarylhydrazine, accompanied by arylhydrazine, arylamine, and aryl azide. Both ^{1}H and ^{13}C CIDNP are observed, and the initial step is second-order, presumably as shown in Equation 4-76.[16,17,166,309]

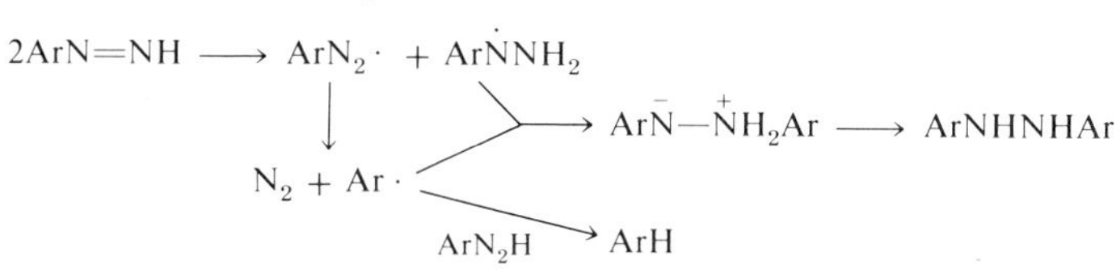

$$(4\text{-}76)$$

Substitution with two aryl groups makes a dramatic increase in thermal stability, and azobenzene, for example, has withstood temperatures up to 600°C without decomposition.[310] Azobenzenes also resist photolysis, and the only change induced by irradiation is isomerization from *trans* to *cis*.[311] The less-stable *cis* isomers are converted cleanly back to *trans* by heat.[312,313]

In between the two foregoing extremes lie azo compounds with one or two alkyl substituents. Benzeneazotriphenylmethane is a simple example; it dissociates reversibly into triphenylmethyl and phenyldiazenyl radicals, and the latter then lose nitrogen readily to leave phenyl radicals. The two radicals undergo further reactions according to conditions, and among them is combination to form tetraphenyl-methane[314,315] (Eq. 4-77). This reaction provided the first synthesis of tetraphenyl-

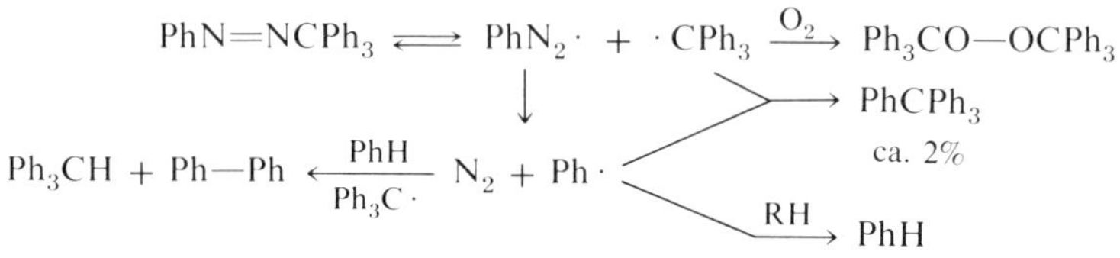

$$PhN\!=\!NCPh_3 \rightleftharpoons PhN_2\cdot + \cdot CPh_3 \xrightarrow{O_2} Ph_3CO\!-\!OCPh_3$$

$$\downarrow \qquad\qquad \longrightarrow PhCPh_3$$

$$Ph_3CH + Ph\!-\!Ph \xleftarrow[Ph_3C\cdot]{PhH} N_2 + Ph\cdot \qquad ca.\ 2\% \qquad\qquad (4\text{-}77)$$

$$\xrightarrow{RH} PhH$$

methane.[316] The reversibility of the first step has been demonstrated by a trapping experiment using another source of phenyldiazenyl radical (Eq. 4-78). The CIDNP

$$PhC(CH_3)_2N\!=\!NPh \longrightarrow Ph\dot{C}(CH_3)_2 + PhN_2\cdot \xrightarrow{Ph_3C\cdot}$$

$$PhN\!=\!NCPh_3 \qquad (4\text{-}78)$$

$$1.3\%$$

effect with ^{15}N-enriched material also supports the path shown.[317] Azotriphenyl-methane fragments even more readily and is so unstable that nitrogen is immediately evolved when an attempt to prepare it is made by oxidizing bis(tri-phenylmethyl)hydrazine (Eq. 4-79); the typical yellow color of triphenylmethyl

$$Ph_3CNHNHCPh_3 \xrightarrow{NaOBr} [Ph_3CNNCPh_3] \longrightarrow N_2 + 2Ph_3C\cdot \qquad (4\text{-}79)$$

radical develops and is discharged on access of oxygen, forming triphenylmethyl peroxide.[318]

The foregoing examples foreshadowed the generalization that azo compounds are decomposed by heat or near ultraviolet light with an ease that varies with the expected stability of the free radicals formed. The products can in general be accounted for satisfactorily on the basis of a simple cleavage into molecular nitrogen and two radicals if the assumption is made that the two radicals have a greatly enhanced probability of combining with each other as a consequence of having been generated in the same solvent cage. Recombination compounds are, indeed, generally the major products,[319,320] barring exceptional radicals such as triphenyl-methyl, and are usually accompanied by disproportionation products and compounds formed by abstracting hydrogen from the medium[321,322] (Eq. 4-80).

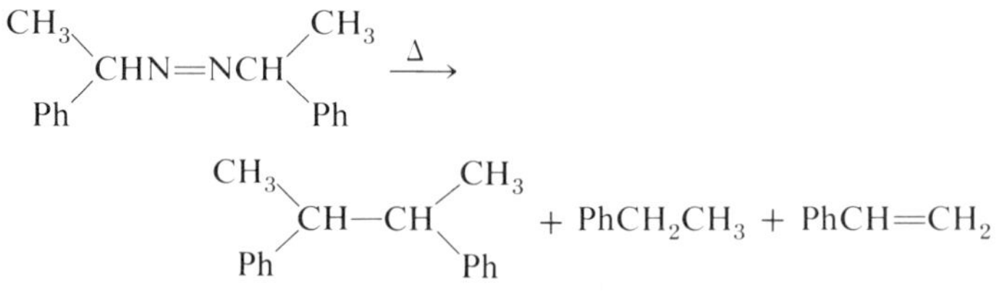

$$\begin{matrix} CH_3 & & CH_3 \\ {}_\diagdown & & {}_\diagup \\ & CHN\!=\!NCH & \xrightarrow{\Delta} \\ {}^\diagup & & {}^\diagdown \\ Ph & & Ph \end{matrix}$$

$$\begin{matrix} CH_3 & CH_3 \\ {}_\diagdown & {}_\diagup \\ & CH\!-\!CH & + PhCH_2CH_3 + PhCH\!=\!CH_2 \qquad (4\text{-}80) \\ {}^\diagup & {}^\diagdown \\ Ph & Ph \end{matrix}$$

Thermal decomposition is first-order[323] and has unusually high activation energy (40–50 kcal/mol is not unusual). There is a large secondary deuterium isotope effect upon α-deuteration of certain symmetrical azo compounds, in which, it has therefore been concluded, both bonds to nitrogen break simultaneously.[324] Further evidence is found in the observation that substituted azoisobutyronitrile derivatives decompose at the same rate whether the meso form or the racemate is used, and the succinonitriles produced (Eq. 4-81) have the same stereoisomeric composition re-

$$CH_3-\underset{\underset{R}{|}}{\overset{\overset{CN}{|}}{C}}-N=N-\underset{\underset{R}{|}}{\overset{\overset{CN}{|}}{C}}-CH_3 \longrightarrow N_2 + CH_3-\underset{\underset{R}{|}}{\overset{\overset{CN}{|}}{C}}-\underset{\underset{R}{|}}{\overset{\overset{CN}{|}}{C}}-CH_3 \qquad (4\text{-}81)$$

gardless of the stereoisomeric composition of the starting material.[325] However, unsymmetrical azo compounds probably decompose in two stages, as implied by the α-deuterium secondary isotope effects with 1-methyl-2-(1-phenylethyl)diazene.

The effect of structure on rates of thermolysis has been much studied,[321,326] especially with cyclic azo compounds.[327] Photolysis has received somewhat less attention,[56,328,329] but it appears that most azo compounds are cleaved by near ultraviolet light, except for six-membered cyclic azo compounds (tetrahydropyridazines), which are resistant but succumb to shorter wavelengths (185 nm).[330]

The high efficiency of recombination of radicals when azo compounds are decomposed in solution and its implication of a cage effect have stimulated intensive investigation.[326] Since radicals that recombine inside the solvent cage in which they were formed are not active in other ways, such as initiation of polymerization or combination with radical scavengers, it is possible to investigate experimentally the relative efficiency of free-radical production and the cage effect. An interesting example is seen in the study of α-azo amidines, which can be studied as the free bases, the monocations, or the dications. Protonation increases the rate of decomposition, but the increase in efficiency of free-radical production is small and not enough to establish firmly whether repulsion of like charges can seriously reduce the cage effect.[331]

The decomposition of azo compounds into free radicals can initiate radical polymerization of olefins,[332] and azoisobutyronitrile and related compounds are widely used industrially for this purpose. Capture of the radicals by olefin must, of course, compete with recombination and other reactions. The recombination of α-cyano radicals shown in Equation 4-81 is only one of two ways in which such species may join; the alternative is C-to-N joining, so as to produce ketenimines (Eq. 4-82). The latter products have actually been observed in several instances,

$$\underset{NC}{\overset{R_2C}{\diagdown}}\underset{\diagup}{\overset{CN}{\diagup}} \underset{\diagup}{\overset{N=N}{\diagdown}} \underset{CR_2}{\overset{}{\diagup}} \xrightarrow{\Delta} [R_2\dot{C}-CN] \rightleftharpoons R_2C=C=N-\overset{\overset{CN}{|}}{C}R_2 \qquad (4\text{-}82)$$

$$\Big\downarrow$$

$$\underset{NC}{\overset{R_2C}{\diagdown}}\underset{\diagup}{\overset{CN}{\diagup}} \underset{CR_2}{\overset{}{\diagup}}$$

such as 1,1-azocyclohexanecarbonitrile. Their rate of formation is decreased by radical scavengers, but their rate of disappearance (to form succinonitriles) is increased by scavengers,[333] and it has therefore been concluded that ketenimines are formed by combination of free radicals outside the solvent cage. The reversal of this process gives back the radicals, which may recombine in either orientation, thus eventually converting all the ketenimine to the more stable succinonitrile structure. Photolysis of azoisobutyronitrile proceeds analogously, and as much as 35% of the ketenimine has been isolated.[334]

Perfluoro azo compounds are particularly amenable to study of their decomposition in the gas phase because of their volatility and the stability of many of their products. An unusual phenomenon is the combination of undecomposed azo compound with perfluoroalkyl radicals to produce tetrasubstituted hydrazines in preparative yield[335] (Eq. 4-83).

$$CF_3N{=}NCF_3 \xrightarrow{325\,°C} (CF_3)_2N{-}N(CF_3)_2, \qquad CF_4, \qquad CF_3N{=}CF_2 \quad (4\text{-}83)$$
$$35\%$$

Tautomerism. Aliphatic azo compounds are in general less stable than the isomeric hydrazones, but isomerization does not occur readily in the absence of acid-strengthening groups on an α-carbon. Tautomerization is catalyzed by hot acid,[336] base,[337] or free radicals[338] (Eq. 4-84). When aqueous acid is used, the hydrazones

$$(CH_3)_2CHN{=}NCH(CH_3)_2 \quad \overset{KOH,\ 180\,°C(fast),\ 25\,°C(slow)}{\underset{aq.\ HCl,\ \Delta}{\longrightarrow}} \quad \begin{array}{l}(CH_3)_2C{=}N{-}NHCH(CH_3)_2 \\[6pt] (CH_3)_2C{=}O + (CH_3)_2CHNHNH_2 \end{array} \qquad (4\text{-}84)$$

are rapidly hydrolyzed to a carbonyl compound and a hydrazine. Tautomerization of azo derivatives of β-keto esters, etc., has been mentioned earlier in this chapter in connection with the Japp-Klingemann reaction.

The tautomerism of benzeneazoalkanes with phenylhydrazones and possibly enehydrazines[339] (Eq. 4-85) has been a matter of controversy for many decades.[340]

$$ArN{=}NCHRCHR_2 \rightleftharpoons ArNHN{=}CRCHR_2 \rightleftharpoons$$
$$ArNHNHCR{=}CR_2 \quad (4\text{-}85)$$

The most recent evidence supports the view that the hydrazone tautomer is the more stable.[336,338] Possible involvement of the enehydrazine is treated in Chapter 2.

Azo/hydrazone tautomerism is also encountered with *o*- or *p*-hydroxy or amino azobenzenes and analogs,[340] which can equilibrate with a quinone hydrazone structure (Eq. 4-86). The position of equilibrium is near enough to the middle that it is

$$ArN{=}N{-}\!\!\!\bigcirc\!\!\!-OH \rightleftharpoons ArNHN{=}\!\!\!\bigcirc\!\!\!={O} \qquad (4\text{-}86)$$

noticeably affected by the medium, but the azo tautomer is generally the preferred form.[341-345] The subject is of great significance to the dye industry, and the emphasis in most investigations has accordingly been placed on naphthalene derivatives. Tautomerism of this sort also plays a role in the behavior of many acid-base indicators. Isomerization to a hydrazone may occur by migration of an acyl group instead of hydrogen.[346]

Acids and other electrophiles. The ability of azo compounds to form salts is so weakly developed that few salts have been isolated.[347,348] A number of hydrohalides and perchlorates of bicyclic azo compounds have been prepared.[349] Their reaction with alcohols is discussed in the section Reduction. Complexes with transition-metal salts, particularly cuprous chloride (Eq. 4-87), are well known.[350] Since they are

$$CH_3N{=}NCH_3 + 2CuCl \rightleftharpoons CH_3N{=}NCH_3 \cdot 2CuCl \qquad (4\text{-}87)$$

usually easily broken up by heat, they serve as useful solid derivatives for the isolation of azo compounds, particularly the volatile aliphatic members.[351] The ratio of azo compound to metal salt is always integral, but it is often other than 1:1.[352]

Azobenzenes without hydroxy or amino substituents cannot undergo the acid-catalyzed tautomerism described in the foregoing section, but they may nevertheless be affected by treatment with hot, strong acid. One such reaction, only rarely encountered, is decoupling to generate a diazonium salt (Eq. 4-88); it has been ob-

$$ArN{=}NAr' + \text{fuming } HNO_3 \longrightarrow ArN_2^+ + Ar'NO_2 \qquad (4\text{-}88)$$

served with amino and alkoxy derivatives of azobenzene,[353-355] and with certain aryl alkyl azo compounds obtained in the Japp-Klingemann reaction. Another type of reaction involves chlorination, reductive cleavage, and benzidine rearrangement (Eq. 4-89) and may begin with addition of hydrogen chloride to the azo group.[356]

$$PhN{=}NPh \xrightarrow[\Delta]{\text{aq. HCl}}$$

$$PhNH_2 \quad p\text{-}ClC_6H_4NH_2 \quad p\text{-}H_2NC_6H_4{-}C_6H_4{-}p\text{-}NH_2 \qquad (4\text{-}89)$$

Apart from Equation 4-88, nitric acid may nitrate the benzene ring of azobenzenes; substitution is at the *para* position if it is free, otherwise *ortho,* and oxidation to an azoxy compound may compete.[357,358]

Azobenzenes are alkylated only by the strongest alkylating agents, such as methyl fluorosulfonate or trifluoromethanesulfonate; diazenium salts are formed in 60 to 90% yields[359] (Eq. 4-90).

$$PhN{=}NPh + CH_3O_3SR \longrightarrow \begin{matrix} Ph \\ \diagdown \\ N{=}NPh \ RSO_3^- \\ \diagup \\ CH_3 \end{matrix} \qquad (4\text{-}90)$$

$$(R = F \text{ or } CF_3)$$

Acylation of azo compounds leads to products in which the azo function has been reduced to the hydrazine stage. Acetyl chloride attacks azobenzene under photolytic conditions to form a ring-chlorinated product[360] (Eq. 4-91). 2,2'-Azopropane has been benzoylated to give a dealkylated hydrazide as product (Eq. 4-92).

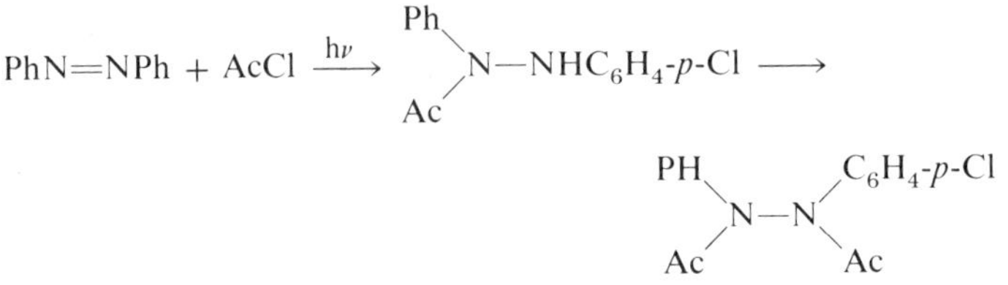

$$PhN=NPh + AcCl \xrightarrow{h\nu} \underset{Ac}{\overset{Ph}{N}}-NHC_6H_4\text{-}p\text{-}Cl \longrightarrow$$

$$\underset{Ac}{\overset{PH}{N}}-\underset{Ac}{\overset{C_6H_4\text{-}p\text{-}Cl}{N}} \qquad (4\text{-}91)$$

$$(CH_3)_2CHN=NCH(CH_3)_2 \xrightarrow[\text{dry } Na_2CO_3]{PhCOCl}$$

$$\underset{PhC\diagdown\!\!\!\diagdown O}{\overset{(CH_3)_2CH}{N}}-NHC\diagdown\!\!\!\overset{Ph}{\diagup}O \qquad (4\text{-}92)$$

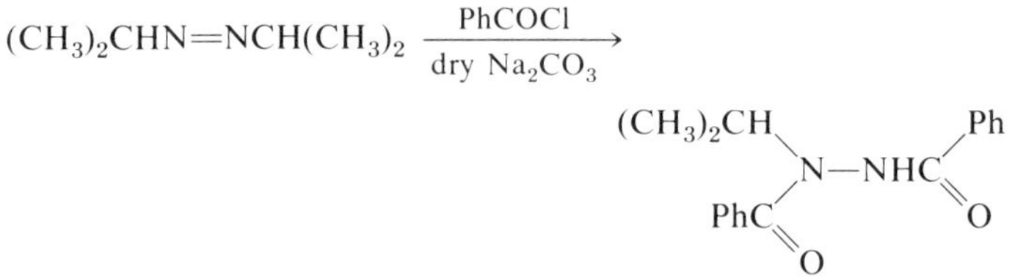

Mercuration of azobenzene with mercuric acetate takes place at the *ortho* position and by subsequent reaction with halogen provides a useful route to *o*-haloazobenzenes.[361] Direct bromination is immeasurably slow in the absence of hydrogen bromide, which catalyzes formation of *p*-bromoazobenzene[362] (Eq. 4-93). It is not

$$PhN=NPh + Br_2 \xrightarrow{HBr} PhN=NC_6H_4\text{-}p\text{-}Br \qquad (4\text{-}93)$$

known whether the species actually brominated is azobenzene or its conjugate acid. A curious reaction, which may or may not involve electrophilic attack on azobenzenes, occurs when they are treated with toluene under Friedel-Crafts conditions (AlCl$_3$); *p*-aminobiphenyls are produced[363] (Eq. 4-94).

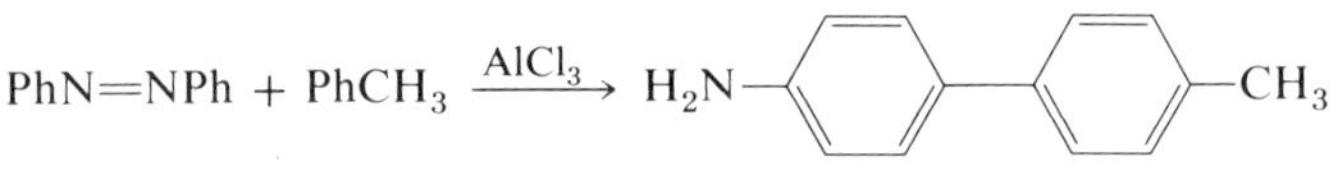

$$PhN=NPh + PhCH_3 \xrightarrow{AlCl_3} H_2N-\!\!\!\langle\;\rangle\!\!\!-\!\!\!\langle\;\rangle\!\!\!-CH_3 \qquad (4\text{-}94)$$

Nucleophilic reagents. Ordinary azoalkanes and azobenzenes show only limited reactivity toward nucleophiles, outside of the base-catalyzed isomerization of azoalkanes to hydrazones already mentioned.[364] Carbanions can either add to azobenzenes or reduce them. Reduction appears to be the favored reaction with Grignard reagents and phenyllithium;[365] the latter reacts with azobenzene to form up to 20% of triphenylhydrazine at depressed temperatures (Eq. 4-95). With other

$$PhN=NPh + PhLi \longrightarrow Ph_2NNHPh \quad \text{and} \quad PhNHNHPh \qquad (4\text{-}95)$$

carbanions, notably benzhydryl, addition to form trisubstituted hydrazines occurs in 49 to 92% yields; some N—N cleavage may occur in certain cases.[366] Sulfinic acids add to azobenzene and some of its derivatives, forming sulfonylhydrazines, which undergo reductive cleavage when refluxed with excess sulfinic acid in ethanol and give rise to sulfones[367] (Eq. 4-96).

In contrast to simple azo compounds, ethyl azoformate (azodicarboxylic ester) has pronounced electrophilic character and reacts readily with a variety of nucleophiles. Dibenzoyldiazene is similar but not so reactive. The type of reactivity shown

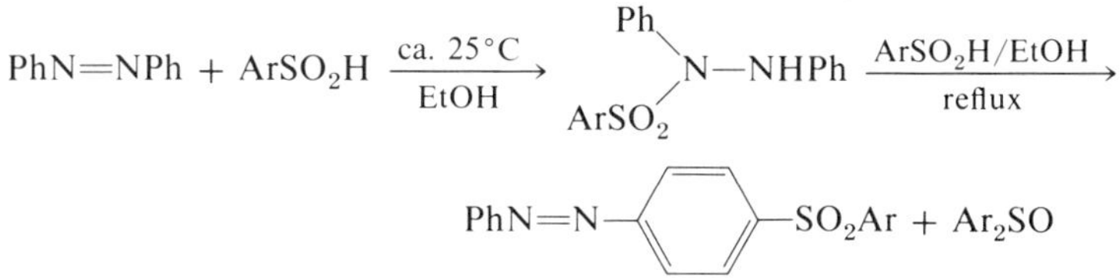

$$PhN{=}NPh + ArSO_2H \xrightarrow[EtOH]{ca.\ 25\,°C} \underset{ArSO_2}{\overset{Ph}{{>}}}N{-}NHPh \xrightarrow[reflux]{ArSO_2H/EtOH}$$

$$PhN{=}N{-}\underset{}{\bigcirc}{-}SO_2Ar + Ar_2SO \qquad (4\text{-}96)$$

markedly resembles that of diazonium salts in coupling reactions and may be expected to be more or less evident in all azo compounds in which both substituents are electron-withdrawing. The subject is an extensive one, and since it has been reviewed in detail elsewhere, only the outline will be given here. The reactions may be classified into three types:[368] simple addition across the N=N bond, addition accompanied or followed by rearrangement in the addend, and addition with rearrangement of the skeleton of the azo reactant such that both substituents become attached to the same nitrogen atom in the product. In some cases, ring closure is a secondary reaction, and attack by the nucleophile on a carbonyl group, especially of esters, is always a potential competing reaction. Simple addition is the most general process; it may be catalyzed by acid or base.

Alcohols and mercaptans form addition products with ethyl azoformate that are believed to be alkoxyhydrazine derivatives or their thio analogs (Eq. 4-97). They are

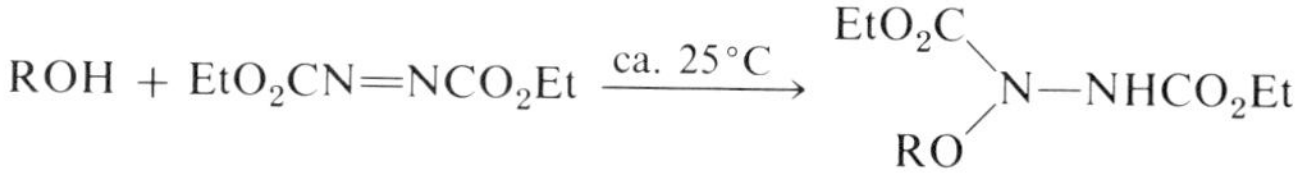

$$ROH + EtO_2CN{=}NCO_2Et \xrightarrow{ca.\ 25\,°C} \underset{RO}{\overset{EtO_2C}{{>}}}N{-}NHCO_2Et \qquad (4\text{-}97)$$

of low stability[369] and are decomposed by acids and even such weak bases as potassium acetate. Sulfides also add to azo formic esters, even though they bear no acidic hydrogen; presumably the initial adduct is a zwitterion having a sulfur-nitrogen bond. Transfer of a proton from an α-carbon to the anionic nitrogen would lead to a sulfur ylide, rearrangement of which to the observed product, an N-(alkylthio-methyl)hydrazine, would be a simple process[370] (Eq. 4-98). Benzyl ethers react at

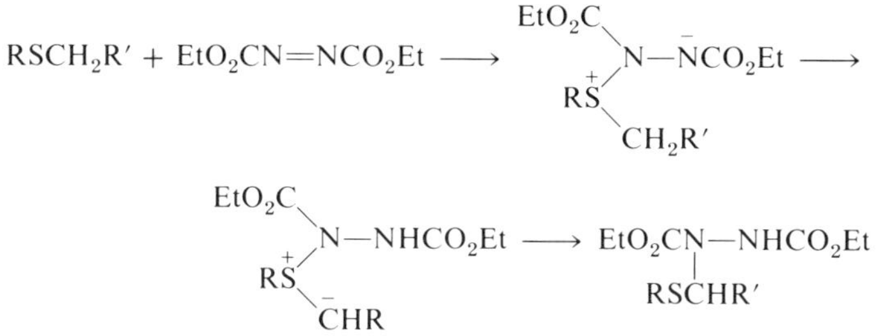

$$RSCH_2R' + EtO_2CN{=}NCO_2Et \longrightarrow \underset{\underset{CH_2R'}{\overset{+}{RS}}}{\overset{EtO_2C}{{>}}}N{-}\overset{-}{N}CO_2Et \longrightarrow$$

$$\underset{\underset{\overset{-}{CHR}}{\overset{+}{RS}}}{\overset{EtO_2C}{{>}}}N{-}NHCO_2Et \longrightarrow EtO_2C\underset{RSCHR'}{\overset{|}{N}}{-}NHCO_2Et \qquad (4\text{-}98)$$

100 °C in the same way.[371] Phenols are oxidized to dihydroxybiphenyls or quinone derivatives.

Primary and secondary aromatic amines add to azoformic esters[372] to form triazane derivatives[373] (Eq. 4-99), but in some cases, such as 2-naphthylamine, which is attacked at the 1 position, substitution at a ring carbon occurs instead. With aliphatic primary and secondary amines, attack at the ester group to form amides is

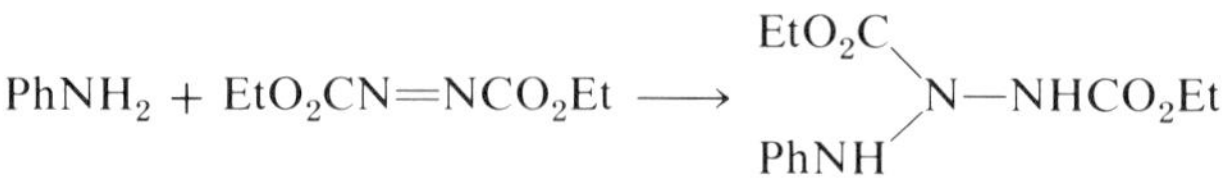

$$PhNH_2 + EtO_2CN{=}NCO_2Et \longrightarrow \underset{PhNH}{\overset{EtO_2C}{\diagdown}}N{-}NHCO_2Et \qquad (4\text{-}99)$$

the dominant reaction, especially with the less-hindered methyl esters.[371,373a] Tertiary amines follow the same reaction course as sulfides and ethers, giving aminomethylhydrazine derivatives (Eq. 4-100), which are readily hydrolyzed to alde-

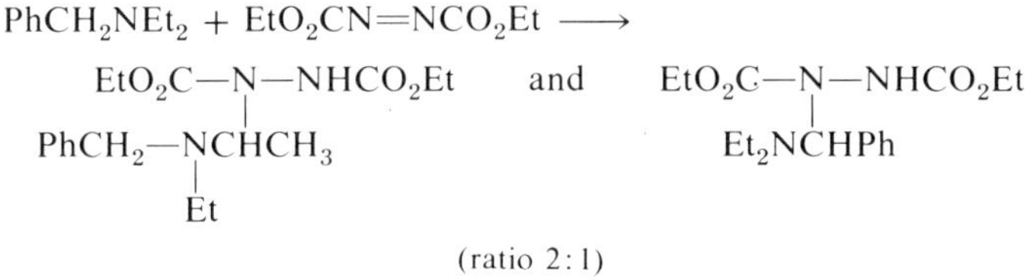

$$PhCH_2NEt_2 + EtO_2CN{=}NCO_2Et \longrightarrow$$

$$\underset{\underset{Et}{\overset{|}{\underset{PhCH_2-NCHCH_3}{}}}}{EtO_2C-N-NHCO_2Et} \quad \text{and} \quad \underset{Et_2NCHPh}{EtO_2C-N-NHCO_2Et} \qquad (4\text{-}100)$$

(ratio 2:1)

hydes.[371,374] Whereas benzyldiethylamine gives a nearly statistical ratio of positionally isomeric products, the *p*-nitro derivative undergoes attachment at the benzylic carbon to the extent of 96%, and benzyldimethylamine becomes attached to the extent of 91% at the benzyl carbon.

Triphenylphosphine reacts in the presumed way that tertiary amines react in the initial step, but because of the lack of an α-hydrogen, reaction proceeds no further.[375] Even azobenzene behaves in this way, and the dipolar products can be isolated as salts[376] (Eq. 4-101).

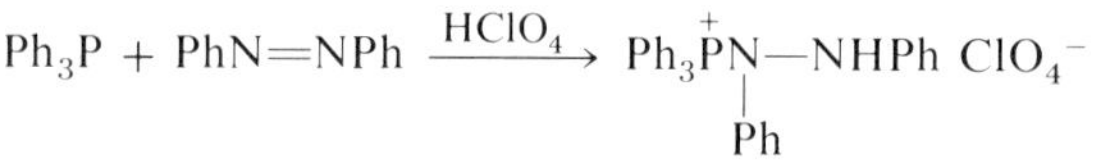

$$Ph_3P + PhN{=}NPh \xrightarrow{\;HClO_4\;} \underset{Ph}{\overset{+}{Ph_3PN}}{-}NHPh\ ClO_4^- \qquad (4\text{-}101)$$

Thiourea, with both nitrogen and sulfur to choose from, reacts at the more nucleophilic site, and the products are isothioureas[377] (Eq. 4-102). The reaction is appar-

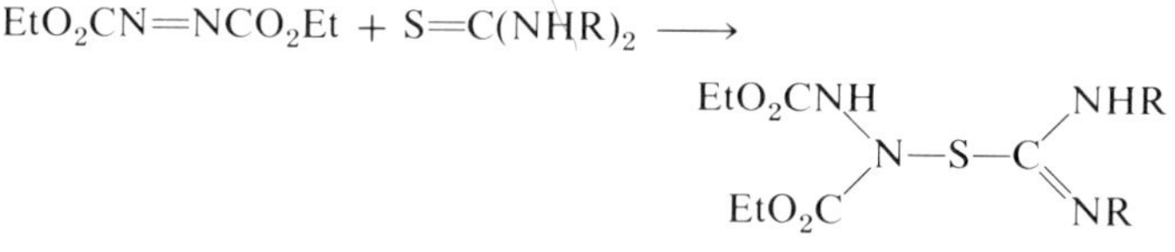

$$EtO_2CN{=}NCO_2Et + S{=}C(NHR)_2 \longrightarrow$$

$$\underset{EtO_2C}{\overset{EtO_2CNH}{\diagdown}}N{-}S{-}C\underset{NR}{\overset{NHR}{\diagup}} \qquad (4\text{-}102)$$

ently general and can be used to prepare carbodiimides by treating the product with triphenylphosphine.

Azoformic esters attack carbon sites in some variety: aliphatic carbanions (including enolates), benzene and naphthalene rings, aldehydes and their phenylhydrazones, and alkenes (including enol ethers). The reactivity toward carbanions is sufficiently great that potassium acetate is sufficient to catalyze the addition of malonic ester to ethyl azoformate at room temperature[378,379] (Eq. 4-103). Ketones, such as cyclohexanone, react at the α position.[374] *t*-Butylmagnesium chloride adds to the N=N bond of *t*-butyl azoformate, providing a route to *t*-butylhydrazine.[379] Triarylsilanes and triarylgermanes also add.[380]

Aldehydes react to give bis(alkoxycarbonyl)hydrazides of the corresponding acid.[381] Aldehyde phenylhydrazones react more readily and exothermically give rise

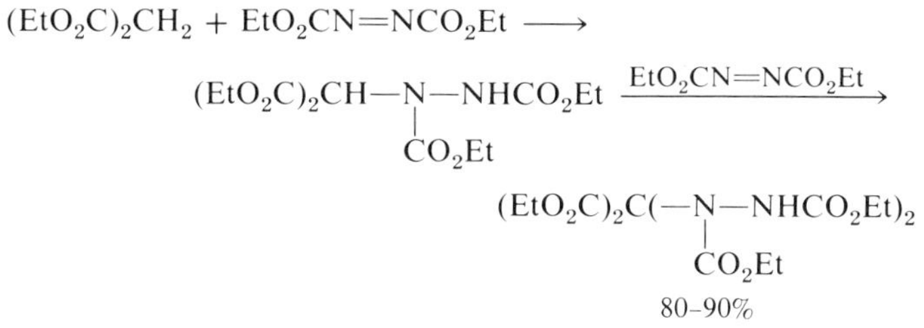

$(EtO_2C)_2CH_2 + EtO_2CN=NCO_2Et \longrightarrow$

$$(EtO_2C)_2CH-\underset{\underset{CO_2Et}{|}}{N}-NHCO_2Et \xrightarrow{EtO_2CN=NCO_2Et}$$

$$(EtO_2C)_2C(-\underset{\underset{CO_2Et}{|}}{N}-NHCO_2Et)_2 \qquad (4\text{-}103)$$

80–90%

to hydrazidines[382] (Eq. 4-104). Since this reaction fails with N,N-disubstituted hydrazines, it has been postulated that it must pass through a cyclic transition state

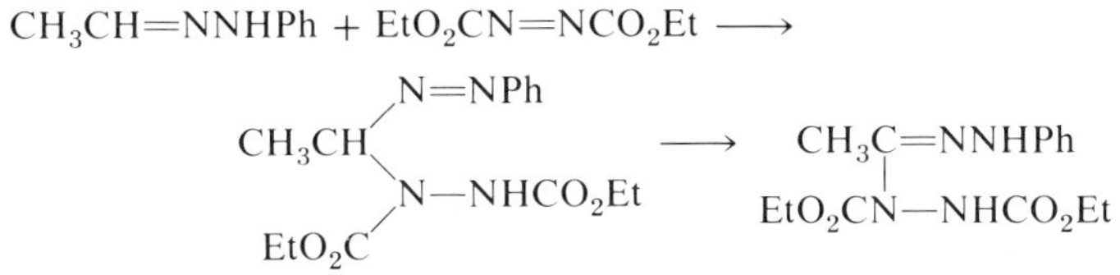

$$CH_3CH=NNHPh + EtO_2CN=NCO_2Et \longrightarrow$$

$$CH_3CH\underset{EtO_2C}{\overset{N=NPh}{\diagup\diagdown}}N-NHCO_2Et \longrightarrow \underset{EtO_2C\overset{}{N}-NHCO_2Et}{CH_3C=NNHPh} \qquad (4\text{-}104)$$

in which the N-attached hydrogen of the hydrazone is transferred to the azo group at the same time as the new N—C bond is formed, analogous to the ene reaction of the same reagent with olefins, about to be discussed. Ketone hydrazones have not been observed to react.

Aromatic rings that are as nucleophilic as toluene are substituted by ethyl azoformate to form N-arylhydrazine-N,N'-dicarboxylic esters[372,373b,383] (Eq. 4-105),

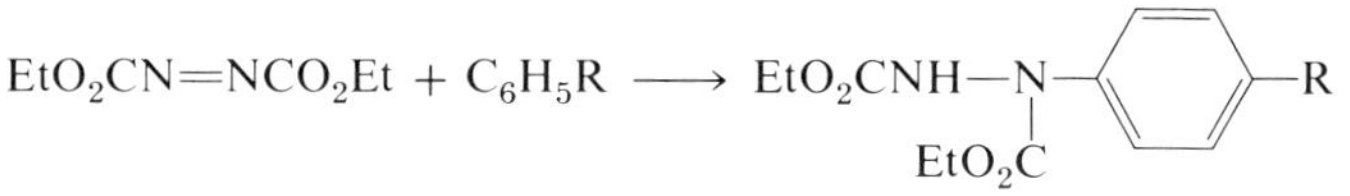

$$EtO_2CN=NCO_2Et + C_6H_5R \longrightarrow EtO_2CNH-\underset{\underset{EtO_2C}{|}}{N}\!\!-\!\!\left\langle\bigcirc\right\rangle\!\!-R \qquad (4\text{-}105)$$

which are easily hydrolyzed to aryl hydrazines, for which the overall process constitutes a useful preparative method. The reaction is acid-catalyzed, but the more reactive aromatic rings do not require catalysis. Enol ethers react analogously, forming alkoxyvinylhydrazinedicarboxylic esters.[371,373a]

Olefins bearing allylic hydrogen react in a distinct manner; substitution occurs with migration of the C=C, bond[384-387] (Eq. 4-106). Yields are only moderate (ca.

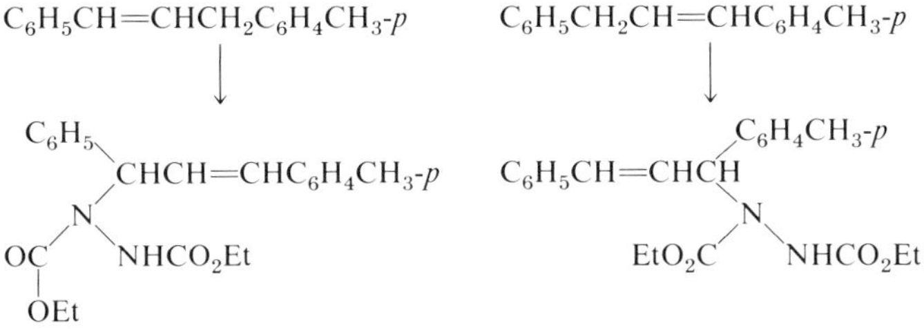

$$C_6H_5CH=CHCH_2C_6H_4CH_3\text{-}p \qquad C_6H_5CH_2CH=CHC_6H_4CH_3\text{-}p$$
$$\downarrow \qquad\qquad\qquad \downarrow$$

$$\underset{\underset{OEt}{|}}{\overset{C_6H_5}{\underset{OC}{\diagup}}}CHCH=CHC_6H_4CH_3\text{-}p \quad C_6H_5CH=CHC\overset{\diagup C_6H_4CH_3\text{-}p}{\underset{\underset{EtO_2C}{}\overset{N}{\diagdown}NHCO_2Et}{}} \qquad (4\text{-}106)$$

40%), and with most olefins, neither free-radical initiators nor acids show catalysis. However, both cyclohexene and cyclopentene show catalysis by peroxide initiators, and it has therefore been inferred that there are two possible paths to product, one being a true ene reaction with a cyclic transition state and the other a free-radical chain reaction. The free-radical path appears to be the only one for fluorene and tetralin. The ordinary reaction shows second-order kinetics, and vinyl deuterium is not disturbed (Eq. 4-107). 1,4-Cyclohexadiene, which gives 80% substitution (Eq. 4-108), undergoes no detectable isomerization to the 1,3 isomer in unconsumed starting material.[388]

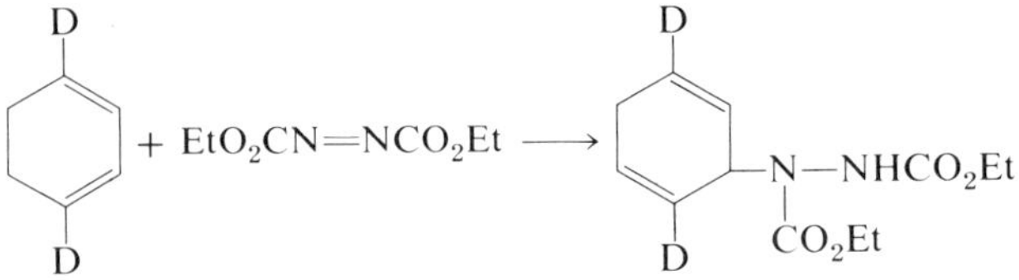

$$ (4\text{-}107) $$

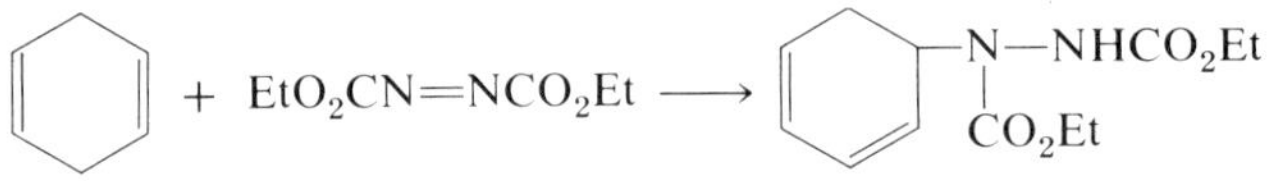

$$ (4\text{-}108) $$

With conjugated dienes, the foregoing substitution reaction must compete with Diels-Alder cycloaddition; substitution commonly predominates, as, for example, with 2,4-dimethyl-1,3-pentadiene, which undergoes substitution to the extent of 87%. In contrast, however, 1,1′-biscyclohexenyl reacts in Diels-Alder fashion to give a tetrahydropyridazine in 80% yield.[389] Styrene reacts in both ways, first undergoing Diels-Alder cycloaddition, followed by substitution with a second equivalent of reagent (Eq. 4-109), probably by the free-radical path.[390]

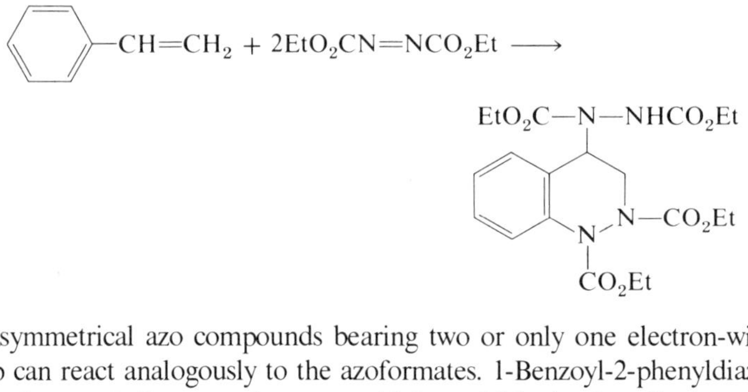

$$ (4\text{-}109) $$

Unsymmetrical azo compounds bearing two or only one electron-withdrawing group can react analogously to the azoformates. 1-Benzoyl-2-phenyldiazene (benzoylazobenzene), for example, reacts with benzaldehyde phenylhydrazone to form an adduct presumably having the hydrazidine structure.[391] *Anti-p*-nitrobenzene-azocyanide attacks 2-naphthol at the 1 position to generate 1,2-naphthoquinone 1-*p*-nitrophenylimine and cyanamide. This reaction presumably starts out according to Equation 4-105, followed by elimination of NCNH$_2$ (Eq. 4-110), which then adds to another molecule of diazocyanide to produce the isolated areneazo-*N*-cyano-formamidine.[392] *p*-Chloro- and bromobenzeneazocyanides react less readily, and the *p*-methyl analog does not react at all; these observations indicate the limits of electron withdrawal required for an electrophilic azo group. However, even azobenzene may react under forcing conditions, for it has been reported to attack

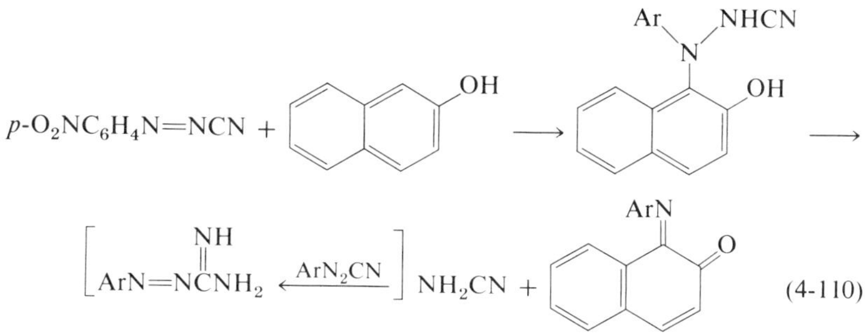

$$p\text{-}O_2NC_6H_4N{=}NCN \ +$$

(with naphthol, intermediate, and product structures) (4-110)

α-naphthol at 180°C to give a small amount of 2-anilino-1,4-naphthoquinone 4-anil.[393]

Cycloaddition. Azo compounds are known to undergo cycloaddition so as to form four-, five-, and six-membered rings. The last is the Diels-Alder reaction, an example of which has already been mentioned. The Diels-Alder reaction was, in fact, discovered in the form of ethyl azoformate acting as a dienophile (Eq. 4-111). The

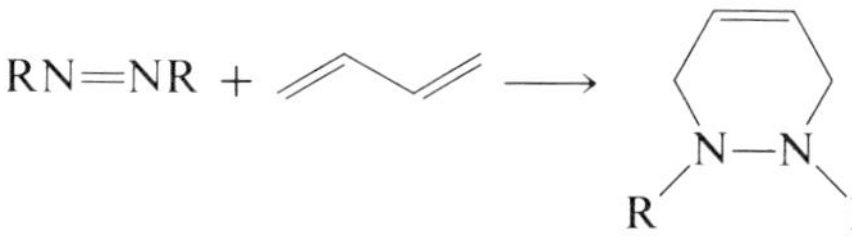

$$RN{=}NR \ + \ \text{(diene)} \ \longrightarrow \ \text{(ring)}$$

(4-111)

reaction goes best when azo compounds bear electron-withdrawing substituents,[394] and perfluoro azo compounds make effective dienophiles, for example.[335] Arenediazocyanides also act as dienophiles.[268A]

Cycloaddition with 1,3-dipolar compounds occurs especially well with azoformic esters. Nitrile ylides, for example, yield 1,2,4-triazolines[395,396] (Eq. 4-112), and ni-

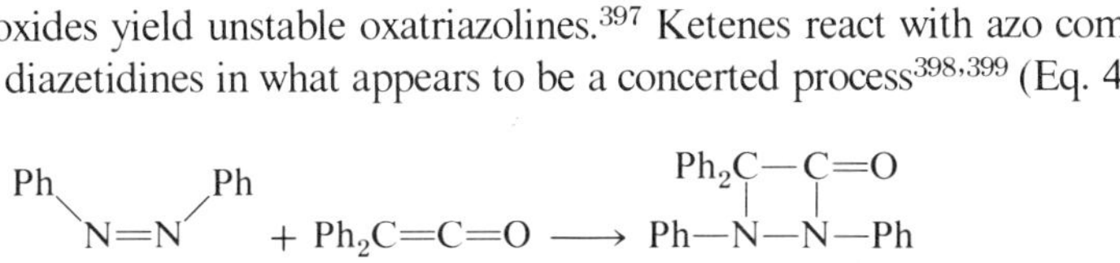

$$R_2\overset{-}{C}{-}\overset{+}{N}{\equiv}CPh \ + \ EtO_2CN{=}NCO_2Et \ \longrightarrow$$

(4-112)

trile oxides yield unstable oxatriazolines.[397] Ketenes react with azo compounds to form diazetidines in what appears to be a concerted process[398,399] (Eq. 4-113). Cis-

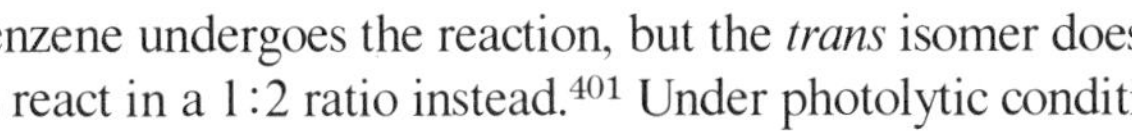

$$\underset{\text{N}{=}\text{N}}{\overset{\text{Ph}\diagdown \quad \diagup \text{Ph}}{}} \ + \ Ph_2C{=}C{=}O \ \longrightarrow \ \underset{Ph-N-N-Ph}{\overset{Ph_2C-C{=}O}{}}$$

(4-113)

azobenzene undergoes the reaction, but the *trans* isomer does not.[135,400] Azoformic esters react in a 1:2 ratio instead.[401] Under photolytic conditions, ethyl azoformate is isomerized to the *cis* isomer, which may then undergo 1,2-addition with olefins to form diazetidines or 1,4-addition (thus involving the carbonyl group) to form dihydrooxadiazines.[402]

Reduction. Azo compounds are in principle reducible in three stages: to hydrazyl radicals (or the corresponding anion radical), to hydrazines (or the corresponding dianions), or finally, to primary amines (Eq. 4-114). Ordinarily reduction is thought

$$RN{=}NR \longrightarrow RNH{-}\overset{.}{N}{-}R \longrightarrow RNHNHR \longrightarrow 2RNH_2 \qquad (4\text{-}114)$$

of in terms of addition of hydrogen and/or removal of oxygen, but addition of atoms other than hydrogen to the $N{=}N$ bond are properly included in the concept of reduction, especially when the atoms attached to nitrogen are more electropositive than nitrogen. Some such reactions have been described in earlier sections of this chapter.

In this section, the presentation is arranged principally according to type of reducing agent. Organization according to type of product should be sought in the preparative sections for the anticipated products (for example, hydrazines in Chap. 1) and in reviews elsewhere.[403] Most of the published work on the subject concerns diaryldiazenes, owing to the importance of the method in synthesis of diarylhydrazines or anilines.

Catalytic hydrogenation[404] over palladium proceeds in well-defined stages and has been used to prepare both hydrazines and amines from azo compounds.[405] More vigorous conditions are usually required to carry the reduction to the amine stage.[406,407] A number of cyclic aliphatic azo compounds have been reduced to hydrazines cleanly with this catalyst.[408] A variant is the use of cyclohexene in place of H_2 as the source of hydrogen; high yields of amine have been obtained.[409]

Among other catalysts for hydrogenation, platinum[337] seems to be similar to palladium, but a systematic comparison does not seem to have been made. 2,2′-Azopropane[337] and a diacyldiazene[410] have been hydrogenated to hydrazines with platinum, and in some other instances, amines have been obtained. Raney nickel is particularly efficient in cleaving hydrazines, so it is not surprising to find that it also cleaves azo compounds to amines.[411] The soluble catalyst bis(dimethylglyoximato)-cobalt has produced almost quantitative yields of hydrazine.[412]

Diimide (diazene), generated in situ from hydrazine in contact with air and a copper catalyst,[413] or from oxidation of hydrazine by sulfur in a separate vessel,[414] or from decarboxylation of azoformic acid,[413] has been shown to be an effective hydrogenating agent for azo compounds, which are converted to hydrazines without further reduction to amine. Reduction reported for hydrazine used alone may take place through diimide.[379] However, substituted hydrazines have also been found to reduce azo compounds in some instances.[415] This type of reduction may be limited to diacyldiazenes, such as azoformic esters.

Among the hydride reagents, lithium aluminum hydride and its alkoxy derivatives are generally sluggish to inert.[405,416] However, when various metal salts (for example, $PbCl_2$) are added, the reduction by $LiAlH_4$ proceeds readily and in high yield to hydrazines.[417] Sodium borohydride also becomes effective when a catalyst (palladium) is used with it.[418] Diborane reduces azobenzenes to hydrazobenzenes at room temperature,[419] but unlike the other hydride reagents, it has been found to produce primary amines.[420] Trialkylboranes have been found to reduce *cis*-azobenzene and ethyl azoformate by adding $R_2B{-}$ and $H{-}$, the third alkyl group being eliminated as alkene; hydrolysis of the adduct gives the hydrazine derivative.[421] A competing reaction occurs with ethyl azoformate in which R_2B- and $R-$ add to the $N=N$ bond; hydrolysis then gives an alkylated hydrazine derivative.

Metals or their lower-valent salts in some variety have been found to reduce azobenzenes. The products may be hydrazines or amines, but unless reduction proceeds rapidly, acidic conditions cause benzidine rearrangement at the hydrazine stage.[422] Reduction by tin or stannous chloride is particularly prone to give this result.[423,424] Titanous chloride reacts quickly and quantitatively[425] and has been manipulated so as to produce hydrazobenzene, benzidine, or aniline.[426,427] Zinc and acetic acid, however, does not appear to result in benzidine rearrangement, and hydrazobenzenes have generally been obtained.[428-430] Electron-withdrawing substituents favor further reduction to the aniline stage. An aliphatic azo compound, 1,1'-diphenylazoethane, has been reduced to primary amine with this combination.[431]

In nearly neutral solution, zinc with aqueous ammonium chloride converts azo compounds to hydrazines in preparative yields.[432,433] In strongly basic solution, with warming, hydrazines are also produced in good yield.[434-436] Aluminum amalgam with water has been reported to produce either amines[437] or hydrazines.[436] Sodium amalgam or sodium and alcohol have reduced 2,2'-azopropane to the hydrazine.[337,436] In aprotic medium, metallic sodium and potassium convert azo compounds successively to anion radicals and dianions.[438,439]

Many types of nonmetallic reducing agents attack azo compounds, such as hydriodic acid, which cleaves azobenzenes to anilines,[432,440] and hydroquinone, which produces hydrazines.[441] Sodium hydrosulfite (sodium dithionite, $Na_2S_2O_4$) has long been used to reduce azobenzenes to anilines in good yields.[442] Sodium sulfide and ammonium hydrogen sulfide have also been used; they lead to hydrazines, but the scattered reports suggest that they are not always effective.[430] Hydrocarbons may act as hydrogen donors in the presence of boron fluoride[443] or under photolytic conditions; the products are hydrazines, but the yields are less than moderate. Hydrazyls appear to be distinct intermediates in the photolytic process,[441,444] for tetrazanes, their dimers, have been found among the products. Heat alone is sufficient at high enough temperatures, for boiling "liquid parraffin" (bp ca. 400 °C) has been used to reduce a group of azobenzenes to anilines in moderate but useful yields; the products distill directly out of the reaction mixture.[445] Alcohols have been reported to reduce azoformic esters and azoalkanes in the form of their salts with strong acids; the products are hydrazine derivatives.

Grignard reagents react with azobenzenes largely as reducing agents, producing the dianions of the hydrazobenzenes that can be obtained by subsequent hydrolysis[364,365] (Eq. 4-115). Addition to the $N=N$ double bond is a significant competing

$$ArN=NAr + RMgX \longrightarrow Ar\bar{N}-\bar{N}Ar(MgX^+)_2 + R-R$$

$$\text{or} \quad \begin{array}{c} Ar \\ \diagdown \\ N-NArMgX \\ \diagup \\ R \end{array} \quad (4\text{-}115)$$

reaction. When electron-withdrawing substituents are present, reduction may proceed all the way to the amine stage.[429] Phenyllithium behaves much like Grignard reagents; lower temperatures favor addition over reduction.[365,411]

Electrochemical reduction, which has been reviewed recently,[296,446] proceeds through all three stages.[447-449] Reduction to hydrazines is reversible,[450] but if the medium is acidic, benzidine rearrangement may occur. In aprotic medium (dimethylformamide), reduction to the anion radical is fast, and it is followed by slower reduction to the dianion.[447,451,452]

Oxidation. Oxidation of azo compounds in general leads to azoxy compounds.[453] Less commonly, azoalkanes may be oxidized to azines.

Peroxycarboxylic acids ($PhCO_3H$, m-$ClPhCO_3H$, $MeCO_3H$, CF_3CO_3H, etc.) have been used widely to convert azoalkanes to azoxyalkanes in preparatively useful yields.[454-456] The reaction is stereospecific, and cis-2,2′-azopropane,[57] for example, is converted to cis-2,2′-azoxypropane, and the $trans$ isomer gives the $trans$ product (Eq. 4-116). Under more strongly acidic conditions, such as hydrogen peroxide

$$\begin{array}{c}R\\ \diagdown \\ N{=}N \quad + \ R'CO_3H \ \longrightarrow \quad N{=}\overset{+}{N}\diagup^{O^-} \quad + \ R'CO_2H\\ \diagup \\ R\end{array}$$

$$\begin{array}{c}N{=}N\\ R\diagup \quad \diagdown R \ + \ R'CO_3H \ \longrightarrow \quad \overset{O^-}{\underset{R}{\overset{+}{N}{=}N}}\end{array}$$

(4-116)

mixed with nitric acid, results are unsatisfactory, owing to acid-catalyzed isomerization to hydrazones.[455] Azobenzenes are also oxidized by peroxy acids stereospecifically, and the cis isomers are oxidized faster than the $trans$.[457] The conditions required are generally mild, and good results have been obtained by carrying out the oxidations at 0 to 10°C in chloroform. With an optically active peroxy acid, (+)-peroxycamphoric acid, asymmetric induction becomes possible, and the cyclic azoalkane 2,3-diazabicyclo[2.2.1]heptane has been converted to an optically active azoxy compound by this means.[454]

The principal alternative to peroxy acids has been hydrogen peroxide (usually 30% aqueous) and acetic acid, which has become popular because of its high yields and versatility; it may be used under refluxing conditions if necessary with less reactive azobenzenes.[458-460] Hydrogen peroxide alone, in concentrations from 30 to 98%, has also been used with success.

The various peroxide reagents seem to give much the same result, all producing mixtures of positionally isomeric azoxy compounds with unsymmetrical azo compounds. The ratios of such isomers vary from nearly equal to almost totally one-sided and are affected by both steric and electronic factors. With simple azoalkanes, oxidation occurs preferentially at the least-hindered nitrogen,[461] but this effect does not always hold if one substituent is unsaturated.[462] Moreover, oxidation at the olefinic site may compete. Methylphenyldiazene is reported[456] to become oxidized at the aryl nitrogen, however, whereas benzylphenyldiazene gives both positional isomers.[463] Some caution is advisable in interpreting the results reported before the widespread use of NMR spectroscopy, however, for it was difficult to establish structure and to assay mixtures before then.

With azobenzenes, electron-withdrawing substituents retard oxidation and direct it to the far nitrogen in general,[464] but there are exceptions, such as *m*-tolyl-*m*-chlorophenyldiazene, which is oxidized adjacent to the chlorophenyl group, whereas *m*-tolyl-*m*-nitrophenyldiazene is oxidized adjacent to the *m*-tolyl group.[465] *Ortho* substituents of nearly all types direct oxidation away from them, except in the case of an *o*-hydroxy group.[466] *p*-Chloro-*p'*-ethoxyazobenzene, in which electronic effects might be expected to be one-sided, is reported to give both isomeric azoxy compounds.[467]

The kinetics of oxidation by peroxy acids are second-order, and the rates of oxidation of substituted azobenzenes follow the Hammett equation with some deviation, *p*-methoxyazobenzene being faster than predicted and *p*-nitro slower.[468] Accumulation of electron-withdrawing groups, as in decafluoroazobenzene, greatly suppresses susceptibility to oxidation, but hot trifluoroperoxyacetic acid will still oxidize them.[407] The mechanism of peroxy acid oxidation might involve simple attack at either azo nitrogen or epoxidation to an oxadiaziridine followed by ring opening[469] (Eq. 4-117). Although a clear decision cannot be made on the basis of the available information, the weight of the evidence favors the direct process.

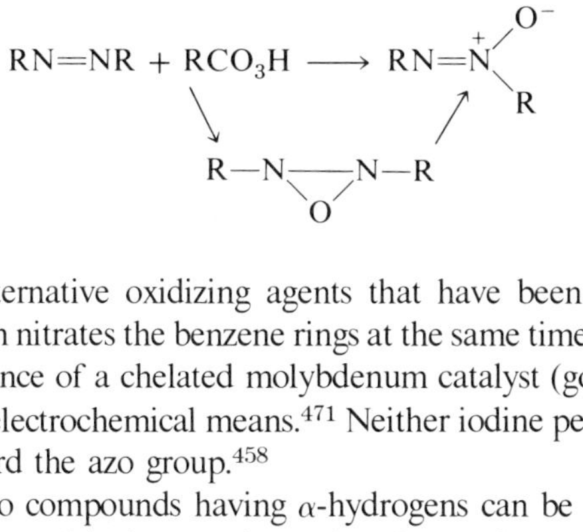

$$RN{=}NR + RCO_3H \longrightarrow RN{=}\overset{+}{N}\overset{O^-}{\underset{R}{\diagup}} \tag{4-117}$$

$$R-N{\underset{O}{\diagdown\diagup}}N-R$$

Alternative oxidizing agents that have been used include fuming nitric acid, which nitrates the benzene rings at the same time,[430] *tert*-butyl hydroperoxide in the presence of a chelated molybdenum catalyst (good yields, but somewhat slow),[470] and electrochemical means.[471] Neither iodine pentafluoride[472] nor ozone is reactive toward the azo group.[458]

Azo compounds having α-hydrogens can be oxidized to azines by exposure to sources of hydrogen-abstracting free radicals, such as RS· and Cl$_3$C·[473] (Eq. 4-118). Yet another type of oxidation is observed during photolysis in the presence of oxygen; methyl hydroperoxide is produced from azomethane (Eq. 4-119), appar-

$$\underset{CH_3}{\overset{Ph}{\diagdown}}CHN{=}NCH\underset{CH_3}{\overset{Ph}{\diagup}} \xrightarrow{RS\cdot} \underset{CH_3}{\overset{Ph}{\diagdown}}C{=}N-N{=}C\underset{CH_3}{\overset{Ph}{\diagup}} \tag{4-118}$$

$$CH_3N{=}NCH_3 \xrightarrow[h\nu,\ 3600\ \text{Å}]{O_2} CH_3OOH, \quad N_2, \quad CH_2{=}O \tag{4-119}$$

ently by combination of methyl radicals with O_2, followed by hydrogen abstraction, perhaps from methoxy radicals.[474]

Effect on neighboring functions.   A carboxyl group attached directly or adjacent to an azo nitrogen is very unstable toward decarboxylation, and the free acids do not appear to have been isolated. Acidification of azoformate salts gives rise to diimide (diazene) in solution, in which it lasts long enough to effect *cis*-hydrogenation of unsaturated compounds.[475] The decarboxylation of α-azo acids is believed to occur by concerted intramolecular hydrogen transfer[476] (Eq. 4-120).

$$\underset{\underset{CO_2H}{|}}{\overset{\overset{CO_2Et}{|}}{Me_2C-N=N-CMe_2}} \longrightarrow Me_2C\overset{\displaystyle N\cdots N}{\underset{\underset{O}{\overset{\|}{C-O}}}{}}\overset{CMe_2CO_2Et}{\underset{H}{\diagdown}} \longrightarrow$$

$$Me_2C{=}NNHCMe_2CO_2Et \qquad (4\text{-}120)$$

A chlorine in a position adjacent to an azo group is very reactive toward nucleophilic displacement by such species as CH_3O^-, CH_3S^-, AcO^-, etc.[321]

Free radicals.   The azo group of azobenzene is not attacked on exposure to phenyl radicals; the ring is attacked instead.[477] However, some areneazoalkanes have been reported to react with phenyl radicals in such a way as to replace the alkyl group by phenyl, probably through initial addition at the alkyl nitrogen.[478]

Formazans

The chemistry of formazans[85,479] parallels that of simple azo compounds, such as reduction to hydrazine derivatives,[480] and that of hydrazines, such as acylation.[481] However, their most important reaction is oxidation to tetrazolium salts (Eq. 4-121).

$$RC\overset{\displaystyle N=NAr}{\underset{\displaystyle N-NHAr}{\diagup}} \underset{[H]}{\overset{[O]}{\rightleftharpoons}} R-C\overset{\displaystyle N-N-Ar}{\underset{\displaystyle N-N-Ar}{\diagup\hspace{-0.3em}\diagdown}}\,\boxed{+} \qquad (4\text{-}121)$$

This reaction passes through an intermediate free-radical stage and is reversible.[482] A variety of reducing agents, particularly those found in biological systems, can accomplish regeneration of formazans. Because of this, tetrazolium salts have acquired a considerable degree of importance in biological work as selective staining agents; the salts penetrate the cell and are reduced to the highly colored formazans at appropriate internal sites.

Among other unusual reactions of formazans may be mentioned that with hydrogen chloride,[483] which entails reductive chlorination, perhaps by way of addition to the N=N bond followed by migration of chlorine (Eq. 4-122). Acid has been observed to bring about hydrolysis to a hydrazine and a diazonium salt[484] or cyclization to a benzo-1,2,4-triazine (the Bamberger benzotriazine synthesis).[481,485]

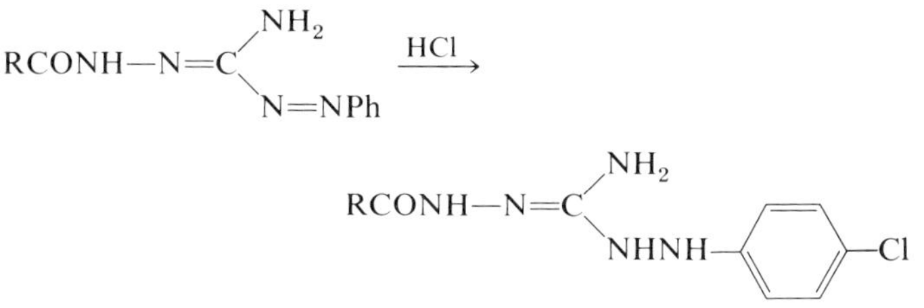

$$RCONH-N=C\begin{smallmatrix}NH_2\\N=NPh\end{smallmatrix} \xrightarrow{HCl}$$

$$RCONH-N=C\begin{smallmatrix}NH_2\\NHNH-\end{smallmatrix}\!\!\!\!\!\!\langle \rangle\!\!-Cl \qquad (4\text{-}122)$$

PREPARATIVE METHODS

Diazonium Compounds[486,487]

Virtually the only preparatively useful method for obtaining diazonium compounds is diazotization of amines, although there are other reactions that also lead to them. The conventional method is to add a solution of sodium nitrite to one of a primary aromatic amine in aqueous mineral acid solution (Eq. 4-123). It has been described

$$ArNH_2 + HX + NaNO_2 \xrightarrow{0 \pm 5^\circ C} ArN_2^+X^- + NaX + 2H_2O \qquad (4\text{-}123)$$

in detail in numerous places,[28,260,269,488-490] and there should be no need to repeat the normal procedure here.

A complication that is likely to arise with large amines is precipitation of an insoluble amine salt before diazotization is begun. If the insolubility is not too pronounced, it may be possible to proceed in spite of it, the salt dissolving as the reaction proceeds. To facilitate this, the insoluble salt should be prepared in as finely divided a state as feasible, commonly by grinding the amine with the mineral acid if necessary. An alternative may be to use an auxiliary solvent, usually acetic acid or ethanol, but the amine salts not infrequently are sparingly soluble in these solvents, too. In extreme cases, such as N-(p-aminophenyl)phthalimide, use of an acid having a very hydrophilic anion in place of the conventional mineral acids can help. Ethanol-2-sulfonic acid (isethionic acid) has been used for this purpose.[491] Still another option is available if the amine is amphoteric, as in the case of sulfanilic acid. This is inverse addition, in which the amine is first dissolved in aqueous base along with the requisite amount of sodium nitrite, and the mixture is then poured into cold mineral acid.

The mechanism[492] of diazotization involves nitrosation in the primary step, and it follows that other nitrosating agents can be used besides nitrous acid (which is, in fact, an equilibrating mixture of different species, including N_2O_3). The very earliest experiments with diazotizing amines were carried out with "nitrous fumes," a mixture of NO and NO_2. Alkyl nitrites may be used instead of sodium nitrite, with certain advantages in special situations.[493] Nitrosylium salts are especially powerful nitrosating agents. They have rarely been used for diazotizations, but the closely related compound nitrosylsulfuric acid, $NOHSO_4$, has seen much use with very weakly basic amines.[494,495] It can be used in concentrated sulfuric acid or glacial acetic acid.

When it is desired to isolate dry diazonium salts, one can often precipitate them from aqueous solution with proper choice of anion; fluoroborates[232,496] and hexafluorophosphates[497] are usually chosen. Alternatively, one can carry out the reaction in a nonaqueous solvent using an alkyl nitrite and precipitate the salt by adding ether, if necessary. Nitrosylium fluoroborate in glacial acetic/propionic acid mixture has been used to prepare some diazonium fluoroborates.[91] Diazonium salts have in many instances been isolated as sparingly soluble salts of large anions, especially metal complexes[256] and arenesulfonates; such salts often have the advantage of improved stability.[28] A neglected method is reaction of thionylanilines (arylsulfilimines), $ArN{=}SO$, with NOCl or N_2O_4 at -70 to $-10°C$ in ether or carbon tetrachloride; conditions are strictly aprotic and yields are good.[498]

Of the other reactions that produce diazonium compounds, probably the most potentially useful is direct introduction of the diazonio group in place of hydrogen.[499] This reaction succeeds only with aromatic rings that are particularly susceptible to electrophilic substitution, and it requires a large excess of nitrous acid (Eq. 4-124). Diazonium compounds can also be made by rearrangement of N-nitroso-

$$(CH_3)_2NPh + NaNO_2 \text{ (15 equiv.)} + aq. \text{ HCl (10 equiv.)} \xrightarrow[0-5°C, \ 14 \ h]{aq. \ acetone}$$

$$(CH_3)_2N{-}\langle \bigcirc \rangle{-}N_2{}^+Cl^- \qquad (4\text{-}124)$$

N-acylanilines, obtained by nitrosation of anilides.[500] Benzaldehyde anils can be converted to diazonium salts by nitrosating agents, a method that might be useful in cases where an amino group has been protected by a benzylidene group during a synthesis.[501,502] Another very limited method is effectively an exchange reaction between one diazonium compound and a different amine, which takes place through an intermediate triazene[503,504] (Eq. 4-125). It proceeds in the direction that

$$ArN_2{}^+ + Ar'NH_2 \longrightarrow ArNH_2 + Ar'N_2{}^+ \qquad (4\text{-}125)$$

converts the more strongly basic amine to a diazonium ion. Nitrosobenzenes can be converted to diazonium nitrates by reaction with excess nitrous acid,[505] but the nitroso compounds are not generally conveniently accessible starting materials.

Diazoates have nearly always been prepared from diazonium salts by adding base, but they can be prepared directly by reaction of nitrosobenzenes with hydroxylamine or by reaction of nitrobenzenes with sodamide.[506]

Aliphatic diazonium ions are prepared in much the same way as their aromatic counterparts, but inasmuch as they are almost invariably fleeting intermediates, they will not be taken up further here. Aliphatic diazoates can be prepared by treating N-alkyl-N-nitrosourethans with strong base.[11]

t-Butyl thionitrite and thionitrate, t-BuSNO and t-BuSNO$_2$, have been applied with good results to diazotization of anilines for *in situ* conversion to aryl halides in a modified Sandmeyer reaction, or to biaryls in a process analogous to the Gomberg-Bachmann synthesis, as well as for Meerwein arylation of alkenes.[507]

Azo Compounds[487,508-509]

There is no completely general method for preparing azo compounds, although oxidation of *sym*-disubstituted hydrazines is widely applicable. There are enough situations in which the hydrazine is only difficulty accessible or, worse still, must be made from the azo compound one seeks, to justify a variety of alternative preparative methods. Many of these have quite limited scope, and many characteristically give yields that in other circumstances would be unacceptably low, but they cannot be dismissed, for they may be the only available methods. Furthermore, some of the reactions that give low yields are quite direct, and can therefore compete with methods that may give high yields in the last step, but require a laborious synthesis of the penultimate compound. The types of reaction involved are oxidations, reductions, condensations (N to N or N to C), isomerization, and elimination. They will be discussed in that order.

Oxidation of hydrazines (Eq. 4-126) has most commonly been accomplished with

$$RNH—NHR \xrightarrow{[O]} RN{=}NR \tag{4-126}$$

mercuric oxide.[46,319,510-512] This has become an expensive reagent, and there is therefore an economic drawback for use on a large scale. N-Bromosuccinimide has also seen substantial use.[513,514] Other oxidizing agents that have been used with some degree of success are hydrogen peroxide,[515] bromine,[516] silver oxide,[44] hypochlorite[517] and hypobromite,[518] alkyl nitrites,[323] hypochlorous acid,[519] nitrogen dioxide,[520] nitric acid,[521] periodic acid,[522] and cupric chloride.[523] Potassium ferricyanide has been shown to be effective when used with a trace of triphenylphenol as an electron phase-transfer agent.[524] Ethyl azoformate takes part in a hydrogen-transfer reaction with hydrazines that is usually so one-sided as to serve as a preparative oxidation of the hydrazine.[525]

The foregoing oxidation methods produce *trans* azo compounds, from which the *cis* isomers must be made by photolysis, with variable success. *Cis*-azobenzenes can be made directly, however, by oxidation of hydrazobenzenes with active manganese dioxide;[526] the temperature must be kept below 70°C to avoid isomerization to the *trans* form, and best results are obtained in the dark and under an argon atmosphere. Yields of 86 to 98% are reported for a variety of examples, except for hydrazobenzenes bearing acetamido, acetoxy, or methoxy groups in a *para* position, in which case only *trans* forms were obtained.

A practical route to some types of cyclic azo compounds makes use of oxidation of the corresponding N,N'-dicarboxylic acids with potassium ferricyanide[527] (Eq. 4-127). The substrates are obtained by Diels-Alder cycloaddition of azoformates to dienes, followed by saponification with lithium methylmercaptide.

$$\left[\begin{array}{c} N—CO_2Et \\ N—CO_2Et \end{array}\right. \xrightarrow[(Me_2N)_3PO]{LiSCH_3} \left[\begin{array}{c} N—CO_2Li \\ N—CO_2Li \end{array}\right. \xrightarrow{K_3Fe(CN)_6} \left[\begin{array}{c} N \\ \| \\ N \end{array}\right. \tag{4-127}$$

Primary amines, both aliphatic and aromatic, can be oxidized to azo compounds in useful yields. With aromatic amines, oxidation is most successful when the ring does not bear electron-donating substituents; with aliphatic amines, the alkyl group should be tertiary to avoid oxidation to imines and nitriles. Iodine pentafluoride, used in chloroform at -10 to $-20°C$, is effective with such amines as α-aminocumene, for example.[472,528,529] Sodium hypochlorite has been used with both aromatic and aliphatic amines.[405,407,530] Peroxyacetic acid used with a Cu^{2+} catalyst works especially well with nitro anilines[531] (Eq. 4-128). Manganese dioxide of the acti-

$$\underset{O_2N}{\overline{}}-NH_2 \xrightarrow[Cu^{2+}]{MeCO_3H} \underset{O_2N}{\overline{}}-N=N-\underset{NO_2}{\overline{}} \qquad (4\text{-}128)$$
$$72\%$$

vated, precipitated variety has given good yields of *trans* azo compounds when used in refluxing benzene.[532] When used below $70°C$, the *cis* isomers are generally formed.[526] Other means are $PhI(OAc)_2$,[533] argentic oxide (AgO),[534] chloramine,[535] anodic oxidation of lithium salts, $RNHLi$, of aliphatic amines,[536] and reaction with ethyl azoformate (p, p'-dimethoxyazobenzene, 28%).[525] Not perhaps unrelated is the oxidation of isocyanates by 100% hydrogen peroxide.[537]

N,N'-Disubstituted sulfamides can be oxidized to azo compounds more satisfactorily than are primary amines, from which they are made by reaction with sulfuryl chloride.[538,539] Sodium hypochlorite is the usual reagent[431,540,541] (Eq. 4-129), but

$$t\text{-BuNH}-SO_2-NH-t-Bu \xrightarrow{NaOCl} [t\text{-BuNHNH-}t\text{-Bu}] \longrightarrow$$
$$t\text{-BuN}=N-t\text{-Bu} \qquad (4\text{-}129)$$

anodic oxidation in methanol in the presence of lithium methoxide has also been used.[542] Closely related is oxidation of disubstituted ureas with *tert*-butyl hypochlorite.[543] The yields are only about 20%, but the ureas are easily made from an aryl isocyanate and a primary amine, and the method seems to be the best general one for alkyl aryl diazenes.

Oxidation of nitriles under fluorinating conditions is apparently confined to the preparation of perfluoroazoalkanes, but in that area is a good method[335] (Eq. 4-130). Chlorination of azines (Eq. 4-131) is a distantly related reaction of prepara-

$$R_F CN + Cl_2 \ (or \ Br_2) + AgF \longrightarrow R_F CF_2 N=NCF_2 R_F \qquad (4\text{-}130)$$

$$R_2C=N-N=CR_2 + Cl_2 \xrightarrow[\text{pet. ether}]{-60°C} R_2C \overset{N=N}{\underset{Cl \ \ Cl}{\diagdown \diagup}} CR_2 \qquad (4\text{-}131)$$

tive value for α-chloroazo compounds,[544] from which a variety of other azo compounds can be made by displacement of the chlorines. Lead tetraacetate oxidizes hydrazones to α-acetoxy azoalkanes.[545]

Among reductive methods, hydrogenation of azines is a useful route to azoalkanes (Eq. 4-132), but it suffers from being erratic.[515,546] Deoxygenation of

$$R_2C=N-N=CR_2 \xrightarrow{H_2/cat.} R_2CH-N=N-CHR_2 \qquad (4\text{-}132)$$

azoxy compounds can be accompished with lithium aluminum hydride[457,547] or phosphorus trichloride, among other reagents; it is perhaps most useful with azoxybenzenes.

Reduction of nitrobenzenes by zinc and alkali is a classical route to symmetrical azobenzenes.[548-550] Some hydrazobenzene is usually formed concurrently, but if oxygen is freely available during the reduction, its formation is suppressed. Lithium aluminum hydride at dry-ice temperatures has also been used to reduce nitro-benzenes to azobenzenes[460,551] (Eq. 4-133).

$$2ArNO_2 \xrightarrow{\text{LiAlH}_4} ArN=NAr \qquad (4\text{-}133)$$

The hydrazone-azoalkane equilibrium, which is generally overwhelmingly on the hydrazone side, has been used for preparation of azoalkanes by shifting the equilibrium, taking advantage of the fact that the boiling points of azo compounds are lower than hydrazones. Potassium hydroxide is used to catalyze the transformation, and the azo compound is distilled out, under vacuum when necessary, at temperatures below 120°C (84–87% yields).[342,552]

The retro-Diels-Alder reaction has been used to prepare some azoalkanes by thermally decomposing tetrahydropyridazines obtained by the Diels-Alder reaction of azoformates and subsequent transformation (reduction of ester function, alkylation) (Eq. 4-134). Yields are high, and *cis*-azoalkanes predominate in some instances.[553]

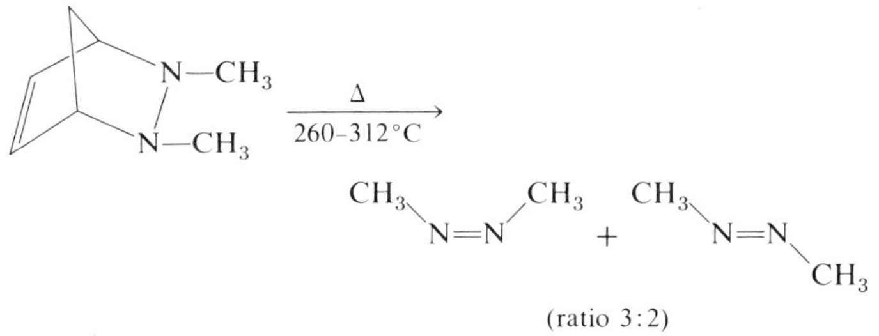

$$(4\text{-}134)$$

(ratio 3:2)

Coupling of diazonium compounds to carbon sites, which has been discussed at some length in the earlier part of this chapter, is a major synthetic method for those structures to which it is applicable. The most general situation is coupling to a benzene ring bearing electron-donating substituents; the orientation is, of course, *ortho/para*, with the *para* isomer generally greatly predominating. Examples of the method can be found in *Organic Syntheses*[489,490] and in the literature of dye chemistry. A variant that allows coupling to unactivated benzene rings is the reaction of diazonium salts, either simple or zinc chloride double salts, with organometallic compounds of magnesium or, better, zinc.[135,554,555] Alkylzinc halides have been thus used to make alkylaryldiazenes.[556] Coupling of diazonium salts to carbanions derived from β-dicarbonyl compounds is not a useful route to azo compounds, for they isomerize or cleave to hydrazones rapidly after formation in the reaction mixtures.[136,142]

The reaction of aryllithium reagents with nitrous oxide is related to the foregoing method in that it apparently proceeds through a lithium diazoate, which couples with more aryllithium[557,558] (Eq. 4-135). Yields are low (less than 30%), and puri-

$$ArLi + N_2O \longrightarrow ArN_2OLi \xrightarrow{ArLi} Li_2O + ArN{=}NAr \qquad (4\text{-}135)$$

fication by chromatography is required, owing to formation of diarylamines and quinone imine derivatives by secondary reactions starting with the addition of aryllithium to the azobenzene, but the method is nevertheless useful where the corresponding nitrobenzene is not easily obtainable. Keeping the diazonium compound in excess may minimize formation of secondary products.

Reductive coupling of diazonium salts to form symmetrical azobenzenes (Eq. 4-36) has preparative value in some situations, such as o,o'-dihydroxyazobenzenes[559,560] and is adaptable to unsymmetrical azobenzenes by first forming a diazosulfonate and then treating it with a different diazonium salt.

There are several reactions that produce N=N bonds by condensation. Perhaps the most useful synthetically is the reaction of nitroso compounds with primary amines[561-563] (Eq. 4-136). Although it has usually been applied with nitrosobenzenes,

$$ArN{=}O + ArNH_2 \xrightarrow[\text{boil}]{PhCH_3} ArN{=}NAr' \qquad (4\text{-}136)$$

trifluoronitrosomethane has been treated with methylamine to prepare 1,1,1-trifluoroazomethane.[564] Conditions for the condensation vary from glacial acetic acid to potassium t-butoxide or $tert$-butyl alcohol and dimethyl sulfoxide.

Nitrosobenzenes also react with thionylanilines to form azobenzenes, but the yields are very low.[565]

Primary aromatic amines have been made to react with nitrobenzenes under rather drastic conditions (solid NaOH, 180°C); yields of unsymmetrical azo compounds are acceptable, although not high (e.g., α-naphthaleneazobenzene, 50%).[566,567]

Unsymmetrical azobenzenes have been made by condensing thionylanilines with arylhydroxylamines; the stoichiometry appears to require three equivalents of arylhydroxylamine for every one of azobenzene produced, and conversions are not high.[568,569]

Chloramines and N,N-dichloramines have been condensed to form symmetrical azoalkanes by treatment with base,[530,570] a reaction that is closely related to the oxidation of primary amines to azo compounds with sodium hypochlorite.

Two methods that depend on elimination from a hydrazone deserve mention. In one, a methoxycyclohexadienone forms an intermediate hydrazone with an arylhydrazine and then loses methanol to form an azobenzene.[571] In the other, which appears to have greater scope, an α-chloro ketone is converted to a monosubstituted hydrazone, which loses the elements of HCl on treatment with base (which may be excess hydrazine) to form a conjugated alkenyldiazene[572,573] (Eq. 4-137).

Lastly, benzeneazoperfluoroalkanes have been made in 41 to 53% yield by reaction of benzenediazonium chloride with perfluoroalkenes in the presence of potas-

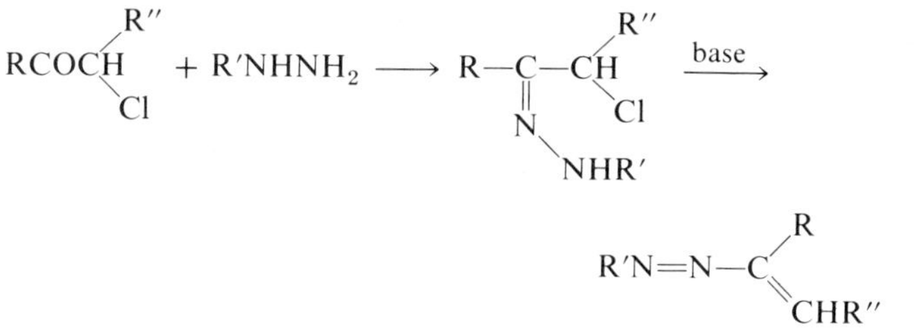

$$RCOCH\overset{R''}{\underset{Cl}{}} + R'NHNH_2 \longrightarrow R-\underset{\underset{NHR'}{\overset{\|}{N}}}{C}-CH\overset{R''}{\underset{Cl}{}} \xrightarrow{\text{base}}$$

$$R'N=N-C\overset{R}{\underset{CHR''}{}} \qquad (4\text{-}137)$$

$$(R,R' = CH_3, R'' = H, 31\%)$$

sium fluoride in dimethylformamide (Eq. 4-138), a reaction that is closely related to coupling with ordinary olefins that occurs in special situations (Eq. 4-26), such as

$$PhN_2{}^+Cl^- + C_nF_{2n} + KF \xrightarrow{\text{DMF}} PhN=N-C_nF_{2n+2} \qquad (4\text{-}138)$$

enamines and the low-yield reaction with some simple alkenes and conjugated dienes.[574,575]

Diazenium compounds are generally prepared by oxidation of hydrazines in acidic medium.[26,27,576]

Formazans[85,577]

Formazans are always made by diazonium coupling reactions. Most commonly, this involves coupling with hydrazones of aldehydes or monohydrazones of α-dicarbonyl compounds[578] (Eq. 4-22). These reactions have been discussed in connection with diazonium reactions earlier in this chapter, and the reader is advised to consult the appropriate references cited there. Alternatively, hydrazidines, prepared by reaction of arylhydrazines with imidic esters, can be oxidized to formazans (Eq. 4-139) (see Chap. 3).

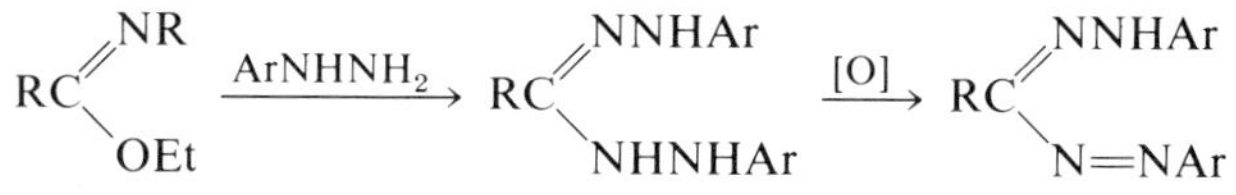

$$RC\overset{NR}{\underset{OEt}{}} \xrightarrow{\text{ArNHNH}_2} RC\overset{NNHAr}{\underset{NHNHAr}{}} \xrightarrow{[O]} RC\overset{NNHAr}{\underset{N=NAr}{}} \qquad (4\text{-}139)$$

ANALYTICAL METHODS

The assay of diazonium solutions[579-581] is important in the dye industry, and various methods have accordingly been developed for their quantitative determination. Volumetric methods using strong reducing agents, such as vanadous salts (VCl_3) or hypovanadous salts (VSO_4), chromous salts, and especially titanous salts ($TiCl_3$) have been used, usually by adding an excess of the reducing solution and back-titrating with a standard oxidizing agent such as a ferric salt. The equivalence ratio is not always obvious and depends on the reagent and conditions; reduction may proceed to a diaryltetrazene, arylhydrazine, or arylamine.

Evolution of nitrogen can be made the basis of quantitative determination, but is not much used today. Coupling reactions can be made the basis for both colorimetric and volumetric analysis. Coulometric analysis has also been used. Spectroscopy is represented for the most part by ultraviolet absorption.

Much the same selection of methods has been used for determination of azo compounds.[582] A curious method not applicable to diazonium compounds has been used with azo compounds: the reduction in weight of a sample of metal, such as copper or tin, that has been exposed to a solution of an azo compound long enough for reduction of the latter to occur. Volumetric titration with standard titanous salts, which take the azo compound to the primary amine, seems to be most widely used,[580,581] but where the absorption maximum and extinction coefficient of the pure compound is known, as in routine industrial processes, colorimetric determination may be the most convenient method.

Qualitative detection of diazonium salts[583] is easily accomplished by means of the intensely colored azo dyes that can be made by a simple coupling reaction. The usual reagent is β-naphthol in alkaline solution; a red dye usually precipitates. It is important that any residual traces of nitrous acid be removed before the test, however, for β-naphthol is easily nitrosated to a colored product that might be mistaken for an azo dye (urea or sulfamic acid are the best scavengers for nitrous acid).

Azo compounds may be detected qualitatively[583] by the fact that they may be reduced to hydrazines, which reduce ammonical silver nitrate (Tollens' reagent). Zinc in aqueous alcohol in the presence of a little acetic acid is the preferred reagent. Nitro groups will respond to the same test, because they are reduced to hydroxylamines, but their presence can be detected by treating the solution with benzoyl chloride and testing for formation of a hydroxamic acid with ferric chloride. Most aliphatic azo compounds can also be detected by virtue of the fact that they do not themselves reduce Tollens' reagent, but can be hydrolyzed to a substituted hydrazine that does.

REFERENCES

1. E. Knoevenagel, *Ber.* **23**, 2994 (1890).

2. J. P. Collman and M. Yamada, *J. Org. Chem.* **28**, 3017 (1963).

3. K. Bott, *Chem. Ber.* **108**, 402 (1975).

4. J. R. Mohrig, K. Keegstra, A. Maverick, R. Roberts, and S. Wells, *J. Chem. Soc. Chem. Commun.*, 780 (1974).

5. K. Bott, *Angew. Chem. Int. Ed.* **10**, 821 (1971).

6. R. A. Bartsch, H. Chen, N. F. Haddock, and P. N. Juri, *J. Am. Chem. Soc.* **98**, 6753 (1976); R. A. Bartsch, in *Progress in Macrocyclic Polyether Chemistry*, R. M. Izatt and J. J. Christensen (eds.), Vol. 2, p. 1, Wiley, New York, 1981.

7. S. H. Korzeniowski, A. Leopold, J. R. Beadle, M. F. Ahern, W. A. Sheppard, R. K. Khanna, and G. W. Gokel, *J. Org. Chem.* **46**, 2153 (1981).

8. E. Müller and H. Haiss, *Chem. Ber.* **96**, 570 (1963).

9. E. Bamberger, *Ber.* **29**, 446 (1896).

10. (*a*) H. Reimlinger, *Angew. Chem.* **73**, 221 (1961); (b) E. Müller, H. Haiss, and W. Rundel, *Chem. Ber.* **93**, 1541 (1960).

11. R. A. Moss, *J. Org. Chem.* **31**, 1082 (1966).

12. H. Gehlen and J. Dost, *Liebig's Ann. Chem.* **665**, 144 (1963).

13. R. N. Butler, T. M. Lambe, J. C. Tobin, and F. L. Scott, *J. Chem. Soc.* [*Perkin 1*], 1357 (1973).

14. T. Kauffmann, H. O. Friestad, and H. Henkler, *Liebig's Ann. Chem.* **634**, 64 (1960).

15. E. Müller and H. Haiss, *Chem. Ber.* **95**, 1255 (1962).

16. E. M. Kosower, *Accts. Chem. Res.* **4**, 193 (1971).

17. W. T. Evanochko and P. B. Shevlin, *J. Am. Chem. Soc.* **100**, 6428 (1978).

18. M. N. Ackermann, J. L. Ellenson, and D. H. Robison, *J. Am. Chem. Soc.* **90**, 7173 (1968).

19. M. N. Ackermann, J. J. Burdge, and N. C. Craig, *J. Chem. Phys* **58**, 203 (1973).

20. N. C. Craig, M. A. Kliewer, and N. C. Shih, *J. Am. Chem. Soc.* **101**, 2480 (1979).

21. M. N. Ackermann, N. C. Craig, R. R. Isberg, D. M. Lauter, and E. P. Tracy, *J. Phys. Chem.* **83**, 1190 (1979).

22. F. Gerson, E. Heilbronner, A. van Veen, and B. W. Wepster, *Helv. Chim. Acta* **43**, 1889 (1960).

23. M. Isaks and H. H. Jaffe, *J. Am. Chem. Soc.* **86**, 2209 (1964).

24. J. H. Collins and H. H. Jaffe, *J. Am. Chem. Soc.* **84**, 4708 (1962).

25. E. Haselbach, A. Henriksson, A. Schmeizer, and H. Berthou, *Helv. Chim. Acta* **56**, 705 (1973).

26. S. Hünig and F. Brühne, *Liebig's Ann. Chem.* **667**, 86 (1963).

27. S. F. Nelsen and R. T. Landis, *J. Am. Chem. Soc.* **96**, 1788 (1974).

28. H. Zollinger, *Diazo and Azo Chemistry,* Interscience Publishers, New York, 1961, p. 47 et seq. and Chap. 7.

29. J. F. McGarrity, in *The Chemistry of the Diazonium and Diazo Groups,* S. Patai (ed.), Wiley, New York, 1978, pp. 221–223.

30. V. Šterba, in *The Chemistry of the Diazonium and Diazo Groups,* S. Patai (ed.) Wiley, New York, 1978, Chap. 2.

31. E. S. Lewis and H. Suhr, *Chem. Ber.* **91**, 2350 (1958).

32. C. Wittwer and H. Zollinger, *Helv. Chim. Acta* **37**, 1954 (1954).

33. B. V. Passet and B. A. Porai-Koshits, *Treatises of the L. T. Institute "Lensoviet"* **11**, 133 (1958).

34. E. S. Lewis and M. P. Hanson, *J. Am. Chem. Soc.* **89**, 6268 (1967).

35. E. S. Lewis and H. Suhr, *J. Am. Chem. Soc.* **80**, 1367 (1958).

36. J. Jahelka, O. Macháčková, and V. Šterba, *Coll. Czech. Chem. Commun.* **38**, 706 (1973).

37. M. Aroney, R. J. W. Le Fevre, and R. L. Werner, *J. Chem. Soc.*, 276 (1955).

38. K. B. Whetsel, G. F. Hawkins, and F. E. Johnson, *J. Am. Chem. Soc.* **78**, 3360 (1956).

39. R. J. Cox and J. Kumamoto, *J. Org. Chem.* **30**, 4254 (1965).

40. N. Sheppard and G. B. B. M. Sutherland, *J. Chem. Soc.*, 453 (1947).

41. (*a*) J. K. Stille, P. Cassidy, and L. Plummer, *J. Am. Chem. Soc.* **85**, 1318 (1963); (*b*) V. V. Ershov, G. A. Nikiforov, and C. R. H. I. De Jonge, *Qunone Diazides,* Elsevier, Amsterdam and New York, 1981.

42. R. J. W. Le Fevre, J. B. Sousa, and R. L. Werner, *J. Chem. Soc.*, 4686 (1954).

43. R. O'Connor, *J. Org. Chem.* **26**, 4375 (1961).

44. M. C. Chaco and N. Rabjohn, *J. Org. Chem.* **27**, 2765 (1962).

45. C. G. Overberger, J.-P. Anselme, and J. R. Hall, *J. Am. Chem. Soc.* **85**, 2752 (1965).

46. L. Spialter, D. H. O'Brien, G. L. Untereiner, and W. A. Rush, *J. Org. Chem.* **30**, 3278 (1965).

47. M. Sukigara and S. Kikuchi, *Bull. Chem. Soc. Japan* **40**, 461, 1077, 1082 (1967).

48. J. Fabian and H. Hartmann, *Light Absorption of Organic Colorants,* Springer-Verlag, New York, 1980, Chap. VII.

49. P. P. Birnbaum, J. H. Linford, and D. W. G. Style, *Trans. Faraday Soc.* **49**, 735 (1953).

50. H. Bock, *Angew. Chem. Int. Ed.* **4**, 457 (1965).

51. S.-J. Yeh and H. Jaffe, *J. Am. Chem. Soc.* **81**, 3274, 3283 (1959).

52. E. Sawicki, *J. Org. Chem.* **22**, 365, 621, 743 (1957).

53. Yu. L. Kaminskii and I. Y. Bernshtein, in *Korrelyatsionniie Uravneniya v Organicheskoi Khimii*, V. Palm (ed.), Tartu University Press, Tartu, Estonia, 1962, p. 338.

54. K. N. Houk, Y.-M. Chang, and P. S. Engel, *J. Am. Chem. Soc.* **97**, 1824 (1975).

55. H. Suhr, *Chem. Ber.* **96**, 1720 (1963).

56. R. F. Hutton and C. Steel, *J. Am. Chem. Soc.* **86**, 745 (1964).

57. J. Swigert and K. G. Taylor, *J. Am. Chem. Soc.* **93**, 7337 (1971).

58. G. A. Olah and J. L. Grant, *J. Am. Chem. Soc.* **97**, 1546 (1975).

59. R. O. Duthaler, H. G. Förster, and J. D. Roberts, *J. Am. Chem. Soc.* **100**, 4974 (1978).

60. E. S. Lewis and H. Suhr, *J. Am. Chem. Soc.* **82**, 862 (1960).

61. H. A. Schoutissen, *Rec. Trav. Chim. Pays Bas* **54**, 381 (1935).

62. J. F. Bunnett, E. Buncel, and K. V. Nahabedian, *J. Am. Chem. Soc.* **84**, 4136 (1962).

63. (*a*) S.-J. Yeh and H. Jaffe, *J. Am. Chem. Soc.* **81**, 3287 (1959); (*b*) B. A. Korolev and B. I. Stepanov, *Zh. Org. Khim.* **5**, 1673 (1969) (English trans. p. 1622).

64. R. J. W. Le Fevre and I. R. Wilson, *J. Chem. Soc.*, 1106 (1949).

65. H. C. Freeman and R. J. W. Le Fevre, *J. Chem. Soc.,* 3128 (1950).

66. R. Dijkstra and J. de Jonge, *Rec. Trav. Chim. Pays Bas* **77**, 538 (1958).

67. G. N. Lewis and M. Calvin, *Chem. Rev.* **25**, 273 (1939).

68. J. J. de Lange, J. M. Robertson, and I. Woodward, *Proc. Roy. Soc. Sec. A* **171**, 398 (1939).

69. G. C. Hampson and J. M. Robertson, *J. Chem. Soc.*, 409 (1941).

70. G. S. Hartley and R. J. W. Le Fevre, *J. Chem. Soc.,* 531 (1939).

71. C. Rømming, *Acta Chem. Scand.* **17**, 1444 (1963).

72. O. Andresen and C. Rømming, *Acta Chem. Scand.* **16**, 1882 (1962).

73. T. N. Polynova, N. G. Bokii, and B. A. Porai-Koshits, *Zh. Strukt. Khim.* **6**, 878 (1965).

74. Ya. M. Nesterova and M. A. Porai-Koshits, *Zh. Strukt. Khim.* **12**, 108 (1971).

75. I. Bø, B. Klewe, and C. Rømming, *Acta Chem. Scand.* **25**, 3261 (1971).

76. Y. M. Nesterova, B. A. Porai-Koshits, N. B. Kupleskaya, and L. A. Kazitsyna, *Zh. Strukt. Khim.* **8**, 1109 (1967).

77. S. Suszko and T. Ignasiak, *Bull. Acad. Polon. Sci.* **18**, 669, 673 (1970).

78. S. Sorriso, in *The Chemistry of the Diazonium and Diazo Groups*, S. Patai (ed.), Wiley, New York, 1978, Chap. 3.

79. R. Huber, R. Langer, and W. Hoppe, *Acta Cryst.* **18**, 467 (1965).

80. A. Angeli, *Ber.* **63**, 1977 (1930).

81. H. Boersch, *Monatsh. Chem.* **65**, 311 (1935).

82. L. D. Vogel, A. M. Rennert, and C. Steel, *J. Chem. Soc. Chem. Commun.*, 536 (1975).

83. P. S. Engel, R. A. Melaugh, M. A. Page, S. Szilgyi, and J. W. Timberlake, *J. Am. Chem. Soc.* **98**, 1971 (1976).

84. C. J. Casewit and W. A. Goddard, III, *J. Am. Chem. Soc.* **104**, 3280 (1982).

85. P. T. S. Lau, Org. *Chem. Bull.* (*Eastman Kodak Co.*) **36**, No. 3 (1964); A. W. Nineham, *Chem. Rev.* **55**, 355 (1955).

86. E. Bamberger and W. Pemsel, *Ber.* **36**, 53, 57, 90, 92, 347, 359 (1903).

87. E. Bamberger, *Ber.* **31**, 2626 (1898); **34**, 574 (1901).

88. A. S. Shawali and B. M. Altahou, *Tetrahedron* **33**, 1625 (1977).

89. B. Levenberg, *Biochim. Biophys. Acta* **63**, 212 (1962).

90. C. D. Ritchie and O. I. Virtanen, *J. Am. Chem. Soc.* **94**, 1589 (1972).

91. A. Ginsberg and J. Goerdeler, *Chem. Ber.* **94**, 2043 (1961).

92. O. Dimroth, H. Leichtle, and O. Friedemann, *Ber.* **50**, 1534 (1917).

93. H. T. Bucherer, *Ber.* **42**, 47 (1909).

94. D. F. De Tar and M. N. Turetzky, *J. Am. Chem. Soc.* **77**, 1745 (1955).

95. C. C. Price and S. Tsunawski, *J. Org. Chem.* **28**, 1867 (1963).

96. A. Hantzsch and H. Freese, *Ber.* **28**, 3237 (1895).

97. L. K. H. van Beek, J.R.G. C. M. van Beek, J. Boven, and C. J. Schoot, *J. Org. Chem.* **36**, 2194 (1971).

98. J. Brokken-Zijp and H. v. d. Bogaert, *Tetrahedron* **29**, 4199 (1973).

99. E. Bamberger and E. Kraus, *Ber.* **29**, 272 (1896).

100. H. H. Hodgson, *Chem. Ind.*, 362 (1945).

101. H. Meerwein, G. Dittmar, G. Kaufmann, and R. Raue, *Chem. Ber.* **90**, 853 (1957); C. D. Ritchie, J. D. Saltiel, and E. S. Lewis, *J. Am. Chem. Soc.* **83**, 4601 (1961).

102. M. F. Ahern, A. Leopold, J. R. Beadle, and G. W. Gokel, *J. Am. Chem. Soc.* **104**, 548 (1982).

103. H. C. Freeman and R. J. W. Le Fevre, *J. Chem. Soc.*, 415 (1951).

104. H. Jonker, T. P. G. W. Thijssens, and L. K. H. Van Beek, *Rec. Trav. Chim. Pays Bas* **87**, 997 (1968); N. Kamigata and M. Kobayashi, *Sulfur Repts.* **2**, 87 (1982).

105. A. Hantzsch and B. Hirsch, *Ber.* **29**, 947 (1896).

106. H. von Pechmann and C. Frobenius, *Ber.* **28**, 170 (1895).

107. L. Wacker, *Ber.* **35**, 3922 (1902).

108. T. P. Ahern, H. Fong, and K. Vaughan, *Can. J. Chem.* **55**, 1701 (1977).

109. E. H. White and H. Scherrer, *Tetrahedron Lett.*, 758 (1961).

110. H. Hansen, S. Hünig, and K. Kishi, *Chem. Ber.* **112**, 445 (1979).

111. P. F. Holt and C. J. McNae, *J. Chem. Soc.*, 1825 (1961).

112. J. P. Horwitz and V. A. Grakauskas, *J. Am. Chem. Soc.* **79**, 1249 (1957).

113. T. Curtius, *Ber.* **26**, 1263 (1893).

114. H. Minato, M. Oku, and H.-P. Chen, *Bull. Chem. Soc. Japan* **39**, 1049 (1966).

115. I. Ugi, H. Perlinger, and L. Behringer, *Chem. Ber.* **92**, 1864 (1959).

116. J. Mai, *Ber.* **25**, 372 (1892).

117. T. Mitsuhashi, Y. Osamura, and O. Simamura, *Tetrahedron Lett.*, 2593 (1965).

118. D. N. Purohit and N. C. Sogani, *J. Chem. Soc.*, 2820 (1964).

119. A. B. Boese, L. W. Jones, and R. T. Major, *J. Am. Chem. Soc.* **53**, 3530 (1931).

120. L. Gattermann and R. Ebert, *Ber.* **49**, 2117 (1916).

121. H. Zollinger, *Chem. Ber.* **51**, 347 (1952).

122. R. Pütter, *Angew Chem.* **63**, 188 (1951).

123. R. Ernst, O. A. Stamm, and H. Zollinger, *Helv. Chim. Acta* **41**, 2274 (1958).

124. H. Zollinger, in *Advances in Physical Organic Chemistry*, V. Gold (ed.), Academic Press, New York, 1964.

125. P. N. Juri and R. A. Bartsch, *J. Org. Chem.* **44**, 143 (1979).

126. J. R. Bourne, E. Crivelli, and P. Rys, *Helv. Chim. Acta* **60**, 2944 (1977).

127. A. Sisti, J. Burgmaster, and M. Fudim, *J. Org. Chem.* **27**, 279 (1962).

128. M. Stiles and A. Sisti, *J. Org. Chem.* **25**, 1691 (1960).

129. K. H. Meyer, A. Irschick, and H. Schlösser, *Ber.* **47**, 1741 (1914).

130. J. F. Bunnett and G. B. Hoey, *J. Am. Chem. Soc.* **80**, 3142 (1958).

131. L. I. Smith and J. H. Paden, *J. Am. Chem. Soc.* **56**, 2169 (1934).

132. K. H. Meyer and H. Tochtermann, *Ber.* **54**, 2283 (1921).

133. L. F. Fieser and W. P. Campbell, *J. Am. Chem. Soc.* **60**, 1142 (1938).

134. E. S. Lewis and H. Suhr, *Chem. Ber.* **92**, 3043 (1959).

135. D. Y. Curtin and J. L. Tveten, *J. Org. Chem.* **26**, 1764 (1962).

136. S. M. Parmenter, *Org. Reactions* **10**, 1 (1959).

137. H. C. Yao and P. Resnick, *J. Am. Chem. Soc.* **84**, 3514 (1962).

138. D. Y. Curtin and M. L. Poutsma, *J. Am. Chem. Soc.* **84**, 4887 (1962).

139. J. Rabischong, *Bull. Soc. Chim. France* **[3|31**, 76, 83 (1904).

140. H. von Pechmann, *Ber.* **25**, 3175 (1892).

141. M. Busch, *J. Prakt. Chem.* **[2|71**, 366 (1905).

142. R. R. Phillips, *Org. Reactions* **10**, 144 (1959).

143. B. Balz and B. Schiemann, *Ber.* **60**, 1186 (1927).

144. E. Bamberger and J. Müller, *Ber.* **27**, 147 (1894).

145. E. Bamberger and J. Frei, *Ber.* **36**, 3833 (1903).

146. F. L. Scott, D. A. O'Sullivan, and J. Reilly, *J. Am. Chem. Soc.* **75**, 5309 (1953).

147. S. Hünig and O. Boes, *Liebig's Ann. Chem.* **579**, 28 (1953).

148. L. Mester and A. Major, *J. Am. Chem. Soc.* **78**, 1403 (1956).

149. K. H. Meyer and H. Hopff, *Ber.* **54**, 2274 (1921).

150. D. J. Cram, *J. Am. Chem. Soc.* **74**, 2159 (1952).

151. A. Quilico and E. Fleischner, *Gaz. Chim. Ital.* **59**, 39 (1929).

152. D. L. De Tar and Y.-W. Chu, *J. Am. Chem. Soc.* **76**, 1686 (1954).

153. S. J. Huang, V. Paneccasio, F. Di Battista, D. Picker, and G. Wilson, *J. Org. Chem.* **40**, 124 (1975).

154. A. D. Ainley and R. Robinson, *J. Chem. Soc.*, 369 (1937).

155. L. Benati and P. C. Montevecchi, *J. Org. Chem.* **42**, 2025 (1977).

156. F. Tröndlin, R. Werner, and C. Rüchardt, *Chem. Ber.* **111**, 367 (1978).

157. N. J. Leonard, *Chem. Rev.* **37**, 269 (1945).

158. J. C. E. Simpson, in *The Chemistry of Heterocyclic Compounds*, Vol. V, A. Weissberger (ed.), Interscience Publishers, New York, 1953.

159. D. Sutton, *Chem. Soc. Rev.* **4**, 443 (1975).

160. D. S. Wulfman, in *The Chemistry of Diazonium and Diazo Groups*, S. Patai (ed.), Wiley, New York, 1978, pp. 274–276.

161. J. F. Bunnett and R. E. Zahler, *Chem. Rev.* **49**, 273 (1951).

162. B. A. Bolto, M. Liveris, and J. Miller, *J. Chem. Soc.*, 750 (1956).

163. R. Meldola and F. Reverdin, *J. Chem. Soc.* **97**, 1204 (1910), and earlier papers cited therein.

164. K. J. P. Orton, *J. Chem. Soc.* **87**, 99 (1905).

165. A. Hantzsch, *Ber.* **36**, 2069 (1903).

166. C. E. McKenna and T. G. Traylor, *J. Am. Chem. Soc.* **93**, 2313 (1971).

167. G. H Coleman, *Org. Synthesis*, Coll. **I**, Ed. 2, 442 (1941).

168. I. M. Hunsberger, E. R. Shaw, J. Fugger, R. Ketcham, and D. Lednicer, *J. Org. Chem.* **21**, 394, 2262 (1956).

169. E. S. Lewis and H. Suhr, *Chem. Ber.* **92**, 3031 (1959).

170. R. Huisgen and R. Lux, *Chem. Ber.* **93**, 540 (1960).

171. R. Walther, *J. Prakt. Chem.* **[2|53**, 433 (1896).

172. H. A. J. Schoutissen, *Rec. Trav. Chim. Pays Bas* **57**, 710 (1938).

173. W. A. Cowdrey and D. S. Davis, *J. Chem. Soc. [Suppl.]*, 48 (1949).

174. E. Pfeil and O. Velten, *Liebig's Ann. Chem.* **565,** 183 (1949).

175. E. Pfeil, *Angew. Chem.* **65,** 155 (1953).

176. D. V. Banthorpe and E. D. Hughes, *J. Chem. Soc.*, 3314 (1962).

177. E. R. Atkinson, C. R. Morgan, H. H. Warren, and T J. Manning, *J. Am. Chem. Soc.* **67,** 1513 (1945).

178. J. I. G. Cadogan, P. G. Hibbert, M. N. U. Siddiqu, and D. M. Smith, *J. Chem. Soc. [Perkin I]*, 2555 (1972).

179. O. A. Stamm and H. Zollinger, *Chimia* **15,** 137, 535 (1961).

180. E. Bamberger, *Ber.* **55,** 3383 (1922).

181. E. Bamberger and O. Baudisch, *Ber.* **42,** 3582 (1909).

182. F. Manisci and A. Portolani, *Gaz. Chim. Ital.* **89,** 1922, 1941 (1959).

183. H. von Pechmann and L. Frobenius, *Ber.* **27,** 672 (1894).

184. E. Bamberger, *Ber.* **27,** 917, 3412 (1894).

185. H. Zollinger, *Angew. Chem. Int. Ed.* **17,** 141 (1978).

186. A. F. Hegarty, in *The Chemistry of Diazonium and Diazo Groups,* S. Patai (ed.), Wiley, New York, 1978, Chap. 12.

187. I. Szele and H. Zollinger, *Helv. Chim. Acta* **61,** 1721 (1978).

188. H. E. Ungnade and E. F. Orwoll, *Org. Syntheses,* Coll. **III,** 130 (1955).

189. J. P. Lambooy, *J. Am. Chem. Soc.* **72,** 5327 (1950).

190. T. Cohen, A. G. Dietz, Jr., and J. R. Miser, *J. Org. Chem.* **42,** 2053 (1977).

191. C. G. Swain, J. E. Sheats, and K. G. Harbison, *J. Am. Chem. Soc.* **97,** 783, 796 (1975).

192. D. F. De Tar and S. K. Wong, *J. Am. Chem. Soc.* **78,** 3916 (1956).

193. D. F. De Tar and A. R. Ballentine, *J. Am. Chem. Soc.* **78,** 3921 (1956).

194. E. A. Lewis, *J. Am. Chem. Soc.* **80,** 1371 (1958).

195. E. S. Lewis and W. H. Hinds, *J. Am. Chem. Soc.* **74,** 304 (1952).

196. K. R. Brown, *J. Am. Chem. Soc.* **82,** 4535 (1960).

197. E. S. Lewis and J. M. Insole, *J. Am. Chem. Soc.* **86,** 32, 34 (1964).

198. R. G. Bergstrom, R. G. M. Landells, G. H. Wahl, Jr., and H. Zollinger, *J. Am. Chem. Soc.* **98,** 3301 (1976).

199. Y. Hashida, R. G. M. Landells, G. E. Lewis, I. Szele, and H. Zollinger, *J. Am. Chem. Soc.* **100,** 2416 (1978).

200. I. Szele and H. Zollinger, *J. Am. Chem. Soc.* **100,** 2811 (1978).

201. C. G. Swain, J. E. Sheats, D. G. Gorenstein, and K. G. Harbison, *J. Am. Chem. Soc.* **97,** 791 (1975).

202. G. A. Olah and J. Welch, *J. Am. Chem. Soc.* **97,** 208 (1975).

203. J. F. Bunnett and C. Yijima, *J. Org. Chem.* **42,** 639 (1977).

204. T. J. Broxton, J. F. Bunnett, and C. H. Paik, *J. Org. Chem.* **42,** 643 (1977).

205. T. J. Broxton and J. F. Bunnett, *Nouv. J. Chim.* **3,** 133 (1979).

206. N. Chatterjee, *J. Ind. Chem. Soc.* **12,** 410 (1935).

207. A. Roe, *Org. Reactions* **4,** Chap. 4 (1949).

208. W. D. Sheppard and C. M. Sharts, *Organic Fluorine Chemistry,* W. A. Benjamin, New York, 1969.

209. M. Hudlicky, *Chemistry of Organic Fluorine Compounds,* MacMillan, New York, 1962.

210. K. G. Rutherford, W. Redmond, and J. Rigamonti, *J. Org. Chem.* **26,** 5149 (1961).

211. C. G. Swain and R. J. Rogers, *J. Am. Chem. Soc.* **97,** 799 (1975).

212. R. C. Petterson, A. Di Maggio III, A. L. Hebert, T. J. Haley, J. P. Mykytka, and I. M. Sarker, *J. Org. Chem.* **36**, 631 (1971).

213. H. J. Lucas and E. R. Kennedy, *Org. Syntheses,* Coll. **II**, 351 (1943).

214. W. A. Waters, *J. Chem. Soc.*, 266 (1942).

215. W. E. Lee, J. G. Calvert, and E. W. Malmberg, *J. Am. Chem. Soc.* **83**, 1928 (1961).

216. H.-W. Schwechten, *Ber.* **65**, 1605 (1932).

217. W. W. Hartman and M. R. Brethen, *Org. Syntheses,* Coll. **I**, Ed. 2, 162 (1941).

218. C. S. Marvel and S. M. McElvain, *Org. Syntheses,* Coll. **I**, Ed. 2, 170 (1941).

219. L. A. Bigelow, *Org. Syntheses,* Coll. **I**, Ed. 2, 135 (1941).

220. H. T. Clarke and R. R. Read, *Org. Syntheses,* Coll. **I**, Ed. 2, 514 (1941).

221. L. Gattermann and A. Cantzler, *Ber.* **25**, 1086 (1892).

222. J. W. Dienske, *Rec. Trav. Chim. Pays Bas* **50**, 176, 407 (1931).

223. M. P. Doyle, B. Siegfried, and F. Dellaria, Jr., *J. Org. Chem.* **42**, 2426 (1977).

224. W. A. Cowdrey and D. S. Davies, *Quart. Rev.* **6**, 358 (1952).

225. J. Kochi, *Tetrahedron* **18**, 483 (1962).

226. J. Kochi, *Organometallic Mechanisms and Catalysis*, Academic Press, New York, 1979, Chap. 9.

227. D. F. De Tar and J. C. Howard, *J. Am. Chem. Soc.* **77**, 4393 (1955).

228. T. Cohen, R. J. Lewarchik, and J. T. Zarino, *J. Am. Chem. Soc.* **96**, 7753 (1977).

229. W. E. Brackman and P. J. Smit, *Rec. Trav. Chim. Pays Bas* **85**, 857 (1966).

230. H.-J. Opgenorth and C. Rüchardt, *Liebig's Ann. Chem.*, 1333 (1974).

231. E. R. Ward, C. D. Johnson, and J. G. Hawkins, *J. Chem. Soc.*, 894 (1960).

232. E. B. Starkey, *Org. Syntheses,* Coll. **II**, 225 (1943).

233. H. H. Hodgson, A. P. Mahadevan, and E. R. Ward, *Org. Syntheses,* Coll. **III**, 341 (1955).

234. H. H. Hodgson and E. Marsden, *J. Chem. Soc.*, 22 (1944).

235. A. N. Frolov, M. S. Pevzner, and L. I. Bagal, *Zh. Org. Khim.* **7**, 1519 (1971) (English trans. p. 1573).

236. H. Meerwein, P. Laasch, R. Mersch, and J. Nentwig, *Chem. Ber.* **89**, 224 (1956).

237. F. Klages and W. Grill, *Liebig's Ann. Chem.* **594**, 21 (1955).

238. E. Börnstein, *Ber.* **34**, 3968 (1901).

239. C. Graebe and W. Mann, *Ber.* **15**, 1683 (1882).

240. J. H. Ziegler, *Ber.* **23**, 2469 (1890).

241. G. E. Hilbert and T. B. Johnson, *J. Am. Chem. Soc.* **51**, 1526 (1929).

242. C. F. H. Allen and D. D. MacKay, *Org. Syntheses,* Coll. **II**, 580 (1943).

243. H. Meerwein, G. Dittmar, R. Göllner, K. Hafner, F. Mensch, and O. Steinfort, *Chem. Ber.* **90**, 841 (1957).

244. A. M. Clifford and J. G. Lichty, *J. Am. Chem. Soc.* **54**, 1163 (1932).

245. R. Leuckart, *J. Prakt. Chem.* **[2]41**, 179 (1890).

246. A. Lustig, *Gaz. Chim. Ital.* **21**, 213 (1891).

247. D. S. Tarbell and D. K. Fukushima, *Org. Syntheses,* Coll. **III**, 809 (1955).

248. K. Hölzle, *Helv. Chim. Acta* **29**, 1883 (1946).

249. I. G. Farbenind., A.G., Brit. Patent No. 384,722; *Chem. Abstr.* **27**, 4251 (1933).

250. A. J. Neale, T. J. Rawlings, and E. B. McCall, *Tetrahedron* **21**, 1299 (1965).

251. R. A. Abramovitch, T. Chellathurai, I. T. McMaster, T. Takaya, C. I. Azogu, and D. P. Vanderpool, *J. Org. Chem.* **42**, 2914 (1977).

252. L. Gattermann, *Ber.* **32**, 1136 (1899).

253. R. H. Bullard, *Org. Syntheses,* Coll. **II,** 494 (1943).

254. C. S. Hamilton and J. F. Morgan, *Org. Reactions* **2,** Chap. 10 (1944).

255. H. M. Leicester, *Org. Syntheses,* Coll. **II,** 238 (1943).

256. A. N. Nesmeyanov, *Org. Syntheses,* Coll. **II,** 433 (1943).

257. D. S. Wulfman, in *The Chemistry of the Diazonium and Diazo Groups,* S. Patai (ed.), Wiley New York, 1978, pp. 296–297.

258. J. C. Clark and R. C. Cookson, *J. Chem. Soc.,* 686 (1962).

259. S. D. Jolad and S. Rajagopal, *Org. Syntheses* **46,** 13 (1966).

260. C. S. Rondestvedt, Jr., *Org. Reactions* **24,** Chap. 3 (1976).

261. I. A. Adel, B. A. Salami, J. Levisalles, and H. Rudler, *Bull. Soc. Chim. France,* 934 (1976).

262. M. P. Doyle, B. Siegfried, R. C. Elliott, and J. F. Dellaria, Jr., *J. Org. Chem.* **42,** 2431 (1977).

263. T. Matsuda, *Chem. Lett.,* 159 (1977).

264. J. K. Kochi, *Free radicals,* Vol. 1, Wiley, New York, 1973, Chap. 11.

265. C. L. Jenkins and J. K. Kochi, *J. Am. Chem. Soc.* **94,** 843, 856 (1972).

266. S. C. Dickerman, D. J. De Souza, and N. Jacobson, *J. Org. Chem.* **34,** 710 (1969).

267. N. I. Ganuschak, V. D. Golik, and I. V. Migaichuk, *Zh. Org. Khim.* **8,** 2356 (1972) (English trans. p. 2403).

268. S. C. Dickerman, K. Weiss, and A. K. Ingberman, *J. Am. Chem. Soc.* **80,** 1904 (1958).

269. D. F. De Tar, *Org. Reactions* **9,** Chap. 7 (1957).

270. K. Kamigata, R. Hisada, H. Minato, and M. Kobayashi, *Bull. Chem. Soc. Japan* **46,** 1016 (1973).

271. A. H. Lewin and R. J. Michl, *J. Org. Chem.* **38,** 1126 (1973).

272. R. Huisgen and W. D. Zahler, *Chem. Ber.* **96,** 736 (1963).

273. R. M. Elofson and F. F. Gadallah, *J. Org. Chem.* **36,** 1769 (1971).

274. M. Stiles and A. J. Sisti, *J. Org. Chem.* **24,** 268 (1959).

275. W. E. Bachmann and R. A. Hoffman, *Org. Reactions* **2,** Chap. 2 (1944).

276. D. E. Rosenberg, J. R. Beadle, S. H. Korzeniowski, and G. W. Gokel, *Tetrahedron Lett.,* 4141 (1980).

277. L. Friedman and J. F. Chlebowski, *J. Org. Chem.* **33,** 1633 (1968).

278. C. Rüchardt and E. Merz, *Tetrahedron Lett.,* 2431 (1964).

279. C. Rüchardt, E. Merz, B. Freudenberg, H.-J. Opgenorth, C. C. Tan, and R. Werner, *Chem. Soc. Spec. Publ.* **24,** 51 (1970).

280. C. Rüchardt and B. Freudenberg, *Tetrahedron Lett.,* 3623 (1964).

281. G. Binsch, E. Merz, and C. Rüchardt, *Chem. Ber.* **100,** 247 (1967).

282. J. I. G. Cadogan, *Accts. Chem. Res.* **4,** 186 (1971).

283. A. V. Dushkin et al., *Zh. Org. Khim.* **13,** 1231 (1977).

284. S. C. Dickerman and K. Weiss, *J. Org. Chem.* **22,** 1070 (1957).

285. S. C. Dickerman and G. B. Vermont, *J. Am. Chem. Soc.* **84,** 4150 (1962).

286. R. Möhlau and R. Berger, *Ber.* **26,** 1994 (1893); cf. also P. A. S. Smith, L. O. Krbechek, and W. Resemann, *J. Am. Chem. Soc.* **86,** 2025 (1964).

287. M. Stiles, R. G. Miller, and U. Burckhardt, *J. Am. Chem. Soc.* **85,** 1792 (1963).

288. N. Kornblum, *Org. Reactions* **2,** Chap. 7 (1944).

289. H. Meerwein, H. Allendörfer, P. Beekman, F. Kundert, H. Morschel, F. Pawellek, and K. Wunderlich, *Angew. Chem.* **70,** 211 (1958).

290. K. G. Rutherford and W. A. Redmond, *J. Org. Chem.* **28,** 568 (1963).

291. M. Schubert and R. Fleischhauer, Ger. Patent No. 905,014; *Chem. Abstr.* **50,** 12111b (1956).

292. M. P. Doyle, J. F. Dellaria, Jr., B. Siegfried, and S. W. Bishop, *J. Org. Chem.* **42,** 3494 (1977).

293. M. S. Newman and W. M. Hung, *J. Org. Chem.* **39**, 1317 (1974).

294. D. F. De Tar and M. N. Turetzky, *J. Am. Chem. Soc.* **78**, 3928 (1956).

295. F. F. Gadallah and R. M. Elofson, *J. Org. Chem.* **34**, 3335 (1969).

296. J. P. Stradins and V. T. Glazer, in *Encyclopedia of the Electrochemistry of the Elements*, Vol. XIII-4, A. J. Bard and H. Lund (eds.), Marcel Dekker, New York, 1979, pp. 164–208.

297. S. H. Korzeniowski, L. Blum, and G. W. Gokel, *J. Org. Chem.* **42**, 1469 (1977).

298. G. D. Hartman and S. E. Biffar, *J. Org. Chem.* **42**, 1468 (1977).

299. G. S. Marx, *J. Org. Chem.* **36**, 1725 (1971).

300. J. B. Hendrickson, *J. Am. Chem. Soc.* **83**, 1251 (1961).

301. N. Kornblum, A. E. Kelley, and G. D. Cooper, *J. Am. Chem. Soc.* **74**, 3074 (1952).

302. F. Tröndlin and C. Rüchardt, *Chem. Ber.* **110**, 2494 (1977).

303. J. F. Bunnett and H. Takayama, *J. Org. Chem.* **33**, 1924 (1968).

304. E. R. Alexander and R. E. Burge, *J. Am. Chem. Soc.* **72**, 3100 (1950).

305. A. F. Levit, L. A. Kiprianova, and I. P. Gragerov, *Zh. Org. Khim.* **11**, 2351 (1975) (English trans. p. 2395).

306. P. J. Zandstra and E. M. Evleth, *J. Am. Chem. Soc.* **86**, 2664 (1964).

307. T. Cohen, A. H. Dinwoodie, and L. D. McKeever, *J. Org. Chem.* **27**, 3385 (1962).

308. T. Cohen, K. W. Smith, and M. D. Swerdloff, *J. Am. Chem. Soc.* **93**, 4303 (1971).

309. R. Galland, A. Heessing, and B. U. Kaiser, *Liebig's Ann. Chem.*, 97 (1976).

310. M. T. Jaquiss and M. Szwarc, *Nature* **170**, 312 (1952).

311. P. D. Wildes, J. G. Pacifici, G. Irick, Jr., and P. D. Whitten, *J. Am. Chem. Soc.* **93**, 2004 (1971); H. Rau and E. Lüddecke, *J. Am. Chem. Soc.* **104**, 1616 (1982).

312. G. Zimmerman, L.-Y Chow, and U.-J. Paik, *J. Am. Chem. Soc.* **80**, 3528 (1958).

313. N. A. Porter and M. O. Funk, *J. Chem. Soc. Chem. Commun.*, 263 (1973).

314. H. Wieland, E. Popper, and H. Seefried, *Ber.* **55**, 1816 (1922).

315. D. H. Hey, M. J. Perkins, and G. H. Williams, *J. Chem. Soc.*, 110 (1965).

316. M. Gomberg, *Ber.* **30**, 2043 (1897).

317. N. A. Porter, G. R. Dubay, and J. G. Green, *J. Am. Chem. Soc.* **100**, 920 (1978).

318. H. Wieland, *Ber.* **42**, 3020 (1909)

319. A. U. Blackham and N. L. Eastough, *J. Am. Chem. Soc.* **84**, 2922 (1962).

320. R. K. Lyon, *J. Am. Chem. Soc.* **86**, 1907 (1964).

321. B. K. Bandlish, A. W. Garner, M. L. Hodges, and J. W. Timberlake, *J. Am. Chem. Soc.* **97**, 5856 (1975).

322. S. E. Scheppele, P. L. Grizzle, and D. W. Miller, *J. Am. Chem. Soc.* **97**, 6165 (1975).

323. S. G. Cohen, F. Cohen, and C.-H. Wang, *J. Org. Chem.* **28**, 1479 (1963).

324. S. Seltzer and F. T. Dunne, *J. Am. Chem. Soc.* **87**, 2628 (1965).

325. C. G. Overberger and M. B. Berenbaum, *J. Am. Chem. Soc.* **73**, 2618, 4883 (1951).

326. G. Koga, N. Koga, and J.-P. Anselme, in *The Chemistry of the Hydrazo, Azo, and Azoxy Groups*, S. Patai (ed.), Wiley, New York, 1975, pp. 862–892.

327. K. MacKenzie, in *The Chemistry of the Hydrazo, Azo, and Azoxy Groups*, S. Patai (ed.), Wiley, New York, 1975, Chap. 11.

328. A. Maschke and B. S. Shapiro, *J. Am. Chem. Soc.* **86**, 1929 (1964).

329. P. S. Engel, D. J. Bishop, and M. A. Page, *J. Am. Chem. Soc.* **100**, 7009 (1978).

330. W. Adam and F. Mazenod, *J. Am. Chem. Soc.* **102**, 7131 (1980).

331. G. S. Hammond and R. C. Neumann, Jr., *J. Am. Chem. Soc.* **85**, 1501 (1963).

332. C. G. Overberger, M. T. O'Shaughnessy, and H. Shalit, *J. Am. Chem. Soc.* **71**, 2661 (1949).

333. G. S. Hammond and J. R. Fox, *J. Am. Chem. Soc.* **86,** 1918 (1964).

334. P. Smith, J. E. Sheats, and P. E. Miller, *J. Org. Chem.* **27,** 4053 (1962).

335. W. J. Chambers, C. W. Tullock, and D. D. Coffman, *J. Am. Chem. Soc.* **84,** 2337 (1962).

336. R. C. Corley and M. J. Gibian, *J. Org. Chem.* **37,** 2910 (1972).

337. H. L. Lochte, W. A. Noyes, and J. R. Bailey, *J. Am. Chem. Soc.* **44,** 2556 (1922).

338. A. J. Bellamy and R. D. Guthrie, *J. Chem. Soc.,* 3528 (1965).

339. B. V. Ioffe and V. S. Stopsky, *Tetrahedron Lett.,* 1333 (1968).

340. R. A. Cox and E. Buncel, in *The Chemistry of the Hydrazo, Azo, and Azoxy Groups,* S. Patai (ed.), Wiley, New York, 1975, pp. 838-849.

341. R. L. Reeves and R. S. Kaiser, *J. Org. Chem.* **35,** 3670 (1970).

342. B. V. Ioffe and L. M. Gershtein, *Zh. Org. Khim.* **5,** 268 (1969) (English trans. p. 257).

343. F. D. Saeva, *J. Org. Chem.* **36,** 3842 (1971).

344. V. Bekarek, J. Dobas, J. Socha, P. Vetesnik, and M. Vecera, *Coll. Czech. Chem. Commun.* **35,** 1406 (1970).

345. L. Skulski, W. Waclawek, and A. Szurowska, *Bull. Acad. Pol. Sci. Ser. Sci. Chim.* **20,** 457 (1972); *Chem. abstr.* **77,** 74610 (1972).

346. D. Y. Curtin and M. L. Poutsma, *J. Am. Chem. Soc.* **84,** 4892 (1962).

347. A. Hantzsch, *Ber.* **42,** 2129 (1909).

348. A. Korczynski, *Ber.* **41,** 4379 (1908).

349. J. P. Snyder, M. L. Heyman, and M. Gundestrup, *J. Chem. Soc. [Perkin I],* 1551 (1977).

350. M. I. Bruce and B. L. Goodall, in *The Chemistry of the Hydrazo, Azo, and Azoxy Groups,* S. Patai (ed.), Wiley, New York, 1975, Chap. 9.

351. O. Diels and W. Koll, *Liebig's Ann. Chem.* **443,** 262 (1925).

352. R. H. Nuttall, E. R. Roberts, and D. W. A. Sharp, *J. Chem. Soc.,* 2854 (1962).

353. R. Meldola and E. S. Hanes, *J. Chem. Soc.* **65,** 834 (1894).

354. G. Charrier and F. Ferreri, *Gaz. Chim. Ital.* **43II,** 148 (1913).

355. F. M. Rowe and W. G. Dangerfield, *J. Soc. Dyers Colorists* **52,** 48 (1936).

356. P. Jacobsen, *Liebig's Ann. Chem.* **367,** 304 (1909).

357. A. Werner and E. Stiasny, *Ber.* **32,** 3256 (1899).

358. C. D. Houghton and W. A. Waters, *J. Chem. Soc.,* 1018 (1950).

359. A. N. Ferguson, *Tetrahedron Lett.,* 2889 (1972).

360. G. E. Lewis and R. J. Mayfield, *Aust. J. Chem.* **19,** 1445 (1966).

361. P. V. Roling, J. L. Dill, and M. D. Rausch, *J. Organomet. Chem.* **69,** C33 (1974).

362. P. W. Robertson, T. R. Hitchings, and G. M. Will, *J. Chem. Soc.,* 808 (1950).

363. R. Pummerer, J. Binapfli, K. Bittner, and K. Schuegraf, *Ber.* **55,** 3095 (1922).

364. H. Gilman and J. C. Baillie, *J. Org. Chem.* **2,** 84 (1937).

365. P. F. Holt and B. P. Hughes, *J. Chem. Soc.,* 764 (1954).

366. E. M. Kaiser and G. J. Bartling, *J. Org. Chem.* **37,** 490 (1972).

367. W. Bradley and J. D. Hannon, *J. Chem. Soc.,* 2713 (1962).

368. E. Fahr and H. Lind, *Angew. Chem. Int. Ed.* **5,** 372 (1966).

369. O. Diels and C. Wulff, *Liebig's Ann. Chem.* **437,** 309 (1924).

370. G. E. Wilson and J. H. E. Martin, *J. Org. Chem.* **37,** 2510 (1972).

371. E. E. Smissman and A. Makriyannis, *J. Org. Chem.* **38,** 1652 (1973).

372. S. H. Schroeter, *J. Org. Chem.* **34,** 4012 (1969).

373. (*a*) O. Diels, *Liebig's Ann. Chem.* **429,** 1 (1922); (*b*) R. B. Carlin and M. S. Moores, *J. Am. Chem. Soc.* **84,** 4107 (1962).

374. R. Huisgen and F. Jakob, *Liebig's Ann. Chem.* **590**, 37 (1954).

375. E. Brun and R. Huisgen, *Angew. Chem. Int. Ed.* **8**, 513 (1969).

376. R. E. Humphrey and E. E. Hueske, *J. Org. Chem.* **36**, 3994 (1971).

377. O. Mitsunobu, K. Kato, and M. Tomari, *Tetrahedron* **26**, 5731 (1970).

378. (*a*) R. Stollé and W. Reichert, *J. Prakt. Chem.* [2]**123**, 74 (1929); (*b*) O. Diels and H. Behnke, *Ber.* **57**, 653 (1924).

379. L. A. Carpino, P. H. Terry, and P. J. Crowley, *J. Org. Chem.* **26**, 4336 (1961).

380. K. H. Linke and H. J. Gohausen, *Chem. Ber.* **106**, 3438 (1973).

381. K. Alder and T. Noble, *Ber.* **56**, 54 (1923).

382. B. T. Gillis and F. A. Daniher, *J. Org. Chem.* **27**, 4001 (1962).

383. R. Huisgen, F. Jakob, W. Siegel, and A. Cadus, *Liebig's Ann. Chem.* **590**, 1 (1954).

384. R. Huisgen and H. Pohl, *Chem. Ber.* **93**, 527 (1960).

385. B. Franzus and J. H. Surridge, *J. Org. Chem.* **27**, 1951 (1962).

386. W. A. Thaler and B. Franzus, *J. Org. Chem.* **29**, 2226 (1964).

387. L. Horner and W. Naumann, *Liebig's Ann. Chem.* **587**, 81 (1954).

388. B. Franzus, *J. Org. Chem.* **28**, 2954 (1963).

389. B. T. Gillis and P. E. Beck, *J. Org. Chem.* **27**, 1947 (1962).

390. G. Ahlgren, B. Åkermark, J. Lewandowska, and R. Wahren, *Acta Chem. Scand.* **B29**, 524 (1975).

391. M. Busch and H. Kunder, *Ber.* **49**, 2347 (1916).

392. Y. Kikuchi, T. Mitsuhashi, O. Simamura, and M. Yoshida, *J. Chem. Soc.* [*C*], 2074 (1971).

393. W. Bradley and L. J. Watkinson, *J. Chem. Soc.*, 319 (1956).

394. A. Rodgman and G. F. Wright, *J. Org. Chem.* **18**, 465 (1953).

395. K. Burger and K. Einhillig, *Chem. Ber.* **106**, 3421 (1973).

396. P. Gilgen, H. Heimgartner, and H. Schmid, *Helv. Chim. Acta* **57**, 1382 (1974).

397. H. Blaschke, E. Brunn, R. Huisgen, and W. Mack, *Chem. Ber.* **105**, 2481 (1972).

398. J. H. Hall and R. Kellogg, *J. Org. Chem.* **31**, 1079 (1966).

399. R. C. Kerber, T. J. Ryan, and S. D. Hsu, *J. Org. Chem.* **39**, 1215 (1974).

400. B. Eistert and M. Regitz, *Chem. Ber.* **96**, 2290 (1963); B. Eistert and K. Schank, *ibid.*, 2304 (1963).

401. L. Horner and E. Spietschka, *Chem. Ber.* **89**, 2765 (1956).

402. E. K. von Gustorf, D. V. White, B. Kim, D. Hess, and J. Leitich, *J. Org. Chem.* **35**, 1155 (1970).

403. B. T. Newbold, in *The Chemistry of the Hydrazo, Azo, and Azoxy Groups*, S. Patai (ed.) Wiley, New York, 1975, pp. 601, 604–614, 631–636.

404. F. W. Whitmore and A. J. Revukas, *J. Am. Chem. Soc.* **59**, 1500 (1937), and papers cited therein.

405. J. Burdon, C. J. Morton, and D. F. Thomas, *J. Chem. Soc.*, 2621 (1965).

406. A. Risaliti and A. Stener, *Ann. Chim.* (*Rome*) **57**, 3 (1967).

407. E. T. McBee, G. W. Calundann, C. J. Morton, T. Hodgins, and E. P. Wesseler, *J. Org. Chem.* **37**, 3140 (1972).

408. M. L. Heyman and J. P. Snyder, *Tetrahedron Lett.*, 2859 (1973).

409. T.-L. Ho and G. A. Olah, *Synthesis*, 169 (1977).

410. D. Y. Curtin and T. C. Miller, *J. Org. Chem.* **25**, 885 (1960).

411. S. Bozzini and A. Stener, *Ann. Chim.* (*Rome*) **58**, 169 (1968).

412. Y. Ohgo, S. Takeuchi, and S. Yoshimura, *Bull. Chem. Soc. Japan* **44**, 283 (1971).

413. E. J. Corey, W. L. Mock, and D. J. Pasto, *Tetrahedron Lett.*, 347 (1961).

414. M. Kira, M. O. Abdel-Rahman, M. N. Tolba, and Z. Nofal, *J. Chem. U.A.R.* **11**, 153 (1968).

415. O. Diels, *Ber.* **56,** 1933 (1923).

416. H. C. Brown and N. M. Yoon, *J. Am. Chem. Soc.* **88,** 1464 (1966).

417. G. Olah, *J. Am. Chem. Soc.* **81,** 3165 (1959).

418. T. Neilson, H. C. S. Wood, and A. G. Wylie, *J. Chem. Soc.,* 371 (1962).

419. H. C. Brown and B. C. Subba Rao, *J. Am. Chem. Soc.* **82,** 681 (1960).

420. H. C. Brown, P. Heim, and N. M. Yoon, *J. Am. Chem. Soc.* **92,** 1637 (1970).

421. A. G. Davies, B. P. Roberts, and J. C. Scaiano, *J. Chem. Soc. [Perkin II],* 803 (1972).

422. P. Jacobson, *Liebig's Ann. Chem.* **428,** 76 (1922).

423. G. P. Warwick, *J. Soc. Dyers Colorists* **75,** 291 (1959).

424. D. J. Byron, G. W. Gray, and B. M. Worrall, *J. Chem. Soc.,* 3706 (1965).

425. J. V. Earley and T. S. Ma, *Mikrochim. Acta,* 685 (1960).

426. N. R. Large and C. Hinshelwood, *J. Chem. Soc.,* 620 (1956).

427. N. R. Large, F. J. Stubbs, and C. Hinshelwood, *J. Chem. Soc.,* 2736 (1954).

428. B. T. Newbold, *J. Chem. Soc.,* 6972 (1965).

429. M. Khalifa, *J. Chem. Soc.,* 1854 (1960).

430. J. Singh, P. Singh, J. L. Boivin, and P. E. Gagnon, *Can. J. Chem.* **40,** 1921 (1962).

431. F. D. Greene, M. A. Berwick, and C. Stowell, *J. Am. Chem. Soc.* **92,** 867 (1970).

432. J. M. Birchall, R. N. Haszeldine, and J. E. G. Kemp, *J. Chem. Soc. [C],* 449 (1970).

433. H. J. Shine and J. T. Chamness, *Tetrahedron Lett.,* 641 (1963).

434. B. T. Newbold and D. Tong, *Can. J. Chem.* **42,** 836 (1964).

435. M. Khalifa and A. A. Abo-Ouf, *J. Chem. Soc.,* 3740 (1958).

436. E. Nölting and E. Fourneaux, *Ber.* **30,** 2930 (1897).

437. M. Hedayatullah and L. Denivelle, *Compt. Rend. Acad. Sci. Paris* **258,** 5467 (1964).

438. A. Zweig and A. K. Hoffmann, *J. Am. Chem. Soc.* **85,** 2736 (1963).

439. W. Schlenk and E. Bergmann, *Liebig's Ann. Chem.* **463,** 281 (1928).

440. R. Meyer, *Ber.* **53,** 1265 (1920).

441. G. O. Schenck and H. Formanek, *Angew. Chem.* **70,** 505 (1958).

442. L. F. Fieser, *Org. Syntheses,* Coll. **II,** 35, 39 (1943).

443. A. N. Nesmeyanov, R. V. Golovnya, and G. A. Mironov, *Issled. Obl. Org. Khim. Izbr. Tr.,* 374 (1971).

444. J. K. S. Wan, L. D. Hess, and J. N. Pitts, *J. Am. Chem. Soc.* **86,** 2069 (1964).

445. L. Bin Din, J. M. Lindley, and O. Meth-Cohn, *Synthesis,* 23 (1978).

446. F. G. Thomas and K. G. Boto, in *The Chemistry of the Hydrazo, Azo, and Azoxy Groups,* S. Patai (ed.), Wiley, New York, 1975, pp. 462–493.

447. J. L. Sadler and A. J. Bard, *J. Am. Chem. Soc.* **90,** 1979 (1968).

448. R. Hazard and A. Tallec, *Bull. Soc. Chim. France,* 2917 (1971).

449. L. Holleck, D. Jannakoudakis, and A. Wildenau, *Electrochim. Acta* **12,** 1523 (1967).

450. B. Nygård, *Arkiv Kem.* **26,** 167 (1967).

451. G. H. Aylward, J. C. Garnett, and J. H. Sharp, *Anal. Chem.* **39,** 457 (1967).

452. G. Pezzatini and R. Guidelli, *J. Chem. Soc. [Faraday Trans.]* **69,** 794 (1973).

453. B. T. Newbold, in *The Chemistry of the Hydrazo, Azo, and Azoxy Groups,* S. Patai (ed.), Wiley, New York, 1975, pp. 557–563, 573–597.

454. F. D. Greene and S. S. Hecht, *Tetrahedron Lett.,* 575 (1969).

455. B. W. Langley, B. Lythgoe, and L. S. Rayner, *J. Chem. Soc.,* 4191 (1952).

456. J. P. Freeman, *J. Org. Chem.* **28,** 2508 (1963).

457. G. M. Badger, R. G. Buttery, and G. E. Lewis, *J. Chem. Soc.*, 2143 (1953).

458. B. T. Newbold, *J. Org. Chem.* **27**, 3919 (1962).

459. P. E. Gagnon and B. T. Newbold, *Can. J. Chem.* **37**, 366 (1959).

460. M. J. S. Dewar and R. S. Goldberg, *Tetrahedron Lett.*, 2717 (1966).

461. S. R. Sandler and W. Karo, *Organic Functional Group Preparations*, Vol. II, Academic Press, New York, 1971, pp. 354*ff*.

462. B. T. Gillis and J. D. Hagarty, *J. Org. Chem.* **32**, 95 (1967).

463. J. N. Brough, B. Lythgoe, and P. Waterhouse, *J. Chem. Soc.*, 4069 (1954).

464. A. Risaliti, *Gaz. Chim. Ital.* **93**, 585 (1963).

465. J. Singh, P. Singh, J. L. Boivin, and P. E. Gagnon, *Can. J. Chem.* **41**, 499 (1963).

466. M. A. Berwick and R. E. Rondeau, *J. Org. Chem.* **37**, 2409 (1972).

467. W. Haug, J. Pelz, and H. Usbeck, *Die Pharmazie* **24**, 442 (1969).

468. G. M. Badger and G. E. Lewis, *J. Chem. Soc.*, 2147 (1953).

469. T. Mitsuhashi, O. Simamura, and Y. Tezuka, *J. Chem. Soc. Chem. Commun.*, 1300 (1970).

470. N. A. Johnson and E. S. Gould, *J. Org. Chem.* **39**, 407 (1974).

471. S. Wawzonek and T. W. McIntyre, *J. Electrochem. Soc.* **114**, 1025 (1967).

472. T. E. Stevens, *J. Org. Chem.* **26**, 2531 (1961).

473. E. C. Kooyman, *Rec. Trav. Chim. Pays Bas* **74**, 117 (1955).

474. N. R. Subbratnam and J. G. Calvert, *J. Am. Chem. Soc.* **84**, 113 (1962).

475. E. E. van Tamelen and R. J. Timmons, *J. Am. Chem. Soc.* **84**, 1067 (1962).

476. E. Benzing, *Liebig's Ann. Chem.* **631**, 1, 10 (1960).

477. J. Miller, D. B. Paul, L. Y. Wong, and A. G. Kelso, *J. Chem. Soc.* [*B*], 62 (1970).

478. H. Lui and J. Warkentin, *Can. J. Chem.* **51**, 1148 (1973).

479. R. Pütter, in *Houben-Weyl, Methoden der Organischen Chemie* 4th Ed., E. Müller (ed.), Georg Thieme-Verlag, Stuttgart, 1965, pp. 682–694.

480. H. Fischer and D. Jerchel, *Liebig's Ann. Chem.* **574**, 85 (1951).

481. E. Bamberger and H. Witter, *J. Prakt. Chem.* [**2**]**65**, 139 (1902).

482. O. W. Maender and G. A. Russell, *J. Org. Chem.* **31**, 442 (1966).

483. H. Gehlen and G. Röbisch, *Liebig's Ann. Chem.* **665**, 132 (1963).

484. E. Bamberger and O. Billeter, *Helv. Chim. Acta* **14**, 219 (1931).

485. G. D. Parkes and S. G. Tinsley, *J. Chem. Soc.*, 1841 (1934).

486. K. Schank, in *Methodicum Chemicum*, Vol. 6, F. Korte (ed.), Academic Press, New York, 1975, pp. 159–178.

487. R. Pütter, in *Houben-Weyl, Methoden der Organischen Chemie*, 4th Ed., Vol. X/3, E. Müller (ed.), Georg-Thieme-Verlag, Stuttgart, 1965, pp. 7–113, 545–626.

488. K. H. Saunders, *The Aromatic Diazo Compounds*, 2d Ed., E. Arnold, London, 1949.

489. J. L. Hartwell and L. F. Fieser, *Org. Syntheses*, Coll. **II**, 145 (1943).

490. H. T. Clarke and W. R. Kirner, *Org. Syntheses*, Coll. **I**, Ed. 2, 374 (1941).

491. P. A. S. Smith, J. H. Hall, and R. O. Kan, *J. Am. Chem. Soc.* **84**, 485 (1962).

492. J. H. Ridd, *Quart. Rev.* **15**, 418 (1961).

493. A. H. Seitz and L. Friedman, *Org. Syntheses*, Coll. **V**, 54 (1973).

494. R. Howe, *J. Chem. Soc.* [*C*], 478 (1966).

495. M. R. Piercey and E. R. Ward, *J. Chem. Soc.*, 3841 (1962).

496. G. Schiemann and W. Winkelmüller, *Org. Syntheses*, Coll. **II**, 299 (1943).

497. K. G. Rutherford and W. Redmond, *Org. Syntheses*, Coll. **V**, 133 (1973).

498. M. Kobayashi and K. Honda, *Bull. Chem. Soc. Japan* **39**, 1778 (1966).

499. H. P. Patel and J. M. Tedder, *J. Chem. Soc.*, 4589, 4593, 4889, 4894 (1963).

550. C. Rüchardt and C. C. Tan, *Chem. Ber.* **103**, 1774 (1970).

501. R. M. Scribner, *J. Org. Chem.* **29**, 3429 (1964).

502. J. Turcan, *Bull. Soc. Chim. France*, 627 (1935).

503. A. Hantzsch and F. M. Perkin, *Ber.* **30**, 1412 (1897).

504. H. Mehner, *J. Prakt. Chem.* [2]**63**, 266 (1901).

505. J. M. Tedder, *J. Chem. Soc.*, 4003 (1957).

506. F. W. Bergstrom and J. S. Buehler, *J. Am. Chem. Soc.* **64**, 19 (1942).

507. S. Oae, K. Shinhama, and Y. H. Kim., *Bull. Chem. Soc. Japan* **53**, 1065, 2023 (1980).

508. S. R. Sandler and W. Karo, *Organic Functional Group Preparations,* Vol. II, Academic Press, New York, 1971, Chap. 14.

509. Azobenzenes: K. H. Schründehütte, in *Houben-Weyl, Methoden der Organischen Chemie,* 4th Ed., Vol. X/3, Georg Thieme-Verlag, Stuttgart, 1965, pp. 219–423. Areneazoalkanes: E. Enders, *ibid.* pp. 471–485.

510. R. Renaud and L. C. Leitch, *Can. J. Chem.* **32**, 545 (1954).

511. J. R. Shelton and C. K. Liang, *Synthesis,* 204 (1971).

512. S. E. Scheppele and S. Seltzer, *J. Am. Chem. Soc.* **90**, 358 (1968).

513. H. Bock, G. Rudolph, and E. Baltin, *Chem. Ber.* **98**, 2054 (1965).

514. L. A. Carpino and P. J. Crowley, *Org. Syntheses,* Coll. **V**, 160 (1973).

515. S. G. Cohen and C. H. Wang, *J. Am. Chem. Soc.* **77**, 2457 (1955).

516. C. G. Overberger, P.-T. Huang, and M. B. Berenbaum, *Org. Syntheses,* Coll. **IV**, 66 (1963).

517. R. Ohme and H. Preuschhof, *Org. Syntheses* **52**, 11 (1972).

518. R. Adams and J. R. Johnson, *Laboratory Experiments in Organic Chemistry,* 4th Ed., Macmillan, New York, 1949, p. 359.

519. N. Rabjohn, *Org. Syntheses,* Coll. **III**, 375 (1955).

520. F. D. Vidal and V. G. Sarli, U. S. Patent No. 3,192,196; *Chem. Abstr.* **63**, 8199a (1965).

521. J. C. Kauer, *Org. Syntheses,* Coll. **IV**, 411, 414 (1963).

522. A. J. Fatiadi, *J. Org. Chem.* **35**, 831 (1970).

523. E. L. Allred and J. C. Henshaw, *J. Chem. Soc. Chem. Commun.,* 1021 (1969).

524. K. Dimroth and W. Tüncher, *Synthesis,* 339 (1977).

525. F. Yoneda, K. Suzuki, and Y. Nitta, *J. Am. Chem. Soc.* **88**, 2328 (1966).

526. J. A. Hyatt, *Tetrahedron Lett.,* 141 (1977).

527. R. D. Little and M. G. Venegas, *J. Org. Chem.* **43**, 2921 (1978).

528. S. F. Nelsen and P. D. Bartlett, *J. Am. Chem. Soc.* **88**, 137 (1966).

529. J. R. Shelton, J. F. Gormish, C. K. Liang, P. L. Samuel, P. Kovacic, and L. W. Haynes, *Can. J. Chem.* **46**, 1149 (1968).

530. J. J. Fuchs, U. S. Patent No. 3,346,554; *Chem. Abstr.* **68**, 2593y (1968).

531. E. Pfeil and K. H. Schmidt, *Liebig's Ann. Chem.* **675**, 36 (1964).

532. O. H. Wheeler and D. Gonzalez, *Tetrahedron* **20**, 189 (1964).

533. K. H. Pauscker, *J. Chem. Soc.*, 1989 (1953).

534. B. Ortiz, P. Villanueva, and F. Walls, *J. Org. Chem.* **37**, 2748 (1972).

535. G. A. Jaffari and A. J. Nunn, *J. Chem. Soc.* [C], 823 (1971).

536. R. Bauer and H. Wendt, *Angew. Chem. Int. Ed.* **17**, 202 (1978).

537. H. Esser, K. Rastädter, and G. Reuter, *Chem. Ber.* **89**, 685 (1956).

538. P. S. Engel and D. J. Bishop, *J. Am. Chem. Soc.* **97**, 6754 (1975).

539. J. W. Timberlake, J. Alender, A. W. Garner, M. L. Hodges, C. Özmeral, S. Szilagyi, and J. O. Jacobus, *J. Org. Chem.* **46**, 2082 (1981).

540. R. Ohme and E. Schmitz, *Angew. Chem. Int. Ed.* **4**, 433 (1965).

541. J. C. Stowell, *J. Org. Chem.* **32**, 2360 (1967).

542. R. Bauer and H. Wendt, *Angew. Chem. Int. Ed.* **17**, 370 (1978).

543. J. S. Fowler, *J. Org. Chem.* **37**, 510 (1972).

544. S. Goldschmidt and B. Acksteiner, *Liebig's Ann. Chem.* **618**, 173 (1958).

545. D. C. Iffland, L. Salisbury, and W. R. Schafer, *J. Am. Chem. Soc.* **83**, 747 (1961).

546. J. Kossanyi, *Compt. Rend. Acad. Sci. Paris* **257**, 929 (1963).

547. J. F. Vozza, *J. Org. Chem.* **34**, 3219 (1969).

548. H. E. Bigelow and D. B. Robinson, *Org. Syntheses*, Coll. **III**, 103 (1955) (azobenzene, 84–86%).

549. H. J. Shine and J. T. Chamness, *J. Org. Chem.* **28**, 1232 (1963).

550. D. A. Blackadder and C. Hinshelwood, *J. Chem. Soc.*, 2898 (1957).

551. R. F. Nystrom and W. G. Brown, *J. Am. Chem. Soc.* **70**, 3738 (1948).

552. J. J. Scheloske, *Wright Air Devel. Center, AD 603684*, 109 pp. (1964); *Chem. Abstr.* **62**, 2718b (1965).

553. S. F. Nelsen, *J. Am. Chem. Soc.* **96**, 5669 (1974).

554. Y. Nomura, H. Azai, R. Tarao, and K. Shoimi, *Bull. Chem. Soc. Japan* **37**, 967 (1964); Y. Nomura and H. Azai, *ibid.*, 970 (1964).

555. Y. Nomura, *Bull. Chem. Soc. Japan* **34**, 1648 (1961).

556. D. Y. Curtin and J. A. Ursprung, *J. Org. Chem.* **21**, 1221 (1956).

557. A. N. Nesmeyanov, *Tetrahedron Lett.*, 1 (1960).

558. R. Meier and W. Frank, *Chem. Ber.* **89**, 2747 (1956).

559. B. M. Bogoslovskii, *Zh. Obshch. Khim.* **16**, 193 (1946).

560. M. Christen, L. Funderbunk, E. A. Halevi, G. E. Lewis, and H. Zollinger, *Helv. Chim. Acta* **49**, 1376 (1966).

561. W. H. Nutting, R. A. Jewell, and H. Rapoport, *J. Org. Chem.* **35**, 505 (1970).

562. M. J. Namkung, N. K. Naimy, C.-A. Cole, N. Ishikawa, and T. L. Fletcher, *J. Org. Chem.* **35**, 728 (1970).

563. H. D. Anspon, *Org. Syntheses*, Coll. **III**, 711 (1955).

564. A. H. Dinwoodie and R. N. Haszeldine, *J. Chem. Soc.*, 2266 (1965).

565. T. M. Pozdnyakova and N. S. Zefirov, *Zh. Org. Khim.* **8**, 1107 (1972).

566. G. M. Badger and G. E. Lewis, *J. Chem. Soc.*, 2151 (1953).

567. M. Martynoff, *Bull. Soc. Chim. France*, 214 (1951).

568. G. E. Lewis and M. A. G. Osman, *Aust. J. Chem.* **17**, 498 (1964).

569. H. E. Fierz-David, L. Blangey, and E. Merian, *Helv. Chim. Acta* **34**, 846 (1951).

570. H. T. Clarke, *J. Org. Chem.* **36**, 3816 (1971).

571. E. C. Taylor, G. E. Jagdmann, and A. McKillop, *J. Org. Chem.* **43**, 4385 (1978).

572. J. van Alphen, *Rec. Trav. Chim. Pays Bas* **64**, 109 (1945).

573. B. T. Gillis and J. D. Hagerty, *J. Am. Chem. Soc.* **87**, 4576 (1965).

574. K. H. Meyer, *Ber.* **52**, 1468 (1919).

575. A. P. Terent'ev and A. A. Demidova, *Zh. Obshch. Khim.* **7**, 2464 (1937); *Chem. Abstr.* **32**, 2094 (1938).

576. G. Cauquis and B. Chebaud, *Tetrahedron* **34**, 903 (1978).

577. (a) K. Schank, in *Methodicum Chemicum,* Vol. 6, F. Korte (ed.), Academic Press, New York, 1975, pp. 198–199; (b) U. Kraatz, S. Linke, E. Wehringer, H. Wollweber, G. Simchen, and W. Walter, *ibid.,* p. 746; (c) R. Pütter, in *Houben-Weyl, Methoden der Organischen Chemie,* 4th Ed., Vol. X/3, E. Müller (ed.), Georg-Thieme-Verlag, Stuttgart, 1965, pp. 622–681 (preparation of formazans, nitrosazones, and nitrazones).

578. G. A. Reynolds and J. A. Van Allan, *Org. Syntheses,* Coll. **IV,** 633 (1963).

579. R. F. Muracca, in *Treatise on Analytical Chemistry,* Part II, Vol. 15, I. M. Kolthoff and P. J. Elving (eds.), Wiley, New York, 1976, pp. 277–337.

580. S. Siggia and J. G. Hanna, *Quantitative Organic Analysis via Functional Groups,* 4th Ed., Wiley, New York, 1979, pp. 654–663 (azo groups), 680–687 (diazonium groups).

581. N. D. Cheronis and T. S. Ma, *Organic Functional Group Analysis,* Wiley, New York, 1964, pp. 262–268.

582. R. F. Muracca, ref. 579, pp. 416–486.

583. N. D. Cheronis and J. B. Entrikin, *Semimicro Qualitative Organic Analysis,* Crowell, New York, 1947.

5.
DIAZO COMPOUNDS, AZAMINES, AND NITRILE IMIDES

The three functional types of compound covered in this chapter have as a common feature the presence of two joined nitrogen atoms in a structure for which no classical octet formula can be written without the use of formal charges.

NOMENCLATURE

The prefix *diazo*, which stands for the divalent substituent $N_2=$, provides the only systematic way for naming compounds of the type $R_2C=N_2$. There is no corresponding suffix, but a near approach to one can be managed if a compound is named as a substituted diazomethane. Thus the compound $CH_3COCH=N_2$ is *diazoacetone* or *diazopropanone*, but $PhCH=N_2$ is usually called *phenyldiazomethane*.

The IUPAC rules do not provide a name for the structure $R_2\overset{+}{N}=N^-$. In *Chemical Abstracts*, such compounds are indexed under "diazenium hydroxide, inner salt," which is an inconveniently long circumlocution. Some authors have called them "1,1-disubstituted diazenes," but this is not without objection. Others have chosen to name them according to the contributing nonoctet structure, $R_2N—N:$, but this is rather like naming an isocyanide as a carbene and necessitates choosing a name for the parent compound NH, a subject of disagreement; the form "aminonitrene," which has no official sanction, has been used in many publications. The coined term "azamine" for the hypothetical parent compound $H_2\overset{+}{N}=\overset{..}{N}{}^-$ is favored by many, although it too does not have official sanction. It will be used in this chapter, however.

Nitrile imides, $R—C\equiv\overset{+}{N}—\overset{-}{N}R$, fall more easily into conventional nomenclature systems, for they are isoelectronic with nitrile oxides, are related to amine oxides, and may be named analogously, as in benzonitrile methylimide for $PH—C\equiv\overset{+}{N}—\overset{-}{N}CH_3$. Some authors have preferred to name them according to a nonoctet contributing structure, as azo carbenes, $R—C—N=N—R$, but this seems artificial. *Chemical Abstracts* defines the parent compound HCNNH as "nitrilimine" and uses *C* and *N* to locate substituents, for example *N*-methyl-*C*-phenylnitrilimine.

PROPERTIES

Diazomethane is an orange gas (bp $-23°C$, mp $-145°C$), and diazoethane boils at $-18°C$ at 89.5 mm pressure. These boiling points are high for compounds of low molecular weight without hydrogen bonding (compare propene, bp $-48°C$) and are the consequence of an appreciable dipole moment (that of diphenyldiazomethane[1a] is 1.42 D, and of diazomethane,[1b] 1.4 D). Diazoalkanes have a high energy content, which is easily evolved explosively when the compound is small, and for this reason, diazomethane is always handled in solution (but even solutions of it have been known to explode). When the carbon skeleton is large enough (more than six carbons) to dissipate internally much of the energy of the decomposing diazoalkane group, the danger of explosion drops off sharply; however, most such compounds decompose readily enough that it is desirable to store them at 0°C or lower. Strongly electron-withdrawing substituents may increase the stability substantially, and, for example, the salts of diazomethanedisulfonic acid, $N_2C(SO_3K)_2$, are relatively inert. Diazomethane solutions become dangerously sensitive on aging and should not be stored, even in a refrigerator.

A compound thought to be the simplest nitrile imide, HCN_2H, was reported[2] in 1954 as a colorless oil that decomposes at room temperature and explodes at 35 to 40°C. It was subsequently[3,4] shown to be isocyanoamine, $CN-NH_2$. Compounds in which the nitrile imide structure is fixed by substitution are apparently formed by base-catalyzed elimination of HCl from hydrazonoyl chlorides; they are too reactive to isolate, but they can be trapped by their capacity to undergo cycloaddition reactions, in which they function as 1,3-dipolar systems.

Azamines were known only as salts or inferred intermediates until 1978,[5] when a six-membered cyclic example (I) was obtained in solution at −78°C. A second cyclic example (II), with a five-membered ring, was reported in 1980.[6] Their solutions are purple. The heat of formation of the pyrrolidine derivative have been estimated at 30.5 kcal/mol, which is about 20 kcal/mol higher than that of the isomeric azo compound (III).

I   II   III

A bond isomer of diazomethane, diazirine, is known[7,8] as a nearly colorless gas (bp −14°C). This compound was only isolated in 1961, but it was anticipated a long time ago, when diazomethane was originally believed to have the diazirine structure. Diazirine is less reactive than diazomethane.

The infrared spectra of diazo compounds[9] show a characteristic absorption at 2050 to 2000 cm^{-1} (about 50 cm^{-1} higher for α-diazo ketones, etc.), owing to N—N stretching.[9,10] The diazo function gives rise to a weak UV maximum near 450 nm ($\varepsilon < 10$) in simple diazoalkanes, with a tail into the visible range that is responsible for the characteristic colors, canary yellow to garnet red. There is also a much stronger maximum at shorter wavelengths, near 240 nm in simple cases.[9,11] The most characteristic feature of the proton NMR of diazo compounds[9] is the downfield position of the α-hydrogens, if present. That of diazomethane[12] is found at 3.08 to 3.20, depending on solvent; anisotropic shielding by the C=N bond is thus weaker than that by carbonyl and ethylenic bonds. The ^{13}C NMR of the α-carbon of diazoalkanes occurs at rather high field (23 to 112 ppm down from TMS), a fact that has been attributed to the importance of the contribution of $R_2\bar{C}-\overset{+}{N}{=}N$ to the total structure[13] (and which would also reduce the effect of anisotropic shielding an α-protons). The ^{15}N NMR of the interior nitrogen atom is found at 69 to 147 ppm upfield from HNO_3, whereas the terminal nitrogen resonates at a much lower position (-66 to $+59$ ppm from HNO_3). The mass spectrometry of diazo compounds has been reviewed,[9] and molecular ions are generally weak or absent, but peaks at 28 (N_2^+), 41 (CHN_2^+), and $M-28$ are commonly seen.

Azamines show as infrared stretching absorption for the $N{=}N$ bond in the region 1595 to 1638 cm^{-1}; their color is due to a maximum in the UV-vis. spectrum at 497 to 550 nm.[5,6]

Diazo compounds are not basic in the sense of being able to form salts, but they nevertheless accept protons readily and are thereby sensitive to acid-catalyzed decomposition, presumably through formation of $R_2CH-N_2^+$. The pK_a of $CH_3N_2^+$ is 10, about the same as nitromethane.[14] They do not ordinarily display acidity, but the α-hydrogens readily undergo exchange with D_2O catalyzed by sodium hydroxide (a phase-transfer catalyst helps).[15,16] Azamines are evidently quite basic, for their salts (azaminium or 1,1-disubstituted diazenium salts) are stable in water solution.

The CNN skeleton of diazomethane (**IV**) has been shown to be linear by diffraction experiments; the hydrogens lie at an angle of about $120°$.[1a,17,18] That the nitrogen atoms are not equivalent has been demonstrated by isotope labeling experiments.[19] In contrast, such experiments have shown the complete equivalence of the nitrogens in the isomeric compound diazirine.[8] The diazo function may evidently interact substantially with adjacent unsaturated groups, for the colors of diazo compounds vary noticeably with the possibility for conjugation. Stronger evidence may be found in the fact that restricted rotation about the $C-C$ bond in α-diazo ketones has been observed; this implies that structure **V** is an important contributor.[20] The preferred geometry has the N_2 *syn* to the oxygen.

$$\underset{\substack{\longleftarrow\longleftarrow \\ 1.300\text{ Å } 1.139\text{ Å}}}{\overset{H}{\underset{H}{>}}C{=}\overset{+}{N}{=}\bar{N}}$$

IV

V

Diazomethane is extremely poisonous, and its inhalation should be carefully avoided. Some chemists become sensitized to it after initial exposure, after which even scarcely detectable traces of the substance put them in acute misery. All diazo compounds should be assumed to be toxic, barring evidence to the contrary.

The only naturally occurring diazo compounds of which I am aware are azaserine, α-amino-β-(diazoacetoxy)propionic acid, a tumor-inhibiting substance isolated from a species of streptomyces,[21] 6-diazo-5-oxo-L-norleucine,[22] and alazopeptin, a tripeptide derived from the latter.[23]

In addition to the diazo compounds that are ordinarily brought to mind by the term *diazoalkane*, there are certain highly conjugated examples that may be considered to be effectively diazonium salts. These are the diazo oxides, or inner salts of phenol-*o* or *p*-diazonium hydroxide, and diazocyclopentadiene (**VI, VII,** and **VIII**). Benzene-1,4-diazooxide,[24] which has sometimes most inappropriately been called "*p*-quinone diazide," is an orange solid that explodes at 92°C and shows infrared absorption at 2080 cm^{-1}, typical of diazo compounds. Diazocyclopentadiene is an

VI   **VII**   **VIII**

unusually stable oil[25] (bp 45°C/40mm) that readily undergoes typical aromatic electrophilic substitution reactions at the 2 and 5 positions. The chemistry of these somewhat ambiguous compounds is considered in this chapter only insofar as it resembles that of other diazo compounds; that part of their chemistry which is of typically diazonium nature is in the province of Chapter 4.

REACTIONS

Diazoalkanes[26-29]

Most of the reactions of diazo compounds can be fitted into a category according to the initial step, which in nearly all cases results in a reactive intermediate that cannot be isolated but leads to the final products by one or more additional steps. The variety of end products may be quite wide, and in some cases, the same end product can be obtained through different initial steps. The structure of the end product is therefore not by itself sufficient indication of the path by which it was formed. The initial steps are (1) cleavage into nitrogen and a carbene, with or without concurrent rearrangement; (2) attack by an electrophilic species at the α-carbon to form an alkanediazonium ion; (3) attack by a nucleophile at the terminal nitrogen atom to form a delocalized carbanion; (4) attack by free radicals; (5) reduction; and (6) cycloaddition to multiple bonds. This section is arranged according to such processes.

Thermolysis and photolysis. The primary action of both heat and light on diazo compounds is to cause fragmentation to molecular nitrogen and a carbene, which is normally formed in the singlet electronic state (unshared electrons paired); photolysis in the presence of triplet sensitizers, such as benzophenone, generates a triplet carbene. The reactions of the two states of the carbene are essentially the same qualitatively, but there may be differences in mechanism, selectivity, and stereochemistry. A complete exposition of the subject is beyond the scope of this book and must be sought among the several excellent reviews of carbenes and their chemistry.[30-35]

The most characteristic reactions of carbenes, whether generated from diazo compounds or in other ways, are insertion into single bonds and addition to multiple bonds. When a suitable reaction partner is not available, the carbene will undergo such reactions intramolecularly if possible. Diazomethane itself is a special case, for no intramolecular reaction can occur with the product, unsubstituted methylene (CH_2), which therefore polymerizes to form ethylene (mostly) and polymethylene (traces).[36,37] If, in addition to the lack of a reaction partner, intramolecular reaction is slow enough, the singlet carbene may undergo intersystem crossing to the triplet state, which is generally more stable, and the ultimate products will be derived from this species. These possibilities are presented in the accompanying scheme, along with some other characteristic reactions:

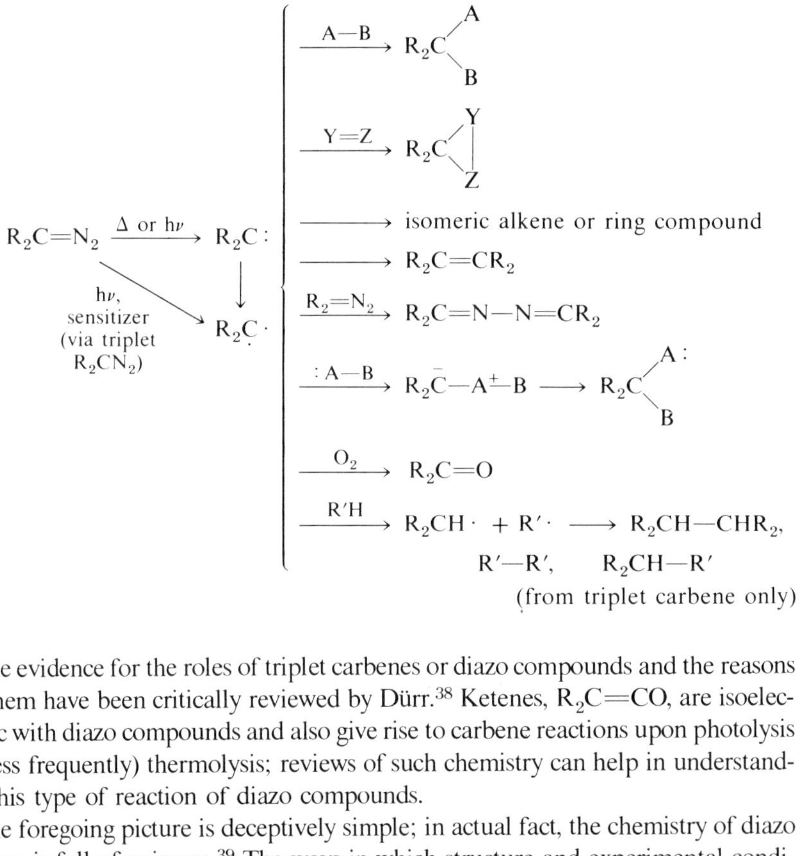

$$R_2C{=}N_2 \xrightarrow{\ \Delta \text{ or } h\nu\ } R_2C\!:\;$$

$$\xrightarrow[\text{(via triplet } R_2CN_2)]{\substack{h\nu, \\ \text{sensitizer}}} R_2C\cdot$$

$$\xrightarrow{\ A{-}B\ } R_2C\!\diagup^{A}_{\diagdown B}$$

$$\xrightarrow{\ Y{=}Z\ } R_2C\!\diagup^{Y}_{\diagdown Z}$$

$$\longrightarrow \text{ isomeric alkene or ring compound}$$

$$\longrightarrow R_2C{=}CR_2$$

$$\xrightarrow{\ R_2{=}N_2\ } R_2C{=}N{-}N{=}CR_2$$

$$\xrightarrow{\ :A{-}B\ } R_2\bar{C}{-}A^{\pm}{-}B \longrightarrow R_2C\!\diagup^{A:}_{\diagdown B}$$

$$\xrightarrow{\ O_2\ } R_2C{=}O$$

$$\xrightarrow{\ R'H\ } R_2CH\cdot + R'\cdot \longrightarrow R_2CH{-}CHR_2,$$
$$R'{-}R', \quad R_2CH{-}R'$$

(from triplet carbene only)

The evidence for the roles of triplet carbenes or diazo compounds and the reasons for them have been critically reviewed by Dürr.[38] Ketenes, $R_2C{=}CO$, are isoelectronic with diazo compounds and also give rise to carbene reactions upon photolysis or (less frequently) thermolysis; reviews of such chemistry can help in understanding this type of reaction of diazo compounds.

The foregoing picture is deceptively simple; in actual fact, the chemistry of diazo alkanes is full of enigmas.[39] The ways in which structure and experimental conditions influence selection among the competing reactions is poorly understood. Decomposition in the gas phase generally gives considerably different results from decomposition in solution. Some reactions appear to be catalyzed by surfaces. Some types of products presumed to be derived from carbenes can also arise directly from the diazo compound. Protic solvents sometimes give different results from aprotic ones, and some reactions are selectively promoted by heavy metals or their salts. The literature abounds with brave attempts to unravel the mysteries, but all too often they succeed only in adding new mysteries.

Intramolecular reactions are mostly of two types: insertion into an adjacent C—H bond, resulting in an alkene, or insertion in a vicinal C—H bond, resulting in a cyclopropane (Eq. 5-1). With simple aliphatic compounds, formation of cyclopro-

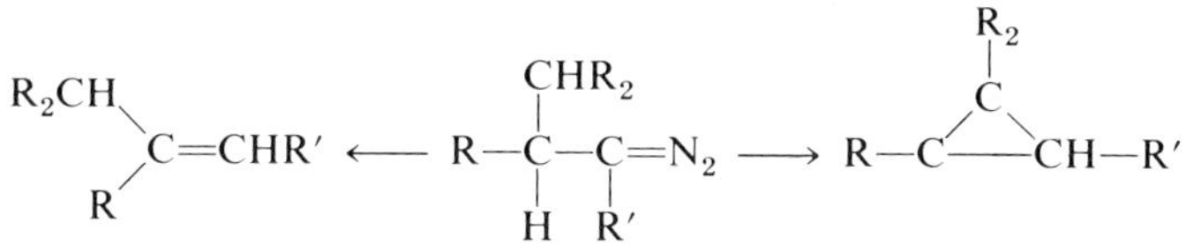

$$
\begin{array}{ccc}
\underset{R}{\overset{R_2CH}{\diagdown}}\!\!C\!\!=\!\!CHR' & \longleftarrow & \underset{\overset{|}{H}\;\overset{|}{R'}}{R\!-\!\overset{|}{C}\!-\!C\!=\!N_2} & \longrightarrow & R\!-\!C\!\!\overset{\overset{\displaystyle R_2}{|}\;\;C}{\diagup\;\diagdown}\!\!CH\!-\!R'
\end{array}
\qquad (5\text{-}1)
$$

panes occurs to a substantial but not major extent; with β-aryl diazo alkanes, formation of alkenes is more strongly favored. Isopropyldiazomethane, for example, gives isobutene and methylcyclopropane in 2:1 ratio.[40] The absence of a β-hydrogen promotes formation of cyclopropanes, for insertion into a C—C bond (i.e., migration of an alkyl group) is more difficult than insertion into a C—H bond. Thus *tert*-butyldiazomethane gives 13 times[40] as much 1,1-dimethylcyclopropane as 2-methylbutene-2 (Eq. 5-2). On the other hand, (1-hydroxycyclobutyl)diazoacetic

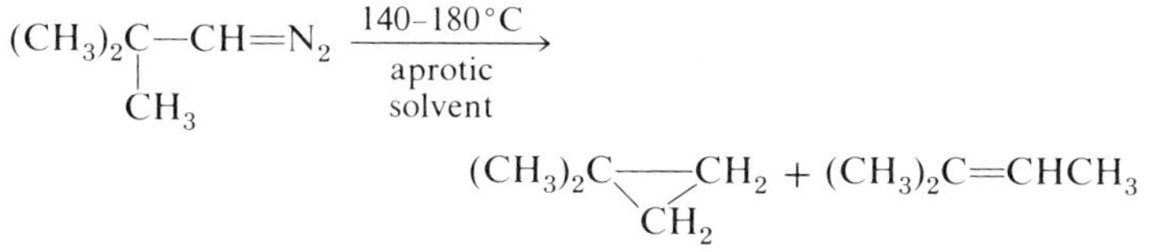

$$
\underset{\overset{|}{CH_3}}{(CH_3)_2C\!-\!CH\!=\!N_2} \xrightarrow[\substack{\text{aprotic}\\\text{solvent}}]{140\text{–}180\,°C}
$$

$$
\underset{\overset{\diagdown}{CH_2}}{(CH_3)_2C\!\!-\!\!\overset{\diagup}{CH_2}} + (CH_3)_2C\!\!=\!\!CHCH_3 \qquad (5\text{-}2)
$$

ester undergoes insertion into a C—C bond exclusively, to form the enol of cyclopentanone-2-carboxylic ester, a course presumably promoted by relief of ring strain[41] (Eq. 5-3). This is not necessarily a carbene reaction, however, for the hydroxy group is a source of potential acid catalysis.

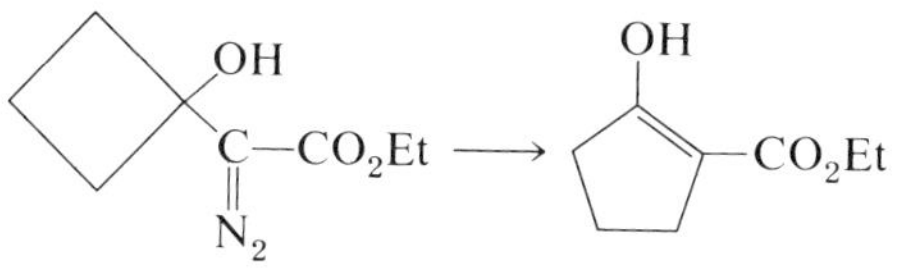

$$
\qquad (5\text{-}3)
$$

The Wolff rearrangement of α-diazo ketones must be considered here, for it is formally a special case of insertion of a carbene into an adjacent C—C bond, resulting in a ketene (Eq. 5-4).

$$
\underset{\overset{\|}{O}}{R\!-\!C\!-\!CH\!=\!N_2} \longrightarrow O\!\!=\!\!CH\!\!=\!\!CHR \qquad (5\text{-}4)
$$

The mechanism of the Wolff rearrangement has been reviewed thoroughly elsewhere.[42-44] The group that migrates from the carbonyl group to the diazomethyl carbon does so with retention of configuration, and there is other evidence that also shows unambiguously that the process is strictly intramolecular.[45] It is less clear whether loss of nitrogen to form a carbene completely precedes the rearrangement step or whether migration of R is concerted with loss of nitrogen. The picture is complicated by the fact that although some α-diazo ketones undergo thermal Wolff rearrangement easily and in good yield, most require catalysis by silver compounds for best results. Copper powder, in contrast, suppresses rearrangement largely or entirely, in favor of intramolecular insertion products.[46] The expected carbene has been intercepted in some cases,[47,48] such as naphthalene-1,2-diazooxide, which rearranges at about 180° in the presence of aniline, but in the presence of benzyl alcohol and an amine at the same temperature, reduction to 2-naphthol takes place[47] (Eq. 5-5). However, the possibility remains that the latter reaction involves

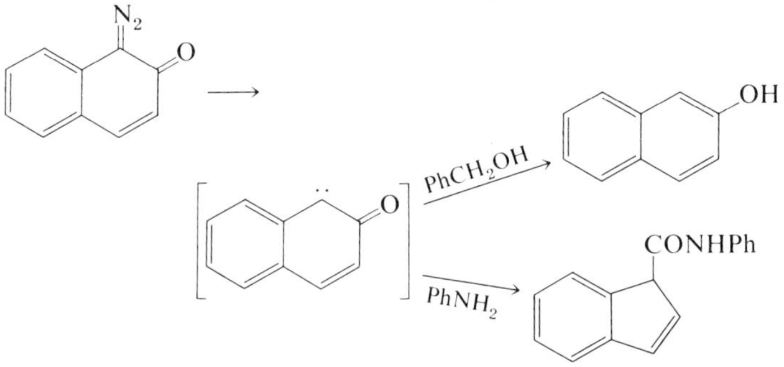

$$ (5\text{-}5) $$

proton catalysis. Possible interception of a carbene from an α-diazo ketone by reaction with oxygen has been reported.[49]

Diazoacetone rearranges to methylketene upon photolysis, and attempts to intercept a carbene have failed.[50] With sensitized photolysis (benzophenone), rearrangement does not occur, and the products are consistent with formation of a triplet carbene. There are many other examples of successful photolytic Wolff rearrangements, such as that of phenyl diazobenzyl ketone, which gives high yields of diphenylketene when photolyzed at room temperature.[51a] At 77 K, however, irradiation produces very little rearrangement; instead, an intermediate is formed that reacts with oxygen to give benzil and with hydrocarbons to give carbene insertion products (Eq. 5-6). The total evidence is fairly conclusive that triplet carbenes do

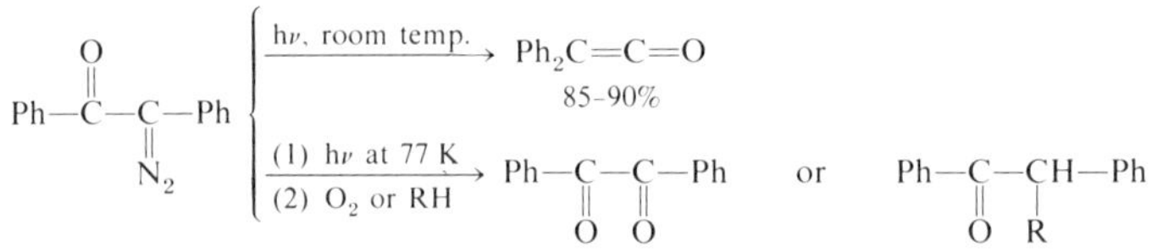

$$ (5\text{-}6) $$

not undergo Wolff rearrangement, but singlets may do so in competition with other carbene reactions. Furthermore, the role of conformation, as shown in structure **V**, has been demonstrated; studies of product composition as influenced by triplet sensitizers and triplet quenchers imply that the main photochemical Wolff rearrangement takes place as a concerted process from the singlet excited state of the s-Z conformation, whereas the s-E conformation fragments to singlet carbene.[51b]

Isotopic labeling has revealed the possibility of two routes for the rearrangement of phenyl diazobenzyl ketone, one of which involves an intermediate in which the two nonring carbons are equivalent.[44,52] The only reasonable candidate is an oxirene **(IX)**. This may be regarded as a product of addition of the α-carbene to the

$$Ph-C\overset{O}{\diagdown\diagup}C-Ph$$
$$\textbf{IX}$$

carbonyl group. This observation raises questions about all the other examples of Wolff rearrangement, the mechanism of which has been assumed not to involve the oxygen atom. If two routes for rearrangement are indeed competitive, it remains to be determined what factors favor one over the other. The conjugation present in IX may make it a particularly favorable case, and α-diazo ketones with only one aryl substituent or none might be expected to be less likely to form an oxirene intermediate.

The Wolff rearrangement is of considerable practical importance as the key step in the Arndt-Eistert homologation of carboxylic acids[53,54] (Eq. 5-7). An acid chlo-

$$RCO_2H \longrightarrow RCOCl \xrightarrow{CH_2N_2, R_3N} RCOCH=N_2 \xrightarrow[EtOH]{Ag_2O}$$

$$RCH=C=O \longrightarrow RCH_2CO_2Et \xrightarrow{[H_2O]} RCH_2CO_2H \qquad (5\text{-}7)$$

ride is made to react with an excess of diazomethane so as to form an α-diazo ketone, which is rearranged by heating, usually with a silver-containing catalyst, in the presence of a substance such as ammonia or alcohol that will combine with the ketene first formed. Experiments with isotopic labeling have shown that the carbon atom of the carboxyl group of the product is the same one that was in the carboxyl group of the starting acid.[55] When a substituted diazomethane is used, an α-substituted acid derivative is produced.[56] The widest application of this synthesis is with diazomethane. Although interesting α-substituted acids can be made, practical difficulties somewhat limit the use of the synthesis with higher diazo alkanes.

Just as the Wolff rearrangement is an analog of the Curtius rearrangement of acyl azides, there is a diazo analog of the Stieglitz rearrangement of triarylmethyl azides. Triphenylmethyldiazomethane losses nitrogen at 70 to 80°C to form triphenylethylene[57] (Eq. 5-8).

$$Ph_3C-CH=N_2 \xrightarrow{70\text{-}80\,°C} Ph_2C=CHPh \qquad (5\text{-}8)$$

Formation of the alkene corresponding to dimerization of a carbene is observed mostly with diazo compounds that cannot readily undergo isomerization at the carbene stage. The commonest examples are the diaryldiazomethanes. Even. so, dimerization must compete with intermolecular reactions, especially reaction with unchanged diazo compound to form an azine[39,58-61] (Eq. 5-9). Whereas diphenyl-

$$Ar_2C=N_2 \xrightarrow{\Delta} Ar_2C=CAr_2, \qquad Ar_2C=N-N=CAr_2 \qquad (5\text{-}9)$$

diazomethane normally forms benzophenone azine in 80 to 90% yield upon thermolysis or photolysis, decomposition in the presence of copper powder causes tetraphenylethylene to become the major product (84–97%).[62]

Formation of azines from diazoalkanes deserves special attention, for it is a competing reaction in many processes, and its mechanism has been a subject of active discussion.[39] Thermal decomposition of diazoalkanes nearly always produces at least traces of azines; the amount may be quite substantial with aryldiazomethanes,[60] but it is often negligible with aliphatic compounds, which give primarily hydrocarbons. Examples of extreme cases are 1-phenyl-1-diazoethane, which when heated or photolyzed in benzene or cyclohexane gives entirely acetophenone azine, and diazocyclohexane, which goes entirely to cyclohexene. By contrast, 1-methyl-1-phenyldiazirine, an isomer of 1-phenyldiazoethane, goes to styrene instead.[61] This fact has been interpreted as evidence that the azine is not formed by combination of a carbene with diazo compound, but results from dimerization before nitrogen is lost. This is not a necessary implication, however, and it is inconsistent with the fact that the kinetics of evolution of nitrogen from diphenyldiazomethane are first-order, consistent with the carbene route.[59] It may, of course, be possible that azines can be formed by either route, depending on structure and conditions. The dimerization route can be formulated as electrophilic attack by the terminal nitrogen of one molecule on the α-carbon of another[58] (Eq. 5-10). Such a

$$ArCOC{=}N_2 + CH_3CH{=}N_2 \longrightarrow ArCOC{-}N{=}N{-}CH{-}N_2^+ \longrightarrow$$
$$\underset{R}{\mid} \qquad\qquad\qquad \underset{R}{\mid}$$

$$ArCOC{=}N{-}N{=}CHCH_3 \qquad (5\text{-}10)$$
$$\underset{R}{\mid}$$

mechanism is supported by the fact that dicyanodiazomethane, an exceptionally electrophilic example, reacts with other diazo compounds even at room temperature, where they are normally stable, producing unsymmetrical azines[63] (Eq. 5-11).

$$(NC)_2C{=}N_2 + Ph_2C{=}N_2 \longrightarrow (NC)_2C{=}N{-}N{=}CPh_2 \qquad (5\text{-}11)$$

A competing factor in thermolysis of diazo compounds is the fact that azines themselves decompose if the temperature is high enough; furthermore, azine decomposition is catalyzed by diazoalkane (see Chap. 2). As a result, conversion of a diazo compound to azine is likely to be higher when the decomposition temperature is minimal. In this connection, it is interesting to consider the reaction of a diazo compound with a carbene generated in another manner. When dichloromethylene is generated from chloroform and base in the presence of diaryldiazomethanes, 1,1-diaryl-2,2-dichloroethylenes are produced in good yields[64] (Eq. 5-12).

$$Ar_2C{=}N_2 + {:}CCl_2 \longrightarrow Ar_2C{=}CCl_2 \qquad (5\text{-}12)$$

The azines are stable under the reaction conditions and thus cannot be intermediates. The mixed azine $Ar_2C{=}N{-}N{=}CCl_2$ is formed in a competing reaction, and its proportion increases with steric hindrance from the aryl groups. Diphenyldiazomethane gives no azine, but bis-(2-naphthyl)diazomethane gives 14%, and dibromomethylene gives more azine than does dichloromethylene. These facts can be reconciled if one assumes that a diazo compound is an ambident reaction partner toward carbenes. If there is little steric hindrance, attack takes place at carbon; increasing steric hindrance directs the attack increasingly to nitrogen (Eq. 5-13). A

$$R_2C=\overset{+}{N}=N^- + R_2'C: \begin{array}{c} \overset{C\ attack}{\nearrow} R_2C=CR_2' + N_2 \\ \\ \underset{N\ attack}{\searrow} R_2C=N-N=CR_2' \end{array}$$

(5-13)

bulky carbene, such as diphenylmethylene, would be more sensitive to this effect, and the overwhelming formation of benzophenone azine from attack of diphenyl-methylene on diphenyldiazomethane can thus be rationalized.

In solvents that are good donors of hydrogen, fragmentation of a diazo compound may produce a dimeric alkane arising from abstraction of a hydrogen atom from the solvent, presumably by a triplet carbene.[65,66] It is not a common reaction, but it may compete with azine formation (Eq. 5-14).

$$Ar_2C=N_2 \begin{array}{c} \overset{benzene}{\nearrow} Ar_2C=N-N=CAr_2 \\ \\ \underset{toluene}{\searrow} Ar_2CH \cdot \longrightarrow Ar_2CH-CHAr_2 \end{array}$$

(5-14)

Carbenes from thermolysis or photolysis of diazo compounds may insert in all sorts of single bonds, but the most commonly observed by far is insertion in C—H bonds. This reaction shows very little discrimination, and with compounds having several different C—H sites for insertion, a statistical distribution is usually found.[67] Pentane, for example, reacts with methylene produced photolytically from diazo-methane to form hexane, 2-methylpentane, and 3-methylpentane.[68] The earliest recognized case of insertion into a saturated C—H bond was that of diethyl ether; photolysis of ethereal diazomethane produces ethyl propyl ether and ethyl isopropyl ether[69] (Eq. 5-15). It was subsequently demonstrated that the former isomer re-

$$CH_3CH_2OCH_2CH_3 + CH_2=N_2 \overset{h\nu}{\longrightarrow}$$
$$CH_3CH_2OCH_2CH_2CH_3, \qquad CH_3CH_2OCH(CH_3)_2 \qquad (5\text{-}15)$$

sulted from insertion into a β-C—H bond rather than a C—O bond.[70] Intermolecular insertions of this kind are best observed with those diazo compounds in which internal insertions cannot compete, such as diazomethane and its halogen derivatives, diazoacetic ester, etc.

Insertion into C—O bonds is rarely observed, and in those instances where it occurs, it is uncertain whether direct insertion was involved or whether the observed product arose from rearrangement of an intermediate ylide. Although tetrahydrofuran does not give rise to tetrahydropyran when diazomethane is photolyzed in it in the liquid phase, 11% of that product, formally derived from insertion in a C—O bond, has been obtained in the gas phase.[71] Alcohols react with diazomethane to give their methyl ethers overwhelmingly, but up to 10% of the homologous alcohol may be formed in a competing reaction involving the C—O bond[72] (Eq. 5-16).

$$ROH + CH_2=N_2 \overset{h\nu}{\longrightarrow} ROCH_3 \text{ (mostly)}, \qquad RCH_2OH \qquad (5\text{-}16)$$

Diazo compounds do not readily give rise to carbene insertions into C—N bonds, and in most instances of decomposition in the presence of amines, only products attributable to C—H insertion have been reported. *N*-Methylpyrrolidine, for example, reacts photolytically with diazomethane to produce *N*-ethylpyrrolidine and 1,2(or 3)-dimethylpyrrolidine, but no detectable *N*-methylpiperidine.[73] However, thermolysis of diazoacetic ester in the presence of benzyldimethylamine produced "some" ethyl 2-dimethylamino-3-phenylpropionate, formally a product of insertion into the benzyl—N bond (Eq. 5-17). With triethylamine, in contrast, the same treat-

$$PhCH_2NMe_2 + N_2{=}CHCO_2Et \xrightarrow{\Delta}$$

$$\underset{\underset{CO_2Et}{|}}{PhCH_2{-}CH{-}NMe_2} + \underset{\underset{CH_2CO_2Et}{|}}{PhCH{-}NMe_2} \qquad (5\text{-}17)$$

ment produced only products of C—H insertion and some ethyl diethylamino-acetate, which probably arose from attack of the carboethoxymethylene on the nitrogen atom to form an ylide, from which ethylene was then eliminated.

The C—C bond is evidently inert to attack by carbenes; I know of no unequivocal example, although products that might have arisen by such a reaction have been reported in rare instances. The weaker S—S bond, however, gives insertion products readily, although here one cannot easily rule out a pathway through an initial ylide. Diphenyldiazomethane in refluxing benzene has been reported to form the diphenylmercaptal of benzophenone in 50% yield[74] (Eq. 5-18).

$$Ph_2C{=}N_2 + PhS{-}SPh \xrightarrow{80\,°C} Ph_2C(SPh)_2 \qquad (5\text{-}18)$$

Photolysis or thermolysis of diazo compounds in the presence of compounds having carbon-halogen bonds may lead to some insertion in the C—X bond, but when insertion in C—H bonds can compete, it usually dominates. Carbon tetrachloride, in which there is no such distraction, is converted by diazomethane all the way to pentaerythrityl tetrachloride, formally a result of insertion of methylene in each C—Cl bond. The reaction is initiated by light, but the quantum yield is so high as to imply a free-radical chain reaction[75] (Eq. 5-19). It should be noted that the

$$CH_2{=}N_2 \xrightarrow{h\nu} CH_2 \xrightarrow{CCl_4} {\cdot}CH_2Cl + {\cdot}CCl_3$$
$${\cdot}CCl_3 + CH_2{=}N_2 \longrightarrow {\cdot}CH_2CCl_3 + N_2 \longrightarrow$$

$$ClCH_2CCl_2 \xrightarrow{CH_2N_2} \text{etc.} \qquad (5\text{-}19)$$

product, a primary alkyl chloride, is apparently immune to attack. However, *tert*-butyl chloride gives rise to substantial amounts of neopentyl chloride, and isopropyl chloride gives some isobutyl chloride, in addition to the products of C—H insertion, *sec*-butyl chloride (mostly) and *tert*-butyl chloride (trace); isopropyl bromide gives similar results.[76] The C—I bond may be more susceptible to insertion, for photolysis of diazomethane in the presence of trifluoroiodomethane produces a compound resulting from double insertion as well as that from a single one[77] (Eq. 5-20). An

$$F_3CI + CH_2{=}N_2 \xrightarrow{h\nu} F_3CCH_2I + F_3CCH_2CH_2I \qquad (5\text{-}20)$$

example of insertion in an aryl-halogen bond has been reported in the case of chlorobenzene, which when photolyzed with diazomethane produces some benzyl chloride, in addition to the main products, chlorotropilidenes and chlorotoluenes.[78]

There are numerous examples of reactions of diazo compounds with inorganic halides to give products of formal insertion reactions, but they do not in general appear to result from initial fragmentation to a carbene, but instead are formed by electrophilic attack on the diazoalkane (see the following section).

Photolysis or thermolysis of diazo compounds in the presence of alkenes gives rise to cyclopropanes formally derived from attack of a carbene on the double bond. For simple alkenes, such a process is probably the actual mechanism, but with more reactive double bonds, an alternative path is cycloaddition to form a pyrazoline, which subsequently loses nitrogen. Photolysis of diazomethane in the presence of *cis*-butene-2 gives dimethylcyclopropane and isomeric olefins. In reasonably concentrated solution, the addition is highly stereospecific, and the dimethylcyclopropane is almost entirely *cis*[79] (Eq. 5-21). When the reaction is carried out highly

$$CH_2{=}N_2 + \underset{CH_3}{\overset{H}{>}}C{=}C\underset{CH_3}{\overset{H}{<}} \longrightarrow$$

$$\underset{\overset{|}{C.{-}H}}{\overset{C{\overset{H}{\nearrow}}}{CH_2}} \underset{CH_3}{CH_3} + C_5H_{10} \text{ isomers} \qquad (5\text{-}21)$$

diluted with perfluoropropane, however, part of the stereospecificity is lost, presumably as a result of increased decay of singlet methylene to triplet.[80] Diazoacetic ester has been used extensively to prepare cyclopropanes.[32]

Benzene and its derivatives give carbene addition products when photolyzed with diazo compounds; the yields may be preparatively useful with those diazo compounds which do not form azines readily. The structure of the products was a source of much confusion in the early years of diazo chemistry, for the situation could not be dealt with adequately without modern spectroscopy. The products are tropilidenes, a rapidly equilibrating mixture of norcaradiene and cycloheptatriene valence tautomers. Benzene itself reacts with diazomethane to give tropilidene and toluene[81] (Eq. 5-22); the possibility that the toluene arose from isomerization of some tropilidene was disproved by labeling experiments.[82] Diazoacetic ester has been studied fairly extensively, since its reaction with benzene produced one of the earliest enigmas of diazo-compound chemistry.[32] It was eventually established by means of nuclear magnetic resonance that a mixture of isomeric cycloheptatrienecarboxylic esters is produced.[83-85] With substituted benzenes, mixtures arise corresponding to attack at different positions (i.e., edges) of the ring; there is weak

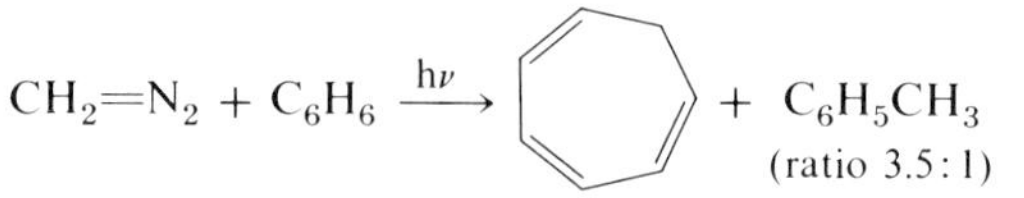

$$CH_2=N_2 + C_6H_6 \xrightarrow{h\nu} \text{(cycloheptatriene)} + C_6H_5CH_3 \qquad (5\text{-}22)$$
$$\text{(ratio 3.5:1)}$$

selectivity, corresponding to electrophilic attack. For diazoacetic ester, the Hammett ρ value is -0.38,[84] but diphenyltriazolyldiazomethane, which thermolyzes at lower temperatures, is more selective ($\rho = 0.8\text{-}1.0$).[86] It attacks toluene twice as fast as it does benzene, and the products correspond to preferred attack at the 3,4 edge (50%) compared with the 2,3 edge (27.3%) and the 1,2 edge (22.7%) (Eq. 5-23).

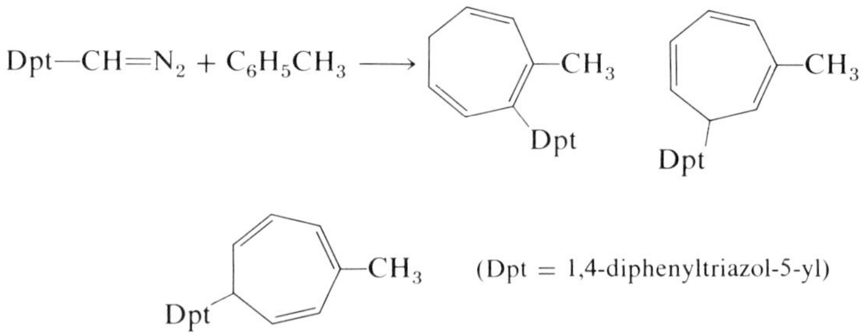

$$Dpt-CH=N_2 + C_6H_5CH_3 \longrightarrow \text{(products)}$$

(Dpt = 1,4-diphenyltriazol-5-yl)   (5-23)

The general reaction path seems to be addition of a singlet carbene to an edge of a benzene ring to produce a norcaradiene structure, which usually has a low energy barrier to ring-opening to a cycloheptatriene, which in turn may undergo rearrangement of the double-bond positions by hydrogen migration if the conditions are sufficiently forceful[86] (Eq. 5-24). Polynuclear and heterocyclic ring systems can be attacked by decomposing diazo compounds in a similar manner.

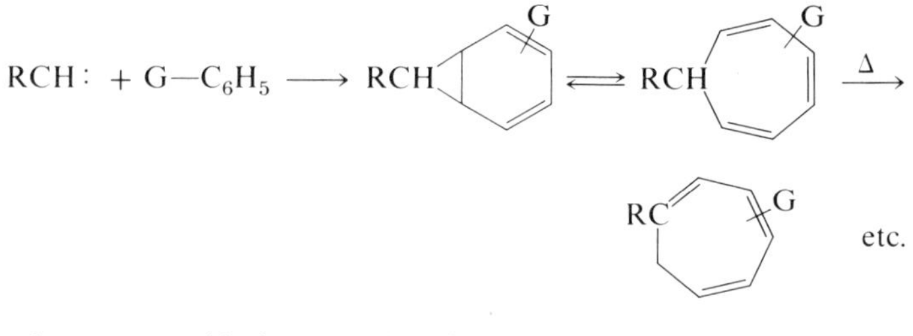

$$RCH: + G-C_6H_5 \longrightarrow RCH \rightleftharpoons RCH \xrightarrow{\Delta}$$

$$RC \qquad \text{etc.} \qquad (5\text{-}24)$$

Acetylenes react with decomposing diazo compounds to form cyclopropenes[87-90] (Eq. 5-25). These are unambiguously the result of carbene reactions rather than

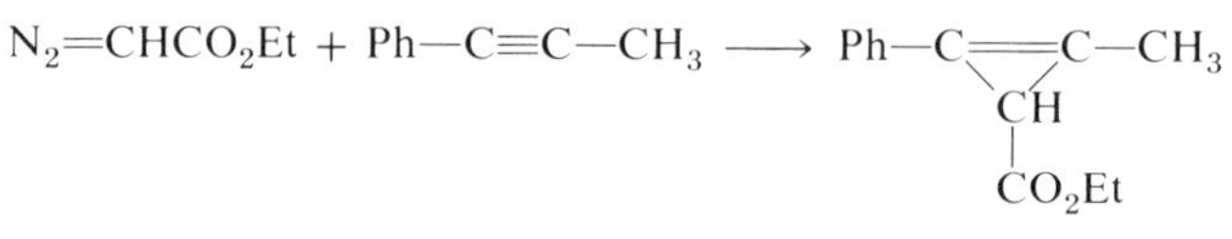

$$N_2=CHCO_2Et + Ph-C\equiv C-CH_3 \longrightarrow Ph-C=\!\!=\!\!C-CH_3 \qquad (5\text{-}25)$$
$$\underset{\displaystyle CO_2Et}{\overset{\displaystyle CH}{|}}$$

cycloaddition of diazo compound, for the pyrazoles that would be formed in the latter process are relatively stable substances.

Formation of ylides by reaction at lone-pair sites has been alluded to in connection with insertion reactions. Such ylides have actually been isolated in situations where the negative charge on carbon is stabilized by substituents, such as a carboethoxy group. Isoquinoline, for example, reacts with thermally decomposing diazoacetic ester to form a quinolinium ylide, and a number of examples of sulfonium ylides have been observed to result from photolysis of diazoalkanes in the presence of dialkyl sulfides[91,92] (Eq. 5-26). Sulfides bearing allylic groups apparently

$$R_2C{=}N_2 + (CH_3)_2S \xrightarrow{h\nu} R_2\overset{-}{C}{-}\overset{+}{S}(CH_3)_2 \tag{5-26}$$

form ylides that rearrange too readily to permit isolation; the allylic group migrates to carbon with allylic inversion[93] (Eq. 5-27). Ylides are not formed when photolysis

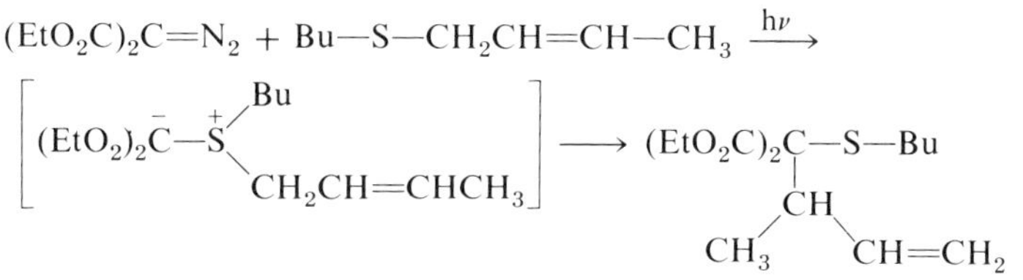

$$(EtO_2C)_2C{=}N_2 + Bu{-}S{-}CH_2CH{=}CH{-}CH_3 \xrightarrow{h\nu}$$

$$\left[(EtO_2)_2\overset{-}{C}{-}\overset{+}{S} \overset{\displaystyle Bu}{\underset{\displaystyle CH_2CH=CHCH_3}{}} \right] \longrightarrow (EtO_2C)_2C{-}S{-}Bu \tag{5-27}$$

is carried out with a triplet sensitizer, such as benzophenone, and it is evident that ylides are formed from singlet carbenes. Formation of ylides at oxygen, and even at halogen, has been invoked to explain products in many instances, but ylides involving these elements have not been isolated. The most convincing evidence for their intermediacy seems to be the occurrence of allylic inversion in the reaction of diazomalonic ester with crotyl chloride[94] (Eq. 5-28); addition to the C—C bond to form a cyclopropane is an important competing reaction.

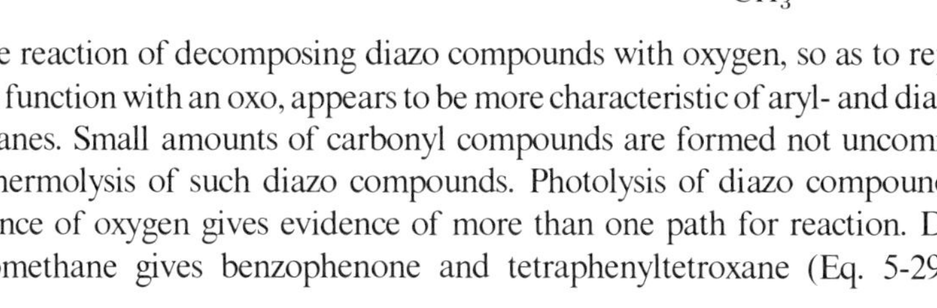

$$(EtO_2C)_2C{=}N_2 + CH_3CH{=}CH{-}CH_2Cl \xrightarrow{h\nu}$$

$$[(EtO_2C)_2\overset{-}{C}{-}\overset{+}{Cl}{-}CH_2CH{=}CHCH_3] \longrightarrow$$

$$(EtO_2C)_2C \overset{\displaystyle Cl}{\underset{\displaystyle \underset{CH_3}{\overset{|}{CHCH=CH_2}}}{}} \tag{5-28}$$

The reaction of decomposing diazo compounds with oxygen, so as to replace the diazo function with an oxo, appears to be more characteristic of aryl- and diaryldiazomethanes. Small amounts of carbonyl compounds are formed not uncommonly in the thermolysis of such diazo compounds. Photolysis of diazo compounds in the presence of oxygen gives evidence of more than one path for reaction. Diphenyldiazomethane gives benzophenone and tetraphenyltetroxane (Eq. 5-29), which

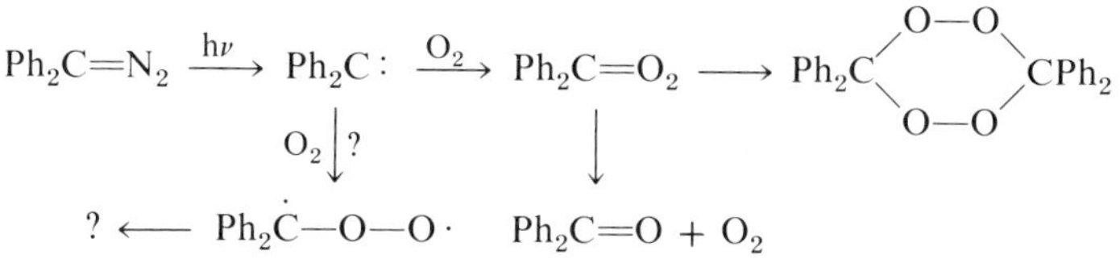

$$Ph_2C{=}N_2 \xrightarrow{h\nu} Ph_2C: \xrightarrow{O_2} Ph_2C{=}O_2 \longrightarrow Ph_2C \overset{\displaystyle O{-}O}{\underset{\displaystyle O{-}O}{}} CPh_2$$

$$O_2 \Big\downarrow ? \qquad\qquad \Big\downarrow \tag{5-29}$$

$$? \longleftarrow Ph_2\dot{C}{-}O{-}O\cdot \qquad Ph_2C{=}O + O_2$$

suggests a peroxo intermediate derived from diphenylmethylene, but the reaction is not clean, and the possibility of a diradical intermediate is real.[95-97] When a singlet sensitizer is used, to promote generation of singlet oxygen, direct attack on the diazoalkane appears to take place, perhaps through cycloaddition, and if an aldehyde is present, a 1,2,4-trioxolane (an "ozonide") as well as the simple carbonyl compound is formed.[98] There is evidence that in unsensitized photolysis, the species that reacts with oxygen may be a photoexcited state of the diazo compound rather than a carbene.[99] However, there is also evidence that oxygen reacts sufficiently more rapidly with triplet carbene than with singlet for it to be used as a triplet carbene trap.[100] Further investigation of this confusing area is obviously desirable.

Metal-catalyzed reactions. The reactions discussed thus far have been purely thermal or photolytic, without the addition of a catalyst. There exists a more or less parallel set of reactions promoted by various transition metal compounds, especially those of copper. In general, not only is the role of the catalyst uncertain, but the actual identity of it may not be known (e.g., whether it is Cu^0, Cu^I, or Cu^{II}). The effect of the catalyst on the rate of decomposition of the diazo compound is seldom explicitly stated, but the examples in which it is mentioned lead one to generalize that the rate is increased (or the temperature for incipient evolution of nitrogen is lowered). It is generally agreed that these reactions do not involve true carbenes, but something closely related to them, perhaps metal-coordinated carbenes, conveniently referred to as *carbenoids*. They may be formed from metal derivatives of the diazo compounds, and some instances are known where metallated diazo compounds, such as the silver[101] and mercury[102] derivatives of diazoacetic ester, have been isolated.

The general effect of metal catalysts is to change the selectivity among competing paths, sometimes to the complete suppression of some of them, by reducing the reactivity of the carbene. Insertions into C—H bonds, for example, are disfavored in the presence of catalysts, which are therefore commonly used when one wishes to prepare a cyclopropane by addition to an alkene. Furthermore, catalyzed additions are highly stereospecific. Many examples of the use of catalytic cyclopropanation for preparative purposes, using diazoacetic ester, diazomethane, etc., may be found in reviews.[32,35,103] The effect may be quite dramatic, as, for example, in the case of allyl diazoacetate, which when heated in the presence of a Cu^{II} catalyst undergoes internal cyclopropanation in good yield, but when photolyzed in the absence of a catalyst forms no cyclopropane at all.[104] Copper has also been used to promote the ring-expansion of aromatic rings with diazomethane.[105] Rhodium complexes promote formation of products of carbene insertion into C—S, C—N, and C—X bonds through intermediate ylides, in preference to cyclopropanation.[106a]

The Wolff rearrangement of α-diazo ketones is suppressed in favor of carbenoid cyclopropanation when copper catalysts are used—an interesting contrast to the effect of silver catalysts, which are commonly used to promote rearrangement.[34]

Attempts to rationalize the role of metal catalysts have generally assumed some sort of coordination of the corresponding carbene to the metal, with or without other ligands, including the olefin to which the carbenoid may ultimately be found added. A determinedly systematic assault on the question[106b,107] has revealed that both the nature of the nonreacting ligands and the ratio of catalyst to diazo compound are significant factors. In the reaction of diazoacetic ester with cyclohexene, high concentrations of catalyst were found to promote formation of the carbene dimer over addition, whereas low concentrations favored formation of norcarane (Eq. 5-30); nickelocene favored maleate over fumarate; whereas nickel tetra-

$$N_2\text{=}CHCO_2Et\ +\ \text{[cyclohexene]} \nearrow EtO_2C\text{—}CH\text{=}CH\text{—}CO_2Et\ \ (cis\ and/or\ trans)$$
$$\searrow \text{[bicyclic]} CHCO_2Et$$

(5-30)

carbonyl favored fumarate. Information seems to be accumulating faster than understanding, but surely the situation will eventually be reversed. A lengthy, comprehensive, critical, and outspoken review of this labyrinthine topic has recently appeared,[108] and more need not be said here.

Proton acids.   Diazo compounds are sensitive to acids, much more so than are the isoelectronic azides, and acids as weak as acetic will react rapidly with diazomethane. In dilute water solution, the principal product is the alcohol corresponding to the diazo compound, although anions in the solution may also be taken up. The initial step is protonation on carbon; there is no convincing evidence for protonation on nitrogen.[109] The intermediate species is thus an aliphatic diazonium ion, RN_2^+, the same as that obtained from the corresponding amine and nitrous acid, and the products are in general the same for the two reactions (Eq. 5-31). The

$$RCH\text{=}N_2 \overset{H^+}{\rightleftharpoons} RCH_2\text{—}N_2^+ \longleftarrow RCH_2NH_2 + HNO_2$$
$$\overset{H_2O}{\swarrow}\qquad \downarrow \qquad \searrow^{A^-}$$
$$RCH_2OH \qquad RCH_2^+ \overset{A^-}{\longrightarrow} RCH_2A$$

(5-31)

diazonium ion may undergo nucleophilic displacement of nitrogen, if it does not first lose nitrogen to form a carbenium ion, according to structure and conditions.

Much of the evidence for intermediate diazonium ions comes from deuterium isotope exchange[110] and kinetic experiments.[14,111] When diazomethane is added to aqueous acid, there is an immediate and dramatic rise in pH, corresponding to protonation at carbon, followed by a drop in pH as the diazonium ion reacts with water to regenerate hydronium ion. In special cases, the reaction has actually been arrested at the diazonium stage, as with hexafluoro-2-diazopropane, which at $-70°C$ reacts with fluorosulfuric acid without evolution of nitrogen to form a solution of $(CF_3)_2CHN_2^+$. When warmed to $-5°C$, decomposition sets in with a rate that depends on the concentration of both the diazonium ion and FSO_3^-.[112] It has also been found that the rate of reaction of diazoacetone and diazoacetic ester with inorganic acids can be correlated with the Swain-Scott nucleophilicity of the anion (for Cl^-, Br^-, I^-).[113-115] At one time it was popular to invoke the concept of "hot" carbenium ions, presumably formed with excess energy by fragmentation of the diazonium ion, but the basis for that hypothesis has been effectively demolished.[116]

In dilute aqueous solution, where the acid is not used up because most of the reaction goes to alcohol, the rate of nitrogen evolution is proportional to the pH, and the reaction has actually been considered as an accurate means of measuring acidities. This situation is compatible with a rapid, reversible protonation of the diazo compound, followed by displacement of nitrogen at a slower rate, because the second step must depend on the concentration of the diazonium ion, which in turn depends on the concentration of the acid taking part in the prior equilibrium, Evidence for this is found in the fact that the rate of reaction of diazoacetic ester is faster with $EtOD_2^+$ than with $EtOH_2^+$, and the product, $EtOCH_2CO_2Et$, contains deuterium in the α position.[117] However, it has been found that the rate of decomposition of diphenyldiazomethane is slower with deuterated acids ($k_H/k_D = 3.5$ for reactions in ethanol). In such a case, the rate of reaction is controlled by the rate of protonation of the diazo compound, for the diazonium ion would be expected to fragment with ease to the unusually stable diphenylmethyl cation. These two situations represent extremes of behavior, according to the type of substitution; intermediate cases are, of course, to be expected.

Carboxylic acids react rapidly with diazomethane in ether solution, evolving nitrogen briskly as they are converted in a few minutes at room temperature to their methyl esters (Eq. 5-32). Purification of the product is especially easy, because the

$$CH_2{=}N_2 + RCO_2H \longrightarrow RCO_2CH_3 + N_2 \qquad (5\text{-}32)$$

only other substance produced is nitrogen, and conversion is essentially quantitative. Other diazoalkanes react more slowly and may require warming. In the absence of water and alcohols, there is no competition for the diazomethane, but in alcoholic solution, substantial amounts of the corresponding methyl ether may be produced. The reaction of diphenyldiazomethane with carboxylic acids in ethanol solution has been found to convert 60 to 70% of the diazo compound to benzhydryl ester, an amount considerably more than one would expect on the basis of the relative nucleophilicities of ethanol and carboxylate ions.[118] The product ratio is little affected by the acid and is unchanged by addition of its salts.[119] These facts imply that in a solvent of lower polarity, such as ethanol, much of the reaction takes place between ion pairs initially formed by reaction of undissociated carboxylic acid with diazo compound; only that fraction of the diphenylmethyl cation which escapes from the ion pair is available to react with the alcohol to form ethyl benzyhydryl ether (Eq. 5-33). Furthermore, the position isomers 1-diazo-2-butene

$$Ph_2C{=}N_2 + RCO_2H \longrightarrow [Ph_2CH^+\ RCO_2^-] \longrightarrow Ph_2CH{-}OCOR$$
$$\phantom{Ph_2C{=}N_2 + RCO_2H \longrightarrow}\! \llcorner\!\!\!\longrightarrow Ph_2CH^+ \xrightarrow{\ EtOH\ } Ph_2CHOEt \qquad (5\text{-}33)$$

and 3-diazo-1-butene react site-specifically with 3,5-dinitrobenzoic acid in ether to form different esters (i.e., without allylic isomerization), whereas in aqueous perchloric acid, either diazo compound produces the same mixture of isomeric alcohols.[118] Stereochemical behavior also supports the concept of reaction at the ion-pair stage in solvents of lower polarity, as opposed to formation of dissociated cations in aqueous or other highly polar media. In solvents of low polarity, chiral alkanediazonium ions produced in other ways, such as reaction of amines with nitrous acid, give products in which the configuration is largely retained, but in highly polar media, racemization and/or inversion occur.[116]

Phenols react with diazo compounds considerably more slowly than do carboxylic acids, but they can nevertheless be converted to their methyl ethers by reaction with diazomethane in practical yields. Hydroperoxides can also be methylated by direct reaction.[120] Alcohols, however, react too slowly for usefulness unless a catalyst is used. Aluminum alkoxides,[121] fluoroboric acid,[122] boron fluoride, aluminum chloride, and even small amounts of toluenesulfonic acid[123] can be used as effective catalysts.[107,124,125] Since these catalyzed reactions take place through oxonium ions, it is not surprising that dialkyloxonium ions can also be alkylated readily. Dimethyloxonium hexachloroantimonate, for example, is converted by diazomethane to the trimethyloxonium salt[126] (Eq. 5-34).

$$Me_2OH^+ \ SbCl_6^- + CH_2{=}N_2 \xrightarrow{\text{liq. } SO_2} Me_3O^+ \ SbCl_6^- \qquad (5\text{-}34)$$

Enols react with diazo compounds almost entirely at oxygen, forming enol ethers[127] (Eq. 5-35). This could be due to the greater acidity of the enol or to a

$$\underset{\displaystyle CH_3\overset{\displaystyle O}{\overset{\|}{C}}CH_2CO_2Et}{} \rightleftharpoons \underset{\displaystyle CH_3\overset{\displaystyle OH}{\overset{|}{C}}{=}CHCO_2Et}{} \xrightarrow{CH_2N_2}$$

$$\underset{\displaystyle CH_3\overset{\displaystyle OCH_3}{\overset{|}{C}}{=}CHCO_2Et}{} \qquad (5\text{-}35)$$

general preference for cations to react at the oxygen atom of enolate ions. There must be some relationship between the position of the keto-enol equilibrium and the site of alkylation, however, for saccharin forms only N-methylsaccharin when alkylated in the solid state, where it is all in the keto form, but undergoes 10% of N-methylation in concentrated ether solution and 24% of O-methylation in dilute ether solution with slow addition of diazomethane.[128]

In another case, a thiolactone reacting with diazoethane, an exceptional side reaction was observed in addition to C- and O-alkylation: diazo coupling occurred at the enolate carbon to produce an N-ethylhydrazone. This would seem to be a reaction of the ion-pair stage before loss of nitrogen.[129]

Amines are ordinarily unreactive toward diazo compounds, but aniline has been induced to react with diazoacetic ester by heating to form ethyl anilinoacetate,[130] and some primary and secondary aliphatic amines have been alkylated by diazoacetic ester in the presence of cuprous cyanide. In the absence of this catalyst, the amines simply catalyze dimerization of the diazoacetic ester to a dihydrotetrazinedicarboxylic ester.[131,132] Acids will catalyze alkylation at nitrogen so long as the anion is of low nucleophilicity, such as ClO_4^- or BF_4^- (chloride salts produce principally alkyl chloride).[133-135] Glycine does not react with dry ethereal diazomethane, but when a trace of water is added, the reaction proceeds virtually quantitatively all the way to betaine[136] (Eq. 5-36).

$$H_3\overset{+}{N}{-}CH_2CO_2^- + CH_2{=}N_2 \longrightarrow (CH_3)_3\overset{+}{N}{-}CH_2CO_2^- \qquad (5\text{-}36)$$

Sulfonic acids are ordinarily converted to methyl esters by diazomethane, although under some circumstances, conversion to polymethylene may compete.[127,137]

Lewis acids.   There is abundant evidence that a large variety of metallic and nonmetallic Lewis acids interact with diazo compounds, but in nearly all cases, the evidence is in the form of kinetics or the nature of products subsequently formed. In two instances, triethylaluminum[138] and a nickel cyclododecatriene complex,[139] visible color changes upon mixing with a diazo compound have been reported at low temperature, unaccompanied by gas evolution; coordination, presumably at carbon, seems probable. The ability of many metal compounds to catalyze carbenoid reactions of diazo compounds has already been mentioned; such reactions may be visualized as commencing by attack of the metal compound as a Lewis acid at the diazo carbon. The adduct might then lose nitrogen to form a coordinated carbene or eliminate a proton from the diazo compound with a ligand to generate a diazoalkylmetal derivative (Eq. 5-37). The formation of mercuric salts of diazoacetic ester

$$Y-M + RCH{=}N_2 \longrightarrow Y-M{\leftarrow}CH(R)N_2 \underset{\searrow}{\overset{\nearrow}{}} \begin{array}{l} MC(R)N_2 \\[4pt] Y-M{\leftarrow}CHR \end{array} \qquad (5\text{-}37)$$

from mercuric acetate or oxide and the formation of Ag_2CN_2 from diazomethane and silver acetate may be considered to follow this course.

Boron compounds (BX_3, BR_3, $B(OR)_3$) convert diazomethane to a high polymer, $(CH_2)_x$ probably by a cationic chain reaction.[140,141] Anhydrous halides of elements of the second and higher periods are converted in wide variety to haloalkyl derivatives by diazomethane[142-146] (Eqs. 5-38, 5-39). Such products correspond to loss of

$$SnCl_4 + CH_2{=}N_2 \text{ (excess)} \longrightarrow Sn(CH_2Cl)_4 \qquad (5\text{-}38)$$
$$57\%$$

$$AsCl_3 + CH_3CH{=}N_2 \xrightarrow[0^\circ C]{\text{benzene}} ClAs(CHClCH_3)_2 \qquad (5\text{-}39)$$
$$40\%$$

nitrogen from an acid-base adduct, accompanied by migration of halogen from metal to carbon. Alkylaluminums behave analogously[142] (Eqs. 5-40, 5-41).

$$R_3Al + CH_2{=}N_2 \longrightarrow R_2AlCH_2R \xrightarrow{CH_2N_2} RAl(CH_2R)_2 \xrightarrow{CH_2N_2}$$
$$(RCH_2)_3Al \qquad (5\text{-}40)$$

$$R_2AlX + CH_2{=}N_2 \longrightarrow R_2AlCH_2X \qquad (5\text{-}41)$$

The haloalkyl organometallic compounds are highly reactive; they react readily with olefins and acetylenes to form cyclopropanes and cyclopropenes, regenerating the metal halide. For this reason, some metal halides can be used as catalysts for the addition of the alkylidene moiety of diazoalkanes to unsaturated compounds[142] (Eq. 5-42). The products are the same as those obtained by thermolysis or photoly-

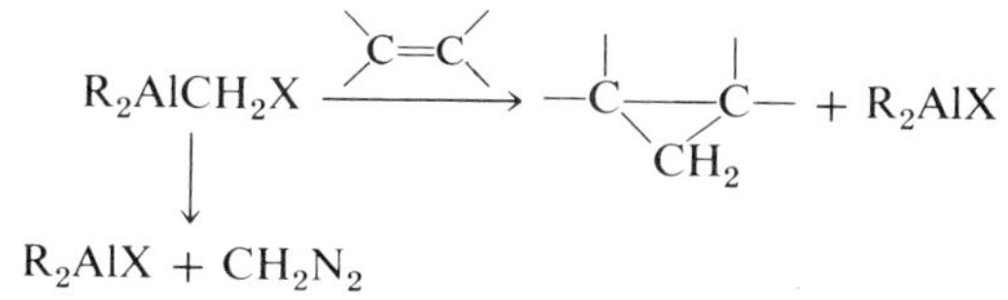

$$R_2AlCH_2X \xrightarrow{\overset{\displaystyle >C=C<}{}} -\overset{\displaystyle |}{\underset{\displaystyle |}{C}}-\overset{\displaystyle |}{\underset{\displaystyle |}{C}}- + R_2AlX$$
$$\text{(with } CH_2 \text{ bridge)}$$

$$\downarrow$$

$$R_2AlX + CH_2N_2 \tag{5-42}$$

sis, but the stereospecificity is much higher.[147] With trialkylboranes and diazoaceto-nitrile or diazoacetic ester, the overall effect[148] is to replace the diazo group by $R + H$ (Eq. 5-43).

$$R_3B + N_2\!=\!CHCN \longrightarrow R_2B\!-\!\underset{\displaystyle R}{\overset{\displaystyle |}{CHCN}} \xrightarrow{[H_2O]} RCH_2CN \tag{5-43}$$

The reactivity of a wide variety of Lewis acids has been compared against diphenyldiazomethane.[149] The relative rates of decomposition of the diazo compound do not strictly follow Lewis acidity. Although $AlCl_3$, $SnCl_4$, and $BF_3 \cdot Et_2O$ are among the fastest catalysts, $FeCL_3$, $CuBr_2$, and ZnI_2 are slower, $ZnCl_2$ and $CuCl_2 \cdot 2H_2O$ are slower still, and $AlCl_3 \cdot 6H_2O$ and $CuCN$ are extremely slow; $CaCl_2$, $LiCl$, $CdCl_2$, and NH_4Cl are inert. Changes from rate-determining addition to rate-determining decomposition of the adduct are encountered, just as with proton-catalyzed reactions.

Electrophilic carbon.   There is abundant evidence for attack of electrophilic carbon on diazo compounds, but most of it concerns carbonyl compounds and their derivatives. The possibility that carbenium ions can react with diazoalkanes is implied by the ionic polymerization of diazomethane with boron fluoride, in which such a reaction is postulated as the chain-propagating step. The reaction can be observed directly with triphenylmethyl perchlorate[150] (Eq. 5-44). The products formed are

$$Ph_3C^+ \ ClO_4^- + CH_2\!=\!N_2 \xrightarrow[Et_2O]{0\,°C} [Ph_3C\!-\!CH_2N_2^+] \longrightarrow$$

$$Ph_2C\!=\!CHPh \tag{5-44}$$

those to be expected from the rearrangement of the transitory diazonium or carbenium intermediates.

Diazoalkanes react with most aldehydes and ketones even in the absence of acids, if hydroxylic substances (water, alcohols) are present. Aldehydes are more reactive than ketones, and electron-withdrawing substituents make the rates faster. There are three types of product: homologated ketones and aldehydes, epoxides (oxiranes), and β-diazo alcohols (Eq. 5-45). Only the first two are commonly encoun-

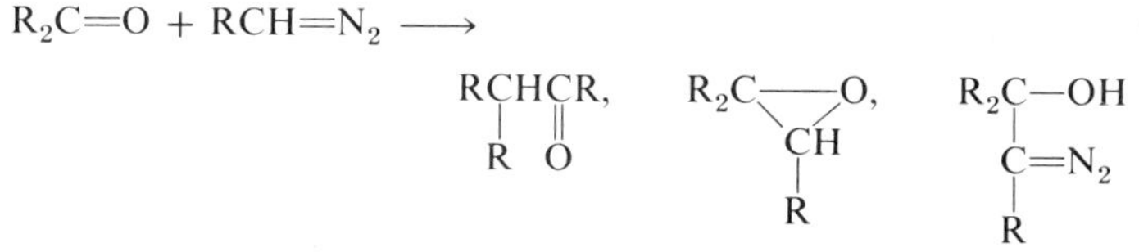

$$R_2C\!=\!O + RCH\!=\!N_2 \longrightarrow$$

$$\underset{\displaystyle R\ \ O}{\overset{\displaystyle |\ \ \ \|}{RCHCR,}} \qquad \underset{\displaystyle \underset{R}{\overset{|}{CH}}}{R_2C\diagdown\!\!\diagup O,} \qquad \underset{\displaystyle \underset{R}{\overset{|}{C\!=\!N_2}}}{R_2C\!-\!OH} \tag{5-45}$$

tered. The relative proportions of the products are very sensitive to the reaction medium; alcohols moderately promote homologation over epoxide formation,[151, 152] and Lewis acids, such as aluminum chloride and boron fluoride, promote homologation to the virtual exclusion of epoxide formation.[124, 153]

Two explanations for these facts have been put forth.[26, 152] In one view, the initial step is always electrophilic attack on the diazo carbon by the carbon of the carbonyl group or its conjugate acid (Eq. 5-46); the resulting transient intermediate cyclizes

$$R_2C=O \xrightarrow{CH_2N_2} \left[R_2C \begin{smallmatrix} O^- \\ \\ CH_2N_2^+ \end{smallmatrix} \right] \longrightarrow R_2C \begin{smallmatrix} O \\ \diagup \\ CH_2 \end{smallmatrix}$$

$$\downarrow H^+$$

$$R_2C=OH^+ \xrightarrow{CH_2N_2} \left[R_2C \begin{smallmatrix} OH \\ \\ CH_2N_2^+ \end{smallmatrix} \right] \longrightarrow \begin{smallmatrix} R-C=OH^+ \\ \\ R-CH_2 \end{smallmatrix} \longrightarrow \begin{smallmatrix} R-C=O \\ \\ R-CH_2 \end{smallmatrix}$$

(5-46)

to an epoxide when the oxygen is neither coordinated nor bonded to hydrogen, but otherwise undergoes rearrangement to the conjugate acid of the homologated carbonyl compound. In the other view, the oxygen of the free carbonyl compound acts as a nucleophilic site and attacks the diazoalkane carbon, producing an intermediate that can only cyclize to epoxide (Eq. 5-47). This reaction is thus uncatalyzed and

$$R_2C=O \xrightarrow{CH_2N_2} [R_2\overset{+}{C}-O-CH_2N_2^-] \xrightarrow{-N_2} R_2C \begin{smallmatrix} CH_2 \\ \diagup \diagdown \\ O \end{smallmatrix}$$

$$\downarrow ROH$$

$$R_2C=O\cdots HOR \xrightarrow{CH_2N_2} [R_2C-OH\cdots\overset{-}{O}R] \longrightarrow R-C=\overset{+}{O}H$$
$$\qquad\qquad\qquad\quad CH_2N_2^+ \qquad\qquad\qquad R-CH_2$$

(5-47)

presumably relatively slow. When the carbonyl group is converted to its conjugate acid, or even hydrogen-bonded to an alcohol, it acts as an electrophilic agent and produces the same β-hydroxy diazonium intermediate shown in Equation 5-46, with the same consequences. The β-hydroxy diazonium ions are the same intermediates that are formed when β-hydroxy amines are treated with nitrous acid, a reaction that also results in rearrangement to a carbonyl compound (semipinacolic deamination).

The zwitterionic adduct of Equation 5-46 may also stabilize itself by internal proton transfer from the diazo carbon to oxygen. This has only been observed when a substituent is present on the diazo carbon that is acid-strengthening, as in diazoacetic ester (Eq. 5-48). On the other hand, the possibility of nucleophilic attack by carbonyl oxygen receives support from the formation of dioxoles from α-diketones, such as benzil and o-quinones[151, 154] (Eq. 5-49). However, since diazo compounds are generally inert toward much stronger oxygen nucleophiles, the mechanism of Equation 5-47 is not so convincing as is cycloaddition oriented C to O and N to C, followed by extrusion of N_2.

The homologation reaction is particularly useful when applied to cyclic ketones,

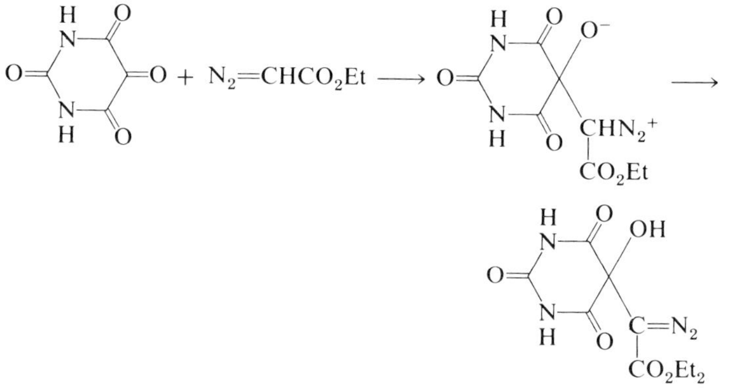

$$\text{(5-48)}$$

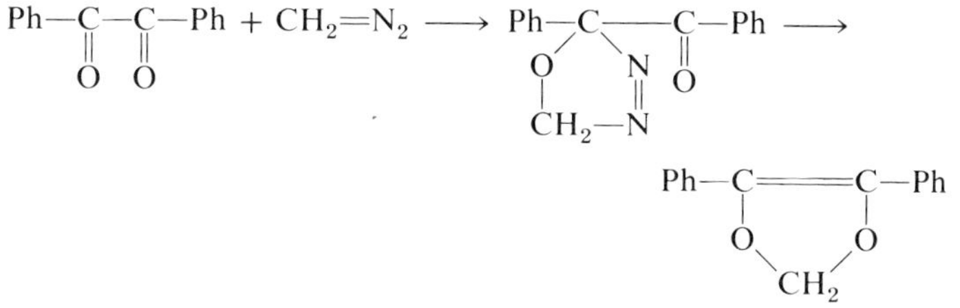

$$\text{(5-49)}$$

for the ring is expanded.[151] Its special value lies in the synthesis of ketones with rings larger than six (Eq. 5-50), where direct ring closure is often difficult. The

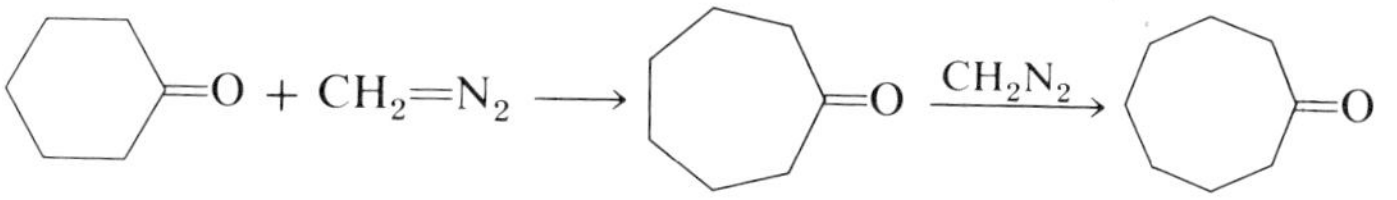

$$\text{(5-50)}$$

products of the diazoalkane ring expansion are, of course, themselves capable of the same reaction, and the conditions must be chosen carefully to achieve the desired result. With higher diazoalkanes, 2-alkylcycloalkanones may be produced. The reaction of diazoethane with 4-substituted cyclohexanones gives high yields of 2-methyl-5-substituted-cycloheptanones in an approximately 1:1 ratio of stereoisomers.[155] Optical activity resident in the migrating carbon in ring expansion is maintained,[156] presumably with retention of configuration. When the diazo and carbonyl groups are in the same molecule, intramolecular ring expansion may be possible, producing bicyclic compounds.[157]

A clever way of converting aldehydes solely to methyl ketones, rather than their mixture with epoxides and homologous aldehydes, has been devised,[158] wherein the aldehyde is first condensed with malononitrile. The resulting dicyano olefins react with diazomethane by methylation at the vinyl hydrogen; alkaline hydrolysis regenerates the carbonyl group (Eq. 5-51).

If the carbonyl group with which a diazo compound may react is also attached to a potential leaving group, elimination may ensue, producing an α-diazo ketone.[52-54] This is the common reaction of acid chlorides[159] and anhydrides,[160,161] as long as there is a base or excess diazoalkane present to take up the acid produced in the

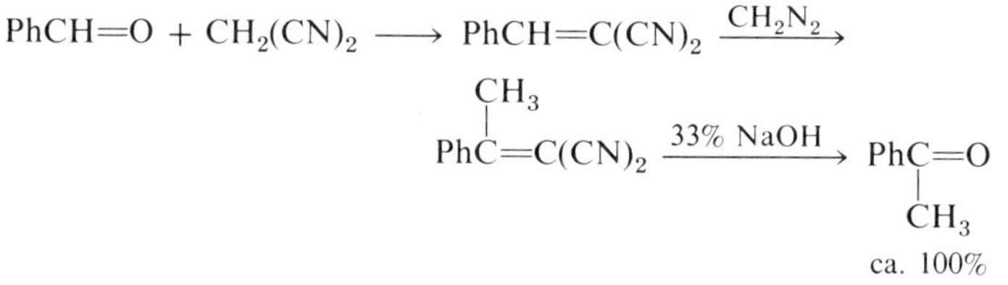

$$PhCH{=}O + CH_2(CN)_2 \longrightarrow PhCH{=}C(CN)_2 \xrightarrow{CH_2N_2}$$

$$\underset{\underset{\displaystyle CH_3}{\textstyle |}}{Ph\overset{\displaystyle}{C}}{=}C(CN)_2 \xrightarrow{33\% \text{ NaOH}} \underset{\underset{\displaystyle CH_3}{\textstyle |}}{Ph\overset{\displaystyle}{C}}{=}O \qquad (5\text{-}51)$$

ca. 100%

elimination step. Otherwise, an α-halo or α-acyloxy ketone is formed, either by reaction of the acid with initially formed α-diazo ketone or by rearrangement of the primary adduct (Eqs. 5-52, 5-53). For the most efficient utilization of diazomethane,

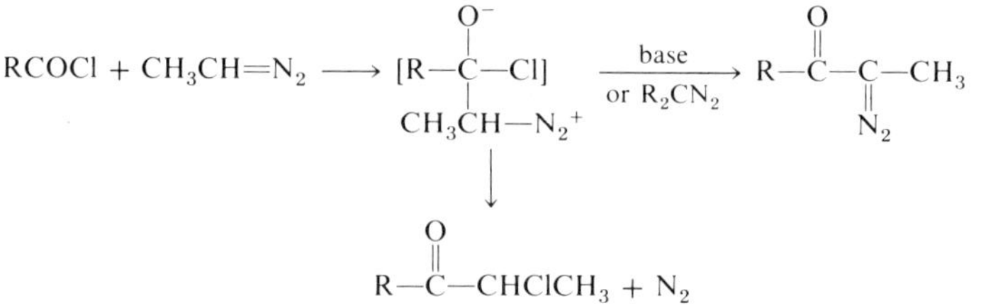

$$RCOCl + CH_3CH{=}N_2 \longrightarrow [R-\overset{\overset{\displaystyle O^-}{|}}{\underset{\underset{\displaystyle CH_3CH-N_2^+}{|}}{C}}-Cl] \xrightarrow[\text{or } R_2CN_2]{\text{base}} R-\overset{\overset{\displaystyle O}{\|}}{C}-\overset{\overset{\displaystyle}{\underset{\underset{\displaystyle N_2}{\|}}{C}}}-CH_3$$

$$\qquad (5\text{-}52)$$

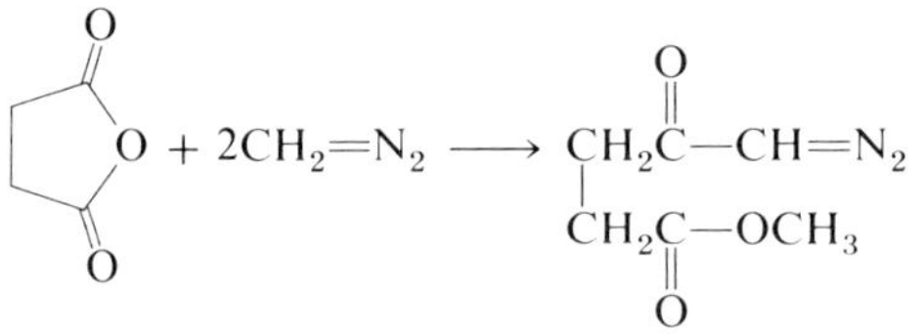

$$\downarrow$$

$$R-\overset{\overset{\displaystyle O}{\|}}{C}-CHClCH_3 + N_2$$

$$(5\text{-}53)$$

$$\overset{\overset{\displaystyle O}{\|}}{CH_2C}-CH{=}N_2$$
$$\underset{\underset{\displaystyle O}{\|}}{CH_2C}-OCH_3$$

it is recommended that reaction be carried out in ethereal triethylamine at -78 to $-25\,°C$ (yields 86–96%).[162, 163] An alternative path for reaction in strongly ionizing media has been suggested in which the acyl cation is the electrophile that attacks the diazo carbon.[164]

Chloroformic esters react like ordinary acyl chlorides, but very slowly. The products are, of course, α-diazo esters rather than α-diazo ketones.[165]

In contrast to acid chlorides and anhydrides, esters are not converted to diazo ketones by diazo compounds. However, diazomethane has long been known to cleave esters, especially aryl esters. It was originally assumed that the esters were hydrolyzed by traces of water, and the acid thereby released was then methylated. Subsequently, it was demonstrated that hydrolysis is unnecessary and that diazomethane actually functions as a transesterification catalyst on mixtures of an ester and an alcohol[166] (Eq. 5-54). The process is most efficient when methanol is used as

$$RCO_2R' + CH_3OH \xrightarrow{CH_2N_2} RCO_2CH_3 + R'OH \qquad (5\text{-}54)$$

the alcohol, although diazomethane will, indeed, catalyze the formation of ethyl esters when ethanol is used. This reaction provides a gentle means of converting a wide variety of esters into their methyl counterparts. Only catalytic amounts of diazomethane (not more than 25%) are required, and amides can also be cleaved if they are derived from amines that are extremely weak bases. It has been suggested that the catalysis results from an increase of nucleophilicity of the alcohol as a result of coordination of its hydrogen atom with the diazomethane carbon. Some esters with strongly electron-withdrawing substituents on the acyl moiety, thus having a more electrophilic carbonyl carbon, react like ketones and aldehydes and form epoxides.[167]

Cleavage of ethers, which has been observed as an intramolecular reaction with some alkoxy-substituted diazomethyl ketones[168] (Eq. 5-55), is catalyzed by strong acid and is evidently unrelated to ester cleavage.

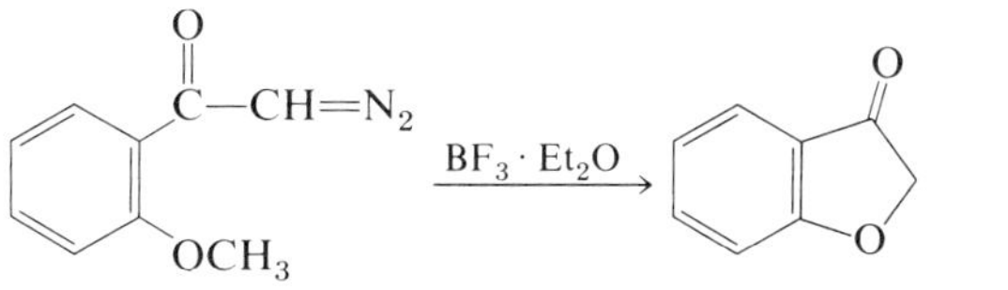

$$\text{(structure)} \xrightarrow{\text{BF}_3 \cdot \text{Et}_2\text{O}} \text{(structure)} \tag{5-55}$$

The reaction of thiocarbonyl compounds with diazo compounds is more readily interpreted in terms of cycloaddition, q.v.

Nitrogen electrophiles. The arenediazonium ion, ArN_2^+, attacks diazo compounds in the same initial manner as other electrophiles, coupling with the carbon. The presumed initial products, $ArN=NCR_2N_2^+$, have not been detected, for they apparently react rapidly in two ways. By loss of nitrogen and combination with the counterion, azo compounds (or the tautomeric hydrazones) are produced, which tautomerize to hydrazonoyl derivatives if the diazo compound has an α-hydrogen[169] (Eq. 5-56). When tautomerization is prevented, α-chloro azo compounds are iso-

$$ArN_2^+ + CH_2{=}N_2 \longrightarrow [ArN{=}NCH_2N_2^+] \xrightarrow{Cl^-} ArN{=}NCH_2Cl \tag{5-56}$$
$$\downarrow \qquad \qquad \downarrow$$
$$H^+ + N_2 + ArNH{-}CN \longleftarrow ArNH{-}N{=}CHN_2^+ \qquad ArNH{-}N{=}CHCl$$

lated.[26] The reaction with the counterion (for example, Cl^-) must, of course, compete with solvolysis if a hydroxylic solvent is used. Indeed, in dilute methanol solution, diazomethane gives rise to the methoxy analogs, N-methyl-N-arylformohydrazonates, $CH_3OCH=NNHAr$, but if the solution is saturated with lithium chloride, the hydrazonoyl chloride again becomes the principal product. The second type of behavior involves rearrangement with loss of nitrogen, producing a cyanamide (which may become alkylated by more diazo compound). By means of tagging experiments with ^{15}N, it has been established that the rearrangement takes place by migration of an anilino group from N to C.[170] Up to 12% of 1-aryltetrazoles have also been isolated from diazonium salts and diazomethane.[171] Such products may arise by cycloaddition to the diazonium group (Eq. 5-57).

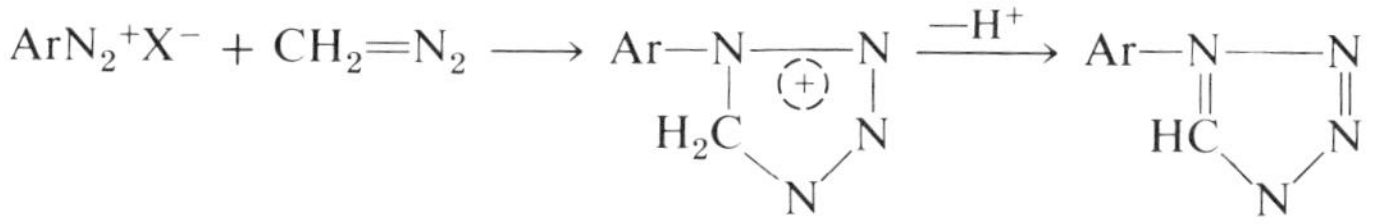

$$ArN_2^+X^- + CH_2{=}N_2 \longrightarrow \text{(structure)} \xrightarrow{-H^+} \text{(structure)} \tag{5-57}$$

Diazoalkanes react spontaneously at room temperature with nitroso compounds to produce nitrones (Eq. 5-58). Nitrogen is evolved with effervescence, and owing to

$$PhN{=}O + Ph_2C{=}N_2 \xrightarrow{\text{spontaneous}} Ph_2C{=}\overset{+}{N}{-}O^- + N_2 \qquad (5\text{-}58)$$
$$\underset{}{\qquad\qquad\qquad\qquad\qquad\qquad} \underset{Ph}{|}$$

| Green | Red | Colorless |

the intense colors of the reactants, the process can be carried out quite nicely as a titration. The reaction probably begins in the same way as with carbonyl groups, but nitrogen is lost without ring closure or rearrangement, owing to the availability of the unshared pair of electrons on the nitroso nitrogen. It is a useful preparative method. The somewhat analogous reaction of diazoalkanes with nitrosyl chloride leads to hydroximoyl chlorides.[172,173]

Azodicarbonyl compounds, such as azodicarboxylic esters, generally function as reagents with electrophilic nitrogen. They react with diazo compounds without catalysis and form azomethine imides or products at least formally derived from them[174-178] (Eq. 5-59). With α-diazo ketones, the hydrazones are the principal prod-

$$R_2C{=}N_2 + EtO_2CN{=}NCO_2Et \longrightarrow [EtO_2N{-}\overset{-}{N}CO_2Et] \longrightarrow$$
$$\underset{R_2\overset{|}{C}{-}N_2{}^+}{}$$

$$EtO_2C\overset{+}{N}{-}\overset{-}{N}CO_2Et \longrightarrow EtO_2C{-}N{-\!\!-\!\!-}N$$
$$\underset{R_2C}{\overset{\|}{}} \qquad\qquad\qquad \underset{R_2C\diagdown\diagup C{-}OEt}{\overset{|}{}\qquad\overset{\|}{}}$$
$$\qquad\qquad\qquad\qquad\qquad\qquad O$$

$$\text{and/or} \qquad (EtO_2C)_2N{-}N{=}CR_2 \qquad (5\text{-}59)$$

uct and are of preparative value as precursors of α-dicarbonyl compounds.[179] With other diazo compounds, hydrazones are favored at higher temperatures and oxadiazolines at lower. All these products can equally well be interpreted as arising from an initial cycloaddition. In one instance, hexafluoroazomethane and diazomethane, the cycloadduct, a tetrazoline, has actually been isolated.[180]

Azodicarbonamide would be expected to behave similarly to the corresponding esters, but its reaction with diazoacetic ester has been an enigma for more than 50 years. It appears now that its reaction is analogous after all; the semicarbazone of ethyl glyoxylate is formed first in the *syn* configuration and is isomerized slowly on heating to the *anti* isomer.[181]

Bases and nucleophiles. Diazoalkanes are essentially inert to strong bases and are in fact often prepared by reactions that require the presence of alcoholic alkali. The latent acidity of the α-hydrogen may lead to proton exchange with the medium, however, and with very strong bases, such as butyllithium, a salt may be formed[182] (Eq. 5-60). Even triphenylmethylsodium will deprotonate diazomethane.[183] The

$$LiCH_3 + CH_2{=}N_2 \longrightarrow LiCH{=}N_2 + LiN(CH_3)NCH_2 \qquad (5\text{-}60)$$
$$\text{ca. 70\%}$$

anions can be captured by aldehydes (not so readily by ketones) so as to form β-diazo alcohols[41] (Eq. 5-61). α-Diazo carbonyl compounds react with themselves

$$\text{EtO}_2\text{CCH}=\text{N}_2 + \text{RCHO} \xrightarrow[\text{EtOH}]{\text{KOH}} \text{EtO}_2\text{C}-\underset{\underset{\text{N}_2}{\parallel}}{\text{C}}-\overset{\overset{\text{OH}}{|}}{\text{C}}\text{HR} \qquad (5\text{-}61)$$

when treated with strong bases, if there is nothing more reactive to capture them; the principal products are dimers, dihydrotetrazines, or tetrazoles, the proportions of which are very sensitive to conditions[26,184] (Eq. 5-62). At low enough tempera-

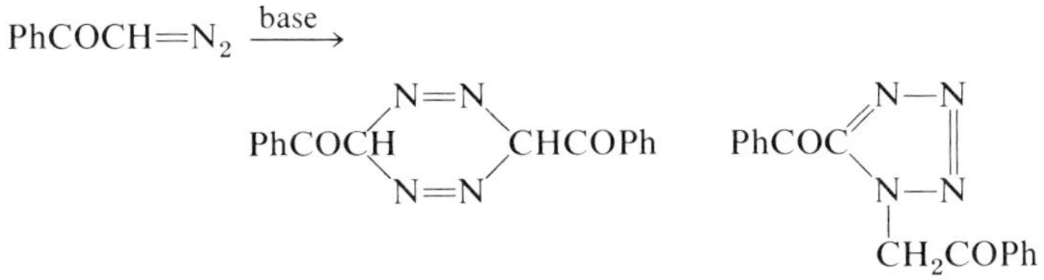

$$\text{PhCOCH}=\text{N}_2 \xrightarrow{\text{base}}$$

(5-62)

tures, derivatives of the monomeric anion can be isolated; the lithium derivative of diazoacetic ester, for example, decomposes above $-50°\text{C}$.[185]

Diazo compounds are susceptible to nucleophilic attack at the terminal nitrogen, and more so than are the isoelectronic azides. Cyanide,[186] other carbanions,[187] Grignard reagents,[188] phosphines,[189] amines,[190] hydrazines,[190,191] sulfites,[192] and sulfide[193] may react in this way (Eqs. 5-63 through 5-69).

$$\text{ArCOCH}=\text{N}_2 + \text{KCN} \longrightarrow \text{ArCOCH}=\text{N}-\text{NCN}^-\text{K}^+ \xrightarrow{\text{H}^+}$$
$$\text{ArCOCH}=\text{NNHCN} \qquad (5\text{-}63)$$

$$\text{Ph}_2\text{C}=\text{N}_2 \xrightarrow[\text{(2) H}_2\text{O}]{\text{(1) } t\text{-BuMgCl}} \text{Ph}_2\text{C}=\text{N}-\text{NH}-t\text{-Bu} \xrightarrow[\text{H}^+]{\text{H}_2\text{O}}$$
$$\text{Ph}_2\text{C}=\text{O} + \text{H}_2\text{N}-\text{NH}-t\text{-Bu} \qquad (5\text{-}64)$$

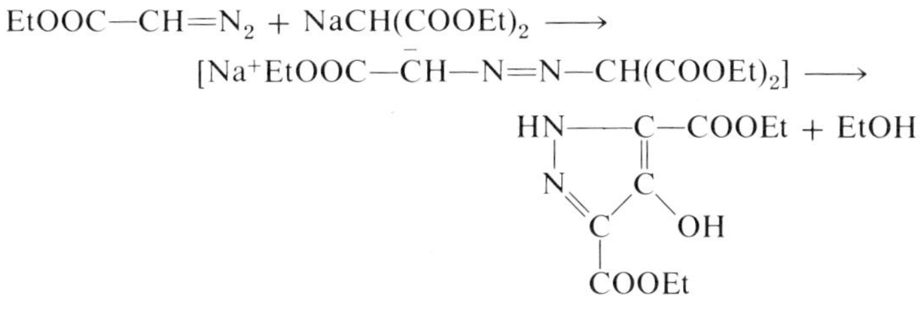

$$\text{EtOOC}-\text{CH}=\text{N}_2 + \text{NaCH(COOEt)}_2 \longrightarrow$$
$$[\text{Na}^+\text{EtOOC}-\overset{-}{\text{CH}}-\text{N}=\text{N}-\text{CH(COOEt)}_2] \longrightarrow$$

(5-65)

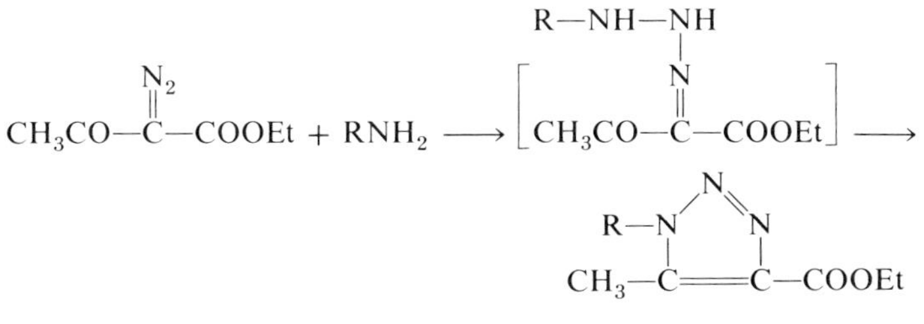

(5-66)

$$R_2C{=}N_2 + R_3P \longrightarrow R_2C{=}N{-}N{=}PR_3 \qquad (5\text{-}67)$$

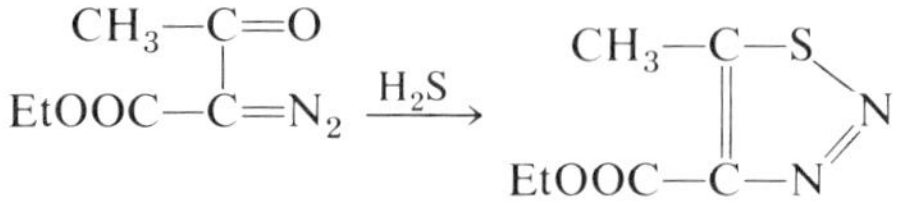

$$(5\text{-}68)$$

$$EtOOC{-}CH{=}N_2 + N_2H_4 \xrightarrow{70\,°C} [H_2N{-}NH{-}N{=}N{-}CH_2COOEt] \longrightarrow$$
$$N_3CH_2COOEt + NH_3 \quad (5\text{-}69)$$

Grignard reagents produce N-monosubstituted hydrazones (Eq. 5-64), which may react further. If the temperature is low ($-70\,°C$), diazoacetic ester is merely metalated by Grignard reagents; at $0\,°C$, the N-alkylation reaction takes over.[185] Hydrolysis of the hydrazones from this reaction constitutes a useful hydrazine synthesis. The reaction with sulfide more often leads to alkylation or reduction.[194] Reaction with amines generally requires strongly electron-withdrawing substituents on the diazo carbon, such as in diazomalonic ester and dicyanodiazomethane; the latter attacks dimethylaniline at the *para* position of the benzene ring.[63] The effects of structure on the reactivity of trivalent phosphorus toward diazomethane have been studied extensively.[195-198] Alkylidenephosphoranes (Wittig reagents) act as alkylidene transfer agents and convert diazo compounds to unsymmetrical azines.

Oxidation. The reaction of diazo compounds with oxygen under carbene-forming conditions has already been mentioned. Ozone appears to react directly with the diazo compound, replacing the diazo group with oxo[199] (Eq. 5-70). Peroxy acids

$$R_2C{=}N_2 + O_3 \longrightarrow \left[R_2C \underset{O_3{}^-}{\overset{N_2{}^+}{\diagup}} \right] \longrightarrow R_2C{=}O + N_2 + O_2 \qquad (5\text{-}70)$$

accomplish the same result,[200,201] and lead tetraacetate forms the diacetate of the corresponding *gem*-diol.[202] An isolated instance of oxidation by an N-oxide has been reported; in the attempted esterification with diazomethane, 4-methoxypicolinic acid N-oxide was converted to 4-methoxypyridine, with loss of CO_2 and formation of formaldehyde.[203]

t-Butyl hypochlorite or hypobromite halogenate the α-carbon, but the products may be difficult to isolate and easily react further[204-206] (Eq. 5-71). The halogen

$$RCH{=}N_2 + t\text{-}BuOX \xrightarrow{-100\,°C} RC \underset{X}{\overset{N_2}{\diagup}} \qquad (5\text{-}71)$$

elements, however, replace the diazo group with X_2, forming *gem*-dihalides[207,208] (Eq. 5-72). This is one of the best methods for preparing such compounds. It seems likely that this reaction involves electrophilic attack to form an intermediate α-halodiazonium intermediate.

Reduction. Diazo compounds can be reduced with or without retention of the diazo nitrogen. Reduction to hydrocarbon has been accomplished by catalytic hy-

$$R_2C{=}N_2 \quad \overbrace{\begin{array}{l} \xrightarrow{X_2} \quad [R_2\overset{\overset{\displaystyle X}{|}}{C}{-}N_2{}^+\ X^-] \longrightarrow R_2CX_2 + N_2 \\[2mm] \xrightarrow{t\text{-BuOCl, AcOH}} \quad R_2C(X)OAc \end{array}}$$

(5-72)

$$(R = R'CO) \qquad\qquad 65{-}91\%$$

drogenation[209] and by sodium naphthalenide.[210] With metallic sodium, diphenyl-diazomethane and diazofluorene develop strong blue colors, believed to be due to radical anions, $Ar_2C{-}N{=}N^-$, detectable by electron spin resonance.[211] With sodium naphthalenide, hydrocarbon is only a minor product, presumed to arise by the radical anion abstracting hydrogen from the solvent; the major product is primary amine (Eq. 5-73). Catalytic hydrogenation may produce hydrazones in addition to

$$Ar_2C{=}N_2 \xrightarrow{NaC_{10}H_8} [Ar_2C^{\mathbf{\cdot}}] \xrightarrow{SoH} Ar_2CH^- \longrightarrow Ar_2CH_2$$

$$\Big\downarrow {Ar_2C{=}N_2}$$

(5-73)

$$[Ar_2C{=}N{-}N{=}CAr_2]^{\mathbf{\cdot}} \xrightarrow{e} Ar_2CHNH_2$$

or instead of hydrocarbons. Hydrogen iodide acts first as an acid in converting diazo compounds to monoiodo compounds, but the latter can be reduced by more HI to form hydrocarbon under appropriate conditions.

Hydrazones, hydrazines, and amines are also produced with other reducing agents, such as amalgams of sodium or aluminum.[209,212-214] Hydrazones have been obtained in good yield using ammonium sulfide[212] or lithium aluminum hydride[26] (Eq. 5-74). Even hydrogen sulfide may reduce diazo compounds in preference to

$$RCOCH{=}N_2 + NH_4HS \longrightarrow RCOCH{=}NNH_2$$

(5-74)

becoming alkylated.[194] This behavior appears to be favored by the presence of electron-withdrawing groups, which would repress the protonation required in the first step of alkylation. Diazoacetophenone and diazoacetic ester undergo reduction by hydrogen sulfide almost exclusively. ortho-Diazooxides are reduced to phenols when heated with benzylamine or benzyl alcohol, but the reaction may involve initial formation of a carbene.[47]

Free radicals. The limited information available suggests that diazo compounds are susceptible to attack by free radicals. Diazomethane reacts with triphenylmethyl to form hexaphenylpropane[215] (Eq. 5-75). The photochemical reaction of diazomethane with carbon tetrachloride, which gives tetrachloroneopentane (Eq. 5-76) has been reported to be a radical-chain reaction.[75]

$$2Ph_3C\cdot + CH_2{=}N_2 \longrightarrow Ph_3C{-}CH_2{-}CPh_3 + N_2$$

(5-75)

$$4CH_2{=}N_2 + CCl_4 \xrightarrow{h\nu} (ClCH_2)_4C + 4N_2$$

(5-76)

The reactions of diazo compounds with nitric oxide and nitrogen dioxide fall in the category of radical reactions, although there is some justification for interpreting reaction with the latter reagent as electrophilic attack by $NO_2{}^+$, actual or potential.

Phenyldiazomethane reacts with nitric oxide to form an azine N,N-dioxide[216] (Eq. 5-77), a structure of unusual interest because of the apparent presence of positive

$$Ph-CH=N_2 + NO \longrightarrow PhCH\overset{+}{=}\overset{\overset{\scriptstyle O^-}{|}}{N}-\overset{\overset{\scriptstyle O^-}{|}}{\overset{+}{N}}=CHPh \qquad (5\text{-}77)$$

charges on adjacent atoms. The azine dioxides react readily with excess nitric oxide to form an aldehyde or ketone, accompanied by the corresponding nitrimine, $R_2C=NNO_2$. Diazo compounds having no α-hydrogen first form iminoxyls, $R_2C=NO\cdot$, detectable by esr; they react with more NO to form nitrimines, probably through Equation 5-78.[217] Monosubstituted diazo compounds react dimerically

$$R_2C=N_2 + NO \longrightarrow R_2C=NO\cdot \xrightarrow{NO} \left[R_2C\overset{+}{=}N\overset{\overset{\scriptstyle O^-}{\diagup}}{\underset{\diagdown NO}{}} \right] \longrightarrow$$

$$R_2C=N-NO_2 \qquad (5\text{-}78)$$

with N_2O_3 to form furoxans.[218] Nitrogen dioxide converts diazo compounds to *gem*-dinitro compounds, as shown by the example of 9-diazofluorene[219] (Eq. 5-79).

$$Ar_2C=N_2 + N_2O_4 \longrightarrow Ar_2C(NO_2)_2 + N_2 \qquad (5\text{-}79)$$

Cycloaddition and related reactions. Diazo compounds act as 1,3-dipoles in adding to a wide variety of dipolarophiles, such as olefins, acetylenes, imines, nitriles, and possibly other types of multiple bonds. Isolated, unstrained carbon-carbon double bonds react very sluggishly with diazoalkanes and for many practical purposes can be considered inert. Consequently, most of the reported examples involve double bonds activated by cyano, carbonyl, or nitro substituents; by conjugation; or by ring strain. The products are pyrazolines, and if the olefin is unsymmetrically substituted, a mixture of isomers is generally to be expected[220] (Eq. 5-80). There appears

$$R_2C=CR_2' + Y_2C=N_2 \longrightarrow$$

$$\begin{array}{ccc} R_2C\!\!-\!\!CR_2' & & R_2'C\!\!-\!\!CR_2 \\ | \quad\quad | & \text{and} & | \quad\quad | \\ N \diagdown \diagup CY_2 & & N \diagdown \diagup CY_2 \\ \;\; N & & \;\; N \end{array} \qquad (5\text{-}80)$$

to be no simple reliable way to predict which isomer will predominate,[107] although there have been massive attacks on the problem from the molecular orbital approach which seem to reveal more about deficiencies in theory than about experimental behavior.[221] The rule of thumb that the diazo carbon behaves as a nucleophile and attacks the double bond so as to generate the most stable carbanion intermediate is fraught with exceptions. Steric hindrance probably also plays a role, and different diazo compounds may react with opposite orientations with the same olefin[222] (Eq. 5-81).

Pyrazolines are not very stable toward extrusion of nitrogen and collapse to cyclopropanes or alkenes, which in many cases are the products actually isolated from reactions of diazo compounds with olefins. If the pyrazolines themselves are de-

$$RCH=CHNO_2 \xrightarrow{CH_2N_2} \begin{matrix} R-CH-CH-NO_2 \\ | \quad\quad | \\ H_2C \quad N \\ \diagdown N \diagup \end{matrix}$$

$$\diagdown \xrightarrow{Ph_2CN_2} \begin{matrix} R-CH-CH-NO_2 \\ | \quad\quad | \\ N \quad CPh_2 \\ \diagdown N \diagup \end{matrix}$$

(R = CH$_3$ or Ph)

(5-81)

sired, the reaction mixture cannot be heated, and in the many situations where the addition is inherently slow, prolonged reaction time, perhaps a week or more, may be required. The rate of addition is substantially increased by high pressures, and a number of examples of improved cycloadditions carried out at 5000 atm have been reported.[107,223] The sensitivity of the rate to structure of the olefin is illustrated by the fact that ethyl acrylate adds about 500 times as fast as styrene, yet ethyl cinnamate is slightly slower than styrene.[224] Vinyl ethers add diazo compounds, but enamines instead convert α-diazo ketones and sulfones to C-alkyl derivatives (α-diazo-β-amino ketones or sulfones).[41,225,226] The solvent exerts a significant effect on the rate of addition, and styrene, for example, gives twice the yield of adduct in wet dioxan as in dry ether at equal times.[227]

Cycloaddition of diazo compounds appears in general to be a concerted process, although bond formation at the two ends may not be equal in the transition state and may be sensitive to structure. The stereochemistry corresponds to *syn*-addition, preserving the geometry of the original olefin, as exemplified by the formation of stereoisomeric pyrazolines from *cis*- and *trans*-α-methylcrotonic esters.[228] In other cases, such as fumaric and maleic esters, the same pyrazoline (*trans* carboethoxy groups) is obtained from either geometric isomer. There is reason to believe that such a result is caused by isomerization of the product under the reaction conditions, owing to the labile α-hydrogens.[229,230]

Ketene reacts with 2 moles of diazomethane to form cyclobutanone[231] (Eq. 5-82), probably by initial cycloaddition to form a pyrazolinone.

$$CH_2=C=O + CH_2=N_2 \longrightarrow \begin{bmatrix} CH_2-C=O \\ | \quad\quad | \\ N \quad CH_2 \\ \diagdown N \diagup \end{bmatrix} \longrightarrow \begin{matrix} CH_2-C=O \\ \diagdown \diagup \\ CH_2 \end{matrix} \xrightarrow{CH_2N_2}$$

$$\begin{matrix} CH_2-C=O \\ | \quad\quad | \\ CH_2-CH_2 \end{matrix}$$

(5-82)

Acetylenes react approximately as readily as olefins with diazo compounds, but the reactions are generally more successful, because the products, pyrazoles, are much more stable than pyrazolines, a fact that allows higher temperatures to be used.

The C=N bond of imines acts as a dipolarophile toward diazo compounds, but most of the reported examples have been benzaldehyde anils[232-235] (Eq. 5-83). In a group of vinylimines of hexafluoroacetone, addition to the imine bond was greatly favored over addition to the olefinic bond, and 1,2,3-triazolines were the major

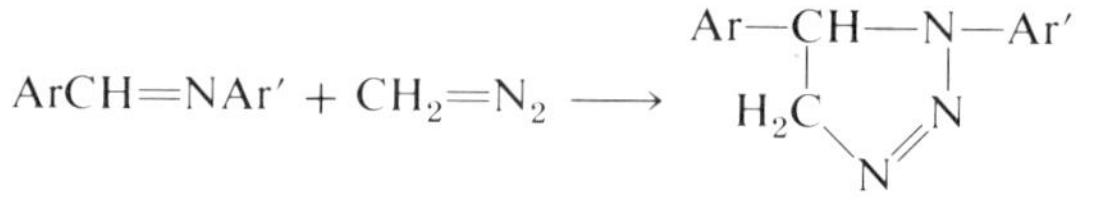

$$\text{ArCH=NAr}' + \text{CH}_2\text{=N}_2 \longrightarrow \begin{array}{c} \text{Ar—CH—N—Ar}' \\ \text{H}_2\text{C} \qquad \text{N} \\ \text{N} \end{array} \qquad (5\text{-}83)$$

product. However, one example gave substantial amounts of a 1,2,4-triazoline[236] (Eq. 5-84). Carbodiimides react to give aminotriazoles[237] (Eq. 5-85), but isocyanates give β-lactams in a reaction analogous to that of ketene.[238]

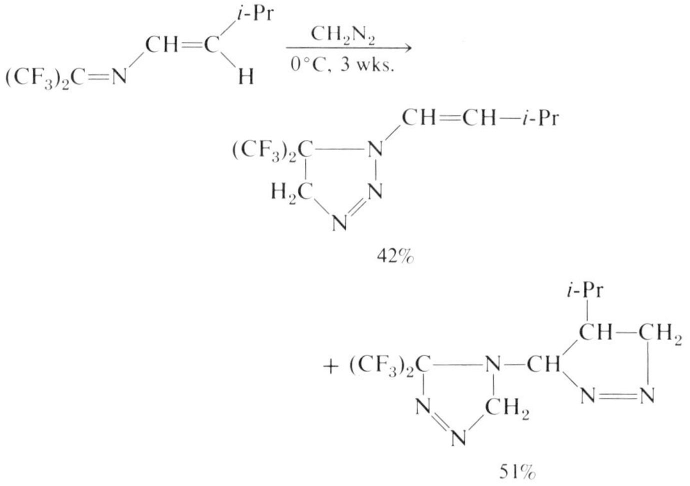

$$(\text{CF}_3)_2\text{C=N} \diagup \overset{i\text{-Pr}}{\underset{\text{H}}{\text{CH=C}}} \xrightarrow[0°C, 3 \text{ wks.}]{\text{CH}_2\text{N}_2}$$

$$\begin{array}{c} \text{CH=CH—}i\text{-Pr} \\ (\text{CF}_3)_2\text{C}\text{—N} \\ \text{H}_2\text{C} \qquad \text{N} \\ \text{N} \end{array}$$

42%

$$+ \ (\text{CF}_3)_2\text{C}\text{—N—CH} \diagup \overset{i\text{-Pr}}{\underset{\text{N=N}}{\text{CH—CH}_2}} \qquad (5\text{-}84)$$

51%

$$\text{RN=C=NR} + \text{CH}_2\text{=N}_2 \longrightarrow \begin{array}{c} \text{R—N}\text{—}\text{C—NHR} \\ \text{N} \qquad \text{CH} \\ \text{N} \end{array} \qquad (5\text{-}85)$$

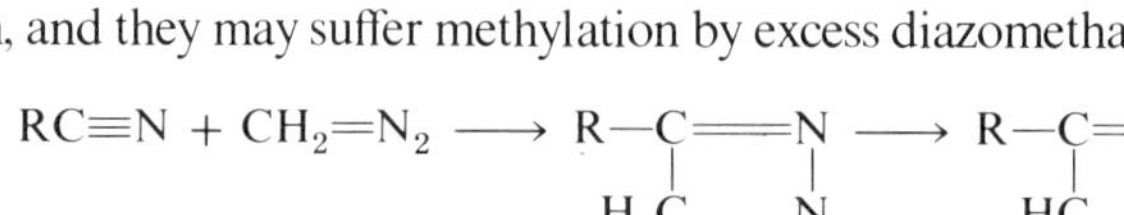

Nitriles do not react readily with diazo compounds unless they bear a strongly electron-withdrawing substituent or unless the cyano and diazo groups are favorably situated in the same molecule; 1,2,3-triazoles are formed by subsequent tautomerism, and they may suffer methylation by excess diazomethane[239] (Eq. 5-86). How-

$$\text{RC≡N} + \text{CH}_2\text{=N}_2 \longrightarrow \begin{array}{c} \text{R—C}\text{==}\text{N} \\ \text{H}_2\text{C} \qquad \text{N} \\ \text{N} \end{array} \longrightarrow \begin{array}{c} \text{R—C}\text{==}\text{N} \\ \text{HC} \qquad \text{NH} \\ \text{N} \end{array} \longrightarrow$$

$$\begin{array}{c} \text{R—C}\text{==}\text{N} \\ \text{HC} \qquad \text{N—CH}_3 \\ \text{N} \end{array} \qquad (5\text{-}86)$$

ever, in the presence of certain organoaluminum compounds, even benzonitrile will react with diazomethane.[240]

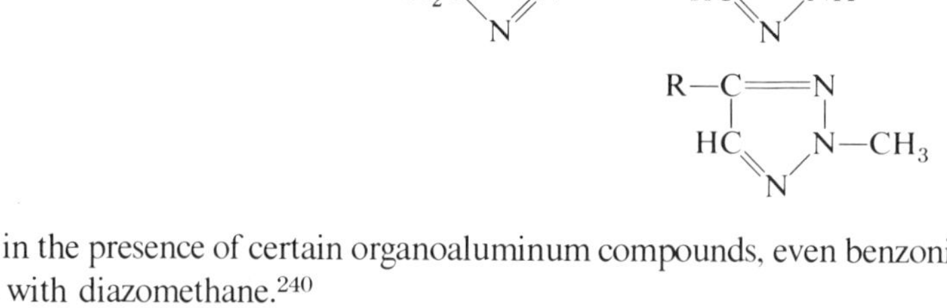

It is not clear whether isocyanides undergo cycloaddition with diazo compounds in general, but some α-diazo ketones have produced low yields of triazoles[241] (Eq. 5-87), and simpler diazo compounds, such as diphenyldiazomethane, form various

$$\text{PhCOCH}{=}\text{N}_2 + \text{PhNC} \xrightarrow[80\,°C]{\text{Cu}} \begin{array}{c} \overset{O}{\overset{\|}{\text{PhC}}}{-}\text{C}{=\!=}\text{CH} \\ \underset{N}{\overset{|}{N}}\diagdown\underset{}{\diagup}\overset{|}{N}{-}\text{Ph} \\ \text{N} \end{array} \qquad (5\text{-}87)$$

12%

products, including ketenimines, that can be conceived of as deriving from an initial cycloadduct, when photolyzed (no reaction occurs in the dark).[242,243]

Thiocarbonyl compounds react with diazo compounds to give an unusual variety of products. For at least some of them, a reasonable starting point is cycloaddition to form a thiadiazoline, which may extrude nitrogen and/or sulfur. α-Diazo ketones, for example, are converted to α,β-unsaturated ketones[244] (Eq. 5-88) or, by

$$R_2C{=}S + N_2{=}\text{CHCOR} \xrightarrow[\text{or } 100\text{-}140\,°C]{\text{Cu, room temp.}} \left[\begin{array}{c} R_2C{-\!-\!-}\text{CHCOR} \\ \underset{S}{|}\diagdown\underset{N}{\overset{\|}{}} \\ \text{N} \end{array} \right] \longrightarrow$$

$$R_2C{=}\text{CHCOR} + N_2 + S \qquad (5\text{-}88)$$

reaction with thiourea, to aminothiazolines.[245] The initial cycloadducts are too labile to have been detected in most cases, but spectroscopy on incompletely purified material has implied the formation of adducts with both orientations: 1,2,3-thiadiazolines and 1,3,4-thiadiazolines.[246] They decomposed easily to thiiranes.[247] In still another variant, diazomethane reacts with thiotropone in 1:2 ratio to form a bis(cycloheptatrieno)dithiepin derivative.[248] Isothiocyanates add diazo compounds to form aminothiadiazoles.[213,249]

Sulfur dioxide reacts with diazo compounds in a mechanistically uncertain manner, apparently giving rise to sulfenes, $R_2C{=}S{=}O_2$, which may react with more diazo compound to form thiirane dioxides or may be hydrolyzed to sulfonic acids. It is possible that the sulfenes are decomposition products of an oxathiadiazoline initially formed by cycloaddition, but evidence is lacking[250-253] (Eq. 5-89). The

$$\text{Ph}_2C{=}N_2 + SO_2 \longrightarrow \left[\begin{array}{c} \text{Ph}_2C{-\!-\!-}N \\ \underset{OS}{|}\diagdown\underset{O}{}\diagup\overset{\|}{N} \end{array} \right] \longrightarrow \text{Ph}_2C{=}SO_2$$

$$\begin{array}{c} \diagup \text{Ph}_2\text{CN}_2 \qquad \diagdown \text{H}_2\text{O} \\ \underset{SO_2}{\overset{}{}} \qquad\qquad \text{Ph}_2\text{CHSO}_3\text{H} \qquad (5\text{-}89) \\ \text{Ph}_2C{-\!-\!-}\text{CPh}_2 \end{array}$$

intermediacy of sulfenes has been confirmed by generating them in a quite different way, dehydrohalogenation of alkanesulfonyl chlorides; the same sorts of products are formed. In some instances, the reaction of the sulfene with diazo compound leads instead to a cycloadduct, a 1,3,4-thiadiazoline dioxide. Both it and thiirane dioxides are converted to olefins on heating.[254]

Nitrile Imides (Nitrilimines)

The chemistry of this class of compounds, which are known only in solution, consists almost entirely of 1,3-addition or cycloaddition. They are generated in situ, usually by dehydrohalogenation of a hydrazonoyl chloride. The diphenyl derivative has been the most extensively investigated.[255]

Phenol, thiophenol, aniline, and benzoic acid add 1,3 to nitrile imides; O-to-N rearrangement follows in the case of phenols and benzoic acid, analogous to the Chapman rearrangement (Eq. 5-90).

$$
\text{PhC}\overset{+}{\equiv}\text{N}\overset{-}{-}\text{NPh}
\begin{cases}
\xrightarrow{\text{PhNH}_2} & \text{PhC}\overset{\nearrow \text{NHPh}}{\underset{\searrow \text{N}-\text{NHPh}}{}} \\[2ex]
\xrightarrow{\text{PhOH}} & \text{PhC}\overset{\nearrow \text{OPh}}{\underset{\searrow \text{N}-\text{NHPh}}{}} \xrightarrow{} \text{Ph}\overset{O}{\overset{\|}{\text{C}}}-\text{NHNPh}_2 \\[2ex]
\xrightarrow{\text{PhSH}} & \text{PhC}\overset{\nearrow \text{SPh}}{\underset{\searrow \text{N}-\text{NHPh}}{}} \\[2ex]
\xrightarrow{\text{PhCO}_2\text{H}} & \text{PhC}\overset{\nearrow \text{N}-\text{NHPh}}{\underset{\searrow \text{OCPh}}{}} \xrightarrow{} \text{PhC}-\text{NH}-\text{NCPh}
\end{cases}
$$

(5-90)

Cycloaddition occurs readily with strained olefins, such as bicyclopentadiene, and less readily with ordinary olefins, to form pyrazolines (Eq. 5-91). Cyclo-

$$
\text{PhC}\overset{+}{\equiv}\text{N}\overset{-}{-}\text{NPh} + n\text{-}\text{C}_5\text{H}_{11}\text{CH}=\text{CH}_2 \xrightarrow{90\,^\circ\text{C}}
$$

$$
\begin{array}{c}
n\text{-}\text{C}_5\text{H}_{11}-\text{CH}-\text{CH}_2 \\
| \qquad\qquad | \\
\text{Ph}-\text{N} \qquad \text{C}-\text{Ph} \\
\diagdown\text{N}\diagup
\end{array}
$$

(5-91)

addition to acetylenes produces pyrazoles. When cycloaddition is slow, dimerization to 1,3,4,6-tetraphenyl-1,4-dihydro-1,2,4,5-tetrazine becomes competitive. Benzaldehyde adds to form an oxadiazoline, and its anil adds to form a 1,2,4-triazoline (Eq. 5-92). Phenyl isocyanate and isothiocyanate also undergo addition at the C=N bond. Nitriles react to form triazoles (Eq. 5-92).

$$
\begin{array}{c}
\text{Ph}-\text{CH}-\text{O} \\
| \qquad\quad | \\
\text{Ph}-\text{N} \quad \text{C}-\text{Ph} \\
\diagdown\text{N}\diagup
\end{array}
\xleftarrow{\text{PhCH}=\text{O}}
\text{PhC}\overset{+}{\equiv}\text{N}\overset{-}{-}\text{NPh}
\xrightarrow{\text{PhCH}=\text{NPh}}
\begin{array}{c}
\text{Ph}-\text{CH}-\text{N}-\text{Ph} \\
| \qquad\quad | \\
\text{Ph}-\text{N} \quad \text{C}-\text{Ph} \\
\diagdown\text{N}\diagup
\end{array}
$$

$$
\Big\downarrow \text{PhCN}
$$

$$
\begin{array}{c}
\text{Ph}-\text{C}=\!=\text{N} \\
| \qquad\quad | \\
\text{Ph}-\text{N} \quad \text{C}-\text{Ph} \\
\diagdown\text{N}\diagup
\end{array}
$$

(5-92)

Azamines

The dominant chemistry of azamines is the competition between dimerization to a tetrazene and fragmentation to nitrogen and hydrocarbon products[256-266] (Eq. 5-93). They undergo these reactions rapidly upon formation by liberation from their

$$N_2 + 2R_2CH \cdot \longleftarrow (R_2CH)_2\overset{+}{N}=\overset{-}{N} \longrightarrow$$

$$(R_2CH)_2N-N=N-N(CHR_2)_2 \quad (5\text{-}93)$$

$$R_2CH-CHR_2 \quad \text{or} \quad \text{disproportionation products}$$

salts, oxidation of hydrazines, etc. There is an optimum pH for formation of tetrazenes, a fact that implies that both azamine and its cation are required for the reaction.[258] It has also been suggested that azamines might behave as aminonitrenes and insert in an N—H bond of the hydrazine from which they were being prepared; the resulting tetrazanes would be easily oxidized to tetrazenes.[262]

Fragmentation of azamines is favored by the presence on the α-carbons of groups that stabilize radicals or charges in the transition state. Oxidation of N-amino-2,6-dimethylpiperidine, for example, leads to a tetrazene through the azamine, but the 2,6-dicyano analog gives 1,2-dicyanocyclopentane as the major product under the same conditions.[267] Fragmentation with recombination of the radicals is sometimes known as the *Busch-Weiss reaction*.

Hydrazones have been reported from some azamine-producing reactions, and it seems probable that they result from tautomerization of an azo compound formed by migration[266] of a substituent from N to N' (Eq. 5-94).

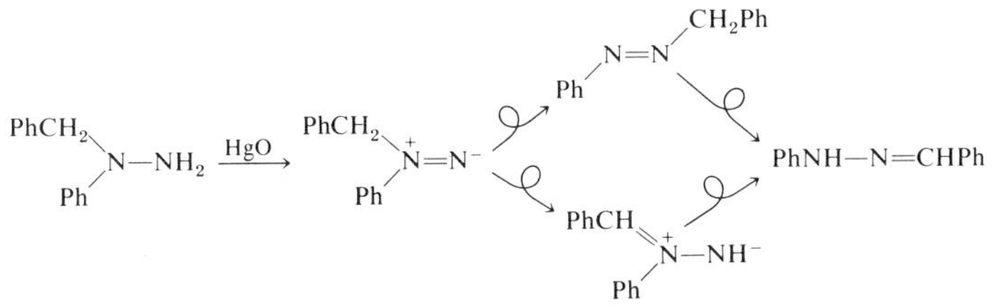

$$(5\text{-}94)$$

Reactions attributable to azaminium cations, $R_2\overset{+}{N}=\overset{-}{N}$, are to be found in Chapter 4 in conjunction with other diazenium ions.

PREPARATIVE METHODS[268]

Diazo Compounds

The most widely used methods for preparing diazo compounds are nitrosation (diazotization) of primary amines, elimination reactions of nitrosoamides and nitrosamino ketones, oxidation of hydrazones and the related Bamford-Stevens reaction of arenesulfonylhydrazones, and the diazo transfer process. In addition, there are a few less widely used reactions, such as the Forster reaction of oximes, the reaction of alkyllithium reagents with nitrous oxide, and the reaction of certain azoles with 2 moles of nitrous acid. Each has its own advantages and limitations, and no one method can be said to be generally the best. The principal variables are the accessibility of the starting materials; their storability, toxicity, and sensitivity; the conditions required to generate the diazo compound from the immediate precursor; the sensitivity of the diazo compound to heat; and its resistance to acid. Even for the most commonly prepared diazo compound, diazomethane, there is no agreement as to what is the best method.

The simplest method is nitrosation of a primary amine, but it is ordinarily successful only when there are acid-strengthening groups on the α-carbon, so that a proton can be ejected faster than nitrogen. Alkoxycarbonyl, sulfonyl, and perfluoroalkyl substituents, for example, allow the conversion to be carried out simply in aqueous solution. The principal use of this method is for preparing α-diazo esters[269a] (Eq. 5-95). Since the products may be destroyed by excessive exposure to

$$\text{EtO}_2\text{CCH}_2\text{NH}_3{}^+\text{Cl}^- + \text{NaNO}_2 \longrightarrow [\text{EtO}_2\text{CCH}_2\text{N}_2{}^+] \longrightarrow$$

$$\text{EtO}_2\text{CCH}{=}\text{N}_2 \quad (5\text{-}95)$$
$$79\text{--}88\%$$

strong acid, the preferred technique is to mix the amine hydrochloride with sodium nitrite without additional acid; the presence of methylene chloride to extract the diazo compound as it is formed improves the yield. Although many perfluoroalkyldi-azomethanes, such as 1-diazo-2,2,3,3,3-pentafluoropropane (58%), can be prepared in a similar way, the reaction fails for perfluorophenyldiazomethane,[269b] and 2,2,2-trichloroethylamine gives 1,1,1,2-tetrachloroethane instead of trichlorodiazo-ethane.[270] The method succeeds with anilines if there is an *ortho* or *para* substituent stituent bearing a sufficiently acidic hydrogen, as in *o*-aminophenol[24,47] (Eq. 5-96) or *p*-(dicyanomethyl)aniline[271] (Eq. 5-97), and with many amino pyrazoles, triazoles, etc.[272]

$$(5\text{-}96)$$

$$(5\text{-}97)$$

Direct diazotization can be applied even to methylamine[273] and octylamine[274] by using nitrosyl chloride at $-80\,°\text{C}$ and treating the resulting nitrosamine with alkali or by treating lithium methylamide with nitrosyl chloride[275] (Eq. 5-98).

$$\text{CH}_3\text{NH}_2 + \text{NOCl} \xrightarrow{-80\,°\text{C}} \text{CH}_2\text{NHNO} \xrightarrow{40\%\ \text{KOH}} \text{CH}_2{=}\text{N}_2 \quad (5\text{-}98)$$

Conversion of primary amines to diazo compounds may be considered as the elimination of water from a primary nitrosamine, R_2CHNHNO, and a closely related reaction is thus elimination of an acid, AOH, from a nitroso amide,

$R_2CHNANO$. This type of reaction is catalyzed by base and has many variants, according to the nature of the acyl group, A. One of the earliest and most widely used examples is the use of N-alkyl-N-nitrosourethans, which at one time was considered the method of choice for making diazomethane[53,276,277] (Eq. 5-99). Alco-

$$CH_3N\binom{CO_2Et}{NO} \xrightarrow{NaOEt} CH_2{=}N_2 + CO_3^{2-} + EtOH \tag{5-99}$$

holic sodium or potassium hydroxides or alkoxides are required, and it is necessary to distill the diazo compound (in the presence of added ether) from the reaction mixture in order to free it of contaminants. This feature is a substantial limitation, as are the instability to storage of the nitrosourethans and their highly irritating characteristics (N-methyl-N-nitrosourethan is also carcinogenic). In order to ameliorate these difficulties, various other types of N-alkyl-N-nitroso amides have been introduced, such as nitrosoureas[53a,278] (Eq. 5-100), nitrososulfonamides[279-281] (Eq. 5-101), nitrosooxamides[282] (Eq. 5-102), and nitrosoterephthalamides[283] (Eq. 5-103).

$$R_2CHN\binom{CONH_2}{NO} \xrightarrow{NaOEt} R_2C{=}N_2 + NaNCO \tag{5-100}$$

$$R_2CHN\binom{SO_2Ar}{NO} \xrightarrow{NaOH} R_2C{=}N_2 + ArSO_3Na \tag{5-101}$$

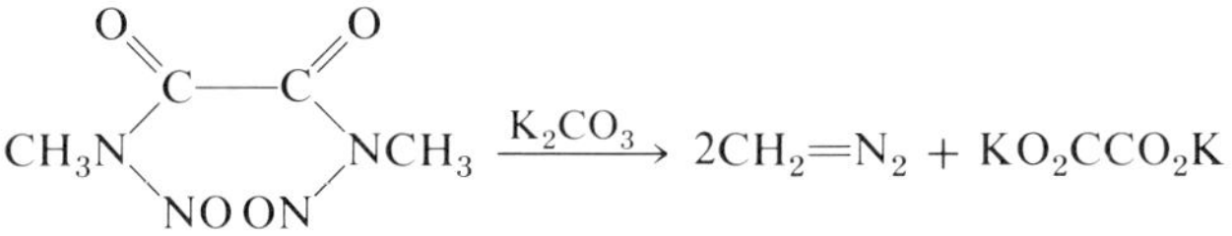

$$CH_3N\overset{O}{\underset{NO}{\|}}C{-}C\overset{O}{\underset{ON}{\|}}NCH_3 \xrightarrow{K_2CO_3} 2CH_2{=}N_2 + KO_2CCO_2K \tag{5-102}$$

$$CH_3N\overset{O}{\underset{NO}{\|}}C{-}\bigodot{-}C\overset{O}{\underset{ON}{\|}}NCH_3 \xrightarrow{NaOH}$$

$$2CH_2{=}N_2 + NaO_2CC_6H_4CO_2Na \tag{5-103}$$

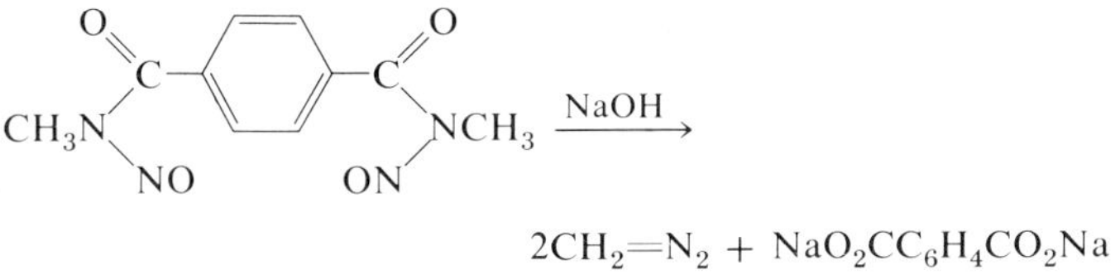

N-Methyl-N-nitroso-p-toluenesulfonamide has been sold commercially as a source of diazomethane under the names *Diazald* and *Diactin*, and N,N'-dimethyl-N,N'-dinitrosoterephthalamide under the name *EXR-101*. All these sources are satisfactory for diazomethane, and N-methyl-N-nitrosourea has the special advantage that ethereal diazomethane can be prepared from it without distillation. Not all these methods work so well for larger diazo compounds, but N-nitrosoacetamides have been found useful for substituted diazo compounds (Eq. 5-104) of some variety[157,284-286] (Eq. 5-105).

Closely related to the foregoing reactions is that of N-methyl-N-nitroso-N'-nitroguanidine, which is a useful source of diazomethane[106,122,287,288] (Eq. 5-106). It

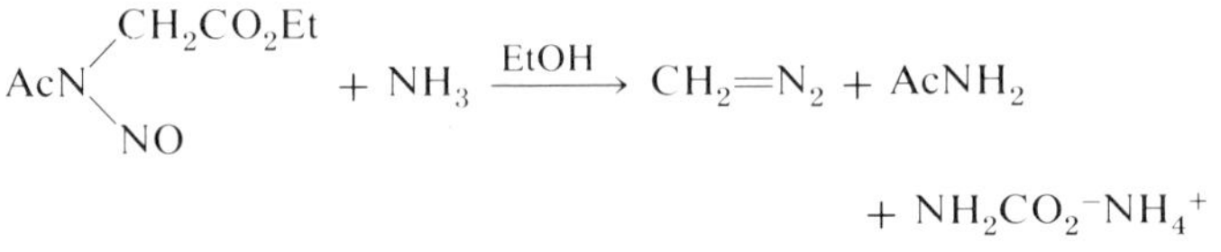

$$AcN\begin{array}{c} {}^{CH_2CO_2Et} \\ {} \\ {}_{NO} \end{array} + NH_3 \xrightarrow{EtOH} CH_2=N_2 + AcNH_2$$

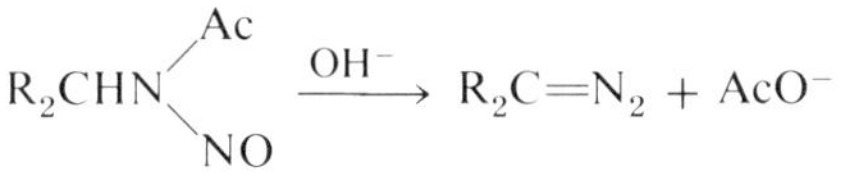

$$+ NH_2CO_2^-NH_4^+ \qquad (5\text{-}104)$$

$$R_2CHN\begin{array}{c} {}^{Ac} \\ {} \\ {}_{NO} \end{array} \xrightarrow{OH^-} R_2C=N_2 + AcO^- \qquad (5\text{-}105)$$

$$CH_3N\begin{array}{c} {}^{NO} \\ {} \\ {}_{C=NNO_2} \\ {}_{\mid} \\ {}_{NH_2} \end{array} \xrightarrow{KOH} CH_2=N_2 + NCNNO_2^-K^+ \qquad (5\text{-}106)$$

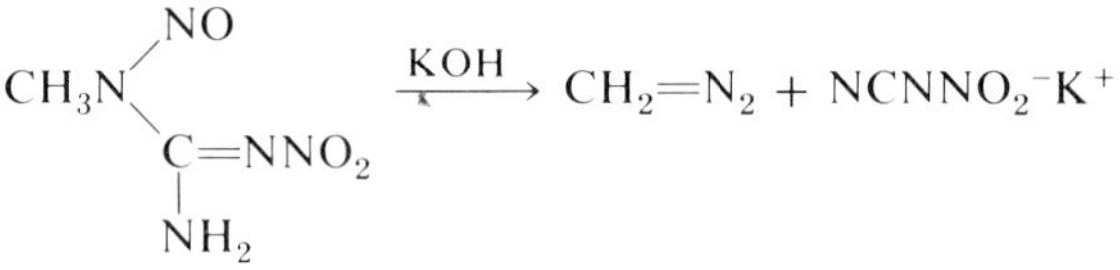

is more stable to heat than nitrosoureas. Some higher diazoalkanes have been made successfully by this method.[289]

Nitrosamines with a labile β-hydrogen can serve as effective sources of diazo compounds. The best-known example is a derivative of mesityl oxide, and it has been widely used to prepare diazomethane[290] (Eq. 5-107). Such compounds are

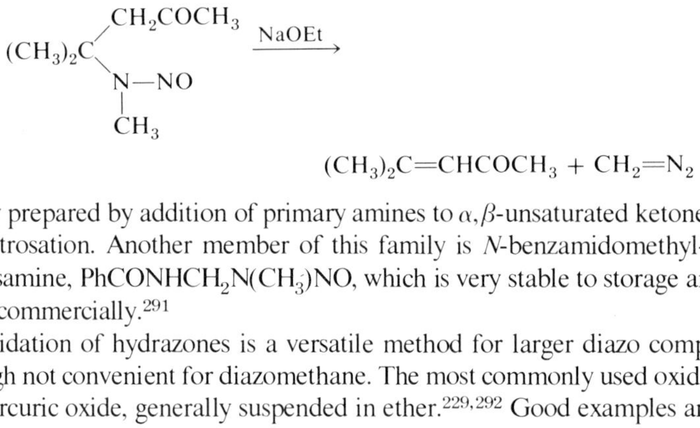

$$(CH_3)_2C\begin{array}{c} {}^{CH_2COCH_3} \\ {} \\ {}_{N-NO} \\ {}_{\mid} \\ {}_{CH_3} \end{array} \xrightarrow{NaOEt}$$

$$(CH_3)_2C=CHCOCH_3 + CH_2=N_2 \qquad (5\text{-}107)$$

easily prepared by addition of primary amines to α,β-unsaturated ketones followed by nitrosation. Another member of this family is N-benzamidomethyl-N-methyl-nitrosamine, $PhCONHCH_2N(CH_3)NO$, which is very stable to storage and is available commercially.[291]

Oxidation of hydrazones is a versatile method for larger diazo compounds, although not convenient for diazomethane. The most commonly used oxidizing agent is mercuric oxide, generally suspended in ether.[229,292] Good examples are benzoyl-phenyldiazomethane ("azibenzil")[293] (87–94%) and 2-diazopropane[294] (Eq. 5-108).

$$(CH_3)_2C=O \longrightarrow (CH_3)_2C=NNH_2 \xrightarrow{HgO/KOH} (CH_3)_2C=N_2 \qquad (5\text{-}108)$$
$$70\text{-}90\%$$

A trace of strong alkali is usually required as a catalyst, but the reaction is even then inclined to be capricious, and different batches of mercuric oxide vary widely in activity for no obvious reason. Other oxidizing agents that have been found of value are manganese dioxide, silver oxide, lead tetraacetate, $1,1$-diacetoxyiodobenzene ("phenyliodoso acetate"), etc. Manganese dioxide is particularly effective, but it must be prepared in an "active" form from potassium permanganate and manganous sulfate.[295] Hydrazones of both aldehydes[296,297] and ketones[298] have been successfully oxidized with this reagent. Silver oxide has the advantage of oxidizing hydrazones rapidly and thus minimizing side reactions. Diazocyclohexane, for example, has been prepared with its use, whereas mercuric oxide failed, giving only azine.[299] Formation of azine may be attributed to disproportionation of the hydrazone into hydrazine and azine before oxidation can take place. Lead tetraacetate is frequently effective, but it has the limitations that excess reagent can cause secondary reactions, and the diazo compound may be sensitive to the acetic acid produced.[202,300] With *1,1*-diacetoxyiodobenzene, a similar reagent, the latter interference has been avoided by having an amine present.[86]

The base-catalyzed decomposition of sulfonylhydrazones (Bamford-Stevens reaction) may be looked upon as a case of internal oxidation of a hydrazone, as well as an elimination of sulfinic acid[301-303] (Eq. 5-109). It is usually carried out with use of

$$\text{PhCH}=\text{NNHSO}_2\text{Ar} \xrightarrow[50-80°C]{\text{NaOCH}_3/\text{C}_5\text{H}_5\text{N}} \underset{55-70\%}{\text{PhCH}=\text{N}_2} + \text{ArSO}_2\text{Na} \qquad (5\text{-}109)$$

sodium alkoxides or potassium hydroxide with mild heating in alcoholic solvents, often ethylene glycol. Purely aliphatic diazoalkanes may not survive these conditions, and for them an alternative method involves isolating the sodium derivative of the tosylhydrazone and heating it in a vacuum.[304] A large amount of work with tosylhydrazones has been carried out in connection with investigations of carbene chemistry and carbenium ion reactions (see Chap. 3). Obviously, the Bamford-Stevens reaction is not suitable for isolation of heat-labile diazo compounds. It can, however, be carried out under only mildly basic conditions, such as with triethylamine[165] (Eq. 5-110) or basic alumina.[305]

$$\text{TosNHN}=\text{CHCOCl} + \text{ROH} + 2\text{Et}_3\text{N} \longrightarrow \text{N}_2=\text{CHCO}_2\text{R} \qquad (5\text{-}110)$$
$$\text{(R = crotyl, 54\%)}$$

The diazo transfer reaction[306] consists of reaction of a nucleophilic carbon site with a bearer of two nitrogens, usually an arenesulfonyl azide, essentially according to Equation 5-111. It may be accompanied by loss of one of the acyl groups (A)

$$\text{TosN}_3 + \text{A}_2\text{CH}^- \longrightarrow \text{A}_2\text{C}=\text{N}_2 + \text{TosNH}^- \qquad (5\text{-}111)$$

from the nucleophilic carbon. The nucleophilic site may be a carbanion, enolate ion, alkene, or alkyne, especially if the latter two are electron-rich, as in enamines and ynamines. The diazo donor may be an azide of various types or even another diazo compound. Most of the azide donors react by forming an intermediate triazene, which will generally not cleave to a diazo compound unless the nucleophilic carbon bears electron-withdrawing groups or has analogous characteristics. Metallocyclopentadienes, for example, are converted to diazocyclopentadienes by tosyl azide,[307,308] but alkyllithium reagents and Grignard reagents are not. The reaction is most suitable for use with active methylene compounds, such as β-keto esters, β-diketones, and their sulfonyl analogs. Diazomalonic ester, for example, can be prepared in one step from diethyl malonate and tosyl azide in 95% yield in acetonitrile containing triethylamine.[309] Acetoacetic esters are converted to the corresponding α-diazo esters in high yields under similar conditions (Eq. 5-112), but with more strongly basic catalysts, such as sodium ethoxide, deacetylation and other

$$CH_3COCH_2CO_2R + TosN_3 \xrightarrow[CH_3CN]{Et_3N} \underset{\underset{N_2}{\parallel}}{CH_3COCCO_2R} \qquad (5\text{-}112)$$

secondary reactions may interfere.[310] In some instances, even triethylamine may bring about deacylation, as with 2-hydroxymethylenecyclohexanone, which affords up to 95% of 2-diazocyclohexanone,[311] and *tert*-butyl acetoacetate, which affords *tert*-butyl diazoacetate[312] (Eq. 5-113).

$$CH_3COCH_2CO_2\text{—}t\text{-}Bu + TosN_3 \xrightarrow{Et_3N} N_2{=}CHCO_2\text{—}t\text{-}Bu \qquad (5\text{-}113)$$

Enamines undergo cleavage at the C=C bond through intermediate amino-triazolines, with formation of diazo compounds and amidines[313] (Eq. 5-114). The

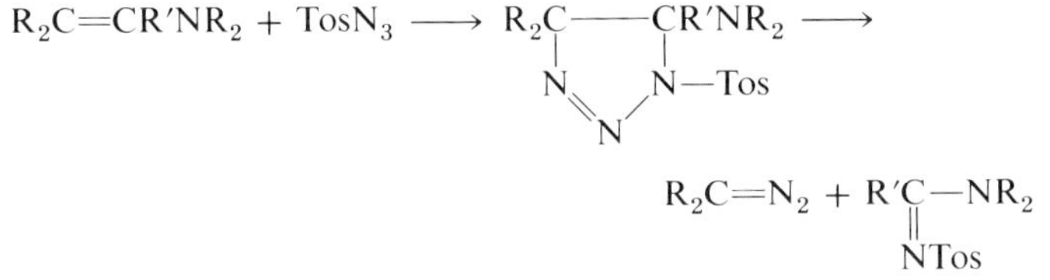

$$R_2C{=}CR'NR_2 + TosN_3 \longrightarrow \underset{\underset{\underset{N}{\diagdown}{\diagup}}{N}}{R_2C}\underset{}{\text{———}}\underset{}{CR'NR_2} \longrightarrow$$

$$R_2C{=}N_2 + \underset{\underset{NTos}{\parallel}}{R'C\text{—}NR_2} \qquad (5\text{-}114)$$

reaction can be carried out in situ with the enamine precursors (secondary amine and carbonyl compound). Activating substituents are not needed on the portion that is to become the diazo compound, and even diazomethane can be prepared in moderately good yields. Much work has been done with keto ynamines; the diazo and amidine functions remain in the same molecule, and α-diazo-β-ketoamidines are produced in good yields[314,315] (Eq. 5-115).

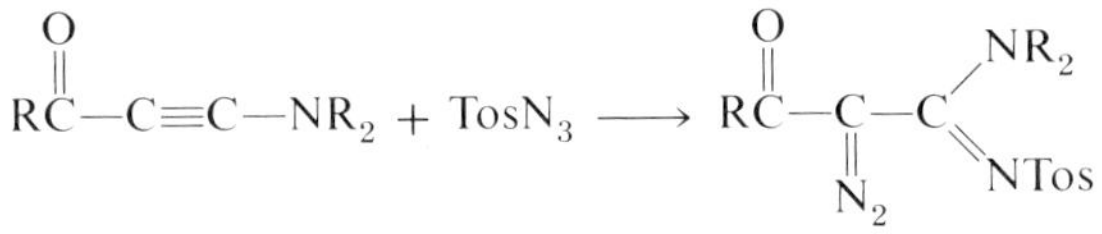

$$\underset{\underset{O}{\parallel}}{RC}\text{—}C{\equiv}C\text{—}NR_2 + TosN_3 \longrightarrow \underset{\underset{O}{\parallel}}{RC}\text{—}\underset{\underset{N_2}{\parallel}}{C}\text{—}C\underset{\diagdown NTos}{\overset{\diagup NR_2}{}} \qquad (5\text{-}115)$$

The principal alternative diazo transfer reagents are the azidinium compounds, such as *N*-ethyl-2-azidopyridinium salts.[316] They react with active methylene compounds under very mild conditions (Eq. 5-116). Azidoiminium compounds react so

$$(EtO_2C)_2CH_2 \xrightarrow[NaOAc, 40°C, 15\ min.]{aq.\ EtOH,\ N_3C_5H_4\overset{+}{N}Et} (EtO_2C)_2C{=}N_2 \qquad (5\text{-}116)$$

$$88\%$$

readily with active methylene compounds, even at room temperature, that a base is not required.[317] Diazoalkanes as diazo transfer agents have been little explored, but their potential is shown by the preparation of a diazo alkene from 2-diazopropane and diphenylacetylene[318] (Eq. 5-117). This method is essentially a photolytic isomerization of nonaromatic triazoles formed by cycloaddition.

Nitrous oxide can be considered as a diazo transfer reagent, and it does indeed react with methyllithium, forming a lithium salt of "isodiazomethane," from which diazomethane can be liberated by alkali[319] (Eq. 5-118). Nitrous oxide will also transfer the diazo group to methylenephosphoranes[320] (5-119).

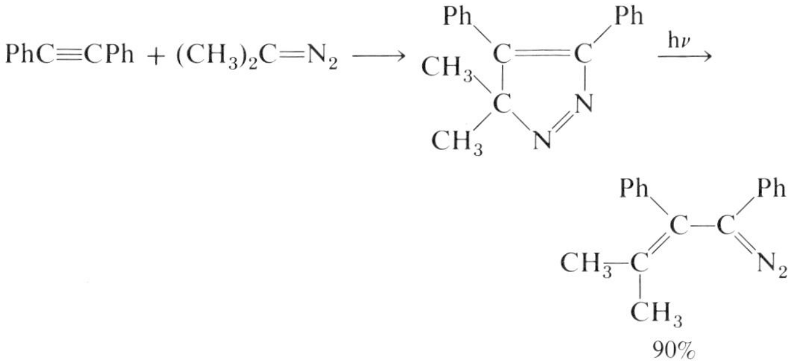

$$PhC{\equiv}CPh + (CH_3)_2C{=}N_2 \longrightarrow \underset{CH_3}{\overset{Ph}{\diagdown}}\underset{}{\text{C}{=}\text{C}}\overset{Ph}{\diagup} \xrightarrow{h\nu}$$

$$\overset{Ph}{\diagdown}\text{C}{-}\text{C}\overset{Ph}{\diagup} \quad (5\text{-}117)$$

90%

$$2CH_3Li + N_2O \xrightarrow{-80\,^{\circ}C} LiCHN_2 \xrightarrow{\text{aq. KOH}} CH_2{=}N_2 \qquad (5\text{-}118)$$

ca. 70%

$$Ph_3P{=}CH_2 + N_2O \xrightarrow[\text{aq. KOH}]{Et_2O,\ 0\,^{\circ}C} CH_2{=}N_2 + Ph_3PO \qquad (5\text{-}119)$$

The reaction of oximes with chloramine or hydroxylamine-O-sulfonic acid in the presence of base, known as the *Forster reaction* (Eq. 5-120), has not been so widely

$$R_2C{=}NOH + NH_2X \xrightarrow{\text{base}} R_2C{=}N_2 + H_2O + X^- \qquad (5\text{-}120)$$

explored as the foregoing methods, but it has the promise of considerable generality.[302,321,322] It has been used to make diazomethane, but it does not appear to work well with aliphatic ketoximes. It is especially well suited for α-diazo ketones, for the required oximes can be made directly from the ketone by nitrosation. In this connection, it has been used in the steroid field to prepare α,α-bisdiazo ketones[323] (Eq. 5-121). A reversal of this reaction, condensation of N,N-dichloroamines with

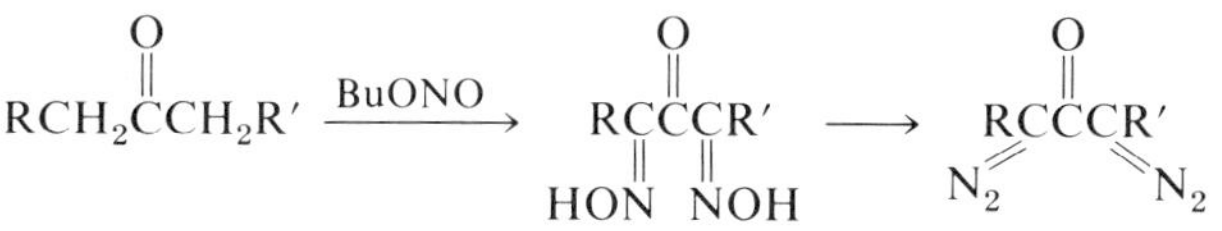

$$\underset{}{RCH_2\overset{\overset{\text{O}}{\|}}{C}CH_2R'} \xrightarrow{\text{BuONO}} \underset{HON\ \ NOH}{R\overset{\overset{\text{O}}{\|}}{C}C\ CR'} \longrightarrow \underset{N_2\ \ \ N_2}{R\overset{\overset{\text{O}}{\|}}{C}C\ CR'} \qquad (5\text{-}121)$$

hydroxylamine in the presence of sodium methoxide, has so far given only poor yields (ca. 25%)[324] (Eq. 5-122).

$$R_2CHNCl_2 + NH_2OH \xrightarrow[CH_3OH]{NaOCH_3} R_2C{=}NOH \qquad (5\text{-}122)$$

Azine N-oxides can be cleaved to diazoalkanes by heat, light, or traces of acid,[325] but the preparative utility of the reaction remains to be established. Reduction of nitrosimines with lithium aluminum hydride (Eq. 5-123) has been found to give

$$Ar_2C{=}N{-}NO + LiAlH_4 \xrightarrow{-30\,^{\circ}C} Ar_2C{=}N_2 \qquad (5\text{-}123)$$

ca. 50%

better yields, especially of hindered diaryl diazomethanes, than more conventional routes,[60] but the method has seen little use.

It has been reported indirectly that *N*-nitroso-*N*-methyl-*p*-nitroaniline is cleaved by alkali to diazomethane,[326] but time has passed the reaction by. *N*-Nitroso-*N*-methylhydrazine has been cleaved to diazomethane[327] (Eq. 5-124), and methylnitramide has been reduced to diazomethane[214] (Eq. 5-125), but neither compound is a practical starting material.

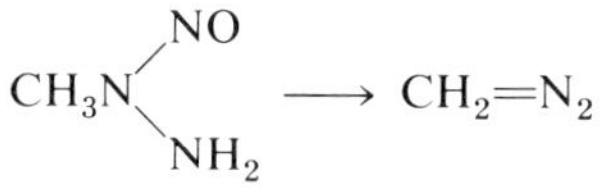

$$CH_3N \begin{array}{c} NO \\ \diagup \\ \diagdown \\ NH_2 \end{array} \longrightarrow CH_2{=}N_2 \qquad (5\text{-}124)$$

$$CH_3NHNO_2 \xrightarrow{[H]} CH_2{=}N_2 \qquad (5\text{-}125)$$

The reaction of chloroform with hydrazine in the presence of alkali affords some diazomethane[328] (Eq. 5-126), but the reaction is probably restricted by mechanism

$$CHCl_3 + N_2H_4 \xrightarrow{OH^-} CH_2{=}N_2 \qquad (5\text{-}126)$$
$$ca.\ 20\%$$

to preparation of diazomethane. It has been adapted to the preparation of $CD_2{=}N_2$ from $CDCl_3$ and N_2D_4.[329]

Nitrile Imides (Nitrilimines)

Dehydrohalogenation of hydrazonoyl chlorides is a mild reaction of some generality, suitable for preparation of nitrile imides in situ for further reaction[255] (Eq. 5-127). Thermolysis of 1,3-disubstituted tetrazoles gives rise to nitrile imides[255,330,331] (Eq. 5-128), but the more drastic conditions required make it a less desirable method.

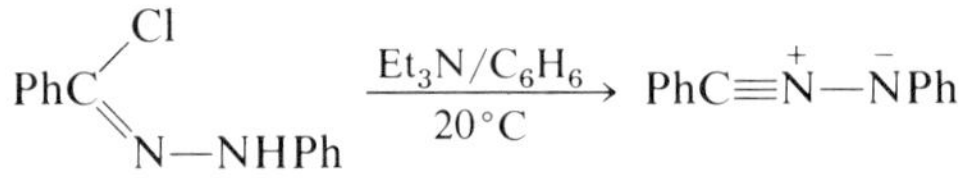

$$PhC \begin{array}{c} Cl \\ \diagup \\ \diagdown \\ N{-}NHPh \end{array} \xrightarrow[20\,°C]{Et_3N/C_6H_6} PhC{\equiv}\overset{+}{N}{-}\overset{-}{N}Ph \qquad (5\text{-}127)$$

$$Ph{-}C \begin{array}{c} N{-}N{-}PH \\ \diagup \quad | \\ \diagdown \quad \\ N{=}N \end{array} \xrightarrow{150\text{-}160\,°C} PhC{\equiv}\overset{+}{N}{-}\overset{-}{N}Ph + N_2 \qquad (5\text{-}128)$$

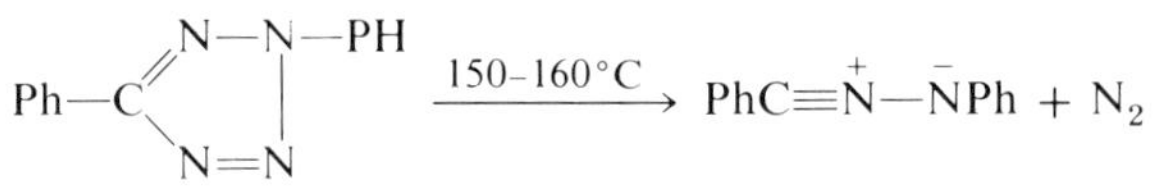

Azamines

Those azamines which have been isolated have been prepared by oxidation of 1,1-disubstituted hydrazines with *tert*-butyl hypochlorite[5,6] (Eq. 5-129).

$$R_2N{-}NH_2 + t\text{-BuOCl} \xrightarrow[-78\,°C]{Et_2N,\ Et_2O} R_2\overset{+}{N}{=}N^- \qquad (5\text{-}129)$$

ANALYTICAL METHODS

Qualitative detection of diazo compounds can usually be accomplished by a combination of infrared and UV-vis spectroscopy,[9,11] as described at the beginning of this chapter; color itself is a useful sign. Diazomethane, for example, has been determined by its absorption at 410 nm.[16] Specific identification can generally be made by means of the ready reaction with acids; if nitrobenzoic acids are used, the resulting esters are usually crystalline, and large numbers of them are to be found in tables of melting points of derivatives of alcohols. The evolution of nitrogen that accompanies the reaction serves as an additional confirmation.

A curious color reaction that has some potential for detecting the presence of the diazo function is the formation of a red suspension of colloidal gold when the compound is brought into contact with aqueous $AuCl_3$; at the point when all the diazo compound is used up, the color is said to change to blue.[332] Another color reaction is that with 4-(p-nitrobenzyl)pyridinium perchlorate, which becomes alkylated at N; base abstracts a benzylic proton and generates an intensely colored quinonoid structure (benzylidene-1,4-dihydropyridines).[333]

Quantitative determination of the diazo function can be managed volumetrically, using a measured excess of an acid such as p-nitrobenzoic and back-titrating the unused portion.[334] If alcohol and water are avoided, conversion of the reactants to ester is essentially quantitative, and the ester can also be isolated and weighed. There have also been attempts to develop a volumetric determination on the basis of the reaction of the diazo function with iodine, but the method has shown itself to be unreliable.[328] Two colorimetric methods have been described. One makes use of 4-(p-nitrobenzyl)pyridinium perchlorate (mentioned earlier). The other is based on the fact that colchiceine produces a strong green color with ferric chloride, but diazomethane reacts with colchiceine to form colchicine, which produces no color.[335] Less conveniently, the volume of nitrogen evolved in reaction with acid or iodine can be measured.[336,337]

A more extensive treatment of the subject of detection and analysis, including tables of spectrographic information, is available elsewhere.[9,11]

REFERENCES

1. (*a*) N. V. Sidgwick, L. E. Sutton, and W. Thomas, *J. Chem. Soc.,* 406 (1933); (*b*) A. P. Cox, L. F. Thomas, and J. Sheridan, *Nature* **181**, 1000 (1958).

2. E. Müller and D. Ludsteck, *Chem. Ber.* **87**, 1887 (1954).

3. E. Müller, P. Kastner, R. Beutler, W. Rundel, H. Suhr, and B. Zach, *Liebig's Ann. Chem.* **713**, 87 (1968).

4. W. P. Feldhammer, P. Baračas, and K. Bartel, *Angew. Chem. Int. Ed.* **16**, 707 (1977).

5. W. D. Hinsberg, III, and P. B. Dervan, *J. Am. Chem. Soc.* **100**, 1608 (1978); W. D. Hinsberg, III, P. G. Schultz, and P. B. Dervan, *J. Am. Chem. Soc.* **104**, 766 (1982).

6. P. G. Schultz and P. H. Dervan, *J. Am. Chem. Soc.* **102**, 878 (1980).

7. W. Graham, *J. Am. Chem. Soc.* **84**, 1063 (1962).

8. E. Schmitz, R. Ohme, and R.-D. Schmidt, *Chem. Ber.* **95**, 2714 (1962).

9. D. A. Ben-Efraim, in *The Chemistry of Diazo and Diazonium Groups*, S. Patai (ed.), Wiley, New York, 1978, Chap. 5.

10. A. Foffani, C. Pecile, and S. Ghersetti, *Tetrahedron* **11**, 285 (1960).

11. R. F. Muracca, in *Treatise of Analytical Chemistry*, Part II, Vol. 15, I. M. Kolthoff and P. J. Elving (eds.) Wiley, New York, 1976, pp. 347–384.

12. A. Ledwith and E. C. Friedrich, *J. Chem. Soc.*, 504 (1964).

13. R. O. Duthaler, H. G. Förster, and J. D. Roberts, *J. Am. Chem. Soc.* **100**, 4974 (1978).

14. J. F. McGarrity and T. Smyth, *J. Am. Chem. Soc.* **102**, 7303 (1980).

15. S. P. Markey and G. J. Shaw, *J. Org. Chem.* **43**, 3414 (1978).

16. P. G. Gassman and W. J. Greenlee, *Org. Syntheses* **53**, 38 (1973).

17. B. L. Crawford, W. H. Fletcher, and D. A. Ramsay, *J. Chem. Phys.* **19**, 406 (1951).

18. H. Boersch, *Monatsh. Chem.* **65**, 311 (1935).

19. K. Clusius and U. Lüthi, *Chimia* **8**, 96 (1954).

20. H. M. Niemeyer, *Helv. Chim. Acta* **60**, 1487 (1977).

21. J. A. Moore, J. R. Dice, E. D. Nicolaides, R. D. Westland, and E. L. Wittle, *J. Am. Chem. Soc.* **76**, 2884 (1954).

22. H. W. Dion, S. A. Fusari, Z. L. Jakubowski, J. G. Zora, and Q. R. Bartz, *J. Am. Chem. Soc.* **78**, 3075 (1956).

23. E. L. Patterson, B. L. Johnson, S. E. De Voe, and N. Bohonos, *Antimicrobial Agents Chemotherapy*, 115 (1965); *Chem. Abstr.* **65**, 10665c (1966).

24. (*a*) V. V. Ershov, G. A. Nikiforov, and C. R. H. I. De Jonge, *Quinone Diazides*, Elsevier, Amsterdam and New York, 1981; (*b*) J. K. Stille, P. Cassidy, and L. Plummer, *J. Am. Chem. Soc.* **85**, 1318 (1963).

25. D. J. Cram and C. K. Dalton, *J. Am. Chem. Soc.* **85**, 1268 (1963).

26. R. Huisgen, *Angew. Chem.* **67**, 439 (1955).

27. F. Egon, *Liebig's Ann. Chem.* **638**, 1 (1960).

28. H. Zollinger, *Diazo and Azo Chemistry*, Interscience Publishers, New York, 1961.

29. S. Patai (ed.), *The Chemistry of Diazonium and Diazo Groups*, Wiley, New York, 1978, Chaps. 4, 6, 7, 8, 9, 10, 12, 13.

30. D. Bethell, "Structure and Mechanism in Carbene Chemistry", in *Adv. Phys. Org. Chem.* **7**, 153 (1969).

31. T. L. Gilchrist and C. W. Rees, *Carbenes, Nitrenes, and Arynes*, Appleton-Century-Crofts, New York, 1969.

32. V. Dave and E. W. Warnhoff, "The Reactions of Diazoacetic Ester with Alkenes, Alkynes, Heterocyclic and Aromatic Compounds", *Org. Reactions*, **18**, Chap. 3 (1970).

33. W. Kirmse, *Carbene Chemistry*, 2d ed., Academic Press, New York, 1971.

34. W. J. Baron et al., in *Carbenes*, Vol. 1, M. Jones, Jr., and R. A. Moss (eds.), Wiley, New York, 1973, Chap. 1.

35. A. P. Marchand and N. M. Brockway, *Chem. Rev.* **74**, 431 (1974) (Carboalkoxy Carbenes).

36. T. Curtius, A. Darapsky, and E. Müller, *Ber.* **41**, 3168 (1908), fn. 2.

37. E. Bamberger and F. Tschirner, *Ber.* **33**, 955 (1900).

38. H. Dürr, *Topics Curr. Chem.* **55**, 87 (1975).

39. G. Cowell and A. Ledwith, *Quart. Rev.* **24**, 119 (1970).

40. L. Friedman and H. Shechter, *J. Am. Chem. Soc.* **81**, 5512 (1959).

41. E. Wenkert and C. A. McPherson, *J. Am. Chem. Soc.* **94**, 8084 (1972).

42. P. A. S. Smith, in *Molecular Rearrangements*, P. de Mayo (ed.), Wiley, New York, 1963, Chap. 8.

43. W. M. Jones, in *Rearrangements in Ground and Excited States*, P. de Mayo (ed.), Academic Press, New York, 1980, Essay 3.

44. H. Meier and K.-P. Zeller, *New Synthetic Methods* (*Verlag Chemie*) **4**, 1 (1979).

45. K. B. Wiberg and T. Hutton, *J. Am. Chem. Soc.* **78**, 1640 (1956).

46. P. Yates, *J. Am. Chem. Soc.* **74**, 5376 (1952).

47. P. A. S. Smith and W. L. Berry, *J. Org. Chem.* **26**, 27 (1961).

48. R. Huisgen, H. König, G. Binsch, and H. J. Sturm, *Angew. Chem.* **73**, 368 (1961).

49. M. Tanaka, T. Nagai, and N. Tokura, *J. Org. Chem.* **38**, 1603 (1973).

50. H. D. Roth and M. L. Manion, *J. Am. Chem. Soc.* **98**, 3392 (1976).

51. (*a*) A. M. Trozzolo, *Accts. Chem. Res.* **1**, 329 (1968); (*b*) H. Tomika, H. Okuno, and Y. Izawa, *J. Org. Chem.* **45**, 5278 (1980).

52. G. Frater and O. P. Strausz, *J. Am. Chem. Soc.* **92**, 6654 (1970).

53. (*a*) W. E. Bachmann and W. A. Struve, *Org. Reactions* **1**, Chap. 2 (1942); (*b*) B. Eistert and F. W. Spangler, *Newer Methods of Preparative Organic Chemistry*, Interscience Publishers, New York, 1948.

54. F. Weygand and H. J. Bestmann, *Angew. Chem.* **72**, 535 (1960).

55. C. Huggett, R. T. Arnold, and T. I. Taylor, *J. Am. Chem. Soc.* **64**, 3043 (1942).

56. A. L. Wilds and A. L. Meader, *J. Org. Chem.* **13**, 763 (1948).

57. L. Hellerman and R. L. Garner, *J. Am. Chem. Soc.* **57**, 139 (1935).

58. P. Yates, D. G. Farnum, and D. W. Wiley, *Tetrahedron* **18**, 881 (1962).

59. H. Reimlinger, *Chem. Ber.* **97**, 339 (1964).

60. H. E. Zimmerman and D. H. Paskovich, *J. Am. Chem. Soc.* **86**, 2149 (1964).

61. C. G. Overberger and J.-P. Anselme, *J. Org. Chem.* **29**, 1188 (1964).

62. H. Nozaki, H. Takaya, S. Moriuti, and R. Noyori, *Tetrahedron* **24**, 3655 (1968).

63. E. Ciganek, *J. Org. Chem.* **30**, 4198 (1965).

64. H. Reimlinger, *Chem. Ber.* **97**, 3503 (1964).

65. W. Kirmse, L. Horner, and H. Hoffmann, *Liebig's Ann. Chem.* **614**, 19 (1958).

66. G. L. Closs and A. D. Trifunac, *J. Am. Chem. Soc.* **92**, 2186, 4549, 4550 (1970).

67. W. von E. Doering, R. G. Buttery, R. G. Laughlin, and N. Chaudhuri, *J. Am. Chem. Soc.* **78**, 3224 (1956).

68. D. B. Richardson, M. C. Simmons, and I. Dvoretzky, *J. Am. Chem. Soc.* **82**, 5001 (1960).

69. H. Meerwein, H. Rathjen, and H. Werner, *Ber.* **75**, 1610 (1942).

70. V. Franzen and L. Fikentscher, *Liebig's Ann. Chem.* **617**, 1 (1958).

71. H. M. Frey and M. A. Voisey, *J. Chem. Soc. Chem. Commun.*, 454 (1966).

72. J. A. Kerr, B. V. O'Grady, and A. F. Trotman-Dickinson, *J. Chem. Soc.* [*A*], 897 (1967).

73. V. Franzen and H. Kuntze, *Liebig's Ann. Chem.* **627**, 15 (1960).

74. A. Schönberg, O. Schütz, and J. Peter, *Ber.* **62**, 440 (1929).

75. W. H. Urry and J. R. Eiszner, *J. Am. Chem. Soc.* **74**, 5822 (1952).

76. V. Franzen, *Liebig's Ann. Chem.* **627**, 22 (1960).

77. M. Hudlicky and J. König, *Coll. Czech. Chem. Commun.* **28**, 2824 (1963).

78. H. Meerwein, H. Disselnkötter, F. Rappen, H. von Rintelen, and H. van der Vloed, *Liebig's Ann. Chem.* **604**, 151 (1957).

79. R. C. Woodworth and P. S. Skell, *J. Am. Chem. Soc.* **81**, 3383 (1959).

80. D. R. Ring and B. S. Rabinovitch, *Int. J. Chem. Kinet.* **1**, 11 (1969).

81. W. von E. Doering and L. H. Knox, *J. Am. Chem. Soc.* **75**, 297 (1953).

82. R. M. Lemmon and W. Strohmeier, *J. Am. Chem. Soc.* **81**, 106 (1959).

83. G. Linstrumelle, *Tetrahedron Lett.*, 85 (1970).

84. J. E. Baldwin and R. A. Smith, *J. Am. Chem. Soc.* **89,** 1886 (1967).

85. J. E. Baldwin and R. A. Smith, *J. Org. Chem.* **32,** 3511 (1967).

86. (*a*) P. A. S. Smith and E. M. Bruckmann, *J. Org. Chem.* **39,** 1047 (1974); (*b*) C. D. Bedford, E. M. Bruckmann, and P. A. S. Smith, *J. Org. Chem.* **46,** 679 (1981).

87. W. von E. Doering and T. Mole, *Tetrahedron* **10,** 65 (1960).

88. T. Terao, N. Sakai, and S. Shida, *J. Am. Chem. Soc.* **85,** 3919 (1963).

89. H. Lind and A. J. Deutschman, *J. Org. Chem.* **32,** 326 (1967).

90. M. Vidal, F. Massot, and P. Arnaud, *Compt. Rend. Acad. Sci. Paris* [*C*] **268,** 423 (1969).

91. W. Ando, T. Yagihara, S. Tozune, and T. Migita, *J. Am. Chem. Soc.* **91,** 2786 (1969).

92. I. Zugravescu, E. Rucinschi, and G. Surpateanu, *Tetrahedron Lett.*, 941 (1970).

93. W. Ando, K. Nakayama, K. Ichibori, and T. Migita, *J. Am. Chem. Soc.* **91,** 5164 (1969).

94. W. Ando, S. Kondo, and T. Migita, *J. Am. Chem. Soc.* **91,** 6516 (1969).

95. H. Staudinger, E. Anthes, and F. Pfenninger, *Ber.* **49,** 1928 (1916).

96. P. D. Bartlett and T. G. Trayler, *J. Am. Chem. Soc.* **84,** 3408 (1962).

97. R. W. Murray and D. P. Higley, *J. Am. Chem. Soc.* **95,** 7886 (1973).

98. D. P. Higley and R. W. Murray, *J. Am. Chem. Soc.* **96,** 3330 (1974).

99. C. Hillhouse, D. S. Wulfman, and B. Poling, unpublished results quoted by D. S. Wulfman, G. Linstrumelle, and C. F. Cooper in ref. 29, Chap. 18.

100. M. Jones, Jr., and K. R. Rettig, *J. Am. Chem. Soc.* **87,** 4013, 4015 (1965).

101. U. Schöllkopf, D. Hoppe, N. Rieber, and V. Jacobi, *Liebig's Ann. Chem.* **730,** 1 (1969).

102. E. Buchner, *Ber.* **28,** 215 (1895).

103. A. P. Marquand, in *The Chemistry of Double-Bonded Functional Groups, Suppl. A,* S. Patai (ed.), Wiley, New York, 1977, Chap. 7.

104. W. Kirmse and H. Dietrich, *Chem. Ber.* **98,** 4027 (1965).

105. E. Müller and H. Kessler, *Liebig's Ann. Chem.* **692,** 58 (1966).

106. (*a*) M. P. Doyle, W. H. Tamblyn, and V Bagheri, *J. Org. Chem.* **46,** 5094 (1981); (*b*) D. S. Wulfman, B. W. Peace, and R. S. McDaniel, Jr., *Tetrahedron* **32,** 1251 (1976).

107. D. S. Wulfman, G. Linstrumelle, and C. F. Cooper, in *The Chemistry of Diazonium and Diazo Groups,* S. Patai (ed.), Wiley, New York, 1978, Chap. 18.

108. D. S. Wulfman and B. Poling, in *Reactive Intermediates,* Vol. 1, R. A. Abramovitch (ed.), Plenum, New York, 1980.

109. K. B. Wiberg and J. M. Lavanish, *J. Am. Chem. Soc.* **88,** 5272 (1966).

110. M. M. Kreevoy and S. J. Thomas, *J. Org. Chem.* **42,** 3979 (1977).

111. W. Kirmse and H. A. Rinkler, *Liebig's Ann. Chem.* **707,** 57 (1967).

112. J. R. Mohrig, K. Keegstra, A. Maverick, R. R. Roberts, and S. Wells, *J. Chem. Soc. Chem. Commun.*, 780 (1974).

113. W. J. Albery and R. P. Bell, *Trans. Faraday Soc.* **57,** 1942 (1961).

114. W. J. Albery, J. E. C. Hutchins, R. M. Hyde, and R. H. Johnson, *J. Chem. Soc.* [*B*], 219 (1968).

115. C. E. McCauley and C. V. King, *J. Am. Chem. Soc.* **74,** 6221 (1952).

116. W. Kirmse, *New Synthetic Methods* (*Verlag Chemie*) **5,** 71 (1979).

117. J. D. Roberts, C. M. Regan, and I. Allen, *J. Am. Chem. Soc.* **74,** 6779 (1952).

118. R. A. M. O'Ferrall, *Adv. Phys. Org. Chem.* **5,** 331 (1967).

119. R. A. M. O'Ferrall, W. K. Kwok, and S. I. Miller, *J. Am. Chem. Soc.* **86,** 5553 (1964).

120. H. Hock and H. Kropf, *Chem. Ber.* **88,** 1544 (1955).

121. H. Meerwein and G. Hinz, *Liebig's Ann. Chem.* **484,** 1 (1930).

122. M. Neeman and W. S. Johnson, *Org. Syntheses,* Coll. **V,** 246 (1973).

123. J. D. Roberts and W. Watanabe, *J. Am. Chem. Soc.* **72,** 4869 (1950).

124. W. S. Johnson, M. Neeman, S. P. Birkeland, and N. A. Fedoruk, *J. Am. Chem. Soc.* **84,** 989 (1962).

125. E. Müller, R. Heischkeil, and M. Bauer, *Liebig's Ann. Chem.* **677,** 55 (1964).

126. F. Klages and H. Meuresch, *Liebig's Ann. Chem.* **592,** 116 (1955).

127. G. S. Hammond and R. M. Williams, *J. Org. Chem.* **27,** 3775 (1962).

128. F. Arndt, B. Eistert, R. Gompper, and W. Walter, *Chem. Ber.* **94,** 2125 (1961).

129. L. Testaferri, M. Tiecco, and P. Zanirato, *J. Org. Chem.* **40,** 3392 (1975).

130. E. Müller and H. Huber-Emden, *Liebig's Ann. Chem.* **649,** 81 (1961).

131. E. Müller, H. Huber-Emden, and W. Rundel, *Liebig's Ann. Chem.* **623,** 34 (1959).

132. T. Saegusa, Y. Ito, S. Kobayashi, K. Hirota, and T. Shimizu, *Tetrahedron Lett.,* 6131 (1966).

133. T. Wieland and H. Wiegandt, *Chem. Ber.* **93,** 1167 (1960).

134. R. Daniels and C. G. Kormendy, *J. Org. Chem.* **27,** 1860 (1962).

135. L. C. King and F. M. Miller, *J. Am. Chem. Soc.* **70,** 4154 (1948).

136. H. Biltz and H. Paetzold, *Ber.* **55,** 1066 (1922).

137. (*a*) J. S. Sherwell, J. R. Russell, and D. Swern, *J. Org. Chem.* **27,** 2853 (1962); (*b*) F. F. Guzik and A. K. Colter, *Can. J. Chem.* **43,** 1441 (1965).

138. H. Hoberg, *Liebig's Ann. Chem.* **695,** 1 (1966).

139. B. Bogdanovic, M. Kröner, and G. Wilke, *Liebig's Ann. Chem.* **699,** 1 (1966).

140. J. Feltzin, A. J. Restaino, and R. B. Mesrobian, *J. Am. Chem. Soc.* **77,** 206 (1955).

141. A. G. Davies, D. G. Hare, O. R. Kahn, and J. Sikora, *Proc. Chem. Soc.,* 172 (1961).

142. H. Hoberg, *Liebig's Ann. Chem.* **656,** 1, 15 (1962).

143. G. Wittig and F. Wingler, *Liebig's Ann. Chem.* **656,** 18 (1962).

144. G. Wittig and K. Schwarzenbach, *Liebig's Ann. Chem.* **650,** 1 (1961).

145. D. Seyferth and E. C. Rochow, *J. Am. Chem. Soc.* **77,** 907, 1302 (1955).

146. A. Y. Yakubovich and V. A. Ginsberg, *J. Gen. Chem. U.S.S.R* **22,** 1534 (1952); *Chem. Abstr.* **47,** 8010, 9254 (1953).

147. G. L. Closs and S. H. Goh, *J. Org. Chem.* **39,** 1717 (1974).

148. J. Hooz and S. Linke, *J. Am. Chem. Soc.* **90,** 6891 (1968).

149. D. S. Crumrine, T. J. Haberkamp, and D. J. Suther, *J. Org. Chem.* **40,** 2274 (1975).

150. H. W. Whitlock, Jr., *J. Am. Chem. Soc.* **84,** 2807 (1962).

151. C. D. Gutsche, *Org. Reactions* **8,** Chap. 8 (1954).

152. J. N. Bradley, G. W. Cowell, and A. Ledwith, *J. Chem. Soc.,* 4334 (1964).

153. E. Müller and M. Bauer, *Liebig's Ann. Chem.* **654,** 92 (1962).

154. H. Biltz and H. Paetzold, *Liebig's Ann. Chem.* **433,** 64 (1923).

155. J. A. Marshall and J. J. Partridge, *J. Org. Chem.* **33,** 4090 (1968).

156. C. D. Gutsche and C. T. Chang, *J. Am. Chem. Soc.* **84,** 2263 (1962).

157. C. D. Gutsche and D. M. Bailey, *J. Org. Chem.* **28,** 607 (1963).

158. J. B. Bastús, *Tetrahedron Lett.,* 955 (1963).

159. M. S. Newman and P. F. Beal, *J. Am. Chem. Soc.* **71,** 1506 (1949).

160. D. S. Tarbell and J. A. Price, *J. Org. Chem.* **22,** 245 (1957).

161. A. Bhati, *J. Org. Chem.* **27,** 1183 (1962).

162. J. N. Bridson and J. Hooz, *Org. Syntheses* **53,** 35 (1973).

163. L. T. Scott and M. A. Minton, *J. Org. Chem.* **42,** 3757 (1977).

164. J. Looker and C. H. Hayes, *J. Org. Chem.* **28**, 1342 (1963).

165. H. O. House and C. J. Blankley, *J. Org. Chem.* **33**, 53 (1968).

166. H. Bredereck, R. Sieber, and L. Kamphenkel, *Chem. Ber.* **89**, 1169 (1956).

167. F. M. Dean and B. K. Park, *J. Chem. Soc. Chem. Commun.*, 162 (1974).

168. H. E. Sheffer and J. A. Moore, *J. Org. Chem.* **28**, 129 (1963).

169. R. Huisgen and H. J. Koch, *Liebig's Ann. Chem.* **591**, 200 (1955).

170. K. Clusius, H. Hürzeler, R. Huisgen, and H. J. Koch, *Naturwissensch.* **41**, 213 (1954).

171. G. S. D. King and M. A. Peiren, *Chem. Ber.* **103**, 2821 (1970).

172. U. S. Seth and S. S. Deshapande, *J. Ind. Chem. Soc.* **29**, 539 (1952).

173. G. S. Skinner, *J. Am. Chem. Soc.* **46**, 731 (1924).

174. L. Horner and E. Lingnau, *Liebig's Ann. Chem.* **591**, 21 (1955).

175. G. F. Bettinetti and P. Grünanger, *Tetrahedron Lett.*, 2553 (1965).

176. E. Fahr, K. Döppert, K. Königsdorfer, and F. Scheckenbach, *Tetrahedron* **24**, 1011 (1967).

177. E. Fahr, J. Markert, and N. Pelz, *Liebig's Ann. Chem.* 2088 (1973).

178. G. F. Bettinetti and L. Capretti, *Gaz. Chim. Ital.* **95**, 33 (1965).

179. E. Fahr, K. Königsdorfer, and F. Scheckenbach, *Liebig's Ann. Chem.* **690**, 138 (1965).

180. V. A. Ginsburg, A. Ya. Yakubovich, et al., *Dokl. Akad. Nauk S.S.S.R.* **142**, 354 (1962); *Chem. Abstr.* **57**, 4518d (1962).

181. M. Pomerantz and S. Bittner, *J. Org. Chem.* **45**, 5390 (1980).

182. E. Müller and W. Rundel, *Chem. Ber.* **90**, 1299 (1957).

183. E. Müller and H. Disselhoff, *Liebig's Ann. Chem.* **512**, 250 (1934).

184. P. Yates, R. G. F. Giles, and D. G. Farnum, *Can. J. Chem.* **47**, 3997 (1969).

185. U. Schöllkopf, B. Banhidai, H. Frasnelli, R. Meyer, and H. Backhaus, *Liebig's Ann. Chem.*, 1767 (1974).

186. L. Wolff and H. Lindenhayn, *Ber.* **36**, 4126 (1903).

187. A. Bertho and H. Nussel, *Liebig's Ann. Chem.* **457**, 278 (1927).

188. G. H. Coleman, H. Gilman, C. E. Adams, and P. E. Pratt, *J. Org. Chem.* **3**, 99 (1938).

189. H. Staudinger and G. Lüscher, *Helv. Chim. Acta* **5**, 75 (1922).

190. L. Wolff and R. Krueche, *Liebig's Ann. Chem.* **394**, 48 (1912).

191. T. Curtius, A. Darapsky, and A. Bockmühl, *Ber.* **41**, 344 (1908).

192. H. von Pechmann, *Ber.* **28**, 1847 (1895).

193. H. Staudinger and J. Siegwart, *Ber.* **49**, 1918 (1916).

194. H. Staudinger, L. Hammett, and J. Siegwart, *Helv. Chim. Acta* **4**, 228 (1921).

195. G. Wittig and W. Haag, *Chem. Ber.* **88**, 1654 (1955).

196. H. Goetz and H. Juds, *Liebig's Ann. Chem.* **678**, 1 (1964).

197. B. H. Freeman, D. Lloyd, and M. I. C. Singer, *Tetrahedron* **30**, 211 (1974).

198. A. M. Pudovik and R. D. Gareev, *Zh. Obshch. Khim.* **45**, 1674 (1975).

199. A. M. Reader, P. S. Bailey, and H. M. White, *J. Org. Chem.* **30**, 784 (1965).

200. A. Schönberg, W. I. Awad, and N. Latif, *J. Chem. Soc.*, 1368 (1951).

201. R. Curci, F. Di Furia, J. Ciabattoni, and P. W. Concannon, *J. Org. Chem.* **39**, 3295 (1974).

202. R. Hensel, *Chem. Ber.* **88**, 527 (1955).

203. E. Profft and W. Steinke, *J. Prakt. Chem.* **[4]13**, 58 (1961).

204. R. J. Bussey and R. C. Neuman, Jr., *J. Org. Chem.* **34**, 1323 (1969).

205. G. L. Closs and J. J. Coyle, *J. Am. Chem. Soc.* **87**, 4270 (1965).

206. K. Schank and R. Blattner, *Chem. Ber.* **114**, 1958 (1981).

207. T. Curtius and A. Darapsky, *Ber.* **39**, 1373 (1906).

208. T. Curtius, *J. Prakt. Chem.* **[2]38**, 396 (1888).

209. H. Staudinger, A. Gaule, and J. Siegwart, *Helv. Chim. Acta* **4**, 212 (1921).

210. R. N. McDonald and K.-W. Lin, *J. Am. Chem. Soc.* **100**, 8028 (1978).

211. T. Kauffmann and S. M. Hage, *Angew. Chem.* **75**, 248 (1963).

212. M. L. Wolfrom and J. B. Miller, *J. Am. Chem. Soc.* **80**, 1678 (1955).

213. H. von Pechmann, *Ber.* **28**, 855 (1895).

214. J. Thiele and C. Meyer, *Ber.* **29**, 961 (1896).

215. W. Schlenk and C. Bornhardt, *Liebig's Ann. Chem.* **394**, 183 (1912).

216. L. Horner, L. Hockenberger, and W. Kirmse, *Chem. Ber.* **94**, 290 (1961).

217. O. L. Chapman and D. C. Heckert, *J. Chem. Soc. Chem. Commun.*, 242 (1966).

218. J. Engbersen and J. B. F. N. Engberts, *Synth. Commun.* **1**, 121 (1971).

219. H. Wieland and C. Reisenegger, *Liebig's Ann. Chem.* **401**, 244 (1913).

220. G. F. Bettinetti, G. Desimoni, and P. Grünanger, *Gaz. Ghim. Ital.* **93**, 150 (1963).

221. K. N. Houk, *Top. Cur. Chem.* **79**, 1 (1979).

222. W. E. Parham, F. D. Blake, and D. R. Theissen, *J. Org. Chem.* **27**, 2415 (1962).

223. H. de Suray, G. Leroy, and J. Weiler, *Tetrahedron Lett.*, 2209 (1974).

224. R. Huisgen, H. Stangl, H. J. Sturm, and H. Wagenhofer, *Angew. Chem.* **73**, 170 (1961).

225. I. Ojima and K. Kondo, *Bull. Chem. Soc. Japan* **46**, 2571 (1973).

226. A. M. van Leusen, B. A. Beith, R. J. Multer, and J. Strating, *Angew. Chem. Int. Ed.* **10**, 271 (1971).

227. P. K. Kadaba and T. F. Colturi, *J. Heterocyc. Chem.* **6**, 829 (1969).

228. C. G. Overerger and J.-P. Anselme, *J. Am. Chem. Soc.* **84**, 869 (1962).

229. R. Baltzly, N. B. Mehta, P. B. Russell, R. E. Brooks, E. M. Grivsky, and A. M. Steinberg, *J. Org. Chem.* **26**, 3669 (1961).

230. J. van Alphen, *Rec. Trav. Chim. Pays Bas* **62**, 210 (1943).

231. P. Lipp, J. Buchkremer, and H. Seeles, *Liebig's Ann. Chem.* **499**, 1 (1932).

232. G. D. Buckley, *J. Chem. Soc.*, 1850 (1954).

233. P. K. Kadaba and J. O. Edwards, *J. Org. Chem.* **26**, 2331 (1961).

234. H. Dehn and M. Süsse, *Z. Chem.* **16**, 102 (1976).

235. P. K. Kadaba, *J. Heterocyc. Chem.* **12**, 143 (1975).

236. K. Burger, J. Fehn, and A. Gieren, *Liebig's Ann. Chem.* **757**, 9 (1972).

237. R. Rotter and E. Schaudy, *Monatsh. Chem.* **58**, 245 (1931).

238. J. C. Sheehan and P. T. Izzo, *J. Am. Chem. Soc.* **70**, 1985 (1948).

239. F. Arndt, H. Scholz, and E. Frobel, *Liebig's Ann. Chem.* **521**, 95 (1936).

240. H. Hoberg, *Liebig's Ann. Chem.* **707**, 147 (1967).

241. M. Muramatsu, N. Obata, and T. Takizawa, *Tetrahedron Lett.*, 2133 (1973).

242. J. A. Green and L. A. Singer, *Tetrahedron Lett.*, 5093 (1969).

243. J. H. Boyer and W. Bevering, *J. Chem. Soc. Chem. Commun.*, 1377 (1969).

244. A. Schönberg and A. Frese, *Chem. Ber.* **96**, 2420 (1963).

245. L. Lardici, C. Battistini, and R. Menicagli, *J. Chem. Soc.* [*Perkin I*], 344 (1974).

246. J. M. Beiner, D. Lecadit, D. Paquer, and A. Thuillier, and J. Vialle, *Bull. Soc. Chim. France*, 1979, 1983 (1973).

247. I. Kalwinsch, Li Xingya, J. Gottstein, and R. Huisgen, *J. Am. Chem. Soc.* **103**, 7032 (1981).

248. T. Machiguchi, Y. Yamamoto, M. Hoshino, and Y. Kitihara, *Tetrahedron Lett.*, 2627 (1973).

249. D. Martin and W. Mucke, *Liebig's Ann. Chem.* **682**, 90 (1965).

250. S. T. Purrington and P. Wilder, Jr., *J. Org. Chem.* **30**, 2070 (1965).

251. G. Hesse and S. Majmudar, *Chem. Ber.* **93**, 1129 (1960).

252. L. V. Vargha and E. Kovacs, *Ber.* **75**, 794 (1942).

253. H. Kloosterziel, M. H. Deinema, and H. J. Backer, *Rec. Trav. Chim. Pays Bas* **71**, 1228 (1952).

254. H. H. Inhoffen, R. Jones, H. Krosche, and U. Eder, *Liebig's Ann. Chem.* **694**, 19 (1966).

255. R. Huisgen, M. Seidel, G. Wallbillich, and H. Knupfer, *Tetrahedron* **17**, 3 (1962).

256. C. G. Overberger, N. P. Mariullo, and R. G. Hiskey, *J. Am. Chem. Soc.* **83**, 1374 (1961).

257. R. L. Hinman and K. L. Hamm, *J. Am. Chem. Soc.* **81**, 3294 (1959).

258. W. R. McBride and E. M. Bens, *J. Am. Chem. Soc.* **81**, 5546 (1959).

259. C. G. Overberger and L. P. Herin, *J. Org. Chem.* **27**, 417 (1962).

260. L. A. Carpino, A. A. Santelli, and R. W. Murray, *J. Am. Chem. Soc.* **82**, 2728 (1960).

261. C. G. Overberger, *Rec. Chem. Prog.* **21**, 21 (1960).

262. D. M. Lemal, T. W. Rave, and S. D. McGregor, *J. Am. Chem. Soc.* **85**, 1944 (1963).

263. L. Horner and H. Ferkeness, *Chem. Ber.* **94**, 712 (1961).

264. L. A. Carpino, *J. Am. Chem. Soc.* **79**, 4427 (1957).

265. R. A. Abramovitch and B. A. David, *Chem. Rev.* **64**, 149 (1964).

266. M. Busch and K. Lang, *J. Prakt. Chem.* **[2]144**, 291 (1936).

267. C. G. Overberger and S. Altscher, *J. Org. Chem.* **31**, 1728 (1966).

268. Reviewed in depth by M. Regitz, in *The Chemistry of Diazonium and Diazo Groups*, S. Patai (ed.), Wiley, New York, 1978, Chaps. 15 and 17.

269. (*a*) N. E. Searle, *Org. Syntheses*, Coll. **IV**, 424 (1963); (*b*) J. H. Atherton, R. Fields, and R. N. Haszeldine, *J. Chem. Soc.* [*C*], 366 (1971).

270. A. Roedig and K. Grohe, *Tetrahedron* **21**, 2375 (1965).

271. H. D. Hartzler, *J. Am. Chem. Soc.* **86**, 2174 (1964).

272. H. Reimlinger, A. van Overstraaten, and H. G. Viehe, *Chem. Ber.* **94**, 1036 (1961).

273. E. Müller, H. Haiss, and W. Rundel, *Chem. Ber.* **93**, 1541 (1960).

274. J. Bakke, *Acta Chem. Scand.* **22**, 1833 (1968).

275. E. Müller and W. Rundel, *Chem. Ber.* **90**, 2673 (1957).

276. W. D. McPhee and E. Klingsberg, *Org. Syntheses*, Coll. III, 119 (1955).

277. R. Huisgen and J. Reinertshofer, *Liebig's Ann. Chem.* **575**, 174 (1952).

278. K. Arndt, *Org. Syntheses*, Coll. **II**, 165 (1943).

279. T. J. De Boer and H. J. Backer, *Org. Syntheses*, Coll. **IV**, 250 (1963).

280. C. G. Overberger and J.-P. Anselme, *J. Org. Chem.* **28**, 592 (1963).

281. M. Hudlicky, *J. Org. Chem.* **45**, 5377 (1980).

282. H. Reimlinger, *Chem. Ber.* **94**, 2547 (1961).

283. J. A. Moore and D. E. Reed, *Org. Syntheses*, Coll. **V**, 351 (1973).

284. H. Reimlinger and L. Skattebøl, *Chem. Ber.* **94**, 2429 (1961).

285. K. Heyns and O. F. Woyrsch, *Chem. Ber.* **86**, 76 (1953).

286. V. Franzen, *Liebig's Ann. Chem.* **614**, 31 (1958).

287. G. A. Akuyunoglu and M. Calvin, *J. Org. Chem.* **28**, 1484 (1963).

288. A. F. McKay, *J. Am. Chem. Soc.* **71**, 1968 (1949).

289. G. Csávássy and Z. A. Györfi, *Liebig's Ann. Chem.*, 1195 (1974).

290. C. E. Redemann, F. O. Rice, R. Roberts, and H. P. Ward, *Org. Syntheses*, Coll. **III**, 244 (1955).

291. M. Sekiya, Y. Okahi, Y. Terao, and K. Ito, *Chem. Pharm. Bull. (Japan)* **24**, 369 (1976).

292. M. S. Newman and A. Arkell, *J. Org. Chem.* **24**, 385 (1959).

293. C. D. Nenitzescu and E. Solomonics, *Org. Syntheses*, Coll. **II**, 496 (1943).

294. S. D. Andrews, A. C. Day, P. Raymond, and M. C. Whiting, *Org. Syntheses* **50**, 27 (1970).

295. J. Attenburrow, A. F. B. Cameron, J. H. Chapman, R. M. Evans, B. A. Hems, A. B. A. Janson, and T. Walker, *J. Chem. Soc.*, 1094 (1952).

296. P. A. S. Smith and J. G. Wirth, *J. Org. Chem.* **33**, 1145 (1968).

297. J. B. F. N. Engberts, G. van Brugger, J. Stratling, and H. Wynberg, *Rec. Trav. Chim. Pays Bas* **84**, 1610 (1965).

298. R. W. Murray and A. M. Trozzolo, *J. Org. Chem.* **29**, 1268 (1964).

299. K. Heyns and A. Heins, *Liebig's Ann. Chem.* **604**, 133 (1957).

300. A. Stojiljkovič, N. Orbovič, and M. L. Mihailovič, *Tetrahedron* **26**, 1101 (1970).

301. D. G. Farnum, *J. Org. Chem.* **28**, 870 (1963).

302. M. P. Cava, R. L. Litle, and D. R. Napier, *J. Am. Chem. Soc.* **80**, 2257 (1958).

303. W. R. Bamford and T. S. Stevens, *J. Chem. Soc.*, 4735 (1952).

304. G. M. Kaufman, J. A. Smith, G. G. Vander Stouw, and H. Shechter, *J. Am. Chem. Soc.* **87**, 935 (1965).

305. J. M. Muchowski, *Tetrahedron Lett.*, 1773 (1966).

306. M. Regitz, in *Newer Methods of Preparative Organic Chemistry*, Vol. VI, W. Foerst (ed.), Verlag Chemie, Weinheim/Bergstr. and Academic Press, New York, 1971, pp. 82–126 ("Transfer of Diazo Groups").

307. W. von E. Doering and C. H. De Puy, *J. Am. Chem. Soc.* **75**, 5955 (1953).

308. M. Regitz and A. Liedhegener, *Tetrahedron* **23**, 2701 (1967).

309. M. Regitz and A. Liedhegener, *Chem. Ber.* **99**, 3128 (1966); cf. also H. J. Ledon, *Org. Syntheses* **59**, 66 (1979) (application of phase-transfer catalysis).

310. M. Regitz, *Liebig's Ann. Chem.* **687**, 101 (1964).

311. M. Regitz, J. Rüter, and A. Liedhegener, *Org. Syntheses* **51**, 87 (1971).

312. M. Regitz, J. Hocker, and A. Liedhegener, *Org. Syntheses*, Coll. **V**, 179 (1973).

313. M. Regitz and G. Himbert, *Liebig's Ann. Chem.* **734**, 70 (1970).

314. G. Himbert and M. Regitz, *Synthesis*, 571 (1972).

315. G. Himbert and M. Regitz, *Chem. Ber.* **105**, 2963 (1972).

316. H. Balli, R. Löw, V. Müller, H. Rempfler, and A. Sezen-Gezgin, *Helv. Chim. Acta* **61**, 97 (1978).

317. B. Kokel and H. G. Viehe, *Angew. Chem. Int. Ed.* **19**, 716 (1980).

318. J. A. Pincock, R. Morchat, and D. R. Arnold, *J. Am. Chem. Soc.* **95**, 7538 (1973).

319. E. Müller and W. Rundel, *Chem. Ber.* **90**, 1302 (1957).

320. W. Rundel and P. Kästner, *Liebig's Ann. Chem.* **686**, 88 (1965).

321. J. Meinwald, P. G. Gassman, and E. G. Miller, *J. Am. Chem. Soc.* **81**, 4751 (1959).

322. W. Rundel, *Angew. Chem.* **74**, 469 (1962).

323. M. P. Cava, E. J. Glanikowski, and P. M. Weintraub, *J. Org. Chem.* **31**, 2755 (1966).

324. E. Bamberger and E. Renault, *Ber.* **28**, 1682 (1895).

325. L. Horner, W. Kirmse, and H. Ferkeness, *Chem. Ber.* **94**, 279 (1961).

326. E. Noelting, quoted by E. Bamberger, *Ber.* **33**, 101 (1900).

327. J. Thiele, *Liebig's Ann. Chem.* **376**, 239 (1910).

328. H. Staudinger and O. Kupfer, *Ber.* **45**, 501 (1912).

329. S. P. McManus, J. T. Carroll, and C. L. Dodson, *J. Org. Chem.* **33,** 4272 (1968).

330. R. Huisgen, M. Seidel, J. Sauer, J. W. McFarland, and G. Wallbillich, *J. Org. Chem.* **24,** 892 (1959).

331. R. Huisgen, *Angew. Chem.* **72,** 359 (1960).

332. A. Ledwith, *Chem. Ind.,* 1310 (1956).

333. R. Preussmann, H. Hengy, and H. Druckrey, *Liebig's Ann. Chem.* **684,** 57 (1965).

334. E. K. Marshall and S. F. Acree, *Ber.* **43,** 2323 (1910).

335. R. F. Raffauf, A. L. Farren, and G. E. Ullyot, *J. Am. Chem. Soc.* **75,** 2576 (1953).

336. N. D. Cheronis and T. S. Ma, *Organic Functional Group Analysis by Micro and Semimicro Methods,* Interscience Publishers, New York, 1964, pp. 265 and 290–291.

337. E. van Hulle, in *Houben-Weyl, Methoden der Organischen Chemie,* 4th Ed., Vol. II, E. Müller (ed.), Georg-Thieme-Verlag, Stuttgart, 1953, pp. 698–699.

6.

AZIDES

NOMENCLATURE

Azides are named in the same way as halides, with the name of the substituent radical followed by the separate word *azide* or with the prefix *azido*. The former method is commonly used for communication, for the azido group is usually the most important part of the molcule to emphasize; the latter method is generally found in indexes, such as those of *Chemical Abstracts* ($C_6H_5N_3$, phenyl azide or azidobenzene; CH_3CON_3, acetyl azide). The equivalent terms *azoimide* (e.g., "azobenzenimide") *azimido*, and *triazo* must be regarded as archaic, for they disappeared from general use over 50 years ago. However, *triazo* inexplicably survives anachronistically in the tables of a popular chemical handbook.

PROPERTIES[1]

Azides generally boil at much the same temperature as the corresponding bromides. Methyl azide (bp 20–21 °C) boils 15° lower than the parent hydrogen azide, owing to lack of hydrogen bonding, and appears to be essentially insoluble in water. Vinyl azide is a clear yellow liquid, (bp 28–30 °C) with a dipole moment of 1.12 D.[2] The simplest *N*-azido compound $(CH_3)_2N$—N_3 is an explosive liquid (bp 32 °C/ll mm)[3] (see Tetrazadienes in Chap. 7). Phenyl azide boils at 73.5 °C under aspirator vacuum (22–24 mm) and is volatile with steam, but it decomposes rapidly, with danger of explosion when one attempts to distill it at atmospheric pressure. It is denser than water, in which it is insoluble; its odor resembles that of benzaldehyde, nitrobenzene, and benzonitrile. Acetyl azide is an explosive liquid,[4] but like most acyl azides, it is too unstable to be distilled at atmospheric pressure. Cyanogen azide, NC—N_3, is a colorless oil that detonates violently.[5] The dipole moments of phenyl azide (1.55 D) and *p*-chlorophenyl azide (0.33 D) show that the azido group is markedly electronegative with respect to carbon.

The structure of the azido group is now known to be linear, although it was for a long time written as a three-membered ring, which represented the only possibility in terms of the classical Kekulé theory and trivalent nitrogen. X-ray and electron-diffraction measurements indicate a linear structure,[6] with substituents lying at an angle of about 120°. Although the azide ion is completely symmetrical, in organic azides, the unsubstituted nitrogens are slightly closer together. These facts may be represented by structural formulas in different ways:[7] resonance hybrid, delocalized molecular orbitals, and as a coordination structure isolectronic with nitrate. On the basis of the last formulation, Franklin[7] referred to hydrogen azide as "ammononitric acid," since the ligands OH^- and $2O^{2-}$ are electrically equivalent to NH^{2-} and N^{3-}.

The infrared spectra of azides show a characteristic absorption[9a] at 2140 to 2240 cm^{-1}, attributable to a triple-bonded or allenic system. The unconjugated azido group absorbs very weakly in the near ultraviolet,[9b,c] but never in the visible range; alkyl, aryl, and acyl azides are colorless. Alkyl azides generally show a band near 287 nm ($\varepsilon \simeq 25$), attributed to a $\pi \rightarrow \pi^*$ transition, and a stronger band near 216 nm.

Azides are not basic enough to form stable salts with acids or to dissolve in aqueous acid, but they are nevertheless capable of accepting protons, as one would expect in view of the feeble but measurably basicity of HN_3. Indeed, they show a general proclivity toward decomposition by strong, concentrated acids. The site of protonation is probably the substituted nitrogen atom, although tautomerism with the terminally protonated form can be expected. Lewis acids also attack azides, and methyl azide has been shown to form an adduct with antimony pentachloride, $CH_3N_3SbCl_5$, of some stability.[10]

Azides are commonly thought of as explosive, because of the properties of hydrogen azide, but when the carbon-containing part of the molecule is sufficiently large, the explosive characteristics are diminished or even removed. It cannot be said precisely at what size violent explosiveness disappears, but compounds with a ratio of $(C + O)/N$ less than 3 should always be regarded as potentially dangerous. Sensitivity is not just a function of size, however, and although methyl azide has been handled in an almost routine manner (but not in the presence of mercury!),[11a] acetyl azide is treacherously sensitive. The heat of formation of the azido group is 205 to 208 kcal/mol; the maximum energy is released when molecular nitrogen is formed. t-Butoxycarbonyl azide, t-BuOCON$_3$, once widely used to introduce the t-butoxycarbonyl protecting group,[11b,c] yields 45% of the energy of nitroglycerine when it explodes (as it has been known to do on distillation), and the 50% point for impact detonation is 2.8 kg cm (cf. nitroglycerine, 1.6).[11b] On the other hand, cis-1,2-diazidocyclohexane has been distilled without incident (bp 54–55 °C/0.005 mm) from a water bath.[12]

The electronic properties of the azido group have been determined from its effect on the strengths of acids and bases to which it is attached and on the reactivity of the benzene ring, as well as by other methods.[13] It has an electron-withdrawing inductive effect intermediate between that of bromine and iodine; α-azido acids have pK_a values of about 3. The inductive effect on the benzene ring is characterized by $\sigma_m = 0.33$–0.37, very much the same as the halogens. However, this effect is counteracted in the *ortho* and *para* positions by a substantial mesomeric electron-releasing effect, very close to that of fluorine; a value of the Hammett substituent constant σ_p of $+0.08$ has been determined. Electron release through polarizability is even more pronounced, and the azido group activates the benzene ring toward electrophilic attack. As measured by the Brown σ_p^+ constant, -0.54, the effect is almost as great as that of the acetamido group (-0.6). *m*-Nitrophenyl azide has been nitrated to 2,4,5-trinitrophenyl azide in 76% yield, for example.[14]

Azides do not appear to be exceptionally toxic, except insofar as they may release hydrogen azide, a powerful vasodilator, or, in the case of aryl azides, may be metabolically reduced to carcinogenic aniline derivatives.

No naturally occurring azide has yet been reported.

REACTIONS

The most prominent reactions of azides are the consequence of fragmentation with loss of nitrogen,[15] brought about by heat, light, strong acids, and sometimes other means. Another important type of azide reaction is cycloaddition, in which the azido group acts as a 1,3-dipolar system. Addition of strong nucleophiles to the terminal nitrogen atom is another general characteristic reaction, and reduction to a primary amino group is a useful step in synthesis (Eqs. 6-1 through 6-5).

$$R-N_3 \xrightarrow{\Delta \text{ or } h\nu} R-N + N_2 \tag{6-1}$$

$$R-N_3 + H^+ \longrightarrow R-NH-N_2^+ \longrightarrow R-NH^+ + N_2 \tag{6-2}$$

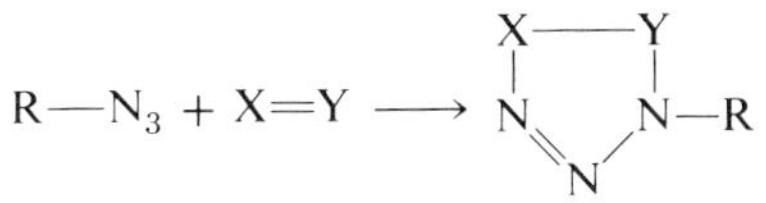

$$R-N_3 + X{=}Y \longrightarrow \begin{array}{c} X-Y \\ | \quad | \\ N \quad N-R \\ \diagdown_N\diagup \end{array} \tag{6-3}$$

$$R-N_3 + B: \longrightarrow R-\bar{N}-N{=}N-B \tag{6-4}$$

$$R-N_3 \xrightarrow{[H]} R-NH_2 \tag{6-5}$$

Heat and light. Ordinary alkyl and aryl azides are indefinitely stable at room temperature, but they begin to decompose at measurable rates not far above 100°C. Other functional groups in the same molecule may greatly accelerate decomposition (e.g., acyl azides, *o*-nitroaryl azides). Thermolysis of undiluted azides can be dangerously explosive, but the reaction is readily controlled in dilute solution. Decomposition may be accompanied or followed by rearrangement or may give rise to an intermediate electron-deficient species, formally a derivative of univalent nitrogen, most commonly called a *nitrene* (but also variously called an *azene, imene, azylene, imidogen,* etc.). (The word *nitrene* is now firmly entrenched, but one should be alert to the fact that it originally referred to an entirely different class of structure, until its meaning was casually subverted in the early 1950s).

The primary process in thermolysis of alkyl azides is believed to be loss of nitrogen to form alkylnitrenes, R—N̈:, but they are so reactive that they have been detected[16] only when generated by photolysis in a frozen glassy matrix at 4°K. Formation of nitrenes from thermolysis is an unproven inference derived from studies of kinetics and products, which are commonly imines arising from migration of a hydrogen (preferred) or an alkyl group from the α-carbon to the retained nitrogen of the azide (Eq. 6-6). The possibility that migration may be concerted with

$$R_3C-N_3 \xrightarrow{\Delta} N_2 + R_2C{=}N-R \tag{6-6}$$
$$(R = H \text{ or alkyl})$$

loss of nitrogen has not been conclusively eliminated. However, there is good evidence for the formation of at least some nitrene from pyrolysis of methyl azide at 350 to 410°C; traces of methylamine are formed and, in the presence of cyclohexane, traces of methylaminocyclohexane.[17] The main product is a polymer apparently derived from the expected rearrangement product, formaldimine. At higher temperatures, some HCN may also be formed.[18] Ethyl azide gives substantial amounts of aziridine (as much as 35%) and small amounts of N-methylformaldimine, in addition to acetaldimine polymer[17] (Eq. 6-7). At lower temperatures,

$$CH_3CH_2N_3 \xrightarrow{400\,°C}$$

$$(CH_3CH{=}NH)_x + \underset{\underset{H}{\overset{|}{N}}}{CH_2{-}CH_2} + CH_2{=}NCH_3 \qquad (6\text{-}7)$$

aziridine becomes the major product, accompanied by piperazine and traces of ethylamine and acetonitrile; the latter two products are at least formally attributable to disproportionation of ethylnitrene.[19]

The behavior of ethyl azide in giving rise to so much aziridine is exceptional; larger alkyl azides give only traces of aziridines or none at all.[17] Aziridine presumably arises from the triplet, or diradical, state of the nitrene, R—N:, by either direct insertion into the C—H bond or more probably through backbiting to form ·CH₂CH₂—NH·. Such a process can become important only if the rates at which the initially formed singlet nitrene, R—N:, relaxes to the lower-energy triplet state and the subsequent backbiting is fast enough to compete with the inherently rapid rearrangement to an imine. This requirement is apparently not met in alkyl azides with longer chains, in which backbiting would be expected to occur at the δ-carbon, for the pyrrolidines that should result have not been observed.[17]

The behavior of alkyl azides on photolysis is similar to that on thermolysis, and the sum of the evidence implies that azides in the singlet state, rather than a hypothetical triplet azide, decompose to singlet nitrenes, which rapidly relax to triplet nitrenes if they do not first rearrange to imines.[15a] Photolysis of sec- and tert-alkyl azides occurs without evidence of intermediate nitrenes; selection of isomeric imine product appears to be determined by rotational equilibrium in a photoexcited azide, which rearranges by a concerted process.[15b]

Triarylmethyl azides constitute a special case, in which the only feasible reaction is rearrangement to an imine, a benzophenone anil. This reaction is known as the *Stieglitz rearrangement*[20] (Eq. 6-8). It has been studied rather thoroughly, from the

$$\underset{\underset{Ar''}{\overset{|}{\underset{|}{C}}}}{\overset{Ar}{\overset{|}{\underset{|}{Ar'{-}C{-}N_3}}}} \xrightarrow{225\,°C} \underset{Ar''}{\overset{Ar'}{C}}{=}N{-}Ar + N_2 \qquad (6\text{-}8)$$

standpoints both of kinetics and product distribution.[15,21] The rates are only weakly sensitive to substituents, but electron-donating groups in a *para* position substantially enhance the migration aptitude of that phenyl group. The evidence generally favors the concept of formation of a discrete nitrene intermediate, with a small amount of participation by an aryl group in the transition state.

Stieglitz rearrangement can also be brought about photolytically.[15,21,22] The results may be somewhat obscured by the fact that the imines are in some cases subject to further photolytic transformation, but in general it appears that photolysis differs from thermolysis in the product ratios. Substituted triarylmethyl azides do not show the influence of substituents on migration aptitudes observed in thermolysis, and isomeric imines are produced in statistically determined ratios; even 1,1-diphenylethyl azide gives a statistical ratio of acetophenone anil and benzophenone methylimine.

At temperatures lower than required for fragmentation, rearrangement of a different kind may occur in special instances. Triarylmethyl azides with the azido group unsymmetrically labeled with ^{15}N undergo some scrambling of the isotope upon heating, through allylic inversion of the azido groups.[23] Allylic azides undergo allylic isomerization (Eq. 6-9); the process is first-order kinetically and is insensitive

$$R—CH=CH—CH_2—N_3 \longrightarrow CH_2=CH—CHR—N_3 \qquad (6\text{-}9)$$

to solvent polarity. These facts and the negative activation entropy imply a cyclic transition state.[24]

Compounds with an azido group attached to an sp^2 carbon, such as vinylic azides, aryl azides, and acyl azides, have distinctive reactions. Vinylic azides[25] appear to form nitrenes by either thermolysis or photolysis; the primary fate of such nitrenes is cyclization to an azirine (Eq. 6-10). The azirines derived from terminal vinylic

$$R—CH=CR'—N_3 \longrightarrow [R—CH=CR'—N] \longrightarrow$$
$$R—\underset{\underset{N}{\diagdown\diagup}}{CH}—CR' \;(+ \; R—CH=C=N—R') \qquad (6\text{-}10)$$
$$<6\%$$

azides are particularly unstable and isomerize to nitriles. Although nitriles are the principal product from terminal vinylic azides, they do not necessarily arise from azirines, for reasonable paths can also be envisaged by direct tautomerization of a nitrene or indirect tautomerization through an intermediate ketene imine (Eq. 6-11).

$$[R—CH=CH—N] \overset{\displaystyle R—CH=C=NH}{\underset{\displaystyle R—CH_2—C\equiv N}{\bigg\langle\;\;\big\downarrow}} \qquad (6\text{-}11)$$

The activation energy for fragmentation of vinylic azides is distinctly lower than that for alkyl azides, however, an observation that has led to the suggestion that the reaction is a concerted process, perhaps proceeding through a triazoline,[26] but it has been objected that the disposition of the orbitals makes such a process impossible in the ground state.[27] In any event, ketene imines have been observed as minor products (5–6%) from the thermolysis of α-azidostyrenes,[28] along with azirines or indoles derived from them by further reaction.[29]

Photolysis of vinylic azides in general gives similar results to thermolysis. Observed differences are due to secondary reactions: further transformations of the primary products by heat or light. α-Azidostyrene, for example, can be photolyzed to 2-phenylazirine in 58% yield, but prolonged photolysis causes dimerization of this product to a dihydropyrazine[30] and a small amount of an azabicyclopentane derived from reaction of some of the azirine with the ketene anil concurrently produced.[31] The azidostilbenes (*cis* or *trans*) also give rise to the pyrazine ring system on photolysis[32] (Eq. 6-12) (compare the Neber reaction of oximes).

$$Ph-CH=\overset{\underset{\displaystyle N_3}{|}}{C}-Ph \xrightarrow{h\nu} \left[Ph-C\overset{\displaystyle N}{\diagdown}CH-Ph \right] \xrightarrow[O_2]{h\nu}$$

$$\begin{array}{c} Ph-\overset{\displaystyle N}{\diagup}-Ph \\ Ph-\diagdown_{\displaystyle N}-Ph \end{array} \qquad (6\text{-}12)$$

Aryl azides without *ortho* substituents (and many that bear them) fragment to nitrenes in the temperature range 100 to 200°C.[15d,33] When fragmentation is effected photolytically at 12 K in a glassy, inert matrix, electron-spin resonance detects a triplet-state nitrene. Thermally generated arylnitrenes combine with carbon monoxide to form isocyanates[35a] (Eq. 6-13) and deoxygenate nitrosamines[35b] (Eq. 6-14). Sufficiently electrophilic arylnitrenes (from aryl azides bearing electron-with-

$$Ar-N_3 \xrightarrow[180°C]{CO} N_2 + Ar-N=C=O \qquad (6\text{-}13)$$

$$Ar-N_3 + (PhCH_2)_2N-NO \xrightarrow{\Delta}$$
$$2N_2 + PhCH_2CH_2Ph + Ar-NO \qquad (6\text{-}14)$$

drawing substituents) can attack other benzene rings to produce diarylamines, but the yields are low.[36]

In the absence of suitable reaction partners, nitrenes from aryl azides may extract hydrogen from other molecules to form amines or give rise to unidentified polymeric materials or azobenzenes.[33] Although azobenzenes are dimers of aryl nitrenes, it seems unlikely that they are formed by simple dimerization in solution reactions, for such reactive species would be present in only very low concentrations. Attack of a nitrene at N-α or N-γ of an undecomposed azide is a more likely possibility.[33,36] In the gas phase, however, direct dimerization is a more probable possibility.[15b]

A variety of thermolytic and photolytic experiments have revealed that aryl nitrenes, whether formed from azides or from other sources, can equilibrate with the isomeric but less-stable pyridyl carbenes, probably by means of a matrix of intermediates[15b,d,37] (Eq. 6-15). The most important manifestation of this phenomenon is the formation of azepines when the azide is decomposed in the presence of a suitable reagent, such as an amine.[38] This reaction was first observed by Wolff in 1912, but the structure of the product was not elucidated until 46 years later, by Huisgen, Vossius, and Appl. In spite of much attention in recent years, the mechanism is only incompletely established, although there is little doubt that it is an intermediate that is trapped. The evidence, which has been critically discussed by Jones,[15b] seems most consistent with the seven-membered ketenimine or its carbene

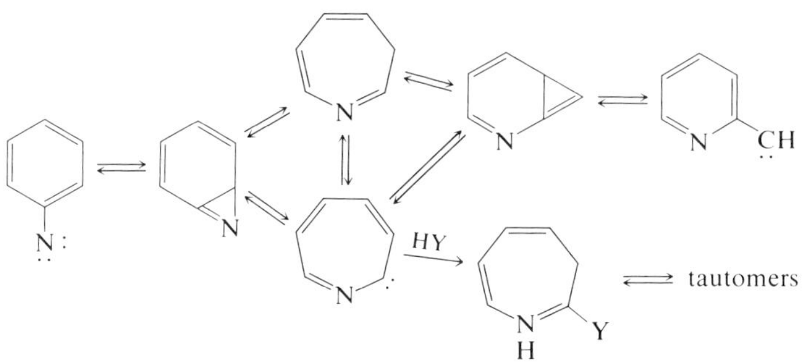

$$(6\text{-}15)$$

$$(HY = NH_3, RNH_2, R_2NH, H_2S, RSH)$$

valence tautomer as the intermediate, as shown, but the benzazirine cannot be ruled out.

High-temperature pyrolysis of aryl azides,[37] extensively investigated by Crow and Wentrup, manifests still another reaction, formation of cyanocyclopentadienes, which are isomers of aryl nitrenes. The reaction is not completely general, for whereas *m*- and *p*-tolyl azides give the same mixture of 2-methyl and 4-methyl-1-cyanocyclopentadiene (Eq. 6-16), *o*-tolyl azide gives only *o*-toluidine and

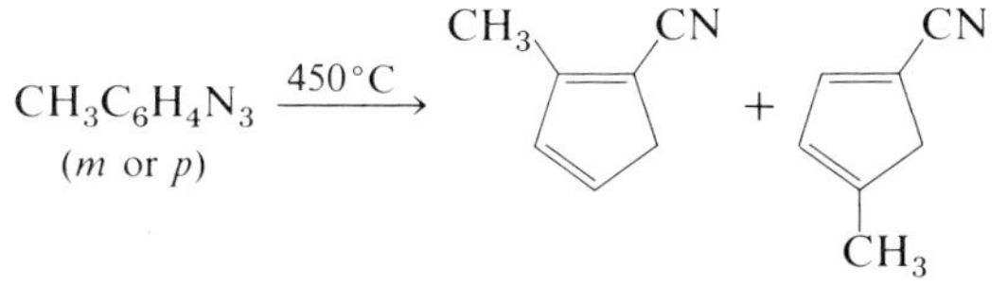

$$(6\text{-}16)$$

o,o'-azotoluene under the same conditions. It has been suggested that the relative rates of intersystem crossing of singlet nitrene to triplet may determine which type of product is formed.[37] An iminocyclohexadienylidene, a carbene tautomer of a phenyl nitrene, has been invoked as the intermediate that undergoes ring contraction.[15b,37]

The foregoing description gives an incomplete picture of a complex situation, for an unassailable rationale to account for the dependence of the products on structure and experimental conditions has not yet emerged. In the case of *ortho*-substituted azides, however, the products are more uniform, and a more coherent view is possible, with some fairly reliable generalizations.[15d,33] Small *ortho*-substituents (e.g., methyl, halo) do not cause new reaction patterns, different from those presented in the preceding paragraphs. Larger groups, containing a β-C—H, undergo intramolecular attack (insertion) by the nitrene, and a heterocyclic ring is formed (Eq. 6-17).

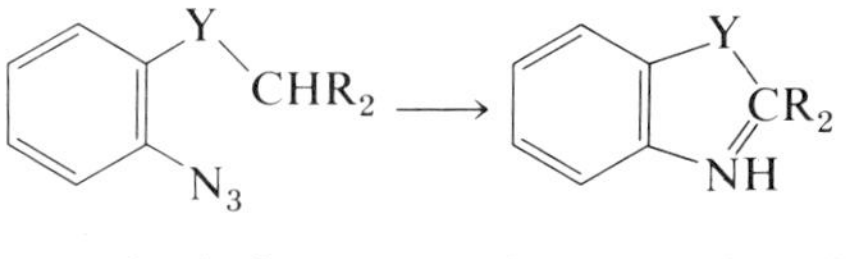

$$(6\text{-}17)$$

Unsaturated substituents carrying an unshared electron pair on a β-atom assist the fragmentation of the azido group, and a heterocyclic ring is formed without the intermediacy of a nitrene (Eq. 6-18).

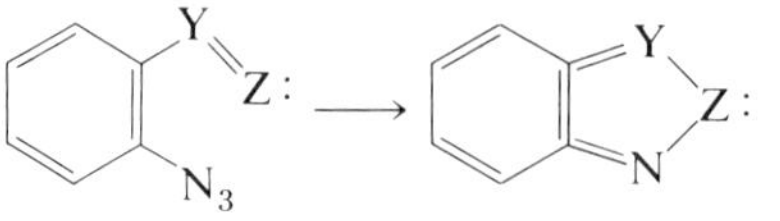

$$(6\text{-}18)$$

o-Alkylphenyl azides require essentially the same temperatures for fragmentation as do azides without *ortho*-substituents. Insertion occurs preferentially at the β position, to give a dihydroindole, but the γ position may also be attacked to form tetrahydroquinolines. *o*-Butylphenyl azide, for example, produces the two types of product in a 4:1 ratio, accompanied by an approximately equal amount of *o*-butylaniline[39] (Eq. 6-19). (The proportion of amine, a product of hydrogen abstrac-

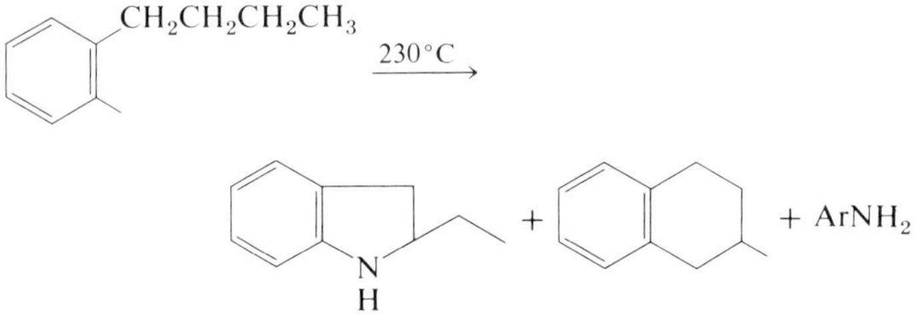

(ratio 43:11:46; 70% total yield)

tion, formed in azide cyclizations varies with the hydrogen-donating property of the solvent, but it cannot be reduced to nothing, because the nitrene can also abstract hydrogen from either the azide or the cyclization product.) There is evidence that such cyclizations may involve either the singlet or the triplet state of the nitrene, for *o*-(2-methylbutyl)phenyl azide gives 2-ethyl-2-methyldihydroindole with 50% retention of configuration. It is believed that retention occurs when the singlet nitrene inserts directly in the C—H bond at the chiral site and that racemization occurs when the triplet nitrene abstracts a hydrogen atom from that site, forming an intermediate diradical that collapses to the cyclization product in a second step.[39] *o*-Dialkylaminophenyl azides give useful yields of benzimidazolines or their oxidation products, benzimidazoles[40] (Eq. 6-20).

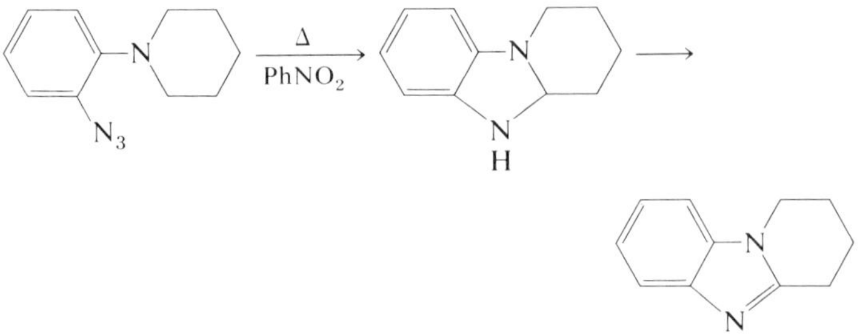

$$(6\text{-}20)$$

Aryl azides bearing an *o*-phenyl group are converted to carbazoles by thermolysis or photolysis in good to excellent yields[33,41] (Eq. 6-21). This is probably the most useful of the nitrene cyclizations, for a variety of substituents can be tolerated, and the *o*-phenyl group can be replaced by other aryl groups, such as thienyl or pyridyl.

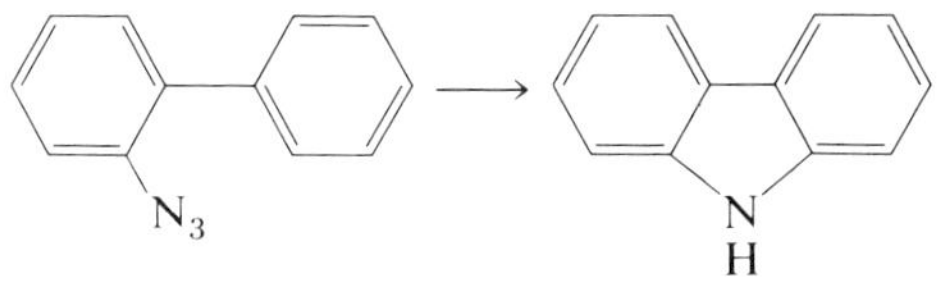

$$(6-21)$$

Cyclization must involve the singlet state of the nitrene, for photolysis with triplet sensitizers (e.g., acetophenone) diverts the reaction to formation azobenzenes, whereas triplet quenchers suppress that side reaction.[42]

Analogous cyclizations can occur when a benzene ring is joined to the *ortho* position through another atom, such as a carbon, nitrogen, or sulfur. In such cases, six-membered rings are commonly formed, but yields are inclined to be low.[43,44] Interesting rearrangements are frequently encountered with systems of this type; unexpected positional isomers of the products may be formed, or the substituent pheny group may undergo expansion to a fused-ring azepine[44-46] (Eq. 6-22). These

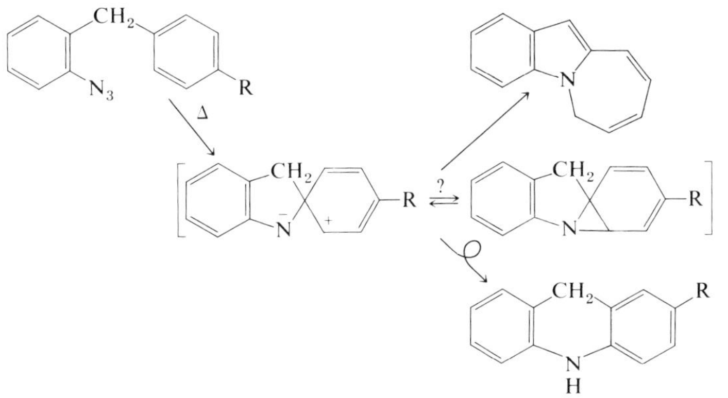

$$(6-22)$$

results can be accounted for if it is assumed that a nitrene is formed and then attacks the 1 position of the substituent phenyl group to form an intermediate that can be represented either as a benzaziridine or a dipolar valence isomer of it. *o*-Benzyl-phenyl azides characteristically give azepine derivatives as the only products, whereas *o*-phenylthiophenyl azides give rise to phenothiazines.

o-Vinylphenyl azides resemble *o*-phenylphenyl azides in that ring closure takes place to form a five-membered ring (indole) and the temperatures required are high, like those required for fragmentation of other aryl azides without anchimeric assistance.[47] There appears to be a steric requirement, however, for *cis-o*-azidostil-bene gives only an 18% yield of 2-phenylindole, whereas the *trans* isomer gives 88%. *o*-Allylphenyl azides also thermolyse to indoles, with concomitant migration of a hydrogen atom and without evidence of anchimeric assistance.[48]

Aryl azides with unsaturated *ortho*-substituents having an unshared electron pair at a β-atom generally fragment at lower temperatures than other aryl azides, in many instances lower than $100°C$, and show a negative activation entropy, consist-ent with a cyclic transition state. Among the examples that have been observed are imines of *o*-azidobenzaldehyde,[45] *o*-azidophenyl ketones,[43,49] *o*-nitrophenyl azides,[50] and *o*-azidoazobenzenes,[51] all of which can be represented by Equation

6-18, to give various fused-ring heterocyclic systems. Even the nitrogen of a pyridine ring may participate, as in 2-(o-azidophenyl)pyridine, forming a pyrid[1,2-b]-indazole[52] (Eq. 6-23). The mechanism and kinetics of these reactions have received much attention.[51,53]

$$\text{(structure)} \longrightarrow \text{(structure)} \tag{6-23}$$

Acyl azides are best known for the Curtius rearrangement, although they are also useful as acylating agents (discussed later in the section Nucleophiles). The Curtius rearrangement is the key step in a widely used method for converting a carboxylic acid to an amine of one less carbon atom.[54] Rearrangement to an isocyanate takes place in high yields on mild heating in solution (Eq. 6-24). Carbamoyl, alkoxy-

$$R-C\overset{\displaystyle O}{\underset{\displaystyle N_3}{\Big\langle}} \longrightarrow R-N=C=O + N_2 \tag{6-24}$$

carbonyl, and sulfonyl azides resist rearrangement and at best give low yields of rearrangement products.

In the Curtius degradation, of which the rearrangement is the central part, an acyl azide is prepared either by reaction of an acyl chloride with sodium azide or of a hydrazide with nitrous acid. More often than not, the acyl azide is not isolated, but is warmed in an inert solvent to convert it to an isocyanate or in alcohol to convert it to a urethan (Eq. 6-25). If an amine is the ultimately desired product, the initial

$$
\begin{array}{c}
\text{RCOCl} + \text{NaN}_3 \xrightarrow[\text{acetone}]{\text{aq.}} \text{RCON}_3 \xrightarrow{\Delta,\ \text{benzene}} \text{R}-\text{NCO} \\
\uparrow \qquad\qquad\qquad\qquad\qquad \Delta, \\
\text{R}-\text{COOH} \qquad\qquad \text{HNO}_2 \quad \text{EtOH} \xrightarrow[\text{reflux}]{\substack{\text{aq.}\\ \text{HCl or}\\ \text{NaOH}}} \text{RNH}_2 + \text{CO}_2 \\
\downarrow \qquad\qquad\qquad\qquad\qquad\qquad \\
\text{RCOOC}_2\text{H}_5 + \text{N}_2\text{H}_4 \xrightarrow{20-80\,^\circ\text{C}} \text{RCONHNH}_2 \longrightarrow \text{R}-\text{NHCOOEt}
\end{array} \tag{6-25}
$$

(Curtius degradation of an acid through its azide)

rearrangement products must be hydrolyzed; refluxing with mineral acid or aqueous alkali is usually quick and effective. The overall yields are usually good, and the conditions can be kept quite mild.

The Curtius rearrangement involves a fragmentation with loss of nitrogen, accompanied by migration of a group from the carbonyl carbon to the remaining nitrogen, a process sometimes called a *1,2-shift*. It has been demonstrated in many ways that the migration component is strictly intramolecular; the migrating group never becomes free of the system to which it is attached at the start and finish, and attempts to intercept it in the form of an ion or radical have all failed.[20] When optically active compounds are used in which the chiral center is attached directly to the carbonyl group, migration occurs with full retention of activity and configuration.[55]

The evidence is conclusive that loss of nitrogen is concerted with the migration step in thermolysis of acyl azides, and a discrete nitrene cannot be an intermediate. Many different methods for intercepting a possible intermediate nitrene have been attempted, but none has succeeded; Curtius rearrangement takes place in high yields in all varieties of solvents, without involving them, except that strongly nucleophilic media may solvolyze the acyl azide faster than it can rearrange. For example, in the presence of water, not even a trace of hydroxamic acid, the product to be expected of reaction of an acylnitrene with water, is detectable.[56] Insertion reactions have not been observed to accompany Curtius rearrangement even in the most favorable circumstances, intra- or intermolecular. The presence of a primary ^{13}C kinetic isotope effect is clear evidence for the involvement of the α-carbon in the transition state.[57]

Acyl azides in which the α-atom is more electronegative than carbon—i.e., carbamoyl azides and carboalkoxy azides—are more resistant to thermal rearrangement. With alkoxycarbonyl azides, fragmentation to a nitrene is the only thermal process;[58] the initially formed singlet nitrene takes part in efficient insertion reactions or crosses over to the triplet state and abstracts hydrogen atoms from the medium[58,59] (Eq. 6-26). The intersystem crossing step appears to be catalyzed by

$$RO-\underset{\underset{O}{\|}}{C}-N_3 \xrightarrow{\Delta} RO-\underset{\underset{O}{\|}}{C}-N^1 \xrightarrow{SoH} RO-\underset{\underset{O}{\|}}{C}-NH-So$$

$$RO-\underset{\underset{O}{\|}}{C}-N^3 \xrightarrow{SoH} RO-\underset{\underset{O}{\|}}{C}-NH\cdot \longrightarrow \qquad (6\text{-}26)$$

$$RO-\underset{\underset{O}{\|}}{C}-NH_2$$

free radicals, such as ROCONH·, and it is retarded by radical scavengers. Carbamoyl azides, however, undergo Curtius rearrangement when pushed and form isocyanatoamines or their further reaction products[60–63] (Eq. 6-27). Sulfonyl azides

$$Et_2N-CON_3 \longrightarrow Et_2N-NCO \xrightarrow{ROH} Et_2N-NH-CO_2R \qquad (6\text{-}27)$$

generally give nitrene reactions on thermolysis,[64] but in 1977, the first example was reported of rearrangement in this class of azide.[64a] Mesitylenesulfonyl azide was found to give mesidine or N,N'-dimesitylsulfamide (43% of the latter) (Eq. 6-28).

$$ArSO_2N_3 \longrightarrow ArNH_2, \qquad ArNHSO_2NHAr \qquad (6\text{-}28)$$

$$(Ar = 2,4,6\text{-trimethylphenyl})$$

Photolysis of acyl azides[58,65] of all types gives rise to products attributable to nitrene intermediates without rearrangement. Curtius rearrangement may occur as well, but the evidence implies that partitioning between the two classes of products occurs at the azide stage and not with the nitrene. Among other facts, the proportion of nitrene to rearrangement products is not affected by the presence of efficient trapping agents for nitrenes, such as cyclohexene. Pivaloyl azide, for example, produces *t*-butyl isocyanate in 40% yield and nitrene-derived products in approximately 45% yield whether photolyzed in methylene chloride (inert to nitrenes) or cyclohexene, with which the nitrene reacts principally by addition to form an aziridine.[66] If the rearrangement products were derived from a nitrene, their proportion would be markedly reduced in the presence of traps, and it has therefore been concluded that rearrangement observed during photolysis is a concerted reaction of an excited state of the azide. Some interesting theoretical calculations of stabilities of acylnitrenes and isomeric species, such as isocyanates, taking into account singlet and triplet states, have been published.[67]

Diarylcarbamoyl azides give typical nitrene reactions on thermolysis; some are attributed to the singlet state (addition to π-bonds, as in conversion of benzenes to azepines) and some to triplet states (abstraction of hydrogen or two-step insertion into C—H bonds).[60a] Whereas diarylcarbamoyl azides give only nitrene products on photolysis, *N*-aryl-*N*-alkylcarbamoyl azides give some rearrangement product as well, and dialkylcarbamoyl azides give Curtius rearrangement products without detectable nitrene products.[61]

Alkoxycarbonyl azides on photolysis give only nitrene-derived products, just as in thermolysis, but whereas in thermolysis, the products appear to be derived entirely from the singlet state of the nitrene, products of both singlet and triplet nitrenes are formed in photolysis.[58] The distribution of products is affected by the solvent; methylene chloride, for example, appears to favor reactions of the singlet state.[68] The most useful reaction of alkoxycarbonyl azides by either thermolysis or photolysis is addition of the singlet nitrene to benzenes to form azepines (Eq. 6-29).

$$\text{EtOCON}_3 + \text{C}_6\text{H}_6 \xrightarrow{\Delta \text{ or } h\nu} \text{EtOCO—N} \hspace{2cm} (6\text{-}29)$$

Cyanogen azide may be thought of as a special case of an acyl azide, related most closely to the carbamoyl azides. It undergoes fragmentation without catalysis at unusually low temperatures (40–60°C) to give products rationally derived from a symmetrical, delocalized nitrene, NCN. It reacts with benzene rings to form azepines and inserts in C—H bonds of alkanes with high efficiency.[69] It can also rearrange to $:\overset{-}{\text{C}}—\overset{+}{\text{N}}\!\!=\!\!\text{N}$, probably by dissociation to atomic C and N_2 and recombination.

Reactions with acids.[15d]   The reactions with Brønsted acids are the most important of the reactions of azides with electrophiles. Although azides are quite generally inert to dilute acids and may confidently be washed with them to remove basic contaminants, almost all of them will react with concentrated sulfuric acid at room temperature with loss of nitrogen. Evolution of nitrogen is usually quantitative and has even been used for estimation of the azido function.[70]

Aliphatic azides (including aralkyl azides) are usually rearranged by treatment with strong acid, an α-substituent (hydrogen, alkyl, or aryl) migrating to the retained nitrogen to form an aldimine or ketimine,[20,71] which in the case of vinyl azides isomerizes to a nitrile[72] (Eqs. 6-30, 6-31). These reactions presumably begin with

$$R_2CR'N_3 \xrightarrow{H_2SO_4}$$

$$N_2 + R_2C{=}NR' \quad or \quad RCR'{=}NR \qquad \left[R_2C \begin{array}{c} \diagup OH \\ \diagdown NHR' \end{array} \right] \qquad (6\text{-}30)$$

$$Ph{-}CO{-}CH{=}CH{-}N_3 \xrightarrow{H_2SO_4} Ph{-}CO{-}CH_2{-}CN \qquad (6\text{-}31)$$
$$82\%$$

protonation of the substituted nitrogen atom to form an aminediazonium ion, $R{-}NH{-}N_2^+$, which may lose nitrogen to leave an aminylium ion, RNH^+, which in turn rearranges to the conjugate acid of the product, or the steps may be synchronous. Since the effects of substituents on rates are consistent with a concerted process and aminylium ions have not been intercepted, it is most reasonable to believe that migration from C to N accompanies loss of nitrogen. With certain tertiary alkyl azides, such as 1-adamantyl, the product may be a carbinolamine rather than an imine, an understandable situation where the formation of an imine double bond would result in excessive strain.[73]

Aryl azides do not rearrange, but usually react with the anion of the acid or with the solvent, forming ring-substituted anilines.[74] Azides that cyclize when heated or irradiated do not cyclize when treated with acid. It appears that the conjugate acids of azides, $ArNH{-}N_2^+ \rightleftarrows Ar{-}\overset{+}{N}{-}N{=}NH$, are first formed, since the kinetics are roughly first-order in ArN_3 and the acidity function h_o.[75] The tautomer having the aminediazonium structure may in principle undergo nucleophilic displacement at the substituted nitrogen atom, producing haloamide or hydroxylamine derivatives that then undergo rearrangement or other reactions, or it might lose nitrogen unimolecularly to produce a delocalized aminylium ion (Eqs. 6-32 through 6-34),

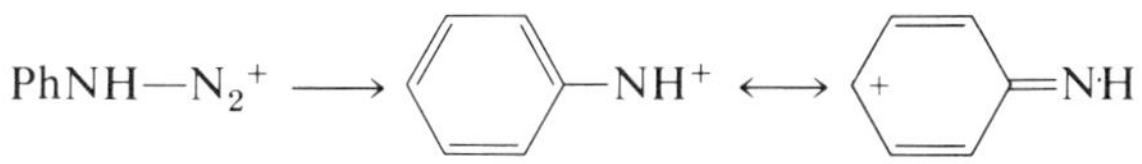

$$PhNH{-}N_2^+ \longrightarrow \langle \text{ring} \rangle{-}NH^+ \longleftrightarrow \langle \text{ring} \rangle{=}NH \qquad (6\text{-}32)$$

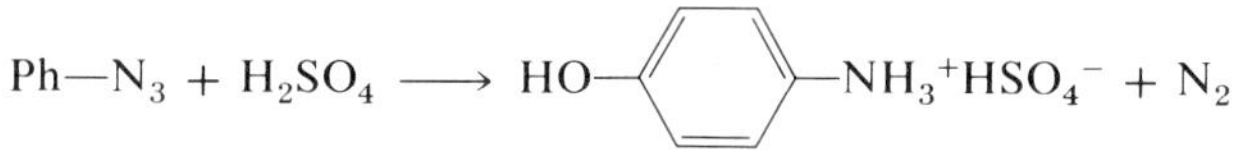

$$Ph{-}N_3 + H_2SO_4 \longrightarrow HO{-}\langle \text{ring} \rangle{-}NH_3^+HSO_4^- + N_2 \qquad (6\text{-}33)$$

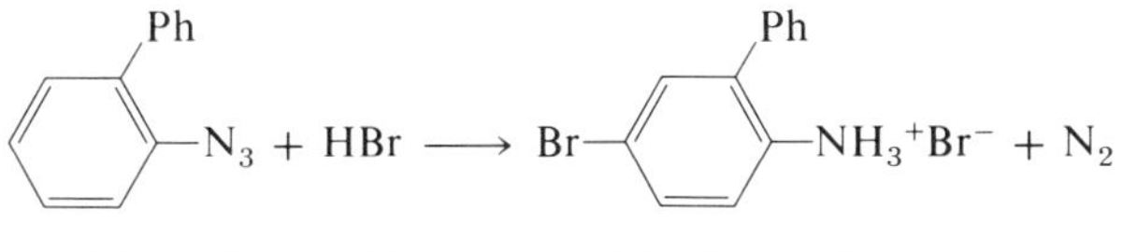

$$\overset{Ph}{\langle \text{ring} \rangle}{-}N_3 + HBr \longrightarrow Br{-}\overset{Ph}{\langle \text{ring} \rangle}{-}NH_3^+Br^- + N_2 \qquad (6\text{-}34)$$

which could combine with a nucleophile in a subsequent step. If an electron-rich benzene ring is present, such as in a phenol, it may be attacked (Eq. 6-35).

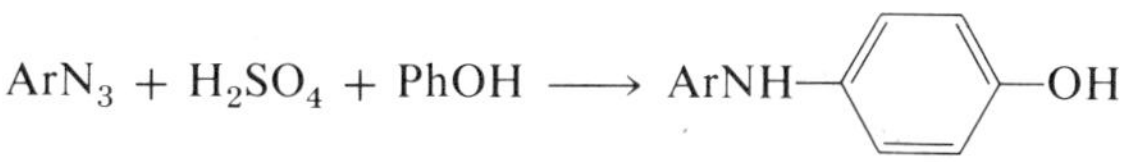

$$ArN_3 + H_2SO_4 + PhOH \longrightarrow ArNH{-}\langle \text{ring} \rangle{-}OH \qquad (6\text{-}35)$$

Aminylium ions may be responsible for the intense colors that are formed in concentrated sulfuric acid, but are destroyed on dilution (compare arylhydroxylamines). *Para*-substituted aryl azides may be converted by strong mineral acids to *ortho*-halo or *ortho*-hydroxy anilines, to anilines without further substitution (Eq. 6-36), or to polymeric substances of the type $(RC_6H_4N)_x$.[76]

$$O_2N\!-\!\!\langle\ \rangle\!-\!N_3 + HBr \longrightarrow$$
$$NO_2$$

$$O_2N\!-\!\!\langle\ \rangle\!-\!NH_2 + Br_2 \qquad (6\text{-}36)$$
$$NO_2$$

The brominative reduction of Equation 6-34 may be converted to the simple reduction of Equation 6-36 by including phenol, a good interceptor for bromine, in the reaction mixture. This fact implies that bromoanilines do not arise by simple combination of Br^- with the aminylium ion. The most reasonable alternative is nucleophilic displacement by bromide on the protonated azide to form nitrogen and an *N*-bromo amine; the latter are known to react with HBr to form bromine and an amine, which would be able to react to form ring-bromoaniline if the bromine is not taken up by a scavenger. If the ring is rendered unreactive by strong electron withdrawal, as in the example in Equation 6-36, bromination would not occur in any case. This interpretation is consistent with the fact that in the example in Equation 6-34, no carbazole is formed; an intermediate aminylium ion should be able to react to some extent with the *ortho*-phenyl group. Displacement at nitrogen would not be expected in decompositions brought about by sulfuric acid, owing to the absence of any effective nucleophile.

More recently,[77] evidence has been obtained that other acid-catalyzed reactions of aryl azides may not involve aminylium ions. When phenyl azide was treated with trifluoroacetic acid in the presence of mesitylene at 85°C, *N*-mesitylaniline was produced in 55% yield; benzene was not attacked. The rate was dependent on the aromatic solvent, a fact that implies that the substrate, ArH, is involved in the transition state for loss of nitrogen. Nevertheless, the overall behavior of aryl azides closely resembles that of the arylhydroxylamines (except for the production of azoxy compounds from the latter), and it is attractive to correlate the various reactions by the hypothesis that aminylium ions are common intermediates when displacement at nitrogen does not preempt them.

Lewis acids, such as anhydrous aluminum chloride, also react with most azides. The products are somewhat different from those from Brønsted acids; phenyl azide, for example,[78] is converted to azobenzene or to aniline and diphenylamine, accompanied by much tar, or it may react with solvent in suitable cases (Eq. 6-37). In the absence of a diluent, an explosion may occur.

Acetic acid and its anhydride react with aryl azides when heated to form derivatives of *o*-aminophenols. Thus boiling acetic anhydride converts phenyl azide to

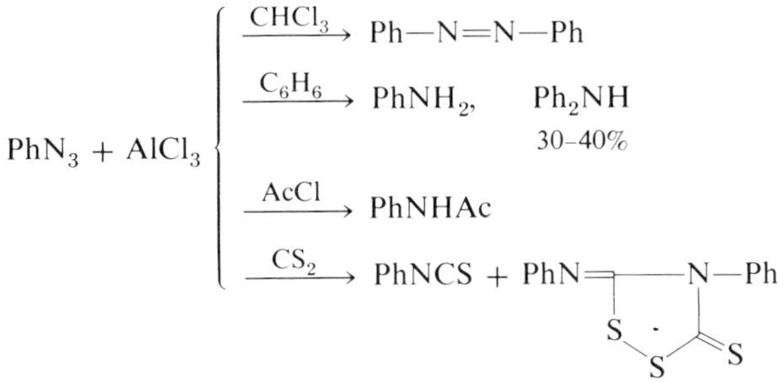

$$PhN_3 + AlCl_3 \begin{cases} \xrightarrow{CHCl_3} Ph-N=N-Ph \\[6pt] \xrightarrow{C_6H_6} PhNH_2, \quad Ph_2NH \\[2pt] \hspace{4cm} 30\text{-}40\% \\[6pt] \xrightarrow{AcCl} PhNHAc \\[6pt] \xrightarrow{CS_2} PhNCS + PhN\overset{\displaystyle}{=\!=\!=}N-Ph \end{cases} \tag{6-37}$$

o-acetamidophenyl acetate, possibly through an intermediate *N*-phenylhydroxamic acid derivative[79] (Eq. 6-38). The same sort of product is also obtained with a

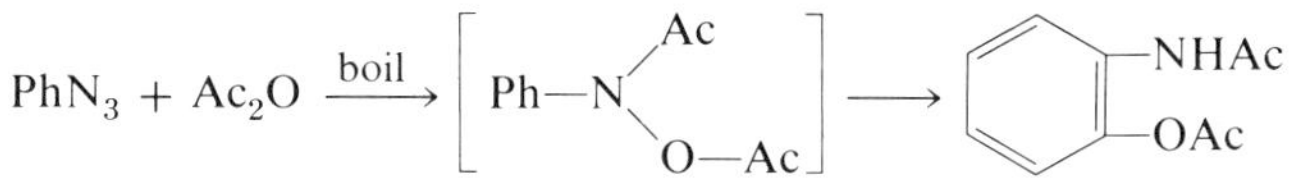

$$PhN_3 + Ac_2O \xrightarrow{boil} \left[Ph-N \begin{array}{c} Ac \\ O-Ac \end{array} \right] \longrightarrow \begin{array}{c} -NHAc \\ -OAc \end{array} \tag{6-38}$$

carboxylic acid in the presence of polyphosphoric acid; the corresponding benzoxazole is in some cases the major product. The oxygen becomes attached to the ring at the *ortho* position most susceptible to electrophilic attack.[80]

Acyl azides undergo either solvolysis or rearrangement at room temperature when treated with strong mineral acid. Such catalysis is so effective that the reaction may be uncontrollably violent unless the diluted azide is brought into contact with the acid in small portions. The most useful application of this phenomenon is the generation of the protonated azide in situ by treating a carboxylic acid with excess hydrogen azide in the presence of concentrated sulfuric acid. This procedure is known as the *Schmidt reaction;*[20,81] it constitutes a one-step conversion of a carboxyl group to an amino group (the isocyanate presumably formed first is usually rapidly hydrolyzed by the reaction medium) (Eq. 6-39).

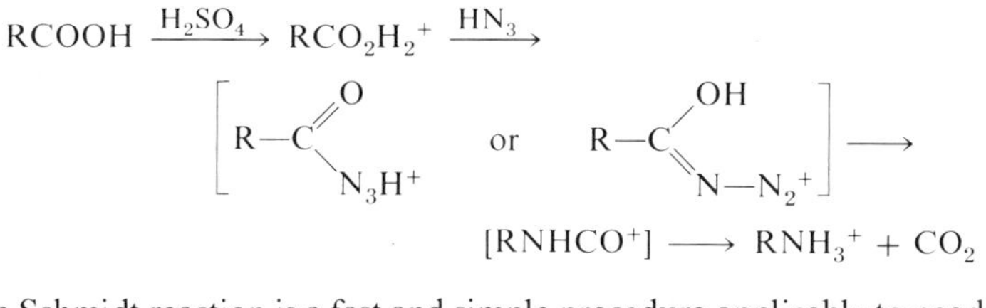

$$RCOOH \xrightarrow{H_2SO_4} RCO_2H_2^+ \xrightarrow{HN_3}$$

$$\left[R-C \begin{array}{c} O \\ N_3H^+ \end{array} \quad \text{or} \quad R-C \begin{array}{c} OH \\ N-N_2^+ \end{array} \right] \longrightarrow$$

$$[RNHCO^+] \longrightarrow RNH_3^+ + CO_2 \tag{6-39}$$

The Schmidt reaction is a fast and simple procedure applicable to nearly all types of carboxylic acids, but its usefulness is obviously limited to acids not having other functional groups that would be affected by strong, concentrated acid. α-Amino acids do not react, so the Schmidt reaction on malonic acids stops at the amino acid stage and is a good preparative method for some of them.

The reactions of HN_3 with aldehydes and ketones are also known as Schmidt reactions. Amides are produced from ketones and nitriles (plus usually minor amounts of formamides) from aldehydes.[20,81] The reactions take place under much milder conditions than the Schmidt reactions of carboxylic acids and often go rapidly at $0°C$ with concentrated hydrochloric instead of sulfuric acid. With unsymmetrical ketones, mixtures may be produced. In the case of aliphatic ketones, the most highly branched group most generally becomes attached to the nitrogen (Eqs. 6-40, 6-41). Acetophenones always give acetanilides in overwhelming proportion, a fact

$$i\text{-Pr}\text{--}CO\text{--}CH_3 \xrightarrow[\text{H}_2\text{SO}_4]{\text{HN}_3} i\text{-Pr}\text{--}NH\text{--}CO\text{--}CH_3 \qquad (6\text{-}40)$$ 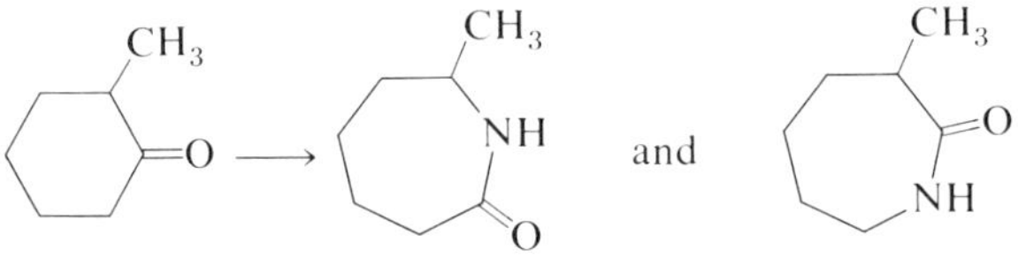
$$\text{(Mostly)}$$

(6-41)

(Mostly)

that makes the combination of Friedel-Crafts acetylation and the Schmidt reaction a usedful alternative to nitration and reduction for the preparation of aromatic amines (Eq. 6-42). Benzophenones, however, show little sensitivity to *meta* and *para*

$$\text{ArH} + \text{Ac}_2\text{O} \xrightarrow{\text{AlCl}_3} \text{Ar}\text{--}\text{Ac} \xrightarrow{\text{HN}_3}$$
$$\text{Ar}\text{--}\text{NH}\text{--}\text{Ac} \xrightarrow{[\text{H}_2\text{O}]} \text{ArNH}_2 \qquad (6\text{-}42)$$

substituents and give mixtures of nearly equal amounts of both possible benzanilides. *Ortho*-substituted benzophenones give mostly the amide derived from migration of the unsubstituted phenyl group to nitrogen, but some *ortho* substituents, such as carboxyl, have the opposite effect, apparently owing to ring formation at an intermediate stage.

The evidence is consistent with a mechanism involving addition of hydrogen azide to the conjugate acid of the carbonyl compound; C-to-N migration with loss of nitrogen may then take place before or after dehydration,[20,82] according to structure and conditions. Amides may thus arise directly or when the iminoalkylium ion resulting from dehydration and rearrangement subsequently combines with water (Eq. 6-43).

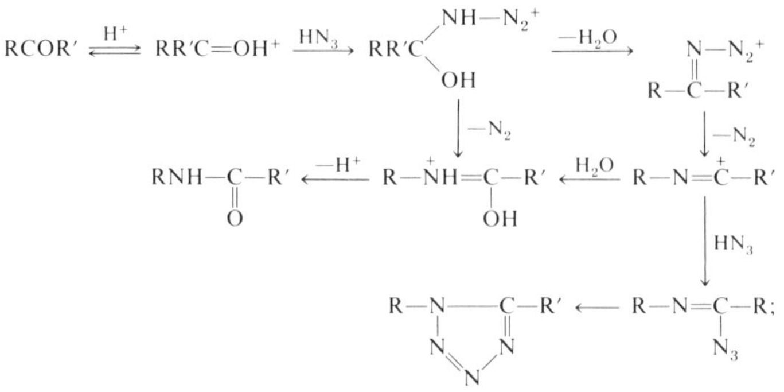

(6-43)

The products from unsymmetrical ketones may be determined by relative migration aptitudes, as appears to be the case with many aliphatic ketones, or by steric effects on the geometric isomerism of the iminoalkylium intermediate, as appears to be the case with most diaryl ketones.[20,82] It is presumed that in the latter case, migration of R from C to N is synchronous with loss of nitrogen and that the group *trans* to the departing N_2 is the one that migrates, in analogy to the Beckmann rearrangement. Migration aptitudes would be expected to operate in rearrangement of the tetrahedral intermediate, but they might also do so in rearrangement of the iminoalkylium ion if the dehydration step is reversible and establishes equilibrium relatively rapidly.

A more recent alternative to the conventional Schmidt procedure on ketones makes use of the reaction of thioketals with iodine azide (Eq. 6-44); yields are

$$R_2C\begin{matrix}SR'\\\\SR'\end{matrix} + I-N_3 \longrightarrow R_2C\begin{matrix}N_3\\\\SR'\end{matrix} \xrightarrow{CF_3COOH} R-CO-NH-R \qquad (6\text{-}44)$$

reported to be 65 to 85%, and tetrazoles are not formed.[83] If sodium azide with stannic chloride, or trimethylsilyl azide, is used instead of IN_3, iminothio ethers are formed in good yield.

Some tertiary alcohols and benzylic alcohols react under the conditions of the Schmidt reaction to give the products to be expected of acid-catalyzed rearrangement of the corresponding azide.[15d] The intermediate azides can, in fact, often be isolated, even in good yield, if the conditions used are not too strenuous.[71]

Sulfonyl azides do not undergo the acid-catalyzed rearrangements of other azides, but it has been reported, somewhat scantily, that they react with benzene rings in the presence of sulfuric acid or aluminum chloride, forming anilines in yields as high as 60%, along with the sulfonic acid and, in the case of $AlCl_3$, sulfones.[84,85]

Electrophilic carbon (carbenium ions, etc.). Azides evidently have very low nucleophilicity, for they are immune to ordinary alkylating agents. Only with trialkyloxonium salts or alkyl halides in the presence of silver perchlorate does attack occur.[86,87] The reaction is then analogous to acid-catalyzed decomposition; the electrophilic species attacks the substituted nitrogen atom, nitrogen is lost, and imines are formed (Eq. 6-45).

$$BuN_3 + (CH_3)_3O^+BF_4^- \longrightarrow \underset{\underset{CH_3}{|}}{Bu-N-N_2^+} \xrightarrow{-N_2}$$

$$PrCH=NCH_3 + BuN=CH_2 + \underset{\underset{CH_3}{|}}{Pr\overset{+}{N}=CH_2}\ BF_4^- \qquad (6\text{-}45)$$

The hydroxyalkylium ions formed by protonation of aldehydes and ketones might be expected to attack azides. 2-Phenylethyl azide and butyl azide have been reported to react with benzaldehyde in the presence of sulfuric acid, forming *N*-alkylbenzamides[88] (Eq. 6-46), but uncertainty clouds the possibility that ketones may react under these conditions. Whereas the yields in the foregoing reaction are low, much higher yields have been reported for the reaction of benzaldehydes with

$$PhCH{=}O + RN_3 \xrightarrow{H_2SO_4} PhCH\underset{\underset{N_2{}^+}{\underset{|}{N}}}{\overset{OH}{\diagup}}R \xrightarrow[\text{PhCH}=O]{\text{in excess}} PhCONHR + N_2$$

$$\text{in benzene} \Big\downarrow$$

$$N_2 + Ph{-}N\underset{CH{=}OH^+}{\overset{R}{\diagup}} \longrightarrow PhNH_2 + CO$$

(6-46)

2-azidoethanol and 3-azidopropanol-1, which give cyclic imino esters, respectively, a 2-aryloxazoline (77%) and a 2-aryldihydrooxazine (82%).[88c]

Two examples, benzaldehyde and cyclohexanone, have been reported of uncatalyzed reaction with phenyl azide.[89] Although these may involve electrophilic attack on the azido group, that does not seem probable, for the experiments were carried out at the boiling points of the reactants, at which temperature phenyl azide actively undergoes thermolytic fragmentation. The imines are formed (Eq. 6-47) by way of a nitrene, but not by way of nitrones, which are stable at the temperatures used.

$$PhN_3 + PhCH{=}O \xrightarrow{\text{boil}} \underset{75\%}{PhN{=}CHPh}$$

(6-47)

Nucleophilic species. Oxy-anion nucleophiles are without effect on azides, and only when the structure holding the azido group is itself susceptible to attack by bases does reaction occur with most of them. Alkyl azides are inert to hydrolysis (azide ion is not a good leaving group) and will even withstand boiling alcoholic alkali, the conditions used to convert 2-haloethyl azides to vinyl azide.[90]

α-Azido ketones lose nitrogen when treated with strong bases, but this reaction is probably an attack on the acidic α-hydrogen rather than the azide. α-Imino ketones or their transformation products are formed.[91a,b] By the same mode of attack, β-azido ketones eliminate HN_3 to form α,β-unsaturated ketones.[91c] Acyl azides are attacked by base at the carbonyl group and are hydrolysed by alkali. Toward amines, they function as acylating agents nearly as active as anhydrides.[54] Since they can be made from esters under mild conditions through the acid hydrazide, they have seen wide application as acylating agents in synthesis, particularly of peptides.[92] However, the rate of acylation of an amino group may not in all cases be greatly different from that of the Curtius rearrangement to an isocyanate, which will react with the amine to form a urea:[92b,d] $RCON_3 \rightarrow RNCO$, $+ R'NH_2 \rightarrow RNHCONHR'$. Alkoxycarbonyl azides,[92b,d] such as t-BuOCON$_3$, do not suffer from this side reaction, and they have accordingly seen use as reagents for protecting amino groups with the t-butoxycarbonyl group[92c] (it is a dangerous reagent, however, and a safer alternative is available[11c]).

Aryl azides bearing electron-withdrawing groups in the *ortho* or *para* positions may be sufficiently susceptible to nucleophilic attack to undergo displacement of the azido group:[93] $YArN_3 + OH^- \rightarrow YArOH + N_3^-$.

Azides appear to be generally inert to amines and hydrazines; aminoaryl azides are stable compounds, and the amino group can be generated from a phthalimide by the Ing-Manske hydrazine cleavage without disturbing an azido group elsewhere in the molecule. Only at temperatures where unimolecular fragmentation of the azido group occurs to a significant extent have amines been reported to react with them. A variety of anilines have been reported to react with several aryl azides to produce azobenzene derivatives[94,95] (Eq. 6-48); the reaction seems to be most suc-

$$Ar—N_3 + Ar'NH_2 \xrightarrow{130-160°C} Ar—N=N—Ar' \qquad (6-48)$$

cessful when the respective aryl groups have opposite electronic features. Some azobenzene derived solely from the aryl azide is always formed. Presumably an aryl nitrene attacks the amino group to form a hydrazo compound, which is then dehydrogenated, perhaps by more nitrene. Trifluoromethanesulfonyl azide, a rather extreme case of an electron-depleted azide, appears to react generally with aliphatic primary amines without involving nitrenes. The amine is converted to an azide, two nitrogen atoms having been acquired from the sulfonyl azide, with retention of configuration, perhaps through a tetrazene intermediate[96] (Eq. 6-49). Amide anions, RNH⁻, will even react with toluenesulfonyl azide in this manner.[97,98]

$$RNH_2 + F_3CSO_2N_3 \xrightarrow{2,6\text{-lutidine}} RN_3 + F_3CSO_2NH_2 \qquad (6-49)$$

In contrast to oxygen and nitrogen nucleophiles, nucleophilic carbon and phosphorus quite generally attack the azido group. Cyanide ion is the simplest example; reaction occurs at the unsubstituted nitrogen atom, producing a triazene,[99,100] the hydrogen of which is acidic owing to the cyano group (Eq. 6-50). These compounds

$$ArN_3 + KCN \longrightarrow Ar—N=N—NCN^-K^+ \begin{cases} \xrightarrow{10\% \text{ KOH}} ArNH_2 \\ \\ \xrightarrow{aq. \text{ HCL}} ArN_2^+ + H_2NCONH_2 \end{cases} \qquad (6-50)$$

are quite reactive and are decomposed to amines by alkali and to diazonium salts and urea by acids.

Grignard reagents, regarded as carbanion sources, react analogously[101-104] and form disubstituted triazenes (although α-azidosuccinamides are claimed to give secondary amines instead[105]). The triazenes are of low stability, and often it is only their decomposition products that are isolated. Sulfonyl and phenylthiomethyl azides, however, give rise to somewhat more stable triazenes, especially with aryl Grignard or aryllithium reagents, in which case they can be converted to amines or aryl azides[106] (Eq. 6-51). This is one of the few preparatively useful methods for

$$ArSO_2N_3 + Ar'MgX \longrightarrow \underset{MgX^+}{ArSO_2\bar{N}—N=NAr'} \xrightarrow{NaN_3/NaOH} Ar'N_3$$

$$(6-51)$$

$$PhSCH_2N_3 + Ar'MgX \longrightarrow \underset{MgX^+}{PhSCH_2\bar{N}—N=NAr'} \underset{\underset{aq. \text{ NH}_3}{\overset{aq. \text{ NaOH}}{\diagdown}}}{\xrightarrow{Ni\text{-Al alloy}}} Ar'NH_2$$

converting a Grignard reagent to an amine. Carbanions of active-methylene compounds, such as β-keto esters, as well as phosphonium ylides, also attack azides readily. The products from sulfonyl azides subsequently cleave to a diazo compound[98,107,108] (Eq. 6-52). This reaction has come to be known as the *diazo transfer*

$$ArSO_2N_3 + (RCO)_2CH^- \longrightarrow ArSO_2NH^- + (RCO)_2C{=}N_2 \qquad (6\text{-}52)$$

reaction;[98] it is a versatile preparative method (see the section Preparative Methods in Chap. 5). With other types of azides, carbanions from active-methylene compounds react to form triazoles; since these can be regarded as deriving from cycloaddition to an enolate double bond, the reaction will be discussed further under Cycloaddition.

Isocyanides may in many respects be considered to be carbon nucleophiles; they react with azides to form carbodiimides, but only when a suitable catalyst is used[109] (Eq. 6-53). Symmetrical carbodiimides are not formed, so an isocyanate intermedi-

$$R{-}N_3 + R'{-}NC \xrightarrow{Fe(CO)_5} R{-}N{=}C{=}N{-}R' + N_2 \qquad (6\text{-}53)$$

ate cannot be involved. The reaction presumably involves a nitrene reacting with a coordinated isocyanide; the conditions used, 90°C for 24 hours, are not inconsistent with initial formation of a nitrene.

Another reagent that may be considered as reacting by nucleophilic carbon is $(CH_3)_2S(O){=}CH_2$, to the extent that the $S{=}CH_2$ bond is semipolar. It reacts in a ratio of 2:1 with alkyl or aryl azides to form triazolines (Eq. 6-54). *p*-Nitrophenyl

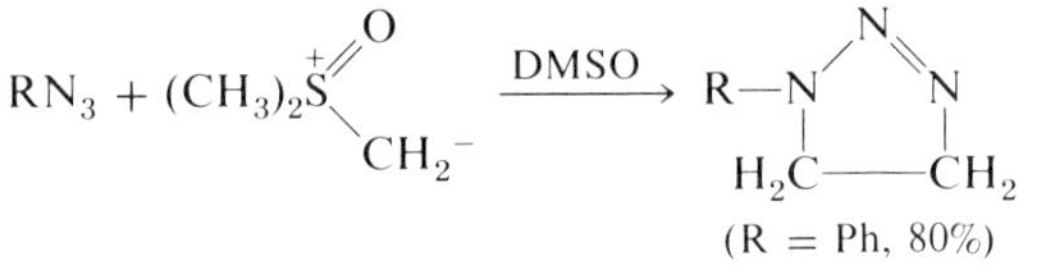

$$(R = Ph, 80\%) \qquad (6\text{-}54)$$

azide and benzoyl azide follow a different course and form vinyltriazenes instead[110] (Eq. 6-55). Vinyltriazenes are also produced by the reaction of alkyllithium or, less

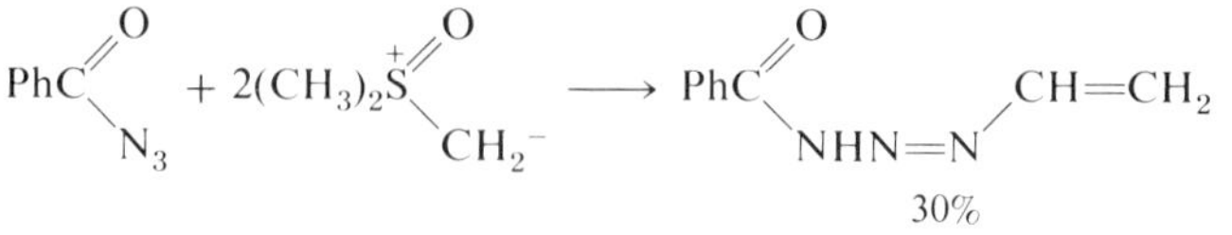

$$30\% \qquad (6\text{-}55)$$

satisfactorily, Grignard reagents to vinyl azides. The alkylvinyltriazenes undergo C-to-N migration of the alkyl group when treated with dilute acid and thereby give rise to ketimines[111] (see Chap. 7).

Phosphines, particularly triphenylphosphine, resemble carbanions in their ability to attack the terminal nitrogen of azides; the initial adducts, called *phosphazenes*, can in some instances be isolated, but they lose nitrogen spontaneously to form phosphine imines[112-114] (Eq. 6-56). In an experiment with toluenesulfonyl azide

$$RN_3 + PR_3' \longrightarrow R{-}N{=}N{-}N{=}PR_3' \longrightarrow R{-}N{=}PR_3' + N_2 \qquad (6\text{-}56)$$

labeled with ^{15}N at the terminal position, none of the label remained in the isolated triphenylphosphine tosylimine.[115] The phosphazenes from triarylmethyl azides are more stable than most; acid decomposes them to unsubstituted triphenylphosphine imine and products derived from the triarylmethyl cation.[114a] The reaction of vinyl azides with triethyl phosphite has been adapted to the synthesis of ketones from trifluoromethyl sulfones[116] (Eq. 6-57).

$$\text{Ph}-\underset{\underset{\text{SO}_2\text{CF}_3}{|}}{\text{CH}}-\text{Et} \xrightarrow[\text{(2) TosN}_3]{\text{(1) NaH}} \text{Ph}-\underset{\underset{\text{N}_3}{|}}{\text{C}}=\text{CHCH}_3 \xrightarrow[\text{(2) H}_2\text{O, H}^+]{\text{(1) P(OEt)}_3}$$

$$\text{Ph}-\underset{\underset{\text{O}}{||}}{\text{C}}-\text{Et} \qquad (6\text{-}57)$$

Two remaining reactions will be taken up in this section, although they cannot be classified unambiguously. One is the reaction of triethylaluminum with phenyl azide, which might be expected to proceed by electrophilic attack by the aluminum or nucleophilic attack by the alkyl groups. At $-70°$C, an addition complex is formed, which breaks up upon warming to form aniline (63.5%), N-ethylaniline (16.3%), o-ethylaniline (9%), and p-ethylaniline (11.2%).[117] The other is the reaction of phenyl azide with phenylhydrazones of aldehydes in the presence of sodium ethoxide.[118] The product is a triazene, formally the result of the 1-carbanion of the hydrazone attacking the terminal nitrogen of the azide, but the product perhaps arose from the delocalized N—N—C anion resulting from removal of the proton from the hydrazone nitrogen, followed by tautomerization (Eq. 6-58). In the case of

$$\text{RCH}=\text{N}-\text{NHPh} + \text{PhN}_3 \xrightarrow[\text{EtOH}]{\text{NaOEt}}$$

$$\text{Ph}-\text{NH}-\text{N}=\text{N}-\text{CR}=\text{N}-\text{NHPh} \qquad (6\text{-}58)$$

benzaldehyde phenylhydrazone, the product is 1,4-diphenyltetrazole and aniline, which might have resulted from cyclization of a triazene first formed.[119]

Cycloaddition.[15a,120-122] Carbon-carbon double bonds, as well as others, react more or less readily with azides to form products that either retain all the nitrogen atoms or have lost two of them as molecular nitrogen. When the conditions are not so drastic as to induce fragmentation of the azide to a nitrene, both products are presumed to arise by initial 1,3-dipolar cycloaddition, which may be followed by extrusion of nitrogen. Simple alkenes with isolated, unstrained double bonds are essentially unreactive toward azides at ordinary temperatures. The strained double bonds in bicyclic systems, such as the dimer of cyclopentadiene, add azides readily (Eq. 6-59). The difference in reactivity is so great that the occurrence of spontaneous reaction has been used as a test for strained double bonds, and phenyl azide has often been used as a reagent to convert terpenes to crystalline derivatives. The products are triazolines, which are characterized by a rather fragile ring system, prone to loss of nitrogen. Whereas with benzoyl azide the triazoline stage can be arrested,[123] with toluenesulfonyl azide decomposition follows so rapidly that the reactants effervesce when brought together.[124-126] The triazoline ring collapses to an aziridine or, when the structure allows, may be converted to an imine. (Eqs. 6-59, 6-60).

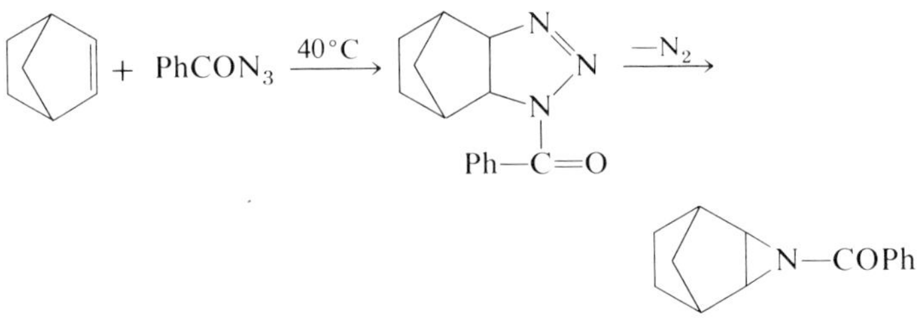

(6-59)

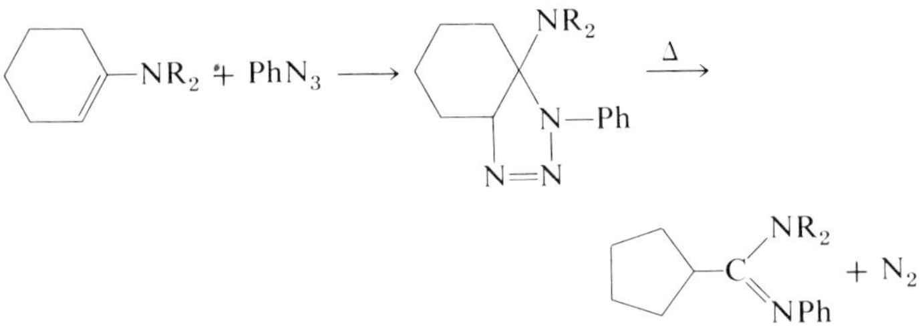

(6-60)

In general, triazolines lose nitrogen more readily than the parent azide. As a result, olefinic solvents can catalyze the decomposition of azides through formation of triazolines. When the alkene and azido functions are suitably situated in the same molecule, the apparent decomposition temperature of the azide is markedly lowered through intramolecular cycloaddition.[122] In such cases, it is not uncommon for cycloaddition to go to completion if the azidoalkene is allowed to stand in solution for some time near room temperature; warming either the cyclization product or the original azide gives the same ultimate products.[127]

In common with other cycloadditions, that of azides to alkenes is stereoselective, and the geometric configuration of the alkene is preserved in the adduct, as shown by the reaction of phenyl azide with the two isomers of β-methylstyrene[128] (Eq. 6-61). These examples also show that the addition is also regioselective. With termi-

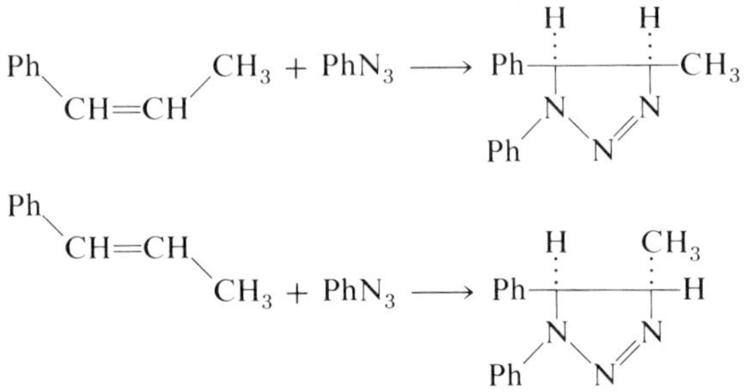

(6-61)

nal alkenes, the terminal nitrogen atom preferentially attaches to the terminal carbon. Electron-withdrawing substituents conjugated to an alkene direct the terminal nitrogen atom to the carbon adjacent to the substituent, as in the example of methyl acrylate, which gives only 1-phenyl-4-carbomethoxytriazoline.[129] Such substituents also increase the reactivity of the C=C bond toward azides. Substituents having a lone electron pair conjugated to the alkene, such as seen in enamines, also increase the reactivity toward azides markedly, but the orientation of the addition is reversed (Eq. 6-60), the substituted nitrogen atom attaching to the carbon closest to the lone-pair substituent. These effects imply the development of some charge separation in the transition state.[120,130] The aminotriazolines decompose rather readily unless the parent azide bore an electron-withdrawing group; the decomposition may involve a rearrangement of the carbon skeleton, as shown in the example of Equation 6-60, in which an amidine structure is formed.[131] Enol ethers are not quite so reactive, but otherwise behave analogously.[132]

Allylic azides constitute a special case. Intramolecular formation of a triazoline is not feasible, and the double bond is not activated. Allylic isomerization occurs instead, and 3-methylallyl azide, for example, is thereby converted to 1-methylallyl azide[133] (Eq. 6-62).

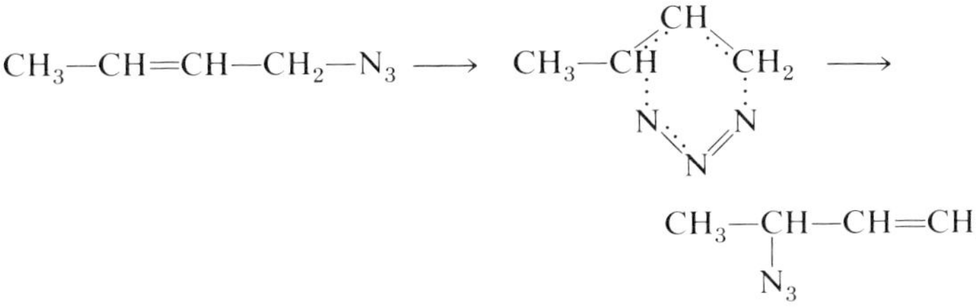

$$CH_3-CH=CH-CH_2-N_3 \longrightarrow CH_3-CH \quad CH_2 \longrightarrow$$

$$CH_3-CH-CH=CH_2 \quad (6\text{-}62)$$

Alkynes undergo cycloaddition with azides to produce triazoles, which, having an aromatic sextet of π-electrons, are far more stable than triazolines. This is a reaction of wide generality and one that gives good yields.[121] Its drawback is that unsymmetrical alkynes are inclined to give mixtures of isomeric triazoles; the regioselectivity is weak, and even methyl propynoate gives two products[134] (Eq. 6-63).

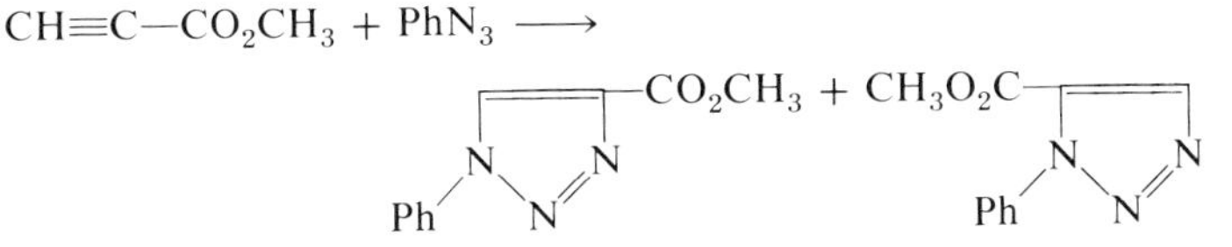

$$CH\equiv C-CO_2CH_3 + PhN_3 \longrightarrow$$

$$\text{—}CO_2CH_3 + CH_3O_2C\text{—} \quad (6\text{-}63)$$

(7:1 ratio)

Compounds that do not themselves have C—C double or triple bonds may nevertheless be induced to manifest the foregoing types of reactions when catalyzed by strong base (usually sodium ethoxide). From carbonyl compounds, enolates are produced, which have an especially reactive double bond as a result of electron donation by the anionic oxygen. Addition is usually followed by elimination so as to generate a triazole (Eq. 6-64). Acetoacetic ester, for example, loses an acetyl group,

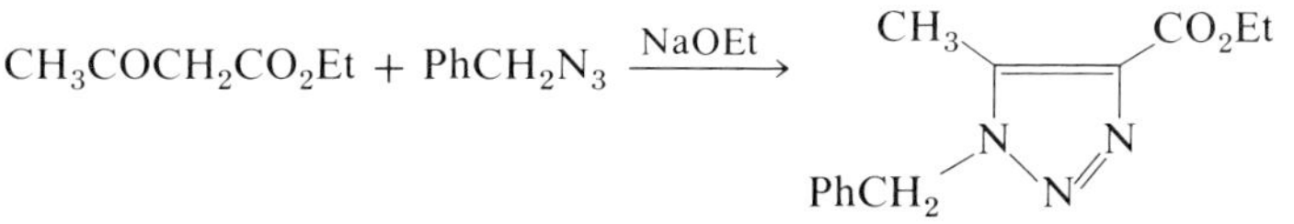

$$CH_3COCH_2CO_2Et + PhCH_2N_3 \xrightarrow{\text{NaOEt}} \quad (6\text{-}64)$$

and diethyl malonate loses a carboethoxy group.[135] This reaction is the classical *Dimroth triazole synthesis*. A variant of it is the use of nitriles, the anions of which can be looked upon as potential ynamine anions; aminotriazoles result[136] (Eq. 6-65). The fact that these reactions can be looked upon as cases of nucleophilic

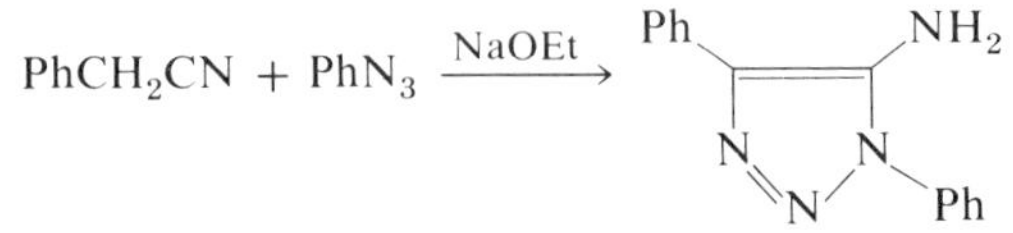

$$PhCH_2CN + PhN_3 \xrightarrow{\text{NaOEt}} \quad\quad\quad\quad (6\text{-}65)$$

attack by a carbanion on the azido group has been mentioned in an earlier section of this chapter.

Azides do not undergo cycloaddition to carbonyl groups, although it has been speculated that such a reaction may play a role in intramolecular reactions of azido ketones, the presumed oxatriazoline decomposing before it is detected. Azides are also inert to the imine function, except in the special case of imidoyl azides, which react intramolecularly to give tetrazoles so rapidly that the open structure can be isolated only in exceptional instances.[98] The reaction of imidoyl chlorides with hydrogen azide, which might be expected to give imidoyl azides, is, in fact, the most general route to 1,5-disubstituted tetrazoles (Eq. 6-66) and is known as the *von Braun-Rudolph synthesis.*[137,138]

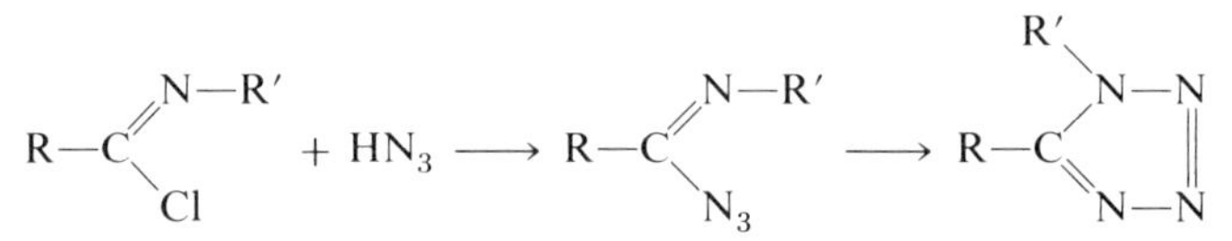

$$ \quad\quad\quad\quad (6\text{-}66)$$

The cyclization of imidoyl azides is reversible, although the equilibrium ordinarily lies far to the tetrazole side. Electron-withdrawing substituents increase the relative stability of the imidoyl azide form, as does protonation by strong acids. Steric strain arising when a tetrazole ring is fused to a previously existing five-membered ring also shifts the equilibrium toward the azide.[98] The position of the equilibrium shows some sensitivity to the polarity of the solvent, and in very polar solvents, such as dimethyl sulfoxide, the tetrazole structure, which has a relatively high dipole moment, is favored markedly.[139] The azide-tetrazole equilibrium is also encountered with hydroximoyl and hydrazonoyl azides.[140-143]

It is interesting to compare the stability toward cyclization of other functions having an azido group joined to an unsaturated carbon. Acyl azides, show no tendency whatsoever to cyclize to oxatriazoles, but thioacyl azides are apparently completely unstable toward the isomeric thiatriazole structure[143] (Eq. 6-67). Vinyl azides

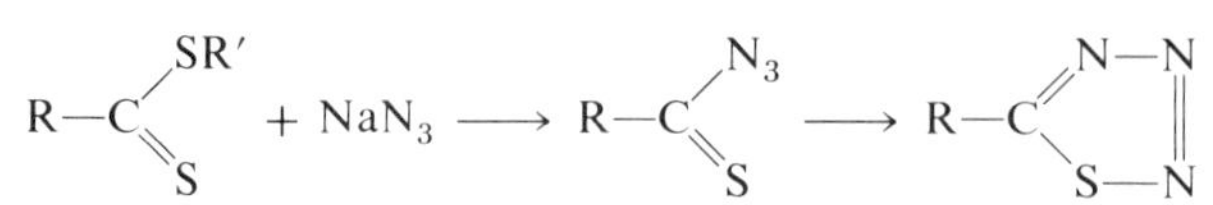

$$R-C \quad + NaN_3 \longrightarrow \quad \longrightarrow \quad\quad (6\text{-}67)$$

show no tendency to cyclize to triazolenines. These observations can be correlated in terms of the relative stabilities of the respective double-bond systems in the open structures ($C=O > C=N > C=S$) together with the fact that the triazolenines, alone of the group of ring systems concerned, do not possess the sextet of π-electrons required for an aromatic structure.

The carbon-nitrogen triple bond of nitriles is ordinarily quite inert toward cycloaddition with azides, although the tetrazoles that would result are thermodynamically favored. Only when the cyano group is strongly polarized by an electron-withdrawing group, such as perfluoroalkyl, does intermolecular reaction take place; intermolecular cycloaddition has been brought about with a few azido nitriles by heat or strong acid,[144] as shown in Equation 6-68.

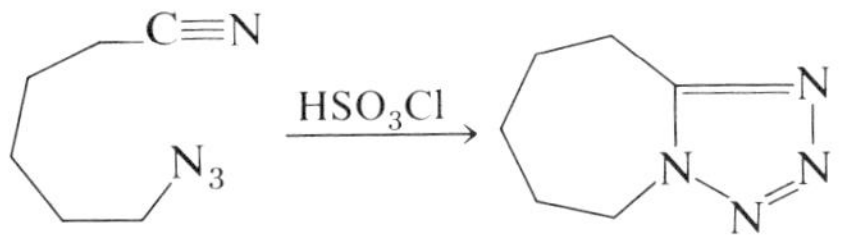

$$\underset{\text{Metrazole}}{} \tag{6-68}$$

Isocyanates undergo cycloaddition with aliphatic azides at their C—N bond to give tetrazolinones[145] (Eq. 6-69).

$$RN_3 + R'N{=}C{=}O \longrightarrow \tag{6-69}$$

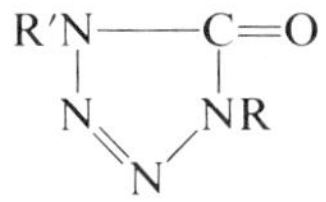

(R' = aryl or acyl)

Neither N=N nor N=O bonds undergo ready reaction with azides, and cyclo-adducts have not been detected. Some aryl azides have been induced to react with nitrosobenzenes in boiling chlorobenzene, conditions sufficient to cause fragmentation of the azides to nitrenes; the products are azoxybenzenes[146] (Eq. 6-70). Only in

$$PhN_3 + Me_2NC_6H_4N{=}O \xrightarrow[\text{ca. } 140°C]{PhCl}$$

$$\underset{32\%}{PhN{=}\overset{O}{\overset{\uparrow}{N}}C_6H_4NMe_2} + \underset{10\text{-}15\%}{Me_2NC_6H_4N{=}\overset{O}{\overset{\uparrow}{N}}C_6H_4HMe_2} \tag{6-70}$$

the special case of o-azidoazobenzenes does the azido group react with the azo function. Nitrogen is lost at a much lower temperature than required for nitrene formation, and a benzotriazole is formed[147] (Eq. 6-71). An intermediate pentazoline is a possibility, but it would be quite strained.

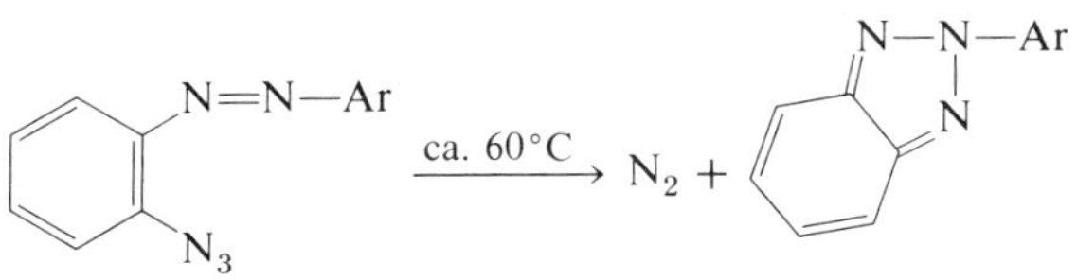

$$\xrightarrow[\text{ca. } 60°C]{} N_2 + \tag{6-71}$$

Free radicals. There is evidence that azides can react with free radicals, but there is insufficient information to give a clear picture. The reduction of phenyl azide to aniline by mercaptans is accelerated by the presence of RS· radicals.[148] Phenyl azide reacts with carbon tetrachloride in the presence of benzoyl peroxide, apparently through attack of Cl_3C· radicals[149] (Eq. 6-72). The same catalyst also induces the reaction of benzenesulfonyl azide with xylene to form a sulfonamide.[114b] How-

$$PhN_3 + CCl_4 \xrightarrow{(PhCO)_2O_2} Cl—\langle\!\!\langle\ \rangle\!\!\rangle—N{=}CCl_2 \qquad (6\text{-}72)$$

ever, the presence of triphenylmethyl radicals does not detectably interfere with the Curtius rearrangement of acyl azides.

Nitrenes appear to be able to attack azides, judging from the formation of low yields of ethyl benzeneazocarboxylate from phenyl azide and a source of carbo-ethoxyaminylidene[150] (Eq. 6-73). A related reaction has been reported from experi-

$$EtO_2C—NHOSO_2Ar + PhN_3 \xrightarrow{base} EtO_2C—N{=}N—Ph \qquad (6\text{-}73)$$

ments in the gas phase in a "flowing afterglow apparatus," in which the anion-radical $PhN^{\cdot}$ is produced; the anion-radical of azobenzene is formed.[36]

Oxidation and reduction.   Oxidizing agents have been reported not to affect the azido group,[151] although the process $RN_3 + [O] \rightarrow RN{=}O + N_2$ should be ther-modynamically favorable. Reduction, however, has been brought about by a wide variety of reagents, most of which convert the azido group to an amino group with evolution of nitrogen. Such reduction has been brought about by catalytic hydroge-bation,[13,152] Raney nickel without added hydrogen,[153] tin or zinc and acid, sodium and alcohol, hydrogen bromide or iodide,[74,154] vanadous and titanous chlorides,[155] diborane,[156] lithium aluminum hydride,[157-159] sodium borohydride in hot isopropyl alcohol,[14] and sodium dithionite,[160] among many others. Electrochemical reduction succeeds even with acyl azides, but an azido group on a saturated carbon bearing an electron-withdrawing substituent is replaced by hydrogen[161] (Eq. 6-74). Hydrogen

$$PhCOCH_2N_3 \xrightarrow{cathodic\ reduction} PhCOCH_3 \qquad (6\text{-}74)$$

sulfide in aqueous pyridine has been recommended as an especially mild and selec-tive means, but the reported examples are all from nucleoside chemistry.[162] With delicate handling, azides can sometimes be reduced to triazenes.[163,164] These may be general intermediates for most azide reductions, for they decompose spontane-ously into an amine and nitrogen (Eq. 6-75). In only one case, that of α-naphthyl

$$PhN_3 \xrightarrow[-20\,°C]{SnCl_2} Ph—N{=}N—NH_2 \longrightarrow PhNH_2 + N_2 \qquad (6\text{-}75)$$

azide, has reduction to a hydrazine been reported.[163] The general ease with which azides can be reduced to amines has made the reaction widely used in synthesis.

PREPARATIVE METHODS[165]

Primary and secondary alkyl azides are most commonly prepared by a displacement reaction between sodium or lithium azide and an alkyl halide,[166] toluenesulfon-ate,[152,157,158] or quaternary ammonium salt[72] in refluxing aqueous alcohol, Carbitol, etc.[166,167] (Eq. 6-76). Displacement occurs with inversion of configuration.[166] Al-though these are generally satisfactory methods, there is sometimes difficulty with

$$R-X + NaN_3 \longrightarrow R-N_3 + NaX \tag{6-76}$$

purification when reaction is incomplete in the case of alkyl bromides, since azides usually boil at temperatures close to the corresponding bromides.[168] Silver azide has also been used, but owing to its explosiveness, it is not to be generally recommended.[169]

A systematic study has been made of the reaction of sulfonates with azide ion in the sugar series.[170] p-Bromobenzenesulfonates react faster than simple sulfonates, but p-nitrobenzenesulfonates give rise to interfering reactions, with production of p-nitrophenyl azide. Hexamethylphosphoramide as solvent gives faster rates than dimethyl sulfoxide or dimethylformamide. Preparation of azides by displacement has also been accelerated by phase-transfer catalysis.[171,172] Tetramethylguanidinium azide has been used as alternative to metal azides;[173] it is soluble in chloroform and gives high yields.

Amino or hydroxy azides can be made by the ring-opening reaction of sodium azide with aziridines and oxiranes.[12] This type of reaction has been developed into a synthetic route to vicinal diazides[12] (Eq. 6-77). Alternatively, vicinal diazides can be made from β-iodo azides, obtained from the addition of IN_3 to alkenes.[174]

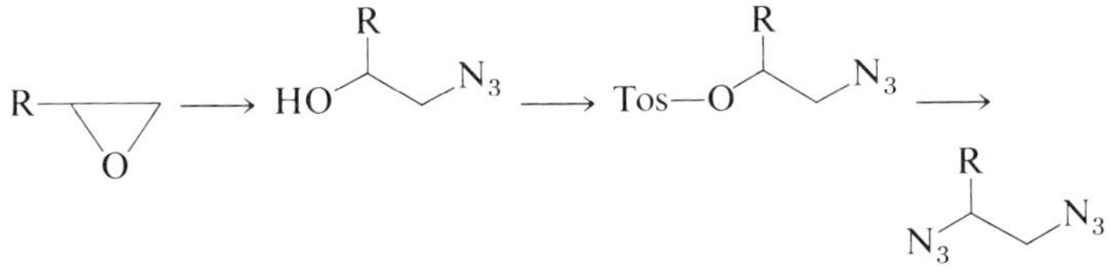

$$\tag{6-77}$$

Tertiary azides are more difficult to prepare. Solvolysis of an alkyl halide in dimethylformamide containing a soluble azide has been used; reaction is slow, and yields are not high.[175] Good yields have been obtained using zinc chloride and sodium azide in carbon disulfide; the reaction takes from 10 to 100 hours at room temperature.[176] By this means, tert-butyl azide has been prepared in 96% yield.

Azides can be prepared from alcohols or alkenes that can be converted to carbonium ions readily. Benzhydrols and triarylmethanols are especially suited to this reaction; a strongly acidic medium is required, and trichloroacetic acid is generally useful[153,177] (Eq. 6-78). With boron fluoride etherate and HN_3, tertiary alcohols can

$$Ar_2CHOH + NH_3 \xrightarrow{Cl_3CCOOH} Ar_2CHN_3 + H_2O \tag{6-78}$$

be converted to azides.[177] The limiting factor in these reactions is the necessity to avoid acidic conditions so strong as to cause decomposition of the azides and yet be able to generate the required carbonium ion. The fact that 2-benzyloxyethyl alcohol has been successfully converted to an azide by this means[153] suggests that an intermediate oxiranium ion, rather than a primary carbonium ion, is involved in this instance.

Addition to alkenes can be used to prepare azides in two ways. Hydrogen azide in the presence of strong acid, or mercuric azide, has been used to prepare tertiary alkyl azides from alkenes.[178,179] Iodine azide adds to produce β-iodo azides, which can be converted to vinyl azides by base-catalyzed elimination; cyclopentenyl azide, for example has been made in 82% overall yield[180] (Eq. 6-79). Another addition

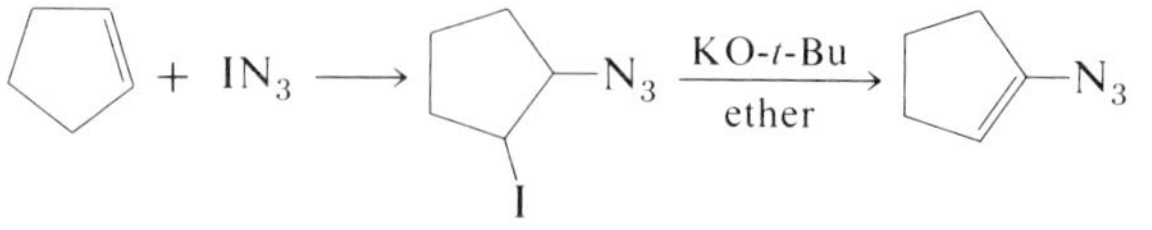

$$\text{(cyclopentene)} + IN_3 \longrightarrow \text{(cyclopentyl–N}_3, I) \xrightarrow[\text{ether}]{\text{KO-}t\text{-Bu}} \text{(cyclopentenyl–N}_3) \qquad (6\text{-}79)$$

(*trans*, 89%)

reaction is an oxidative process that transfers two azido groups to an alkene to form a vicinal diazide[181] (Eq. 6-80). The reagent is prepared from trimethylsilyl azide and

$$\begin{array}{c} \diagdown \\ C=C \\ \diagup \end{array} \xrightarrow[-40 \text{ to } -20°C]{Me_3SiN_3, \ Pb(OAc)_4} \begin{array}{c} | \quad | \\ -C-C- \\ | \quad | \\ N_3 \ N_3 \end{array} \qquad (6\text{-}80)$$

lead tetraacetate or *1,1*-diacetoxyiodobenzene and is presumably a mixed acetate/azide. The reaction is not stereospecific, and some β-acetoxy azide is also produced. Closely related to this is the reaction of alkenes with sodium azide in the presence of *tert*-butyl hydroperoxide and ferrous sulfate[182] (Eq. 6-81).

$$PhCH{=}CH_2 + NaN_3 \xrightarrow[FeSO_4]{t\text{-BuOOH}}$$

$$\begin{array}{cc} PhCH{-}CH_2 + PhCH{-}CH_2 & (6\text{-}81) \\ | \quad | \quad\quad | \quad\quad | \\ N_3 \ N_3 \quad\quad N_3 \ O{-}t\text{-Bu} \end{array}$$

A totally different approach to the preparation of azides is found in the reaction of carbanions with tosyl azide. If the carbanion bears no α-hydrogen, the intermediate triazene loses a sulfinate ion to form the azide (Eq. 6-82). Phenylmalonic ester,

$$R^- + N_3SO_2C_7H_7 \longrightarrow R{-}N{=}N{-}\bar{N}{-}SO_2C_7H_7 \longrightarrow$$

$$R{-}N_3 + C_7H_7SO_2^- \qquad (6\text{-}82)$$

for example, is converted to ethyl α-azido-α-phenylmalonate in 77% yield.[183] When an α-hydrogen is present, the competing cleavage to diazoalkane and sulfonamide anion dominates. Triflyl azide can be used to prepare azides by a diazo transfer process by reaction with primary amines[96] (Eq. 6-49) or their anions, easily generated by reaction with a Grignard reagent[97,98] (Eq. 6-83).

$$RNH^- + C_7H_7SO_2N_3 \longrightarrow R{-}N_3 + C_7H_7SO_2NH^- \qquad (6\text{-}83)$$

Alkylhydrazines can be used as sources of azides if they are first acylated; otherwise, nitrosation takes place on the alkylated nitrogen, and no azide results. Blocking with the carbamoyl group is convenient[184] (Eq. 6-84).

$$\begin{array}{c} i\text{-Bu}{-}N{-}CONH_2 + HNO_2 \longrightarrow i\text{-Bu-N}_3 + N_2 + CO_2 \qquad (6\text{-}84) \\ | \\ NH_2 \end{array}$$

Two electrolytic methods for preparing azides have been reported. Anions of carboxylic acids are converted to the azide with one less carbon atom at a platinum anode in the presence of azide ions[185] (Eq. 6-85). The anions of nitroalkanes also give azides under these conditions[186] (Eq. 6-86).

$$C_5H_{11}CO_2^- + N_3^- \xrightarrow{\text{Pt anode}} C_5H_{11}N_3 \ (+ \ C_{10}H_{22}) \qquad (6\text{-}85)$$

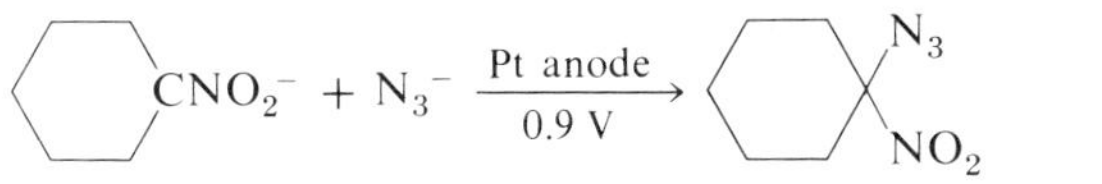

$$(6\text{-}86)$$

There are a few methods of potential preparative value for alkyl azides that do not quite fit into the foregoing categories and which have not yet seen substantial application. One is the reaction of peroxide adducts of ketones with sodium azide in the presence of ferrous sulfate[182] (Eq. 6-87); it is apparently a free-radical process.

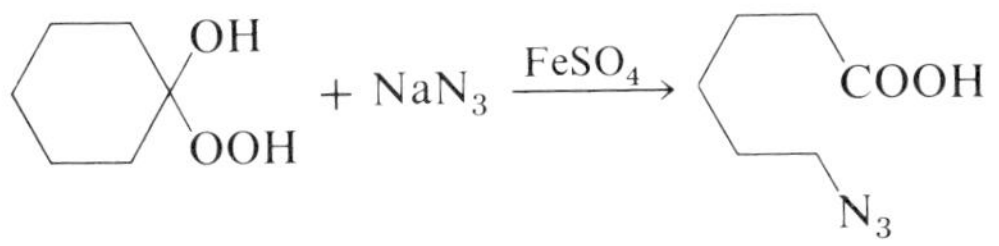

$$(6\text{-}87)$$

Somewhat similar is the reaction of carboxylate ions with cupric azide, which appears to involve ligand transfer of radicals[187] (Eq. 6-88).

$$BuCO_2^- + Cu(N_3)_2 \xrightarrow[\text{AcOH/MeCN}]{Cu^I} \underset{\text{ca. 50\%}}{Bu-N_3} \qquad (6\text{-}88)$$

Another reaction that appears to involve free radicals is the reaction of trialkylboranes with sodium azide in the presence of ferric sulfate and hydrogen peroxide, which takes place at room temperature.[188] For synthetic purposes, the starting materials are alkenes, which are easily converted to trialkylboranes; terminal alkenes thus produce primary alkyl azides (Eq. 6-89). Yields as high as 100% are

$$EtCH=CH \longrightarrow Bu_3B \xrightarrow[Fe_2(SO_4)_3]{NaN_3, \ H_2O_2}$$

$$\underset{56\%}{CH_3CH_2CH_2CH_2N_3} + Bu_2BOH \qquad (6\text{-}89)$$

reported, but this is on the basis of consuming only one of the three boron-attached alkyl groups. Finally, a reaction that seemingly stands alone is the conversion of β-hydroxyalkyl-N-nitrosoacetamides to vinyl azides by reaction with base and sodium azide.[189]

Aryl azides are most generally and conveniently prepared by the reaction of diazonium salts with sodium azide; reaction occurs rapidly in the cold, and the yields are good[190] (Eq. 6-90). The azide nitrogen is derived largely from the diazo-

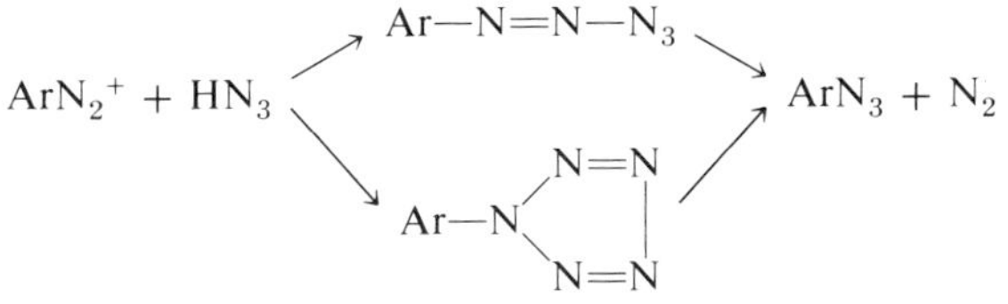

$$(6\text{-}90)$$

nium function, which first couples to form a mixture of a diazoazide and a pentazole.[191] Other reagents, such as hydrazine,[192] hydroxylamine,[193] ammonia and bromine,[194] and sulfonamides,[195] also convert diazonium compounds to azides through coupling and cleavage (see Chap. 4).

Aryl azides may in some instances be prepared by a nucleophilic displacement reaction when there are sufficiently activating substituents present (Eq. 6-91). A

$$\text{Ar—X} + \text{NaN}_3 \xrightarrow{\text{DMSO}} \text{Ar—N}_3 + \text{NaX} \qquad (6\text{-}91)$$

single nitro group is sufficient if dimethyl sulfoxide is used as the medium; reaction takes place between room temperature and 100°C, and yields run as high as 94%.[196]

Aryl azides are also easily prepared from aryl hydrazines by reaction with nitrous acid[197] (Eq. 6-92). However, the required hydrazines must usually be made from the

$$\text{ArNHNH}_2 + \text{HNO}_2 \longrightarrow \text{Ar—N}_3 + 2\text{H}_2\text{O} \qquad (6\text{-}92)$$

corresponding diazonium salt, which is usually more efficiently converted to an azide directly by reaction with sodium azide.

Otherwise unreactive aryl halides can be converted to azides in moderate yields through their Grignard reagents. Addition to tosyl azide forms a triazene, which is easily decomposed to an aryl azide with aqueous sodium azide[106] (Eq. 6-51).

Nitrosobenzenes can be converted into azides by reaction with hydroxylamine[198] or hydrogen azide,[199] often very smoothly (Eq. 6-93), but these reactions have sel-

$$\text{ArNO} + 2\text{HN}_3 \xrightarrow{20\text{-}40\,°\text{C}} \text{ArN}_3 + \text{H}_2\text{O} + 2\text{N}_2 \qquad (6\text{-}93)$$

dom been used for synthesis, owing to the difficulty in obtaining the nitrosobenzenes.

Acyl azides are most generally prepared from acid chlorides by reaction with sodium azide, anhydrous or in aqueous solution[54,200] (Eq. 6-94). A possible im-

$$\text{RCOCl} + \text{NaN}_3 \xrightarrow[\text{or benzene}]{\text{aq. acetone}} \text{RCON}_3 + \text{NaCl} \qquad (6\text{-}94)$$

provement is the use of tetramethylguanidinium azide in chloroform solution, which has given yields of 84 to 97% at 0°C.[201] Acyl azides can equally well be prepared by the reaction of hydrazides with nitrous acid. The hydrazides are readily prepared from esters by reaction with hydrazine, a process often more convenient than preparation of an acid chloride[54,202] (Eq. 6-95).

$$\text{RCOOR}' + \text{NH}_2\text{NH}_2 \longrightarrow \text{RCONHNH}_2 \xrightarrow{\text{HNO}_2} \text{RCON}_3 \qquad (6\text{-}95)$$

ANALYTICAL METHODS

Infrared spectroscopy provides the simplest general method for detecting azides. Strong absorption in the range 2160 to 2100 cm^{-1} is always exhibited, along with less useful absorption at 1340–1180 and 680 cm^{-1}. Carbodiimides and acetylenes, however, also absorb in the same general high-frequency range. Ultraviolet absorption[9b,203] by alkyl azides falls in the 264 to 284-nm range, and can be used to distinguish azides in the presence of the corresponding alcohols. The usefulness of mass spectroscopy is limited by the fact that nitrogen is lost so easily that the parent peak is rarely detectable; a peak at M − 28 is the highest mass seen.

Most azides can reliably be reduced to primary amines, which can then be detected and identified by conventional means.[13,152,204] Two-thirds of the azide nitrogen is usually evolved more or less quantitatively (see Eq. 6-75), and measurement of its volume may provide a quantitative assay. Many azides react quantitatively with concentrated hydriodic acid, not only with liberation of nitrogen, but also with formation of an equivalent of iodine (compare Eq. 6-36), which can be titrated with thiosulfate.[74,114b,205] Nonreducing acids, exemplified by sulfuric acid, also liberate nitrogen more or less quantitatively from azides of all types[206] other than triaryl-methyl azides, which are inclined to lose hydrogen azide when treated with concentrated sulfuric acid.[207] Acyl azides liberate nitrogen quantitatively on warming even without a strong acid,[208] owing to the Curtius rearrangement (Eq. 6-24). Analysis of acyl azides by this means is particularly useful, because combustion analysis is quite unreliable, owing to explosion in the combustion tube.

Acyl azides can also be determined by means of hydrolysis or aminolysis. Cold alkali hydrolyses them to azide ion, which can be determined by titration with ceric ion or by precipitation with silver nitrate.[209] The latter reagent is no longer much favored, for the silver azide that must be collected, dried, and weighed is dangerously explosive. With aminolysis, the azide ion is ignored, but the amide formed by acylation of the amine can be easily collected if it is a solid.[210] For this purpose, aniline or nitroanilines are usually suitable. (Eq. 6-96). It is essential that the reac-

$$RCON_3 + ArNH_2 \longrightarrow RCONHAr + HN_3 \qquad (6\text{-}96)$$

tion not be allowed to become warm, for Curtius rearrangement can intervene, and give rise to a urea.

Alkyl and aryl azides have been converted to solid derivatives for characterization by reaction with triphenylphosphine[211] (Eq. 6-56) or by cycloaddition with strained olefins or acetylenes[184,212] (Eqs. 6-59 and 6-63). Although the triazoles that are formed from acetylenes are usually stable and crystalline, the fact that mixtures of differently oriented triazoles are likely to be formed limits the usefulness of the method.[134] Propynoic acid is the favored reagent.

REFERENCES

1. Reviews: (a) S. Patai, ed., *The Chemistry of the Azido Group*, Interscience Publishers, Wiley, New York, 1971; (b) J. H. Boyer and F. C. Canter, *Chem. Rev.* **54**, 1 (1954); (c) C. G. Overberger, J.-P. Anselme, and J. G. Lombardino, *Chemistry of Organic Compounds with Nitrogen-Nitrogen Bonds*, Ronald Book Co., New York, 1966; (d) carbamoyl azides: E. Lieber, R. L. Minnis, and C. N. R. Rao, *Chem. Rev.* **65**, 377 (1965).

2. G. Favini and I. R. Bellobono, *Gazz. Chim. Ital.* **92**, 468 (1962).

3. H. Bock and K. L. Kompa. *Angew. Chem.* **74**, 327 (1962).

4. Y. Tsuno, *Mem. Inst. Sci. Ind. Res. Osaka Univ.* **15**, 83, 187 (1958); *Chem. Abstr.* **53**, 1116 (1959).

5. F. D. Marsh, *J. Org. Chem.* **37**, 2966 (1972).

6. (*a*) Methyl azide: L. Pauling and L. O. Brockway, *J. Am. Chem. Soc.* **59**, 13 (1937); (*b*) cyanuryl azide: I. E. Knaggs, *Proc. Roy. Soc. [A]*, **150**, 348 (1935); (*c*) E. W. Hughes, *J. Chem. Phys.* **3**, 1 (1935).

7. Z. Iqbal, *Structure and Bonding* **10**, 25 (1972).

8. E. C. Franklin, *The Nitrogen System of Compounds*, Reinhold, New York, 1935.

9. (*a*) E. Lieber and E. Oftedahl, *J. Org. Chem.* **24**, 1014 (1959); (*b*) E. Lieber and E. Oftedahl, *Spectrochim. Acta* **19**, 1135 (1963); (*c*) W. D. Closson and H. B. Gray, *J. Am. Chem. Soc.* **85**, 290 (1963).

10. J. Goubeau, E. Allenstein, and A. Schmidt, *Chem. Ber.* **97**, 884 (1964).

11. (*a*) C. L. Currie and B. de B. Darwent, *Can. J. Chem.* **41**, 1048 (1963); (*b*) *Chem. Eng. News* **54** (No. 22), 3 (1976); (*c*) M. Itoh, D. Hagiwara, and T. Kamiya, *Tetrahedron Lett.* 4393 (1975).

12. G. Swift and D. Swern, *J. Org. Chem.* **32**, 511 (1967).

13. P. A. S. Smith, J. H. Hall, and R. O. Kan, *J. Am. Chem. Soc.* **84**, 485 (1962); ref. 1*a*, Chap. 4, by M. E. C. Biffin, J. Miller, and D. B. Paul.

14. A. S. Bailey and J. R. Case, *Tetrahedron* **3**, 113 (1958).

15. (*a*) G. L'abbe, *Chem. Rev.* **69**, 345 (1969); (*b*) W. M. Jones, in *Rearrangements in Ground and Excited States*, P. de Mayo (ed.), Academic Press, New York, 1980, Essay III; (*c*) E. P. Kyba and R. A. Abramovitch, *J. Am. Chem. Soc.* **102**, 735 (1980); (*d*) ref. 1*a*, Chap. 5; (*e*) W. Lwowski, in *Reactive Intermediates*, Vol. 2, M. Jones, Jr., and R. A. Moss (eds.), Wiley, New York, 1981, Chap. 8.

16. (*a*) F. D. Lewis and W. H. Saunders, Jr., in *Nitrenes*, W. Lwowski (ed.), Wiley, New York, 1970, Chap. 2; (*b*) E. Wasserman, G. Smolinsky, and W. A. Yager, *J. Am. Chem. Soc.* **86**, 3166 (1964), **84**, 3220 (1962).

17. W. Pritzkow and D. Timm, *J. Prakt. Chem.* **32**, 178 (1966).

18. F. O. Rice and C. J. Grelecki, *J. Phys. Chem.* **61**, 830 (1957).

19. G. Geiseler and W. König, *Z. Phys. Chem.* **227**, 81 (1964).

20. P. A. S. Smith, in *Molecular Rearrangements*, P. de Mayo (ed.), Wiley, New York, 1963, Chap. 8.

21. F. D. Lewis and W. H. Saunders, Jr., *J. Am. Chem. Soc.* **89**, 645 (1967).

22. W. H. Saunders, Jr., and E. Caress, *J. Am. Chem. Soc.* **86**, 861 (1964).

23. F. D. Lewis and W. H. Saunders, Jr., *J. Am. Chem. Soc.* **90**, 3828 (1968).

24. C. A. VanderWerf and V. L. Heasley, *J. Org. Chem.* **31**, 3534 (1966).

25. (*a*) G. Smolinsky, *Trans. N.Y. Acad. Sci*, **30**, 511 (1968); (*b*) G. Smolinsky and C. A. Pryde, *J. Org. Chem.* **33**, 2411 (1968); ref. 1*a*, Chap. 10; (*c*) reviewed in detail by G. L'abbé, *New Synthetic Methods (Verlag Chemie)* **5**, 1 (1979).

26. G. L'abbé and G. Mathys, *J. Org. Chem.* **39**, 1778 (1974).

27. G. R. Harvey and K. W. Ratts, *J. Org. Chem.* **31**, 3907 (1966).

28. G. Smolinsky, *J. Org. Chem.* **37**, 3557 (1962); *J. Am. Chem. Soc.* **83**, 4483 (1961).

29. K. Isomura, S. Kobayashi, and H. Taniguchi, *Tetrahedron Lett.*, 3499 (1968).

30. L. Horner, A. Christmann, and A. Gross, *Chem. Ber.* **96**, 399 (1963).

31. H. Reimlinger, F. P. Warner, and D. R. Arnold, *Angew. Chem. Int. Ed.* **7**, 130 (1968).

32. F. W. Fowler, A. Hassner and L. A. Levy, *J. Am. Chem. Soc.* **89**, 2077 (1967).

33. (*a*) P. A. S. Smith, in *Nitrenes*, W. Lwowski (ed.), Wiley, New York, 1970, Chap. 3; (*b*) E. F. V. Scriven, in *Reactive Intermediates*, Vol. 2, R. A. Abramovitch (ed.), Plenum, New York, 1982, Chap. 1.

34. (*a*) F. J. Weigert, *J. Org. Chem.* **38**, 1316 (1973); (*b*) K. Nishiyama and J.-P. Anselme, *J. Org. Chem.* **42**, 2636 (1977).

35. R. A. Abramovitch, S. R. Challand, and E. F. V. Scriven, *J. Org. Chem.* **37**, 2705 (1972).

36. R. N. McDonald and A. K. Chowdhury, *J. Am. Chem. Soc.* **102**, 5118 (1980).

37. (*a*) W. D. Crow and C. Wentrup, *Tetrahedron Lett.*, 5569 (1968); (*b*) C. Wentrup, *Top. Curr. Chem.* **62**, 173 (1976); (*c*) O. L. Chapman, R. S. Sheridan, and J.-P. Leroux, *J. Am. Chem. Soc.* **100**, 6245 (1978).

38. (*a*) L. Wolff, *Liebig's Ann. Chem.* **394**, 59 (1912); (*b*) R. Huisgen, D. Vossius, and M. Appl, *Chem. Ber.* **91**, 1 (1958); (*c*) A. Albini, G. F. Bettinetti, E. Fasani, and S. Pietra, *Gazz. Chim. Ital.* **112**, 13 (1982).

39. (*a*) G. Smolinsky and B. I. Feuer, *J. Am. Chem. Soc.* **86**, 3085 (1964); (*b*) G. Smolinsky and B. I. Feuer, *J. Org. Chem.* **29**, 3097 (1964), **31**, 3882 (1966).

40. (*a*) K. H. Saunders, *J. Chem. Soc.*, 3275 (1955); (*b*) O. Meth-Cohn, R. K. Smalley, and H. Suschitzky, *J. Chem. Soc.*, 1666 (1963); (*c*) J. Schmutz and F. Kunzle, *Helv. Chim. Acta* **39**, 1144 (1956).

41. P. A. S. Smith and J. H. Hall, *J. Am. Chem. Soc.* **84**, 480 (1962).

42. J. S. Swenton, T. J. Ikaler, and B. H. Williams, *J. Am. Chem. Soc.* **92**, 3103 (1970).

43. P. A. S. Smith, B. B. Brown, R. K. Putney, and R. K. Reinisch, *J. Am. Chem. Soc.* **75**, 6335 (1953).

44. J. I. G. Cadogan, *Accts. Chem. Res.* **5**, 303 (1972); M. Messer and D. Farge, *Bull. Soc. Chim. France*, 2832 (1968), 4955 (1969).

45. L. O. Krbechek and H. Takimoto, *J. Org. Chem.* **29**, 1150 (1964).

46. G. R. Cliff and G. Jones, *J. Chem. Soc.* [*C*], 3418 (1971).

47. P. A. S. Smith and C. D. Rowe, unpublished results; C. D. Rowe, doctoral dissertation, University of Michigan, 1968.

48. P. A. S. Smith and S.-S. P. Chou, *J. Org. Chem.* **46**, 3970 (1981); S.-S. P. Chou, doctoral dissertation, University of Michigan, 1979.

49. R. Kwok and P. Pranc, *J. Org. Chem.* **33**, 2880 (1968); K. H. Wünsch and A. J. Boulton, *Adv. Heterocyc. Chem.* **8**, 303 (1967).

50. (*a*) T. F. Fagley, J. R. Sutter and R. L. Oglukian, *J. Am. Chem. Soc.* **78**, 5567 (1956); (*b*) A. S. Bailey and J. R. Case, *Tetrahedron* **3**, 113 (1958); (*c*) F. B. Mallory and C. S. Wood, *J. Org. Chem.* **27**, 4109 (1962).

51. J. H. Hall and F. W. Dolan, *J. Org. Chem.* **43**, 4608 (1978).

52. (*a*) R. A. Abramovitch and K. A. H. Adams, *Can. J. Chem.* **39**, 2516 (1961); (*b*) P. J. Bunyan and J. I. G. Cadogan, *J. Chem. Soc.*, 42 (1963).

53. L. K. Dyall, *Aust. J. Chem.* **28**, 2147 (1975).

54. P. A. S. Smith, *Organic Reactions* **3**, Chap. 9, (1947).

55. F. Bell, *J. Chem. Soc.*, 835 (1934).

56. C. R. Hauser and S. W. Kantor, *J. Am. Chem. Soc.* **72**, 4284 (1950).

57. Arthur Fry and J. C. Wright, *Chem. Eng. News*, 28 (Jan. 1, 1968).

58. W. Lwowski, in *Nitrenes*, W. Lwowski (ed.), Wiley, New York, 1970, Chap. 6.

59. D. S. Breslow, T. J. Prosser, A. F. Marcantonio, and C. A. Genge, *J. Am. Chem. Soc.* **89**, 2384 (1967); D. S. Breslow and E. I. Edwards, *Tetrahedron Lett.*, 2123 (1967).

60. (*a*) W. Lwowski, R. De Mauriac, T. W. Mattingly, Jr., and E. Scheiffele, *Tetrahedron Lett.*, 3285 (1964); (*b*) W. S. Wadsworth and W. D. Emmons, *J. Org. Chem.* **32**, 1279 (1967).

61. W. Lwowski, R. A. de Mauriac and M. Thompson, *J. Org. Chem.* **40**, 2608 (1975).

62. W. Reichen, *Helv. Chim. Acta* **59**, 2601 (1976).

63. R. Richter, H. Ulrich and D. J. Duchamp, *J. Org. Chem.* **43**, 3060 (1978).

64. (*a*) R. A. Abramovitch, T. Chellathurai, W. D. Holcomb, I. T. McMaster and I. P. Vanderpool, *J. Org. Chem.* **42**, 2920 (1977); (*b*) D. C. Appleton, J. McKenna, J. M. McKenna, L. B. Sims, and A. R. Walley, *J. Am. Chem. Soc.* **98**, 292 (1976).

65. R. S. Berry, D. Cornell, and W. Lwowski, *J. Am. Chem. Soc.* **85,** 1199 (1963).

66. (*a*) G. T. Tisue, S. Linke and W. Lwowski, *J. Am. Chem. Soc.* **89,** 6303 (1967); (*b*) S. Linke, G. T. Tisue, and W. Lwowski, *ibid.* **89,** 6308 (1967).

67. (*a*) D. Poppinger and L. Radom, *J. Am. Chem. Soc.* **100,** 3674 (1978); (*b*) A. Rauk and P. F. Alewood, *Can. J. Chem.* **55,** 1498 (1977).

68. P. Casagrande, L. Pellacani and P. A. Tardella, *J. Org. Chem.* **43,** 2725 (1978).

69. (*a*) F. D. Marsh and H. E. Simmons, *J. Am. Chem. Soc.* **87,** 3529 (1965); (*b*) A. G. Anastassiou, H. E. Simmons, and F. D. Marsh, in *Nitrenes,* W. Lwowski (ed.), Wiley-Interscience, New York, 1970.

70. R. D. Richmond, *Analyst* **33,** 179 (1908).

71. C. H. Gudmundsen and W. E. McEwen, *J. Am. Chem. Soc.* **79,** 329 (1957).

72. A. N. Nesmeyanov and M. I. Rybinskaya, *Izvest. Akad. Nauk, Otdel. Khim. Nauk,* 816 (1962) (English trans. p. 761).

73. T. Sasaki, S. Eguchi, T. Katada, and O. Hiroaki, *J. Org. Chem.* **42,** 3741 (1977).

74. P. A. S. Smith and B. B. Brown, *J. Am. Chem. Soc.* **73,** 2438 (1951).

75. P. A. S. Smith and J. H. Carter, unpublished results; J. H. Carter, doctoral thesis, University of Michigan, 1962.

76. E. Bamberger, *Liebig's Ann. Chem.* **443,** 192 (1925).

77. R. J. Sundberg and K. B. Sloan, *J. Org. Chem.* **38,** 2052 (1973).

78. W. Borsche and H. Hahn, *Ber.* **82,** 260 (1949).

79. (*a*) R. K. Smalley, *J. Chem. Soc.,* 5571 (1963); (*b*) H. Suschitzky, *ibid.,* 1663 (1963).

80. (*a*) E. B. Mullock and H. Suschitzky, *J. Chem. Soc.* [*C*], 1937 (1968); (*b*) R. Garner, E. B. Mullock, and H. Suschitzky, *ibid.,* 1980 (1966).

81. H. Wolff, *Org. Reactions* **3,** Chap. 8 (1946).

82. (*a*) P. A. S. Smith and E. P. Antoniades, *Tetrahedron* **9,** 210 (1960); (*b*) D. L. Fishel and M. J. Maximovich, *Abstracts of Papers, Am. Chem. Soc. National Meeting, Chicago, Ill.,* Aug., 1964, p. 85S.

83. B. M. Trost, M. Vaultier, and M. L. Santiago, *J. Am. Chem. Soc.* **102,** 7929 (1980).

84. G. S. Sidhu, G. Thyagarajan and U. T. Bhalero, *Chem. Ind.* (*Lond.*), 1301 (1966).

85. R. Kreher and J. Jager, *Angew. Chem. Int. Ed.* **4,** 706 (1965).

86. N. Wiberg and K. H. Schmid, *Angew. Chem.* **76,** 381 (1964).

87. W. Pritzkow and G. Pohl, *J. Prakt. Chem.* [4]**20,** 132 (1963).

88. (*a*) J. H. Boyer and L. R. Morgan, Jr., *J. Am. Chem. Soc.* **81,** 3369 (1959); (*b*) J. H. Boyer, W. E. Krueger, and R. Modler, *J. Org. Chem.* **34,** 1987 (1969); (*c*) J. H. Boyer, F. C. Canter, J. Hamer, and R. K. Purney, *J. Am. Chem. Soc.* **78,** 325 (1956).

89. A. Neiman, V. I. Maimind, and M. M. Shemyakin, *Izvest. Akad. Nauk,* 1498 (1962) (Consultants Bureau English translation, p. 1418).

90. (*a*) M. O. Forster and S. H. Newman, *J. Chem. Soc.* **99,** 1277 (1911); (*b*) R. H. Wiley and J. Moffatt, *J. Org. Chem.* **22,** 995 (1957).

91. M. O. Forster and S. H. Newman, *J. Chem. Soc.* **97,** 2570 (1910); (*b*) E. Oliveri-Mandalá and G. Caronna, *Gazz. Chim. Ital.* **71,** 182 (1941); (*c*) J. H. Boyer, *J. Am. Chem. Soc.* **73,** 5248 (1951).

92. (*a*) J. Meienhofer, in *The Peptides,* E. Gross and J. Meienhofer (eds.), Academic Press, New York, 1979, Chap. 4; (*b*) K. Inouye, K. Watanabe, and M. Shin, *J. Chem. Soc.* [*Perkin I*], 1905 (1977); (*c*) L. A. Carpino et al., *Org. Syntheses* **44,** 15 (1964); (*d*) J. Suh and B. H. Lee, *J. Org. Chem.* **45,** 3103 (1980).

93. J. F. Bunnett and R. E. Zahler, *Chem. Rev.* **49,** 278, 283 (1951).

94. R. E. Banks and A. Prakash, *Tetrahedron Lett.,* 99 (1973).

95. E. F. V. Scriven, H. Suschitzky, and G. V. Garner, *Tetrahedron Lett.*, 103 (1973).

96. C. J. Cavender and V. J. Shiner, Jr., *J. Org. Chem.* **37**, 3567 (1972); J. Zaloom and D. C. Roberts, *ibid.* **46**, 5173 (1981).

97. (*a*) J.-P. Anselme and W. Fischer, *Tetrahedron* **25**, 855 (1969); *J. Am. Chem. Soc.* **89**, 5284 (1967); (*b*) J. B. Hendrickson and W. A. Wolf, *J. Org. Chem.* **33**, 3610 (1968).

98. M. Regitz, "Transfer of Diazo Groups," in *Newer Methods of Preparative Organic Chemistry* Vol. VI, W. Foerst (ed.), Academic Press, New York, 1971.

99. H. Bretschneider and H. Rager, *Monatsh. Chem.* **81**, 970, 981 (1950).

100. L. Wolff and G. K. Grau, *Liebig's Ann. Chem.* **394**, 68 (1912).

101. J. H. Boyer and F. C. Canter, *Chem. Rev.* **54**, 1 (1954).

102. K. Clusius and H. R. Weisser, *Helv. Chim. Acta* **35**, 1548 (1952).

103. O. Dimroth, *Ber.* **39**, 3905 (1906).

104. V. Ya. Pochinok, *J. Gen. Chem.* (*U.S.S.R.*) **16**, 1303, 1306 (1946); *Chem. Abstr.* **41**, 3066f,h (1947).

105. W. I. Awad, F. G. Baddar, M. A. Omara and S. M. A. R. Omran, *J. Chem. Soc.*, 2040 (1965).

106. (*a*) P. A. S. Smith, L. B. Bruner, and C. D. Rowe, *J. Org. Chem.* **34**, 3430 (1969); (*b*) B. M. Trost and W H. Pearson, *J. Am. Chem. Soc.* **103**, 2483 (1981).

107. T. Weil and M. Cais, *J. Org. Chem.* **28**, 2472 (1963).

108. (*a*) G. Harvey, *J. Org. Chem.* **31**, 1587 (1966); (*b*) M. Rosenberger, P. Yates, and W. Wolf, *Tetrahedron Lett.*, 2285 (1964).

109. T. Saegusa, Y. Ito, and T. Shimizu, *J. Org. Chem.* **35**, 3995 (1971).

110. G. Gaudiano, C. Ticozzi, A. Umani-Rouchi, and P. Bravos, *Gazz. Chim. Ital.* **97**, 1411 (1967).

111. A. Hassner and B. A. Belinka, Jr., *J. Am. Chem. Soc.* **102**, 6185 (1980).

112. H. Staudinger and E. Hauser, *Helv. Chim. Acta* **4**, 861 (1921).

113. (*a*) L. Horner and A. Gross., *Liebig's Ann. Chem.* **591**, 117 (1955); (*b*) K. Brass and F. Albrecht, *Ber.* **61**, 983 (1928).

114. (*a*) J. E. Leffler, V. Honsberg, Y. Tsuno, and I. Forsblad, *J. Org. Chem.* **26**, 4810 (1961); (*b*) J. E. Leffler and Y. Tsuno, *ibid.* **28**, 190 (1963).

115. H. Bock and M. Schnöller, *Angew. Chem.* **80**, 167 (1968).

116. J. B. Hendrickson, K. W. Blair, and P. M. Keehn, *J. Org. Chem.* **42**, 2935 (1977).

117. K. Hoegerle and P. E. Butler, *Chem. Ind.* (*Lond.*), 933 (1964).

118. O. Dimroth and S. Merzbacher, *Ber.* **43**, 2899 (1910).

119. F. D. Chattaway and G. D. Parkes, *J. Chem. Soc.*, 113 (1925).

120. R. Huisgen, R. Grashey, and J. Sauer, *The Chemistry of Alkenes*, S. Patai (ed.), Wiley-Interscience, New York, 1964, pp. 806ff; R. Huisgen, *J. Org. Chem.* **41**, 403 (1976).

121. T. Sheradsky, in ref. 1*a*, Chap. 6.

122. A. Padwa, *New Synthetic Methods* (*Verlag Chemie*) **5**, 43 (1979).

123. R. Huisgen and G. Müller, quoted by R. Huisgen, *Angew. Chem.* **72**, 359 (1960).

124. L. H. Zalkow and A. C. Oehlschlager, *J. Org. Chem.* **28**, 3303 (1963).

125. L. H. Zalkow and C. O. Kennedy, *J. Org. Chem.* **28**, 3309 (1963).

126. J. E. Franz and C. Osuch, *Tetrahedron Lett.*, 827 (1963).

127. A. L. Logothetis, *J. Chem. Soc.*, 749 (1965).

128. P. Scheiner, *Tetrahedron* **24**, 349 (1968).

129. R. Huisgen, G. Szeimies, and L. Moebius, *Chem. Ber.* **99**, 475 (1966); W. Broeck, N. Overbergh, D. Samyn, G. Smets, and G. L'abbé, *Tetrahedron* **27**, 3527 (1971).

130. M. C. Meilahn, B. Cox, and M. E. Munk, *J. Org. Chem.* **40**, 819 (1975).

131. R. Fusco, G. Bianchetti and D. Pocar, *Gazz. Chim. Ital.* **92**, 849, 933 (1962); G. Bianchetti, P. Dalla Croce and D. Pocar, *Tetrahedron Lett.*, 2043 (1965).

132. R. A. Wohl, *Helv. Chim. Acta* **56**, 1826 (1973); *Tetrahedron Lett.*, 3111 (1973).

133. A. Gagnieux, S. Winstein, and W. G. Young, *J. Am. Chem. Soc.* **82**, 5956 (1960).

134. R. Huisgen, R. Knorr, L. Möbius and G. Szeimies, *Chem. Ber.* **98**, 4014 (1965).

135. J. R. E. Hoover and A. R. Day, *J. Am. Chem. Soc.* **78**, 5832 (1956).

136. E. Lieber, T. S. Chao, and C. N. R. Rao, *J. Org. Chem.* **22**, 654 (1957); E. Lieber, C. N. R. Rao, and T. V. Rajkumar, *ibid.* **24**, 134 (1959).

137. P. A. S. Smith, *J. Am. Chem. Soc.* **76**, 436 (1954).

138. R. M. Herbst, C. W. Roberts, H. T. Givens, and E. K. Harvill, *J. Org. Chem.* **17**, 262 (1952).

139. C. Temple, Jr., W. C. Coburn, M. D. Thorpe, and J. A. Montgomery, *J. Org. Chem.* **30**, 2395 (1965).

140. M. S. Chang and A. J. Matuszko, *J. Org. Chem.* **28**, 2260 (1963).

141. F. Eloy, *J. Org. Chem.* **26**, 952 (1961).

142. H. Behringer and H. J. Fischer, *Chem. Ber.* **95**, 2546 (1962).

143. E. Lieber, E. Oftedahl, and C. N. R. Rao, *J. Org. Chem.* **28**, 194 (1963).

144. W. R. Carpenter, *J. Org. Chem.* **27**, 2085 (1962).

145. J.-M. Vandensavel, G. Smets and G. L'abbé, *J. Org. Chem.* **38**, 675 (1973).

146. G. V. Garner, K. B. Niewiadomsky, and H. Suschitzky, *Chem. Ind. (Lond.)*, 462 (1972).

147. R. A. Carboni and J. E. Castle, *J. Am. Chem. Soc.* **84**, 2453 (1962).

148. T. Shingaki, *Sci. Rep. Coll. Gen. Educ. Osaka Univ.* **11**, 67, 81 (1963); *Chem. Abstr.* **60**, 6733d (1964).

149. J. E. Leffler and H. H. Gibson, Jr., *J. Am. Chem. Soc.* **90**, 4117 (1968).

150. H. H. Gibson, Jr., H. R. Gaddy, III, and C. S. Blankenship, *J. Org. Chem.* **42**, 2443 (1977).

151. S. G. Fridman, *Men. Inst. Chem., Acad. Sci. Ukrain. U.S.S.R.*, **4**, No. 3, 351 (1937); *Chem. Abstr.* **32**, 5373⁴ (1938).

152. J. P. Horwitz, A. J. Tomson, J. A. Urbanski, and J. Chua, *J. Org. Chem.* **27**, 3045 (1962).

153. D. Balderman and A. Kalir, *Synthesis*, 24 (1978).

154. T. Wieland and H. Urbach, *Liebig's Ann. Chem.* **613**, 84 (1958).

155. B. Stanovnik, M. Tisler, S. Polanc, and M. Gracner, *Synthesis*, 65 (1978).

156. G. J. Matthews and A. Hassner, *Tetrahedron Lett.*, 1833 (1969).

157. A. K. Bose, J. F. Kistner, and L. Farber, *J. Org. Chem.* **27**, 2925 (1962).

158. J. H. Boyer, *J. Am. Chem. Soc.* **73**, 5865 (1951).

159. D. Lednicer and D. E. Emmert, *J. Org. Chem.* **40**, 3839 (1975).

160. R. Adams and L. Whitaker, *J. Am. Chem. Soc.* **78**, 658 (1956).

161. P. E. Iversen, in *Encyclopedia of the Electrochemistry of the Elements*, Vol. XIII, A. J. Bard and H. Lund (eds.), Marcel Dekker, New York, 1979, pp. 209–212.

162. T. Adachi, Y. Yamada, and I. Inouye, *Synthesis*, 45 (1977).

163. O. Dimroth and K. Pfister, *Ber.* **43**, 2757 (1910).

164. H. Rathsburg, *Ber.* **54**, 3183 (1921).

165. S. R. Sandler and W. Karo, *Organic Functional Group Preparation*, Vol. II, Academic Press, New York, 1971, Chap. 13.

166. E. Lieber, T. S. Chao, and C. N. R. Rao, *J. Org. Chem.* **22**, 238 (1957).

167. (a) P. Brewster, F. Hiron, E. D. Hughes, C. K. Ingold, and P. A. Rao, *Nature* **166**, 178 (1950); (b) L. A. Freiberg, *J. Org. Chem.* **30**, 2476 (1965).

168. E. Lieber, C. N. R. Rao, T. S. Chao, and W. H. Wahl, *J. Sci. Ind. Res.* **16B,** 95 (1957).

169. H. Böhme and D. Morf, *Chem. Ber.* **91,** 660 (1958).

170. M. C. Wu, L. Anderson, C. W. Slife, and L. J. Jensen, *J. Org. Chem.* **39,** 3014 (1974).

171. Y. Nakajima, R. Kinishi, J. Oda, and Y. Inouye, *Bull. Chem. Soc. Japan* **50,** 2025 (1977).

172. E. Manhart and K. von Werner, *Synthesis,* 705 (1978).

173. A. J. Papa, *J. Org. Chem.* **31,** 1426 (1966).

174. T. Sasaki, K. Kanematsu and Y. Yukimoto, *J. Org. Chem.* **37,** 890 (1972).

175. R. A. Abramovitch and E. P. Kyba, *J. Am. Chem. Soc.* **96,** 430 (1974).

176. J. A. Miller. *Tetrahedron Lett.,* 2959 (1975).

177. A. Astier, A. Pancrazi and Q. Khuong-Hun, *Tetrahedron* **34,** 1487 (1978).

178. A. Pancrazi, *Tetrahedron* **30,** 2337, 2579 (1974).

179. C. H. Heathcock, *Angew. Chem. Int. Ed.* **8,** 134 (1969).

180. A. Hassner and F. W. Fowler, *J. Org. Chem.* **33,** 2686 (1968).

181. E. Zbiral, *Synthesis,* 285 (1972).

182. F. Minisci and R. Galli, *Tetrahedron Lett.,* 533 (1962); F. Minisci, *Gazz. Chim. Ital.* **85,** 1300 (1955).

183. S. J. Weininger, S. Kohen, S. Mataka, G. Koga, and J.-P. Anselme, *J. Org. Chem.* **39,** 1591 (1974).

184. P. A. S. Smith, J. M. Clegg, and J. Lakritz, *J. Org. Chem.* **23,** 1595 (1958).

185. An example of a rare exception is α-hydrazinophenylacetic acid: A. Darapsky et al., *J. Prakt. Chem.* **[2]146,** 268 (1936).

186. C. M. Wright (Hercules Powder Company), personal communication.

187. C. L. Jenkins and J. K. Kochi, *J. Org. Chem.* **36,** 3095 (1971).

188. A Suzuki, M. Ishidoya and M. Tabata, *Synthesis,* 687 (1966).

189. M. S. Newman and W. C. Liang, *J. Org. Chem.* **38,** 2438 (1973).

190. P. A. S. Smith and J. H. Boyer, *Org. Syntheses,* Coll. **IV,** 75 (1963).

191. I. Ugi, *Tetrahedron* **19,** 1901 (1963).

192. A. Wohl and H. Schiff, *Ber.* **33,** 2741 (1900).

193. L. Gatterman and R. Ebert, *Ber.* **49,** 2117 (1916).

194. M. O. Forster, *J. Chem. Soc.* **107,** 260 (1915).

195. H. Bretschneider and H. Rager, *Monatsh. Chem.* **81,** 970 (1950); P. K. Dutt, *J. Chem. Soc.* **125,** 1463 (1924).

196. P. A. Grieco and J. P. Mason, *J. Chem. Eng. Data* **12,** 623 (1967).

197. R. O. Lindsay and C. F. H. Allen, *Org. Syntheses,* Coll. **III,** 710 (1955).

198. E. Bamberger, *Liebig's Ann. Chem.* **424,** 233 (1921).

199. S. Maffei and L. Coda, *Gazz. Chim. Ital.* **85,** 1300 (1955).

200. J. Munch-Petersen, *Org. Syntheses,* Coll. **IV,** 715 (1963).

201. K. Sakai and J.-P. Anselme, *J. Org. Chem.* **36,** 2387 (1971).

202. P. A. S. Smith, *Org. Syntheses,* Coll. **IV,** 819 (1963).

203. E. Lieber and A. F. Thomas III, *Appl. Spectrosc.* **15,** 144 (1961).

204. (*a*) W. R. Carpenter, *Anal. Chem.* **36,** 2352 (1964); (*b*) J. Cleophax, S. D. Gero and J. Hildesheim, *J. Chem. Soc. Chem. Comm.* 94 (1968).

205. K. A. Hoffman, H. Hoch, and H. Kirmreuther, *Liebig's Ann. Chem.* **380,** 131 (1911).

206. J. S. Fritz and G. S. Hammond, *Qualitative Organic Analysis,* Wiley, New York, 1957, pp. 110–111.

207. M. J. Coombs, *J. Chem. Soc.* 4200 (1958).

208. W. I. Awad, Y. A. Gawargious, and S. S. M. Hassan, *Talanta* **14**, 1441 (1967).

209. C. V. Hart, *J. Am. Chem. Soc.* **50**, 1922 (1928).

210. P. P. T. Sah and W.-H. Yin, *Rec. Trav. Chim. Pays Bas* **59**, 238 (1940).

211. W. Fischer and J.-P. Anselme, *J. Am. Chem. Soc.* **89**, 5284 (1967).

212. P. Scheiner, J. H. Schumaker, S. Deming, W. J. Libbey, and G. P. Nowack, *J. Am. Chem. Soc.* **87**, 306 (1965).

7.
OTHER FUNCTIONS WITH CHAINS OF THREE OR MORE NITROGENS

NOMENCLATURE

Chains of nitrogen atoms joined only by single bonds are named by means of a prefix corresponding to the number of nitrogen atoms in the chain and the generic root -*azane*; H_2N—NH—NH—NH_2 is thus *tetrazane*. The chains are numbered in the same way as carbon chains. Unsaturation is indicated by changing the last syllable to -*ene, -diene,* etc.; $(CH_3)_2NN$=NNHEt is thus called *1,1-dimethyl-4-ethyltetraz-2-ene*. This system is sometimes extended to chains of only two nitrogens, allowing azo compounds to be named alternatively as diazenes, and it has been proposed that *azane* be accepted as a systematic alternative for ammonia. The names "prozane" and "buzane" instead of triazane and tetrazane were originally proposed by Curtius, but their use has long since died out. Names ending in -*azone* were once widely used for some types of unsaturated nitrogen chains, but their meaning degenerated into such ambiguity that they too, have long ago been abandoned. The word *tetrazene* is also sometimes used for the explosive, 4-amidino-1-nitrosaminoamidino-1-tetrazene. Dipolar isomers of the triazene structure $R\overset{-}{N}$—$\overset{+}{N}(R)$=NR are commonly called *azimines*.

GENERAL REMARKS[2]

The considerable variety of compounds of different structural types, each with more or less distinct chemistry, evokes a departure from the form of the other, more homogeneous chapters. Furthermore, the literature concerning most of the types is quite limited. In view of this state of affairs, the properties, reactions, and preparation will be dealt with together in most cases, and the chapter has been subdivided according to length and degree of saturation of the nitrogen chains.

It appears in general that chains of three or more nitrogens are easily broken by elimination of an amine when a hydrogen is present on at least one nitrogen and by homolytic dissociation when fully substituted. Some degree of stabilization seems to be conferred by unsaturation, either N=N or C=N, and the known examples of chains of three or more nitrogens are almost all unsaturated in one way or another. Although some structural types that might be stable remain unrepresented because no feasible synthetic route has been contrived, others remain unknown in spite of a variety of fully credible preparative routes. Such structural types are therefore inferred to be inherently unstable. The types taken up in this chapter are those of which at least one isolable example has been reported.

TRIAZANES

Very few triazanes have been reported, and for many of them the supporting evidence is inadequate; the analyses in several instances are not acceptable by present-day standards.

The simplest triazane to have been isolated is actually a quaternary compound, 2,2-dimethyltriazanium chloride, prepared by the action of chloramine on *unsym*-dimethylhydrazine[3,4] (Eq. 7-1). It is stable to boiling ethanol and decomposes at about 135°C.

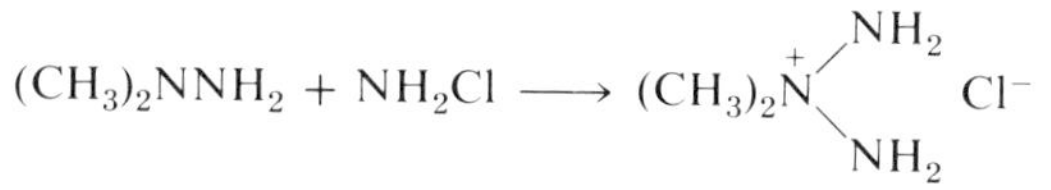

$$(CH_3)_2NNH_2 + NH_2Cl \longrightarrow (CH_3)_2\overset{+}{N}\overset{NH_2}{\underset{NH_2}{\diagup}} \quad Cl^- \qquad (7\text{-}1)$$

1-Acyl-2-phenyl-3-benzylidenetriazanes have been prepared by reduction of N'-nitroso-N'-phenyl hydrazides with sodium amalgam and treatment of the reaction mixture with benzaldehyde[5] (Eq. 7-2). The intermediate 1-acyl-2-phenyltriazanes

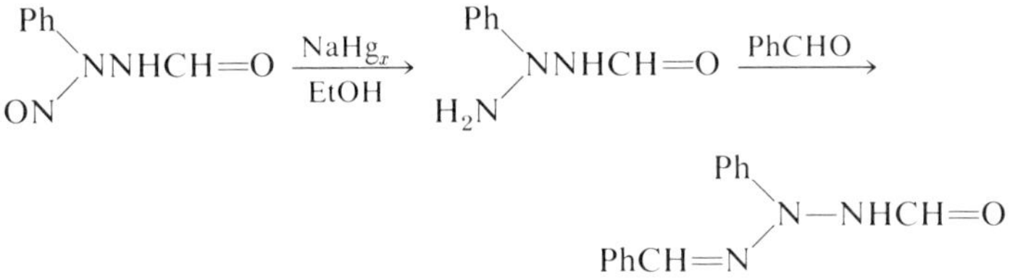

$$\underset{ON}{\overset{Ph}{\diagdown}}NNHCH=O \xrightarrow[\text{EtOH}]{NaHg_r} \underset{H_2N}{\overset{Ph}{\diagdown}}NNHCH=O \xrightarrow{PhCHO}$$

$$\underset{PhCH=N}{\overset{Ph}{\diagdown}}N-NHCH=O \qquad (7\text{-}2)$$

can be handled in solution, but are too unstable to isolate, and they easily revert to the original hydrazide. The benzylidene derivatives are colorless, crystalline substances, soluble in aqueous alkali, from which they are reprecipitated unchanged by acid. Hot, dilute hydrochloric acid breaks up the molecule into benzaldehyde and phenylhydrazine. The formyl member has been acetylated by a mixture of acetic anhydride and acetyl chloride; alkali removes the formyl group from the product, leaving the acetyl derivative (Eq. 7-3).

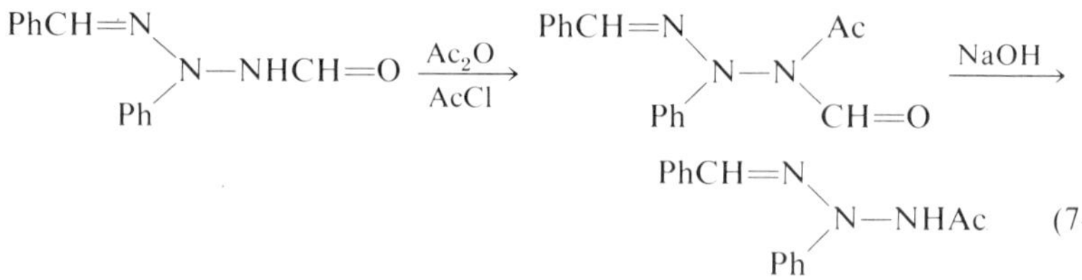

$$\underset{Ph}{\overset{PhCH=N}{\diagdown}}N-NHCH=O \xrightarrow[\text{AcCl}]{Ac_2O} \underset{Ph}{\overset{PhCH=N}{\diagdown}}N-N\overset{Ac}{\underset{CH=O}{\diagup}} \xrightarrow{NaOH}$$

$$\underset{Ph}{\overset{PhCH=N}{\diagdown}}N-NHAc \qquad (7\text{-}3)$$

Dicarboethoxytriazanes, such as 1-phenyl-2,3-dicarboethoxytriazane (mp 138°C) have been obtained in some instances by the addition of amines to azoformic esters[6] (Eq. 7-4) (see Chap. 4). They decompose easily into the original reactants.

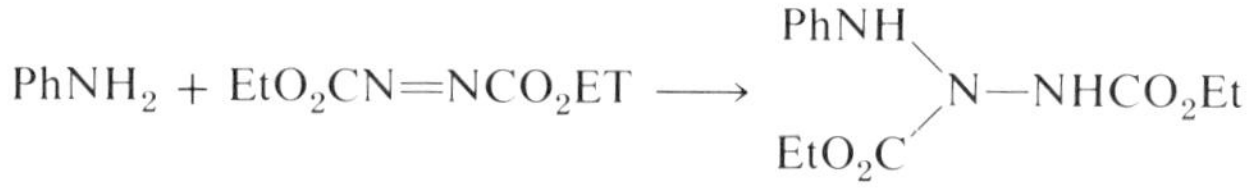

$$PhNH_2 + EtO_2CN=NCO_2ET \longrightarrow \underset{EtO_2C}{\overset{PhNH}{\diagdown}}N-NHCO_2Et \qquad (7\text{-}4)$$

1,3-Diphenyl-1,2,3-tribenzoyltriazane (mp 160–161°C) has been obtained by treating diphenyltriazene with excess Grignard reagent, which acts as a reducing agent, and benzoylating the reaction mixture[7] (Eq. 7-5). Another triazene,

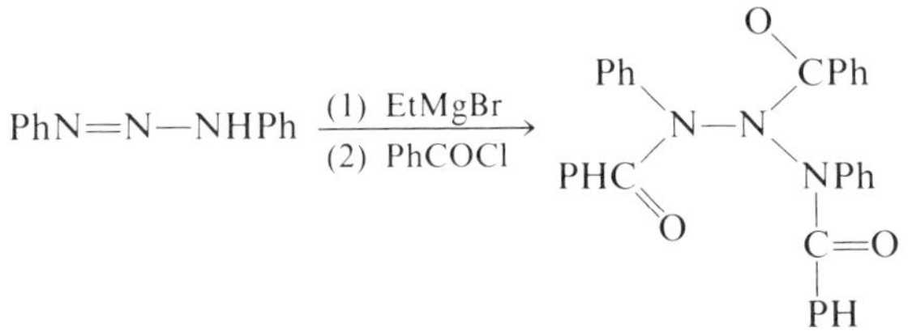

$$PhN\!\!=\!\!N\!\!-\!\!NHPh \xrightarrow[\text{(2) PhCOCl}]{\text{(1) EtMgBr}}$$

(7-5)

guanylcarbamyltriazene, has also been reduced to what may have been a triazane; the unstable product was obtained only in solution and was easily oxidized back to the triazene.[8]

Reduction of nitrosohydrazine derivatives offers a logical route to triazane, yet it has had little success, and except for the nitrosohydrazides, even the mildest reduction cleaves the nitroso group off. All attempts to prepare trimethyltriazane by reducing trimethylnitrosohydrazine have failed, for example; only products of N—N cleavage were obtained.[9] Several simple triazanes were originally reported to have been obtained by reduction of the compounds resulting from the condensation of N-nitrosophenylhydrazine with acetaldehyde ammonia (α-hydroxyethylamine), but the entire series was subsequently shown to have been misidentified.[10]

TRIAZENES

The triazenes are certainly the best known of all the chains of three or more nitrogens, other than azides. The diaryl derivatives, such as 1,3-diphenyltriazene (diazoaminobenzene) (mp 98–99°C) are fairly stable substances, as are the 1-aryl-2,2-dialkyl derivatives, many of which have been distilled at temperatures as high as 158°C/0.7 mm.[11] The 1-aryl-3-alkyl derivatives, such as 1-phenyl-3-methyltriazene (mp 37–37.5°C) are somewhat less stable and generally decompose below 100°C. The only simple monoaryl triazene, 1-phenyltriazene[12] (mp 50°C), is extremely unstable; more complicated monoaryl triazenes, such as 1-(1-anthraquinonyl)triazene, are known, however.[13] Purely aliphatic triazenes are represented by 1,3-dimethyltriazene, a colorless, water-miscible liquid that boils with decomposition at 92°C (explodes on rapid heating), and several trialkyl triazenes with methyl and benzyl groups.[14] 1-Methyl groups are found at δ 2.86–2.93 in the NMR spectrum, and 3-methyl groups at δ 3.05–3.46.[14b]

Disubstituted triazenes are distinctly acidic and usually form easily isolated cuprous salts. The triazenes evidently possess some basic character also, for they are decomposed by mineral acids, in some instances extremely easily. Some 1-aryl-3,3-dialkyltriazenes form isolable picrates, however.

1,3-Disubstituted triazenes are obviously capable of tautomerism, but separate tautomers of unsymmetrical examples have not yet been isolated. It is true that some 1,3-diaryl triazenes have been obtained in different forms, but these could as well be geometric isomers or polymorphic forms on the basis of the available evidence.[15] Indeed, the molecular weights of such triazenes indicate that they form hydrogen-bonded aggregates, in which tautomeric proton transfer would be so easy that it is doubtful that separate tautomers could ordinarily be isolated.[16] Attempts to prepare isomeric 1,3-diaryltriazenes have always given the same product[17] (Eq. 7-6).

$$ArN_2^+ + Ar'NH_2 \longrightarrow \begin{array}{c} ArN{=}N{-}NHAr' \\ \text{or} \\ ArNH{-}N{=}NAr' \end{array} \longleftarrow Ar'N_2^+ + ArNH_2 \qquad (7\text{-}6)$$

A spectroscopic study of a group of 1,3-arylmethyltriazenes has demonstrated that a dynamic equilibrium (on the NMR time scale) exists between the tautomers when there is an electron-withdrawing *para* substituent, but *p*-tolylmethyltriazene exists entirely as the conjugated tautomer, whereas examples with *ortho* substituents (nitro, carboethoxy, or acetyl) exist only as the nonconjugated tautomers[18] (by reason of chelate hydrogen bonding?). 1,3-Diaryltriazenes at room temperature give an NMR spectrum that is an average of the tautomers, but below $-80°C$, the tautomers can be separately resolved.[19]

The geometric configuration of triazenes is probably usually *trans*. Although dipole moments do not conclusively settle the point,[20] x-ray crystallography of 1-phenyl-3-(2,5-dibromophenyl)triazene clearly shows *trans* geometry, with an $N{-}N{=}N$ angle of 107.7° (the $N{-}N$ distance is 1.45 Å, and the $N{=}N$ distance is 1.25 Å).[21] Delocalization of the triazene double bond causes restricted rotation about the $N{-}N$ bond; 1-phenyl-3,3-dimethyltriazene shows a barrier of 12.7 kcal/mol.[22]

The diaryltriazenes are commonly light yellow or golden when pure, but traces of the isomeric aminoazobenzene, which are difficult to remove, may make the color darker.[23] 1-Aryl-3,3-dialkyltriazenes are likely to be yellowish, but aryl alkyl triazenes, as well as phenyltriazene and 1,3-dimethyltriazene, are colorless. The $N{=}N$ stretching frequency of triazenes has been identified with absorption at 1410 to 1418 cm^{-1} in the infrared[24] and thus resembles that of other azo compounds. 1,3-Diphenyl-3-acetyltriazene, a colorless solid, shows carbonyl absorption at 1700 cm^{-1}, and ultraviolet λ_{max} at 290 nm. Absorption in the near ultraviolet is strong for most triazenes and usually appears as more than one band.[20]

Alkylidenetriazenes are known.[25]

Although many triazenes are fairly stable to heat (especially diaryltriazenes), most of then decompose at temperatures below 300°C. Nitrogen is lost and amines are formed[19,26] (Eq. 7-7), but the mechanism is not clearly understood. It has been

$$ArN{=}N{-}NHAr' \xrightarrow[\text{Bu}_2 \text{ phthalate}]{235{-}300°C} \underset{\text{ca. 50\%}}{ArNHAr'} \qquad (7\text{-}7)$$

(Ar = anthraquinon-1-yl)

suggested that it is heterolytic, with $ArNH^-$ and R^+ intermediates. In the mass spectrometer, the molecular ion can lose N_2, CH_3, CH_3N_3H, CH_2N_2, or CH_3NH, in the case of arylmethyltriazenes.[27]

A study of aryl alkyl triazenes has shown that there is a competing path of decomposition leading to an aniline and other products, not all of which have been identified.[27] This course becomes more important in protic solvents, such as ethanol, and is believed to start with cleavage to an amine and a diazonium ion, but systematic investigation has been hampered by difficult reproducibility.

The reaction leading to secondary amine has also been brought about by treating triazenes with aluminum chloride, boron trifluoride/ether, or silica.[28,29] Under these conditions, the benzene ring may also be alkylated; benzyl-p-nitrophenyltriazene, for example, gives roughly equal quantities of N-benzyl- and o-benzyl-p-nitroaniline (Eq. 7-8). These products appear to derive from quite different reac-

$$p\text{-}O_2NC_6H_4N{=}NNHCH_2Ph \xrightarrow{\;BF_3/Et_2O\;} p\text{-}O_2NC_6H_4NHCH_2Ph$$

$$\text{and} \quad O_2N{-}\!\!\left\langle\!\!\bigcirc\!\!\right\rangle\!\!{-}NH_2 \quad (7\text{-}8)$$
$$\overset{|}{CH_2Ph}$$

tions, for when the reaction is conducted in benzene, the solvent undergoes substantial alkylation, presumably by a carbocation, but when the alkyl group is attached at a chiral carbon, its configuration is retained in the secondary amine. It does not appear to have been determined which nitrogen atom of the triazene is retained in the secondary amine.

Diaryltriazenes decompose rather easily in acidic media to form anilines and diazonium salts; this is simply the reversal of the reaction by which the triazenes are most generally made. Owing to the tautomeric equilibrium in unsymmetrical 1,3-diaryltriazenes, both possible amines and both possible diazonium salts are formed (Eq. 7-9). Recombination can occur if the medium is not strongly acidic and can

$$ArN_3HAr' \xrightarrow{\;H^+\;} \begin{matrix} ArN_2^+ + Ar'NH_2 \\[4pt] ArNH_2 + Ar'N_2^+ \end{matrix} \longrightarrow ArN{=}N{-}HNAr \quad \text{and}$$

$$Ar'N{=}N{-}NHAr' \quad (7\text{-}9)$$

lead to disproportionation to a pair of symmetrical triazenes as well as the original one.[30] When thermal and protolytic decomposition take place simultaneously, the products may be more complex; the secondary amine from the thermal process may combine with a diazonium ion to produce a trisubstituted triazene. 1-Aryl-3-methyltriazenes, when warmed in ethanol, thus give rise to some 1,3-diaryl-3-methyltriazene.[27] N-Areneazoaziridines cleave at 60 to 75°C to aryl azide and alkene with retained geometry.[31]

Acid also causes a slower rearrangement to an aminoazobenzene structure[32] (Eq. 7-10); this almost certainly occurs to a significant extent through the aforemen-

$$PhN{=}N{-}NHPh \xrightarrow{\;H^+,\,\Delta\;} H_2N{-}C_6H_4{-}N{=}N{-}Ph \quad (7\text{-}10)$$

tioned cleavage, followed by slow electrophilic attack on the ring of the amine. The *para* isomer is generally formed unless the *para* position is blocked. This rearrangement is accelerated by the presence of free aromatic amine,[33] an effect that is ascribed to a concerted process in which the N—N bond of the triazene is cleaved at the same time as the new N—C bond is formed.[34]

Alkylvinyltriazenes undergo a different sort of rearrangement when treated with dilute acid. The alkyl group migrates to the β-carbon of the vinyl group and nitrogen is lost, forming an imine that is quickly hydrolyzed to the corresponding carbonyl compound[35] (Eq. 7-11). The reaction is presumed to proceed by cleavage of

$$CH_2=C\underset{NHN=NCH_2CH_2CH_2CH_3}{\overset{Ph}{{<}}} \xrightarrow{H_3O^+} CH_3CH_2CH_2CH_2CH_2C\underset{O}{\overset{Ph}{{<}}}$$

$$\text{and} \quad CH_3CH_2\underset{CH_3}{\underset{|}{C}}HCH_2C\underset{O}{\overset{Ph}{{<}}} \qquad (7\text{-}11)$$

the protonated triazene into a primary enamine and an alkanediazonium ion, followed by alkylation of the enamine by the latter or the carbenium ion derived from it.

Homolytic decomposition of triazenes is also catalyzed by acids. Acetic acid or hydrogen chloride is the most effective agent, and the free radicals are believed to arise from the diazoacetate or diazonium chloride first formed.[36] The aryl radicals produced can be captured by olefins, coumarin, etc., which are thus arylated. 1,3-Diphenyltriazene has been observed to catalyze the polymerization of acrylonitrile.[37]

Triazenes may act as useful alkylating agents through the acid-catalyzed cleavage reaction if temperatures below that at which homolytic cleavage is favored are used. 1-Phenyl-3-methyltriazene has been claimed to rival diazomethane as a methylating agent, and 1-p-chlorophenyl-3-butyltriazene has been used to butylate not only carboxylic acids, but also phenols, mercaptans, and some alcohols[38,39] (Eq. 7-12).

$$p\text{-ClC}_6\text{H}_4\text{N}=\text{N}-\text{NHBu} + \text{ArCO}_2\text{H} \longrightarrow \text{ArCO}_2\text{Bu} \qquad (7\text{-}12)$$
$$67\text{-}73\%$$

The alkanediazonium ion, identical to that obtained from the corresponding diazoalkane, may be an intermediate, but kinetic evidence for a concerted process has been presented.[39]

Disubstituted triazenes can be alkylated in alkaline solution. The anion of an unsymmetrical triazene is, of course, an ambident system and can become alkylated at either terminal nitrogen (Eq. 7-13). Both isomeric products have indeed been isolated.[14b,40,41] Phase-transfer catalysis facilitates alkylation of triazenes.[41]

$$\overline{\text{Ar}-\text{N}\cdots\text{N}\cdots\text{N}}-\text{Ar}' + \text{RX} \longrightarrow$$

$$\text{Ar}-\underset{R}{\underset{|}{N}}-\text{N}=\text{N}-\text{Ar}' \quad \text{and} \quad \text{Ar}-\text{N}=\text{N}-\underset{R}{\underset{|}{N}}-\text{Ar}' \qquad (7\text{-}13)$$

Trisubstituted triazenes are apparently susceptible to alkylation only by exceptionally strong reagents. Trimethyloxonium fluoroborate attacks 1-p-tolyl-3,3-diisopropyltriazene mostly at the 3 position, causing cleavage to diazonium salt and amine, but a small amount of attack takes place at N-1, to form a triazenium salt in 3% yield (Eq. 7-14), but 3-methyl-1,3-di-p-tolyltriazene, lacking the steric hindrance

$$ArN{=}N{-}NR_2 + (CH_3)_3O^+BF_4^- \longrightarrow [ArN{=}N{-}\overset{+}{\underset{CH_3}{N}}R_2] \quad \text{or} \quad Ar\overset{+}{\underset{CH_3}{N}}{-}N{-}NR_2 \ BF_4^-$$

$$\downarrow \qquad\qquad\qquad\qquad (7\text{-}14)$$

$$ArN_2^+ + CH_3NR_2 \longrightarrow (CH_3)_2\overset{+}{N}R_2 \ BF_4^-$$

of the two isopropyl groups, gives no triazenium salt. When, however, one of the 3-substituents bears a good leaving group (Br—, CH_3SO_3—) on the β position, intramolecular alkylation at the 1 position of the triazene is the dominant reaction, producing cyclic triazenium salts.[42]

The common acylating agents attack mono- and disubstituted triazenes normally to give monoacyl derivatives; phenyl isocyanate has been particularly widely used[12,43] (Eq. 7-15). Since only one of the theoretically possible isomeric acylation

$$Ph{-}N{=}N{-}NH{-}CH_3 \xrightarrow{\ PhNCO\ } Ph{-}N{=}N{-}N\overset{\displaystyle CH_3}{\underset{\displaystyle CONHPh}{\big<}} \xrightarrow{\ HCl\ }$$

$$PhN_2^+ + PhNHCONHCH_3 \qquad (7\text{-}15)$$

products is usually obtained, acylation was at one time believed to be a reaction peculiar to the saturated nitrogen and thus diagnostic of the tautomeric constitution. This hypothesis was vitiated when it was shown for phenylcamphoryltriazene that spontaneous migration of an acyl group occurs on standing[44] (Eq. 7-16). Nitrosation

$$RN_3HPh \xrightarrow{\ PhNCO\ } RN{=}N{-}N\overset{\displaystyle Ph}{\underset{\displaystyle CONHPh}{\big<}} \xrightarrow{\ \text{spont.}\ } RN{-}N{=}NPh$$
$$\underset{\displaystyle CONHPh}{|}$$

$$\downarrow HCl \qquad\qquad\qquad\qquad\qquad \downarrow HCl \qquad (7\text{-}16)$$

$$PhNHCONHPh \qquad\qquad\qquad RNHCONHPh$$

is accomplished by reaction of triazenes with amyl nitrite[19,40,45,46] or triazene salts with nitrosyl chloride[47] (Eq. 7-17). The resulting nitrosotriazenes are isomeric with

$$Ar_2N_3^- \ M^+ + NOCl \longrightarrow Ar{-}\underset{\displaystyle NO}{\underset{\displaystyle |}{N}}{-}N{=}N{-}Ar \qquad (7\text{-}17)$$

the diazoanhydrides (see Chap. 4) and not identical with them. However, the products obtained by their decomposition are similar, and it is believed that an important path is homolytic cleavage into $ArN_2O\cdot$ and $ArN_2\cdot$, with or without prior rearrangement into diazoanhydrides. Radicals have been detected by CIDNP, and the products, which are complex, are consistent with the formation of aryl radicals.

Diazonium salts couple with disubstituted triazenes to form pentazadienes,[27] but the reaction does not occur readily with diaryltriazenes except in alkaline solution (see the section Pentazadienes).

An instance of a triazene acting as a partner in a cycloaddition reaction has been reported.[48,49] Diaryltriazenes react with tetracyanoethylene to form anils and arylhydrazones of mesoxalonitrile (Eq. 7-18). 1-Phenyl-3,3-dimethyltriazene has been reported to undergo [2 + 2] addition with diphenylketene.[50,51]

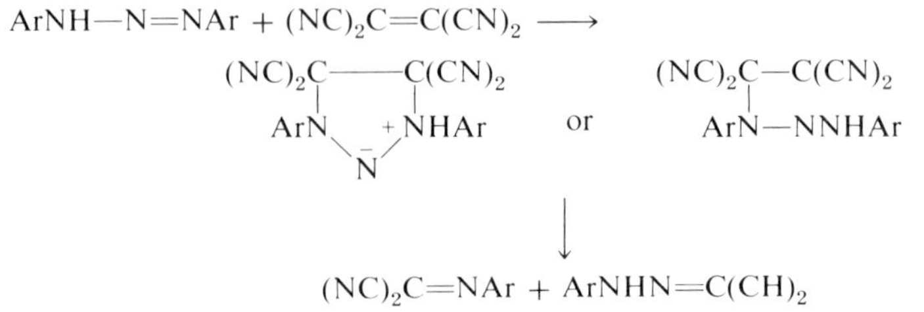

$$ArNH—N=NAr + (NC)_2C=C(CN)_2 \longrightarrow$$

$$
\begin{array}{ccc}
(NC)_2C\!-\!-\!-\!-C(CN)_2 & & (NC)_2C\!-\!C(CN)_2 \\
| \qquad\qquad | & & | \\
ArN \quad +NHAr & \text{or} & ArN\!-\!NNHAr \qquad (7\text{-}18) \\
\diagdown \quad \diagup & & \\
N & &
\end{array}
$$

$$\downarrow$$

$$(NC)_2C=NAr + ArNHN=C(CH)_2$$

Triazenes are readily reduced, but all attempts to stop the reduction at the triazane stage have failed; amines and hydrazines are the only products isolated[52] (Eq. 7-19). Oxidation of disubstituted triazenes with permanganate produces hexazadienes[53] (Eq. 7-20). This may be a case of dimerization of intermediate free

$$ArN=N—NHAr \xrightarrow{[H]} ArNH_2 \quad \text{and} \quad ArNHNH_2 \qquad (7\text{-}19)$$

$$ArN=N—NHAr \xrightarrow[\text{acetone}]{KMnO_4} ArN=N—\underset{\underset{Ar}{|}}{N}—\underset{\underset{Ar}{|}}{N}—N=NAr \qquad (7\text{-}20)$$

radicals, for such radicals, triazaallyl radicals, have been detected by ESR when 1,3-dimethyltriazene is photolyzed in the presence of *tert*-butyl peroxide at −100°C. The species is somewhat stable at that temperature, but disappears when the temperature is raised.

Triazenes are prepared for the most part either by coupling of diazonium salts to primary or secondary amines[11,31,39,54] (Eq. 7-21), or by the reaction of azides with Grignard reagents[52,55,56] (Eq. 7-22). Experimental conditions, especially order of

$$ArN_2^+ + R_2NH \longrightarrow ArN=N—NR_2 \qquad (7\text{-}21)$$

$$RN_3 + R'MgX \longrightarrow [RN\!\cdots\!\overset{-}{N}\!\cdots\!NR']MgX \xrightarrow{H^+}$$
$$RN=NNHR' \rightleftharpoons RNHN=NR' \qquad (7\text{-}22)$$

addition, can be of great importance in the diazonium reaction. The success of the preparation of 1-*p*-tolyl-3-methyltriazene[54] depends on adding the diazonium salt to the methylamine, but with diazonium salts bearing electron-withdrawing substituents, the reverse addition may be used.[27] Under unfavorable conditions, pentazadienes or diaryltriazenes may be formed. Coupling of diazonium salts to amides in ether (but not water),[55,56] or to sodium derivatives of amides,[57,58] is a preparative route to acyltriazenes. The Grignard reaction is the only suitable method for preparing dialkyltriazenes. Releasing the acid-sensitive triazenes from their magnesium salts is not a trivial step, and it is usually advisable first to prepare the more stable cuprous salts by treating with ammoniacal cuprous salt solution. In the instance of the especially sensitive compound dimethyltriazene, the free substance was eventually obtained by grinding the cuprous derivative with diphenyltriazene;[14a] even ammonium chloride was too acidic to use.

Diaryltriazenes have also been obtained by treating aryl amines with amyl nitrite in aprotic solvents,[19] from treating diazonium salts with sodium acetate,[27] and from warming certain heterocyclic nitrosamines in methanol solution.[59]

Monosubstituted triazenes are the most difficult to prepare. Reduction of an azide to a triazene, which is only exceptionally successfull, is exemplified by the preparation of phenyltriazene[12] (Eq. 7-23). Coupling of diazonium salts with ammonia

$$Ph-N_3 \xrightarrow[\text{ether}]{\text{SnCl}_2} Ph-N=N-NH_2 \qquad (7\text{-}23)$$

cannot usually be stopped short of formation of pentazadiene, but there are some exceptions, such as the preparation of 1-anthraquinonyltriazene.[13]

Trisubstituted triazenes can be prepared by alkylating salts of disubstituted triazenes.[14] Vinyltriazenes have been prepared by the reaction of $(CH_3)_2S(O)=CH_2$ with p-nitrophenyl or benzoyl azide, but simple aryl and alkyl azides give triazolines instead[60] (Eq. 7-24).

$$p\text{-}O_2N-C_6H_4N_3 + 2(CH_3)_2S\overset{\displaystyle O}{\underset{\displaystyle CH_2}{\big\backslash\!\!\big\backslash}} \longrightarrow$$

$$p\text{-}O_2N-C_6H_4NH-N=N-CH=CH_2 \qquad (7\text{-}24)$$
$$45\%$$

AZIMINES

The azimines are triazenes with a substituent on each nitrogen, such that only semipolar structures, analogous to the nitro group, can be written for them:

$$\overset{-}{R}N-\overset{+}{N}=NR \longleftrightarrow RN=\overset{+}{N}-\overset{-}{N}R$$
$$\underset{R}{|} \qquad\qquad \underset{R}{|}$$

They are, in fact, isoelectronic with nitro and azoxy compounds. Purported examples in the earlier literature have been reinvestigated, and the open-chain examples were found to have other structures.[61] Acyclic examples apparently decompose in situ.[62] Stable representatives are known in which part of the azimine chain is incorporated into an aromatic ring, as in the benzocinnoline N-imides formed by the action of alkyl nitrites on 2,2'-diaminobiphenyl or of hydroxylamine-O-sulfonic acid on benzocinnoline.[63] The closest approach to isolable acyclic examples are actually tetrazene derivatives,[64] and cis and $trans$ forms are separable. They are phthalimido derivatives and are pale yellow when the substituents are alkyl; orange when they are aryl. They decompose at their melting points (all above 100°C) to form phthalhydrazides (Eq. 7-25).

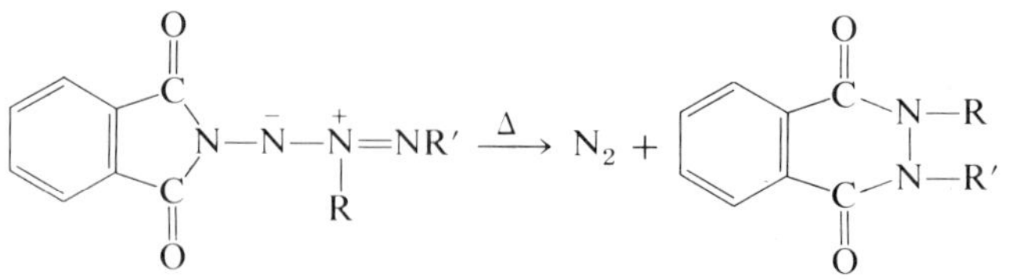

$$(7\text{-}25)$$

TETRAZANES

A number of fully substituted tetrazanes and several partially substituted tetrazanes are known. Hexaphenyltetrazane is formed by oxidation of triphenylhydrazine (Eq. 7-26); it is a white solid when very cold, but green at room temperature, and it

$$\varnothing_2N\text{---}NH\varnothing \xrightarrow[-80°C]{PbO_2 \text{ or } K_3Fe(CN)_6}$$

$$(7\text{-}26)$$

Colorless   Blue

dissociates reversibly in solution into blue triphenylhydrazyl radicals.[65] Many other hexaaryltetrazanes have been reported, and nearly all are highly dissociated into hydrazyl radicals, some, such as tripicrylhydrazyl, apparently even in the solid state. For this reason, these compounds are referred to as *hydrazyls* at least as often as they are called *tetrazanes* (see Chap. 1). 1,2,3,4-Tetraaryl-1,4-diacyltetrazanes, however, dissociate very little. The dissociation of 1,1,4,4-tetraaryl-2,3-dibenzoyltetrazane has been thoroughly investigated from both equilibrium and kinetic standpoints.[66] Both rate and equilibrium constants obey the Hammett linear free energy relation ($\rho = -1.52$ for equilibrium), and there is a linear relation between entropy and enthalpy.

1,2,4-Trimethyl-1,3,4-*tris*(trifluoromethyl)tetrazane (bp 75°C/135 mm) is formed by photolysis of methyltrifluoromethyldiimide[67] (Eq. 7-27); it apparently does not dissociate detectably under ordinary conditions.

$$CH_3\text{---}N\text{=}N\text{---}CF_3 \xrightarrow{h\nu} CF_3\text{---}N\text{---}N\text{---}N\text{---}N\text{---}CF_3 \qquad (7\text{-}27)$$

with substituents CH_3, CF_3, CH_3, CH_3

A partially unsubstituted tetrazane has been reported to result from the oxidation of 1-hydrazinophthalazine by oxygen in the presence of ferrous ion; the tetrazane forms a deep red, presumably chelate, complex with ferrous ion.[68]

1,2,3,4-Tetracarboethoxytetrazane (mp 80°C) has been reported to result from reductive coupling of azodicarboxylic ester[69] (Eq. 7-28). Above its melting point, this tetrazane apparently dissociates into hydrazyl radicals, for products corresponding to their disproportionation are then found.

A number of alkylidene tetrazanes have been reported in the older literature as products of the oxidation of arylhydrazones or of the addition of arylhydrazones to

$$\text{EtOOC}-\text{N}{=}\text{N}-\text{COOEt} \xrightarrow[\text{i-PrOH}]{\text{$h\nu$}}$$

$$\text{EtOOC}-\text{NH}-\underset{\underset{\text{COOEt}}{\vert}}{\overset{\overset{\text{COOEt}}{\vert}}{\text{N}}}-\underset{\underset{\text{COOEt}}{\vert}}{\text{N}}-\text{NH}-\text{COOEt} \qquad (7\text{-}28)$$

80%

azo compounds. More recent evidence, however, makes it appear very unlikely that any of these products were tetrazanes (see Chap. 2 for references and further discussion).

TETRAZENES

Both 1-tetrazenes, $\text{R}-\text{N}{=}\text{N}-\text{NRNR}_2$, and 2-tetrazenes, $\text{R}_2\text{N}-\text{N}{=}\text{N}-\text{NR}_2$, are known in some variety in both partially and completely substituted examples. Tetramethyltetrazene-2[70] is a liquid distillable at 74°C/110 mm; 1,4-dimethyl-1,4-diphenyltetrazene-2 forms colorless leaflets (mp 137°C dec.), stable to boiling water; tetraphenyltetrazene-2 forms yellowish crystals (mp 123°C dec)[71]; and 1,4-bis(carboethoxy)-1, 4-dimethyltetrazene-2 has a melting point of 186 to 187°C.[72] Presumably the tetrazenes that have been reported have the *trans* configuration about the double bond, but only one example, 1,4-di-*tert*-butyl-1,4-diphenyltetrazene-2, has been investigated by x-ray crystallography.[73]

The 1-tetrazenes are prepared by the coupling of diazonium salts to hydrazines[74,75] (Eq. 7-29) (see Chap. 4). Support for the assignment of the 1,3-disubstituted struc-

$$\text{Ar}-\text{N}_2^+ + \text{RNHNH}_2 \longrightarrow \text{Ar}-\text{N}{=}\text{N}-\underset{\underset{\text{R}}{\vert}}{\text{N}}-\text{NH}_2 \qquad (7\text{-}29)$$

ture for the tetrazenes derived from monosubstituted hydrazines is adduced from the formation of hydrazones with aldehydes, identical to the products obtained by coupling diazonium salts to preformed hydrazones[74] (Eq. 7-30).

$$\text{PhCH}_2\text{NHNH}_2 + p\text{-O}_2\text{NC}_6\text{H}_4\text{N}_2^+ \longrightarrow$$

$$\text{PhCH}_2\text{N}\overset{\displaystyle{\diagup}\text{NH}_2}{\underset{\displaystyle\diagdown}{\text{N}{=}\text{N}-\text{C}_6\text{H}_4\text{NO}_2\text{-}p}} \xrightarrow{\text{PhCHO}}$$

$$\text{PhCH}_2\text{N}\overset{\displaystyle{\diagup}\text{N}{=}\text{CHPh}}{\underset{\displaystyle\diagdown}{\text{N}{=}\text{N}-\text{C}_6\text{H}_4\text{NO}_2\text{-}p}} \qquad (7\text{-}30)$$

$$\diagup$$

$$\text{PhCH}_2\text{NH}-\text{N}{=}\text{CHPh} + p\text{-O}_2\text{NC}_6\text{H}_4\text{N}_2^+$$

The 1-tetrazenes are presumably weakly basic and acidic, but they are decomposed by both alkalies and acids. The acid-induced decomposition evidently proceeds through initial decoupling; the products are a phenol, a hydrazine, and nitrogen.[76] Reduction gives a pair of hydrazines (Eq. 7-31) or azide and amine derived from the 1,4-diaryltetrazene. The only report of oxidation concerns 1,3-diaryltetrazenes, which have been converted to octazatrienes.[77]

$$
\text{Ph}-\text{N}=\text{N}-\underset{\underset{\text{Et}}{|}}{\text{N}}-\text{NH}_2
\begin{array}{c}
\xrightarrow{\text{H}^+} \text{PhOH} + \text{EtNHNH}_2 + \text{N}_2 \\
\xrightarrow[\text{AcOH}]{\text{Zn}} \text{PhNHNH}_2 + \text{EtNHNH}_2
\end{array}
\tag{7-31}
$$

1-Aryl-4-acyltetrazenes are cyclized to tetrazoles by alkali[78] (Eq. 7-32). However, a report of another type of base-catalyzed reaction of tetrazenes has been shown to

$$
\text{PhNHN}=\text{N}-\text{NHCOPh} \xrightarrow{\text{NaOH}}
\begin{array}{c}
\text{N}-\text{N} \\
\parallel \quad \diagdown \\
\quad \quad \text{C}-\text{Ph} \\
\text{N}-\text{N} \diagup \\
\quad \quad \quad \diagdown \text{Ph}
\end{array}
\tag{7-32}
$$

be in error. It was once thought that diazonium salts condensed with benzaldehyde arylhydrazones to give benzylidenetetrazenes, which underwent base-catalyzed rearrangement to formazans. The compounds originally thought to be tetrazenes have since been shown to be geminal bis-azo compounds resulting from coupling at carbon rather than nitrogen; their conversion to formazans is a simple tautomerization.[79]

The 2-tetrazenes are prepared from *unsym*-disubstituted hydrazines by oxidation in various ways,[2,70-72,80] including mercuric oxide,[81,82] sodium hypochlorite,[82] etc., and decomposition of salts of the corresponding arenesulfonyl derivatives. These methods apparently give rise to an intermediate azamine, $\text{R}_2\overset{+}{\text{N}}=\overset{-}{\text{N}}$, which dimerizes (Eq. 7-33) (see Chap. 5). They may also be generated from secondary amines by

$$
\text{Ph}_2\text{NNH}_2 + \text{KMnO}_4 \longrightarrow \text{Ph}_2\overset{+}{\text{N}}=\overset{-}{\text{N}} \longrightarrow \text{Ph}_2\text{N}-\text{N}=\text{N}-\text{NPh}_2 \tag{7-33}
$$

reaction with Angeli's salt[83] (sodium nitrohydroxamate) (Eq. 7-34) or hydroxylamine-*O*-sulfonic acid.[84]

$$
\text{R}_2\text{NH} \xrightarrow{\text{Na}_2\text{N}_2\text{O}_3} \text{R}_2\overset{+}{\text{N}}=\overset{-}{\text{N}} \longrightarrow \text{R}_2\text{N}-\text{N}=\text{N}-\text{NR}_2 \tag{7-34}
$$

Oxidation of α-alkylcarbazate esters with bromine in dilute hydrochloric acid produces 1,4-bis(carboalkoxy)-2-tetrazenes in moderately good yields, but the reaction fails with with the corresponding semicarbazides or hydrazides.[72] The fact that this reaction takes place in an acidic medium seems to preclude azamines as intermediates.

2-Tetrazenes are weakly basic; tetramethyl-2-tetrazene has a pK_b value of 7.78, and tetraethyl, 6.45.[81] Salts have not been isolated, for evolution of nitrogen takes place when 2-tetrazenes are acidified, and amines and other products are formed[71] (Eq. 7-35). Initial formation of secondary amine cation-radicals, $\text{Ph}_2\overset{+}{\text{N}}\text{H}\cdot$, followed by disproportionation, could account for the observed products, which are benzidines or their oxidation products if *para* positions are open, or 9,10-diarylphenazine

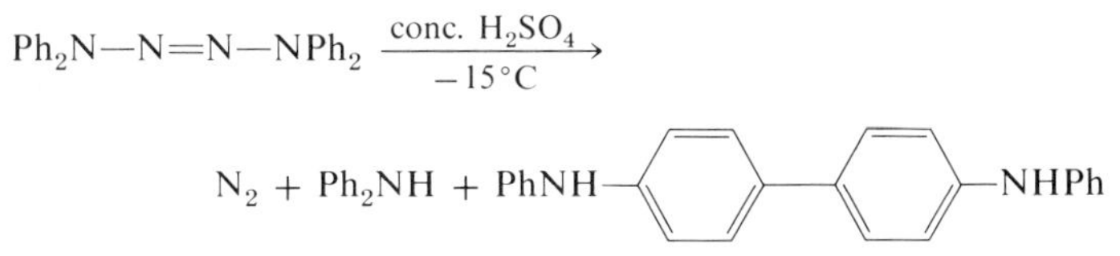

$$Ph_2N-N=N-NPh_2 \xrightarrow[-15°C]{conc. \; H_2SO_4}$$

$$N_2 + Ph_2NH + PhNH-\!\!\!\raisebox{-1ex}{\bigcirc}\!\!\!-\!\!\!\raisebox{-1ex}{\bigcirc}\!\!\!-NHPh \qquad (7\text{-}35)$$

derivatives[71,72] (Eqs. 7-36, 7-37). When hydrochloric acid is used, the products may contain chlorine.

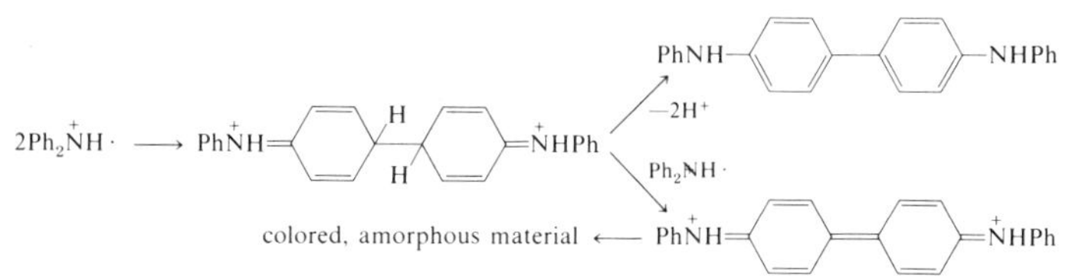

$$ (7\text{-}36) $$

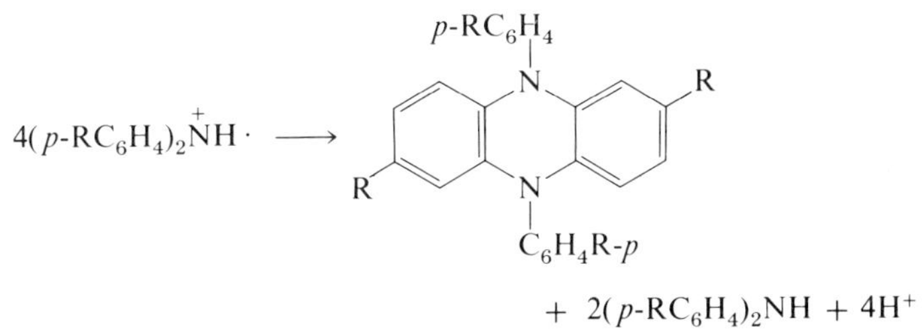

$$4(p\text{-}RC_6H_4)_2\overset{+}{N}H\cdot \longrightarrow$$

$$+ \; 2(p\text{-}RC_6H_4)_2NH + 4H^+$$

2-Tetrazenes are also decomposed by photolysis[85-88] and by mild heating (generally 80–120°C),[89,90] which cause loss of nitrogen and initial formation of secondary amino radicals (Eq. 7-38). The amino radicals then either dimerize to form

$$Me_2N-N=N-NMe_2 \xrightarrow{\Delta \; or \; h\nu} N_2 + Me_2N\cdot \xrightarrow{[H]} Me_2NH$$

$$\downarrow \qquad\qquad (7\text{-}38)$$

$$Me_2N-NMe_2$$

hydrazines, inside or outside of the solvent cage, or abstract hydrogen from the medium. Unsymmetrical tetrazenes give both the unsymmetrical hydrazine and the two possible symmetrical ones. A violet solid that appears to be composed of free dimethylamino radical has actually been isolated from the decomposition of tetramethyltetrazene followed by rapid quenching in liquid nitrogen.[91] At −160°C, the color fades and tetramethylhydrazine and other products are formed. This overall behavior is consistent with the low dissociation energy (16.5 kcal/mol) reported for the N—N bond in 2-tetrazenes.[92]

The behavior of decomposing 2-tetrazenes in the presence of olefins is not completely understood, but it appears that the amino radicals can be trapped under some circumstances. Thermolysis of tetramethyl-2-tetrazene initiates polymerization.[70] Photolysis in acidic medium in the presence of olefin and oxygen gives β-hydroperoxyamines, but it is stated that thermolysis under the same conditions does not, nor does photolysis in neutral medium.[93]

The kinetics of decomposition of tetrazenes have been studied. The activation energy for tetramethyltetrazene is moderately high,[79,94] and viscosity effects on the decomposition of tetraphenyl-2-tetrazene are consistent with sequential rather than simultaneous cleavage of the N—N bonds.[95]

Reduction of 2-tetrazenes catalytically or by dissolving metals always gives a secondary amine[96,97] (Eq. 7-39); neither tetrazanes nor hydrazines have been re-

$$R_2N-N{=}N-NR_2 \xrightarrow[\text{or Zn/HCl}]{H_2/Pd} R_2NH \qquad (7\text{-}39)$$

ported. Stannous chloride, which will reduce azamines, the monomers corresponding to tetrazenes, does not reduce tetrazenes themselves.[98] Tetraalkyltetrazenes are oxidizable, and their $E°$ values have been compared.[99] Permanganate attacks the 1- and 4-alkyl groups, converting them successively to acyl groups[100] (Eq. 7-40);

$$(RCH_2)_2N-N{=}N-N(CH_2R)_2 \xrightarrow{KMnO_4}$$

$$\begin{array}{c} \text{O} \\ \parallel \\ RC \\ \diagdown \\ N-N{=}N-N(CH_2R)_2 \longrightarrow \\ \diagup \\ RCH_2 \end{array}$$

$$\begin{array}{c} \text{O} \qquad\qquad \text{O} \\ \parallel \qquad\qquad \parallel \\ RC \qquad\qquad CR \\ \diagdown \qquad\qquad \diagup \\ N-N{=}N-N \qquad (7\text{-}40) \\ \diagup \qquad\qquad \diagdown \\ RCH_2 \qquad\qquad CH_2R \end{array}$$

tetramethyl-2-tetrazene gives 65% of 1,4,4-trimethyl-1-formyltetrazene (mp 55°C, pK_b 14.57) or 50% of 1,4-dimethyl-1,4-diformyltetrazene (mp 166°C).

Little has been reported about the reaction of nucleophiles with 2-tetrazenes. 1,4-Bis(carboalkoxy)tetrazenes have been found to react with secondary amines or with alkali to give tetrazolinones or azide and urethan, depending on the other substituents.[85] Electrophilic attack has received a little more attention. Alkylation appears to take place at N-1, but the nitrogen chain is broken and nitrogen is evolved.[101] Acylating agents give similar results.[102,103] Phenyl isocyanate, for example, converts tetramethyl-2-tetrazene to N-phenyl-N',N'-dimethylurea and N-methylformaldimine. Nitrosating agents (N_2O_3 or N_2O_4) cleave a methyl group from tetramethyl-2-tetrazene to give yellow, explosive 1-nitroso-1,4,4-trimethyl-2-tetrazene[100] (Eq. 7-41).

$$Me_2N-N{=}N-NMe_2 \xrightarrow[-20°C,\ CHCl_3]{N_2O_3\ or\ N_2O_4} \begin{array}{c} ON \\ \diagdown \\ N-N{=}N-NMe_2 \qquad (7\text{-}41) \\ \diagup \\ Me \end{array}$$

One cannot help admiring the courage of the investigators who treated tetramethyl-2-tetrazene with tetranitromethane. Condensation with elimination of the elements of nitrous acid occurs to give a dinitrovinyltetrazene[101] in 30% yield at room temperature in hexane. The properties of the product (mp 129°C) suggest that it exists as a zwitterion, $Me_2N-N{=}N-\overset{+}{N}Me{=}CH-\overset{-}{C}(NO_2)_2$.

TETRAZADIENES

Whereas no simple tetrazadienes have been reported, substances that can be considered as formally derived from such a structure are known in the form of N-azidoamines, such as N-azido-bis(trimethylsilyl)amine:[102]

$$(Me_3Si)_2N-N_3 \longleftrightarrow (Me_3Si)_2\overset{+}{N}=N-N=\overset{-}{N}$$

N-Azidodimethylamine[103] is a liquid (bp 32°C/11 mm), and azidodibenzylamine appears to have been isolated as an unstable oil from the reaction of N,N-dibenzylhydrazide anion, $(PhCH_2)_2N-NH^-$, with p-toluenesulfonyl azide.[104] It slowly loses nitrogen and forms bibenzyl.

Diaryl tetrazadienes have been considered in the attack of arylnitrenes on aryl azides during thermolysis, but the evidence is unfavorable.[105]

PENTAZADIENES

Although no authentic pentazanes or pentazenes have been reported, 1,4-pentazadienes are known in considerable variety from coupling of diazonium salts with primary amines[106,107] or triazenes.[107] 1,5-Diphenylpentazadiene forms yellow prisms that explode on heating or rubbing.[107] It is acidic enough to dissolve in dilute alkali, but it does not dissolve in cold, dilute acids. Hot acid brings about decomposition into aniline, nitrogen, and phenol (Eq. 7-42), presumably through initial

$$PhN=N-NH-N=NPh \xrightarrow[\Delta]{aq. H^+} PhNH_2 + PhOH + N_2 \qquad (7-42)$$

decoupling to a diazonium salt and phenyltriazene. 1,5-Diphenyl-3-methylpentazadiene forms bright yellow needles (mp 112–113°C); cold acid breaks the compound into aniline, methylamine, nitrogen, and a diazonium salt.[108] Reduction gives both phenylhydrazine and methylhydrazine and, presumably, aniline and methylamine (Eq. 7-43).

$$PhN=N-\underset{\underset{CH_3}{|}}{N}-N=NPh \xrightarrow[AcOH]{Zn} PhNHNH_2 + CH_3NHNH_2 \qquad (7-43)$$

Although pentazadienes decompose on mild heating, often violently, little has been done to investigate the reaction. Free radicals are evidently formed, for pentazadienes have been used as polymerization initiators.[109]

HEXAZADIENES

The only isolated example of a disubstituted hexazadiene appears to be the 1,6-bis(5-tetrazolyl) derivative, a yellowish solid that is stable at room temperature but explodes at 90°C or when pressed.[110,111] It is formed by coupling of 2 moles of 5-diazotetrazole with hydrazine (Eq. 7-44), a reaction that has not been duplicated with benzenediazonium compounds apparently because the intermediate mono-

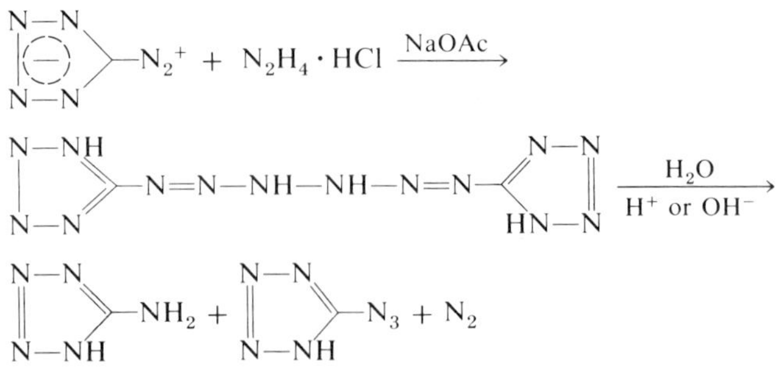

$$\text{(tetrazolyl)}-N_2^+ + N_2H_4 \cdot HCl \xrightarrow{\text{NaOAc}}$$

$$\text{(tetrazolyl)}-N{=}N{-}NH{-}NH{-}N{=}N-\text{(tetrazolyl)} \xrightarrow[\text{H}^+ \text{ or OH}^-]{\text{H}_2\text{O}}$$

$$\text{(tetrazolyl)}-NH_2 + \text{(tetrazolyl)}-N_3 + N_2 \qquad (7\text{-}44)$$

substituted tetrazenes break up before they can be engaged by a second diazonium ion. This hexazadiene is acidic enough to form a sodium salt (yellow) with concentrated alkali, but it cannot be said with certainty whether the acidity is due to the hexazadiene hydrogens or the tetrazole hydrogens.

The hardly surprising extreme sensitivity of ditetrazolylhexazadiene precluded analyzing it in the ordinary manner, but its constitution is more or less adequately supported by a determination of its degradation products (Eq. 7-44).

The simplest tetrasubstituted hexazadienes are 1,6-diaryl-3,4-dimethyl derivatives, obtained by the coupling of negatively substituted diazonium salts with sym-dimethylhydrazine.[111] Cold, concentrated sulfuric acid reverses the coupling. 1,6-Bis(p-chlorophenyl)-3,4-dimethyl-1,5-hexazadiene thus obtained is a slightly greenish solid (mp 86–87°C); although it is stable to prolonged storage in a refrigerator and can apparently be handled "safely" when protected from shock, it explodes when heated or struck.

A group of hexaarylhexazadienes has been prepared by oxidizing diaryltriazenes.[53] They are not very stable, and only those bearing electron-withdrawing substituents have been isolated. 1,3,4,6-Tetrakis(p-chlorophenyl)hexazadiene is a yellow solid that decomposes at 110°C; even at room temperature, slow decomposition to nitrogen and an azobenzene is noticeable (Eq. 7-45).

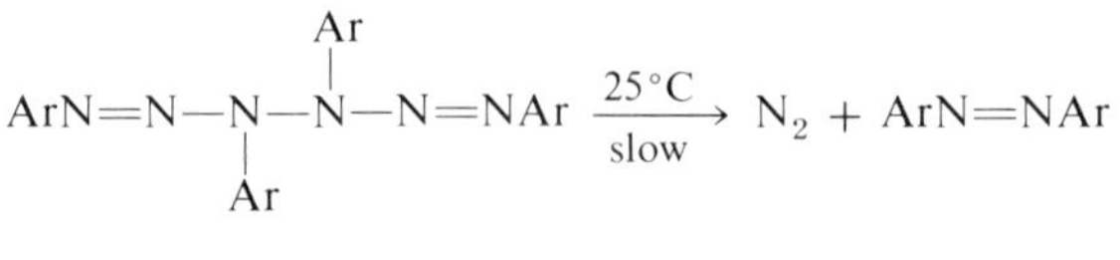

$$\underset{\underset{Ar}{|}}{ArN{=}N{-}N}{-}\underset{\underset{Ar}{|}}{N}{-}N{=}NAr \xrightarrow[\text{slow}]{25°C} N_2 + ArN{=}NAr \qquad (7\text{-}45)$$

Although the coupling of diazonium salts to hydrazine is not a generally successful route to hexazadienes, secondary hydrazides give 3,4-diacylhexazadienes more readily[111] (Eq. 7-46). They are colorless, crystalline solids whose ultraviolet spectra

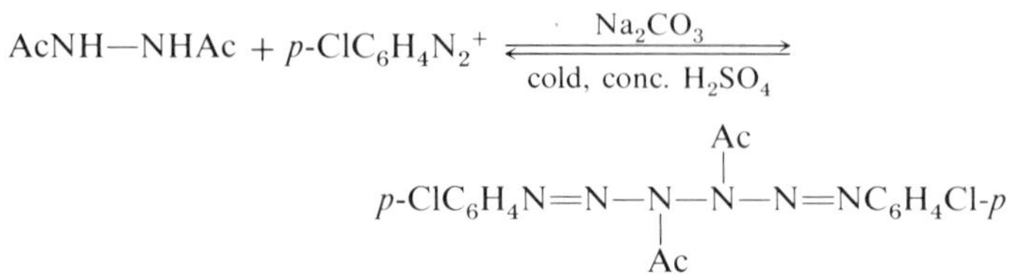

$$\text{AcNH}{-}\text{NHAc} + p\text{-ClC}_6\text{H}_4\text{N}_2^+ \underset{\text{cold, conc. H}_2\text{SO}_4}{\overset{\text{Na}_2\text{CO}_3}{\rightleftharpoons}}$$

$$p\text{-ClC}_6\text{H}_4\text{N}{=}N{-}\underset{\underset{Ac}{|}}{N}{-}\underset{\underset{Ac}{|}}{N}{-}N{=}NC_6H_4Cl\text{-}p \qquad (7\text{-}46)$$

(70–80%; mp 132–133°C dec.)

have a strong maximum near 290 nm; they melt with decomposition at temperatures well above 100°C. Alcoholic potassium hydroxide first cleaves the diacylhexazadienes into a mole each of aryl azide, aryl acyl triazene, and nitrogen, the expected decomposition products of the presumed intermediate monoacylhexazadiene (Eq. 7-47). Hot hydrochloric acid is almost without action on the diacyl-

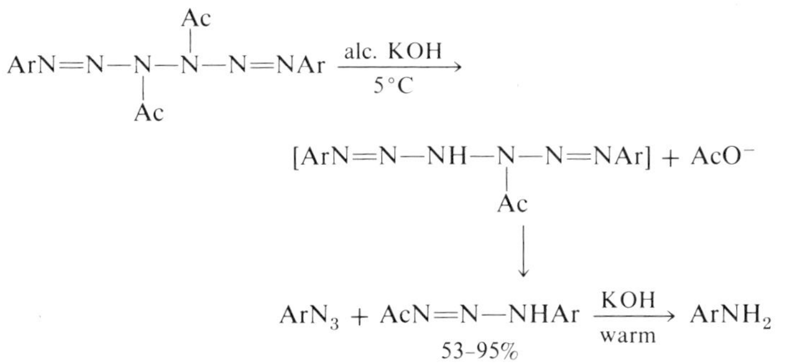

$$\underset{\underset{\overset{|}{Ac}}{\overset{\overset{Ac}{|}}{}}}{ArN=N-N-N-N=NAr} \xrightarrow[5°C]{alc. KOH}$$

$$[ArN=N-NH-\underset{\overset{|}{Ac}}{N}-N=NAr] + AcO^-$$

$$\downarrow \qquad\qquad (7\text{-}47)$$

$$ArN_3 + AcN=N-NHAr \xrightarrow[warm]{KOH} ArNH_2$$
$$53\text{-}95\%$$

hexazadienes, but cold, concentrated sulfuric acid effects decoupling in a few minutes, regenerating the diazonium salt and hydrazide from which the compound was prepared.

OCTAZATRIENES

A small group of 1,3,6,8-tetraaryloctazatrienes has been reported as products in low yield of the oxidation of 1,3-diaryltetrazenes[5] (Eq. 7-48). The tetraphenyl derivative,

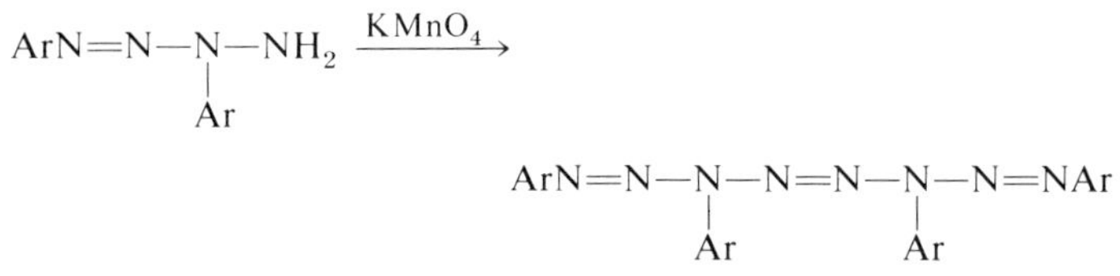

$$\underset{\underset{Ar}{\overset{|}{}}}{ArN=N-N-NH_2} \xrightarrow{KMnO_4}$$

$$\underset{\underset{Ar}{\overset{|}{}}\quad\underset{Ar}{\overset{|}{}}}{ArN=N-N-N=N-N-N=NAr} \qquad (7\text{-}48)$$

for which no analysis was reported, is an explosive, difficultly soluble, sulfur-yellow solid (mp 51-52°C). The 3,6-diphenyl-1,8-bis(p-tolyl) analog is a yellow solid (mp 61.5-64°C dec.) that is very sensitive to shock.[77] When allowed to decompose in cumene at 30°C, it gave rise to toluene, benzene, azoarenes, and p,p'-dimethylbiphenyl.[77] These substances have been claimed to be initiators for radical-induced polymerization.[109]

CHAINS LONGER THAN N_8

The only reported instance of a nitrogen chain longer than eight atoms appears to be the product of irradiation of 1,2,4-triazoledione, which forms a low polymer with a nitrogen backbone believed to be 40 nitrogens long.[112] However, the structure has not been firmly established.

REFERENCES

1. S. H. Patinkin, J. P. Horwitz, and E. Lieber, *J. Am. Chem. Soc.* **77**, 562 (1955).

2. C. G. Overberger, J. G. Lombardino, and J.-P. Anselme, *Chemistry of Organic Compounds with Nitrogen-Nitrogen Bonds,* Ronald Press, New York, 1965.

3. R. Gösl, *Angew. Chem.* **74**, 470 (1962).

4. K. Utvary and H. G. Sisler, *Inorg. Chem.* **7**, 698 (1968).

5. (*a*) A. Wohl, *Ber.* **33**, 2759 (1900); (*b*) A. Wohl and H. Schiff, *Ber.* **35**, 1900 (1902).

6. (*a*) K. E. Cooper and E. H. Ingold, *J. Chem. Soc.,* 1894 (1926); (*b*) I. G. S. Maara and S. B. Srinavastava, *J. Ind. Chem. Soc.* **37**, 177 (1960); (*c*) R. B. Carlin and M. S. Moores, *J. Am. Chem. Soc.* **84**, 4107 (1962).

7. H. Gilman and R. M. Pickens, *J. Am. Chem. Soc.* **47**, 2406 (1925).

8. J. Thiele and W. Osborne, *Ber.* **30**, 2867 (1897).

9. A. F. Graefe, *J. Org. Chem.* **23**, 1230 (1958).

10. H. Voswinckel, *Ber.* **35**, 3271 (1902); E. Bamberger, *Ber.* **35**, 756, 1896 (1902).

11. C. S. Rondestvedt, Jr., and S. J. Davis, *J. Org. Chem.* **22**, 200 (1957).

12. O. Dimroth, *Ber.* **40**, 2376 (1907).

13. L. Wacker, *Ber.* **35**, 3922 (1902).

14. (*a*) O. Dimroth, *Ber.* **36**, 909 (1903); (*b*) D. H. Sieh, D. J. Wilbur, and C. J. Michejda, *J. Am. Chem. Soc.* **102**, 3883 (1980).

15. T. W. Campbell and B. F. Day, *Chem. Rev.* **48**, 299 (1951).

16. C. K. Ingold and H. A. Piggott, *J. Chem. Soc.* **121**, 2381 (1922).

17. O. Dimroth, *Ber.* **38**, 638, 2328 (1905).

18. K. Vaughan, *J. Chem. Soc.* [*Perkin II*], 17 (1977).

19. G. Vernin, C. Siv, J. Metzger, J. Elguero, and A. Archavlis, *Helv. Chim. Acta* **60**, 495 (1977).

20. R. J. W. LeFevre and T. H. Liddicoot, *J. Chem. Soc.,* 2743 (1951).

21. Y. D. Kondrashev, *Soviet Phys.-Crystallog.* **6**, 413 (1962); Y. A. Amalchenko and Y. D. Kondrashev, *ibid.* **10**, 690 (1966); **12**, 359 (1967).

22. C. B. Mayfield and E. H. Wagoner, *J. Am. Chem. Soc.* **90**, 510 (1968).

23. F. Dwyer, *J. Am. Chem. Soc.* **63**, 78 (1941).

24. R. Kübler, W. Lüttke, and S. Weicherlin, *Z. Elektrochem.* **64**, 650 (1960); S. Weicherlin and W. Lüttke, *Tetrahedron Lett.,* 1711 (1964).

25. H. Balli and F. Kersting, *Liebig's Ann. Chem.* **663**, 103 (1963).

26. V. A. Puchkov, *Zh. Obshch. Khim.* **29**, 3058 (1959).

27. T. P. Ahern, H. Fong, and K. Vaughan, *Can. J. Chem.* **55**, 1701 (1977).

28. R. Kreher and K. Goth, *Z. Naturforsch.* **B31**, 217 (1976).

29. M. Kawanisi, I. Otani, and H. Nozaki, *Tetrahedron Lett.,* 5575 (1968).

30. A. Hantzsch and F. M. Perkin, *Ber.* **30**, 1412 (1897).

31. R. E. Clark and C. D. Clark, *J. Org. Chem.* **42**, 1136 (1977).

32. H. Zollinger, *Diazo and Azo Chemistry,* Interscience Publishers, New York, 1963, pp. 182*ff.*

33. H. Goldschmidt, S. Johnson, and E. Overwien, *Z. Physik. Chem.* **110**, 25 (1924), and earlier papers.

34. E. D. Hughes and C. K. Ingold, *Quart. Rev.* **6**, 34 (1952).

35. A. Hassner and B. A. Belinka, Jr., *J. Am. Chem. Soc.* **102**, 6185 (1980).

36. O. Vogl and C. S. Rondestvedt, Jr., *J. Am. Chem. Soc.* **77**, 3067 (1955).

37. C. Konongsberger and G. Salomon, *J. Polymer Sci.* **1**, 200 (1946).

38. E. H. White, A. A. Baum, and D. E. Eitel, *Org. Syntheses*, Coll. **V**, 797 (1975).

39. N. Isaacs and E. Rannala, *J. Chem. Soc.* [*Perkin II*], 899 (1974).

40. C. Smith and C. H. Watts, *J. Chem. Soc.* **97**, 562 (1910).

41. G. Vernin and J. Metzger, *Synthesis*, 921 (1978).

42. H. Hansen, S. Hünig, and K. Kishi, *Chem. Ber.* **112**, 445 (1979).

43. O. Dimroth, *Ber.* **38**, 677 (1905).

44. M. O. Forster and C. S. Garland, *J. Chem. Soc.* **95**, 2051 (1909).

45. J. I. G. Cadogan, R. G. M. Landells, and J. T. Sharp, *J. Chem. Soc.* [*Perkin I*], 1841 (1977).

46. L. Fisera, J. Kovac, E. Komanova, and J. Lesko, *Tetrahedron* **30**, 4123 (1974).

47. E. Müller and H. Haiss, *Chem. Ber.* **95**, 1255 (1962).

48. C. M. Camaggi, R. Leardini, and C. Chatgilialoglu, *J. Org. Chem.* **42**, 2611 (1977).

49. R. C. Kerber and T. J. Ryan, *Tetrahedron Lett.*, 703 (1970).

50. E. Nölting and F. Binder, *Ber.* **20**, 3004, 3014 (1887).

51. K. Clusius and H. R. Weisser, *Helv. Chim. Acta* **35**, 1548 (1952).

52. W. Theilacker and E. C. Fintelmann, *Chem. Ber.* **91**, 1597 (1958).

53. F. Bernardi, M. Guerra, L. Lunazzi, G. Panciera, and G. Placucci, *J. Am. Chem. Soc.* **100**, 1607 (1978).

54. W. W. Hartman and J. B. Dickey, *Org. Syntheses*, Coll. **II**, 163 (1943).

55. O. Dimroth, M. Eble, and W. Gruhl, *Ber.* **40**, 2390 (1907).

56. V. Ya. Pochinok, *J. Gen. Chem.* (*U.S.S.R.*) **16**, 1303, 1306 (1946); *Chem. Abstr.* **41**, 3066f,h (1947).

57. T. Ignasiak, J. Suszko, and B. Ignasiak, *J. Chem. Soc.* [*Perkin I*], 2126 (1975).

58. D. Y. Curtin and J. D. Druliner, *J. Org. Chem.* **32**, 1552 (1967).

59. J. Goerdeler and M. Roegler, *Chem. Ber.* **103**, 112 (1970).

60. G. Gaudiano, C. Ticozzi, A. Umani-Rouchi, and P. Bravos, *Gazz. Chim. Ital.* **97**, 1411 (1967).

61. R. C. Kerber, *J. Org. Chem.* **37**, 1587 (1972); R. C. Kerber and J. Haffron, *ibid.* **37**, 1592 (1972).

62. K.-H. Koch and E. Fahr, *Angew. Chem. Int. Ed.* **9**, 634 (1970).

63. S. F. Gait, C. W. Rees and R. C. Storr, *Chem. Commun.*, 1545 (1971).

64. L. Hoesch, M. Karpf, E. Dunkelblum, and A. S. Dreiding, *Helv. Chim. Acta* **60**, 816 (1977); C. Lauenberger, M. Karpf, L. Hoesch, and A. S. Dreiding, *ibid.* **60**, 831 (1977).

65. S. Goldschmidt, *Liebig's Ann. Chem.* **473**, 137 (1929).

66. W. K. Wilmarth and N. Schwartz, *J. Am. Chem. Soc.* **77**, 4543, 4551 (1955).

67. V. A. Ginsburg, A. Ya. Yakubovich, A. S. Filatov, V. A. Shpanskii, E. S. Vlasova, G. E. Zelenin, L. F. Sergienko, L. L. Martynova, and S. P. Marakov, *Doklady Akad. Nauk S.S.S.R.* **142**, 88 (1962); *Chem. Abstr.* **57**, 642e (1962).

68. D. Walz and S. Fallab, *Helv. Chim. Acta* **43**, 540 (1960).

69. G. O. Schenck and C. Foote, *Angew. Chem.* **70**, 505 (1958).

70. B. L. Erusalimsky, B. A. Dolgoplosk, and A. P. Kravunenko, *J. Gen. Chem.* (*U.S.S.R.*) **27**, 267 (1957).

71. (*a*) E. Fischer, *Liebig's Ann. Chem.* **190**, 167 (1878); (*b*) H. Wieland, *Ber.* **41**, 3498 (1908).

72. W. S. Wadsworth, *J. Org. Chem.* **34**, 2994 (1969).

73. S. F. Nelsen, R. T. Landis, II, and J. C. Calabrese, *J. Org. Chem.*, **42**, 4192 (1977).

74. A. Wohl and H. Schiff, *Ber.* **33**, 2741 (1899).

75. E. Fischer, *Ber.* **43**, 3500 (1910).

76. J. P. Horwitz and V. A. Grakauskas, *J. Am. Chem. Soc.* **80**, 296 (1958).

77. H. Minato, M. Oku and S. H.-P. Chan, *Bull. Chem. Soc. Japan* **39**, 1049 (1966).

78. O. Dimroth and G. de Montmollin, *Ber.* **43**, 2904 (1910).

79. A. F. Hegarty and F. L. Scott, *J. Org. Chem.* **32**, 1957 (1967).

80. G. S. Hammond, B. Seidel, and R. E. Pincock, *J. Org. Chem.* **28**, 3275 (1963).

81. W. R. McBride and W. E. Thun, *Inorg. Chem.* **5**, 1846 (1966).

82. S. F. Nelsen and R. Fibiger, *J. Am. Chem. Soc.* **94**, 8497 (1972).

83. D. M. Lemal and T. W. Rave, *J. Am. Chem. Soc.* **87**, 393 (1965).

84. R. Stradi, *Atti Accad. Naz. Lincei, Cl. Sci. Fis. Mat. Nat. Rend.* **43**, 350 (1967).

85. J. S. Watson, *J. Chem. Soc.*, 3677 (1956); R. G. Child, G. Morton, C. Pidacks, and A. S. Tomcufcik, *Nature* **201**, 391 (1964); D. Mackay and W. A. Waters, *J. Chem. Soc.* [*C*], 813 (1966).

86. B. G. Gowenlock, P. P. Jones, and D. R. Snelling, *Can. J. Chem.* **41**, 1911 (1963).

87. J. C. McGowan and T. Powell, *Rec. Trav. Chim. Pays Bas* **81**, 1061 (1962).

88. D.-H. Bac and H. J. Shine, *J. Org. Chem.* **45**, 4448 (1980).

89. B. R. Cowley and W. A. Waters, *J. Chem. Soc.*, 1228 (1961).

90. W. Schlenk and E. Bergmann, *Liebig's Ann. Chem.* **463**, 281 (1927).

91. F. O. Rice and C. J. Grelecki, *J. Am. Chem. Soc.* **79**, 2679 (1957).

92. J. R. Majer, *Trans. Faraday Soc.* **57**, 23 (1961).

93. L. J. Magdjinski and Y. L. Chow, *J. Am. Chem. Soc.* **100**, 2444 (1978).

94. A. Good and J. C. J. Thynne, *J. Chem. Soc.* [*B*], 684 (1967).

95. K. Sugiyama, T. Nakaya and M. Imoto, *Bull. Chem. Soc. Japan* **48**, 941 (1975).

96. L. Birkofer, *Ber.* **75**, 429 (1942).

97. C. Paal and W.-N. Yao, *Ber.* **63B**, 57 (1930).

98. W. R. McBride and E. M. Bens, *J. Am. Chem. Soc.* **81**, 5546 (1959).

99. S. F. Nelsen, V. E. Peacock, and C. R. Kassel, *J. Am. Chem. Soc.* **100**, 7017 (1978); S. F. Nelsen, L. A. Grezzo, and V. A. Peacock, *J. Org. Chem.* **46**, 2402 (1981).

100. W. E. Thun and W. R. McBride, *J. Org. Chem.* **34**, 2997 (1969).

101. E. Stöldt and R. Kreher, *Angew. Chem. Int. Ed.* **17**, 203 (1978).

102. N. Wiberg and A. Gieren, *Angew. Chem.* **74**, 942 (1962).

103. H. Bock and K. L. Kompa, *Angew. Chem.* **74**, 327 (1962).

104. G. Koga and J.-P. Anselme, *J. Org. Chem.* **35**, 960 (1970).

105. P. A. S. Smith, in *Nitrenes*, W. Lwowski (ed.), Wiley, New York, 1970, Chap. 3; R. N. McDonald and A. K. Chowdhury, *J. Am. Chem. Soc.* **102**, 5118 (1980).

106. A. N. Howard and F. Wild, *Biochem. J.* **65**, 651 (1957).

107. H. von Pechmann and L. Frobenius, *Ber.* **27**, 651, 703 (1894); **28**, 170 (1895).

108. H. Goldschmidt and V. Badl, *Ber.* **22**, 934 (1889).

109. J. C. McGowan and L. Seed, British Patent No. 834,332; *Chem. Abstr.* **54**, 2343 (1960).

110. K. A. Hofmann and H. Hock, *Ber.* **44**, 2946 (1911).

111. (*a*) J. P. Horwitz and V. A. Grakauskas, *J. Am. Chem. Soc.* **79**, 1249 (1957); (*b*) V. A. Grakauskas, Doctoral dissertation, Illinois Institute of Technology, 1955.

112. W. H. Pirkle and J. C. Stickler, *J. Am. Chem. Soc.* **92**, 7497 (1970).

INDEX